T0280563

This book is concerned with the involvement of the cerebellum in learning and remembering how to carry out skilled movements such as walking, riding a bicycle, and speaking. Processes of plasticity or change have been identified at the cellular level in the cerebellum that could underlie the learning and memory of skilled movements, but whether these processes actually have such a role is controversial.

The book brings together cellular studies of synaptic plasticity with behavioral and systems level studies focused on the learning and memory of motor tasks. Four target articles address the phenomenology and mechanisms of synaptic plasticity at excitatory and inhibitory synapses in the cerebellar cortex, and another four target articles address the evidence for cerebellar involvement in motor learning in animals and man. Both experimental and theoretical approaches are included. The target articles are followed by informed commentaries by scientists from related subspecialties, and by replies to these commentaries by the authors of the target articles. Thus this book is unique in bringing together studies of plasticity at the cellular level with studies of plasticity or learning at the behavioral level and in attempting to build bridges between the two.

The book will appeal to neuroscientists, neurologists, and physiologists interested in the neural control of movement.

Motor learning and synaptic plasticity in the cerebellum

Motor learning and synaptic plasticity in the cerebellum

Edited by

PAUL J. CORDO
Robert S. Dow Neurological Sciences Institute

CURTIS C. BELL
Robert S. Dow Neurological Sciences Institute

STEVAN HARNAD
University of Southampton

CAMBRIDGE UNIVERSITY PRESS
Cambridge, New York, Melbourne, Madrid, Cape Town, Singapore, São Paulo

Cambridge University Press
The Edinburgh Building, Cambridge CB2 8RU, UK

Published in the United States of America by Cambridge University Press, New York

www.cambridge.org
Information on this title: www.cambridge.org/9780521592864

Originally published in 1996 by Cambridge University Press as a special issue of the
interdisciplinary journal *Behavioral and Brain Sciences*

A catalogue record for this publication is available from the British Library

Library of Congress Cataloguing in Publication data
Motor learning and synaptic plasticity in the cerebellum / edited by
Paul J. Cordo, Curtis C. Bell, Stevan Harnad.
p. cm.
Includes bibliographical references and index.
ISBN 0-521-59286-0. – ISBN 0-521-59705-6 (pbk.)
1. Cerebellum – Physiology. 2. Motor learning.
3. Neuroplasticity. I. Cordo, Paul J. II. Bell, Curtis C. (Curtis
Calvin) III. Harnad, Stevan R.
QP379.M68 1997
152.3′34 – dc21 97-1229
CIP

ISBN 978-0-521-59286-4 hardback
ISBN 978-0-521-59705-0 paperback

Transferred to digital printing 2008

Contents

Preface

The histological simplicity and organization of the cerebellar cortex have fascinated neuroscientists for more than 150 years. These structural features made it possible for early anatomists to establish the basic connectivity among the different cellular elements of the cerebellum; and the same features facilitated the physiological work of J. Eccles, M. Ito, R. Llinas, and others in the 1960s that established the polarity and other aspects of the synaptic connections. The cerebellum thus became the first central nervous system structure in a vertebrate for which a wiring diagram could be drawn showing the morphology of the different elements, their connectivity, and their physiological interactions. This knowledge generated a great deal of excitement in the late 1960s and convinced many neuroscientists that a fundamental understanding of a major central nervous system structure was near at hand. It seemed that only a few years' work would be necessary to establish "what the cerebellum does and how it does it," in the phrase of the time. The excitement and promise were reflected in the title of the 1967 book by J. Eccles, M. Ito, and J. Szentagothai that summarized the anatomical and physiological findings, *The Cerebellum as a Neuronal Machine.* The circuitry of the cerebellum and its promise still fascinate many neuroscientists, but a good functional understanding continues to elude us.

Fascination with the cerebellum was heightened by the addition of a second theme to that of circuitry in our conceptual approach to cerebellar function. This is the theme of the cerebellar cortex as a site of learning – and in particular of motor learning. This theme began with theoretical work in the late 1960s and early 1970s by D. Marr and J. Albus which hypothesized that a temporal association between climbing fiber and parallel fiber inputs to Purkinje cells could alter the synaptic efficacy of the parallel fiber synapse and that this could serve as the substrate for motor learning. Interest in this theme was strengthened by the discovery of vestibulo-ocular plasticity, a type of motor learning, and the work of M. Ito and others suggesting that such plasticity might be explained by a Marr-Albus type of *learning* process in the cerebellar cortex. A possible role for the cerebellum in motor learning was further supported by the finding that the temporal association between parallel fiber and climbing fiber inputs does indeed lead to a change, in fact a depression, in the synaptic effect of the paired parallel fiber input. This phenomenon of long-term depression (LTD) was first identified *in vivo* by Tongroach, Sakurai, and Ito and later in *in vitro* preparations by Sakurai, Crépel, Hirano, Linden, and their colleagues.

The phenomenon of LTD is now well established, but the connection between this and other types of synaptic plasticity in the cerebellum and motor learning remains unresolved and controversial. In addition, although most would agree that the cerebellum plays some role in motor learning, the nature of that role is also quite unresolved and controversial. These controversies concerning the role of the cerebellum in motor learning and the connection between motor learning and synaptic plasticity were the subjects of the symposium "Controversies in Neuroscience IV: Motor Learning and Synaptic Plasticity in the Cerebellum," upon which this book is based. The symposium brought together scientists working on the cerebellum at the cellular level with scientists working at the systems and behavioral levels. The goals were to build some bridges across these separate levels of analysis and to help clarify the controversies surrounding the role of the cerebellum in motor learning.

This symposium, held in Portland, Oregon, on 24–26 August 1993, was the fourth of five with the theme Controversies in Neuroscience. These symposia were organized by the Robert S. Dow Neurological Sciences Institute and supported by NIH, NSF, and Good Samaritan Foundation. Early symposia in the series focused on movement control, neural transplantation, and g-proteins in the nervous system. Each conference has resulted in the publication of a dedicated issue of the Cambridge University Press scientific journal *Behavioral and Brain Sciences.* This book reproduces one of those issues of *BBS.*

The papers on plasticity at the cellular level are presented first, followed by the papers on learning at the systems or behavioral levels. The first paper ("Cerebellar long-term depression as investigated in a cell culture preparation"), by D. Linden, and the second paper ("Cellular mechanisms of long-term depression in the cerebellum"), by F. Crépel, N. Hemart, D. Jaillard, and H. Daniel, are concerned with the phenomenology of LTD and the contributions of various cellular processes to the generation of LTD. Linden's work was done in cultured Purkinje cells, whereas Crépel et al.'s was done in the *in vitro* slice preparation. Both papers show that LTD is dependent on calcium influx through voltage-gated calcium channels and on activation of protein kinase C via metabotropic glutamate receptors. The two sets of studies obtained opposite results, however, with regard to the role of nitric oxide (NO). Blockade of NO synthesis had no effect on LTD in the culture preparation but had a clear effect in slices. The third paper ("Long-lasting potentiation of GABAergic inhibitory synaptic transmission in cerebellar Purkinje cells: Its properties and possible mechanisms"), by M. Kano, is concerned with plasticity at the synapses between inhibitory interneurons and Purkinje cells. Plasticity at inhibitory synapses has received much less attention than plasticity at excitatory synapses but appears to be clearly present in the cerebellum and must have an important role in the plastic phenomena that take place there. The fourth paper ("Nitric oxide and synaptic plasticity: NO news from the cerebellum"), by S. Vincent, discusses the biochemical pathways responsible for the synthesis and action of nitric acid and argues against an essential role for NO in LTD.

The fifth paper ("Models of the cerebellum and motor learning"), by J. Houk, J. Buckingham, and A. Barto, is the first of four papers concerned with the systems or behavioral levels of analysis. The paper by Houk et al. addresses the question "How can modeling studies help us understand the role of the cerebellum in motor learning?" and reviews several important models of the cerebellum, including the authors' own model. The authors' own model includes certain cellular properties of Purkinje cells as well as properties of the cerebellum and systems levels. The sixth paper ("On climbing fiber signals and their consequence(s)"), by J. Simpson, D. Wylie, and C. De Zeeuw, reviews what is known about the messages conveyed by climbing fibers and about the effects of climbing fibers on Purkinje cell activity. The paper also reviews some of the theories regarding the role of the climbing fiber in cerebellar function. Much controversy exists regarding this role. Such controversy is not surprising, however, given the similar controversy regarding cerebellar function and the likelihood that the climbing fiber is a key which, once understood, might unlock the problem of cerebellar function. The seventh paper, by A. Smith ("Does the cerebellum learn strategies for the optimal time-varying control of joint stiffness?"), is directly concerned with motor learning and argues that the cerebellum stores and shapes the time-varying patterns of muscle activation that control posture and movement. The author further argues that teloceptive and proprioceptive sensory stimuli could serve as learned cues for associated patterns of muscle activation. The eighth and final paper, by W. Thach ("On the specific role of the cerebellum in motor learning and cognition: Clues from PET activation and lesion studies in man"), is similar to the paper by Smith in arguing for the cerebellum's role in motor learning. Thach's paper makes extensive use of information from human studies and extends the role of the cerebellum into the realm of cognition, where planning can be viewed as a form of "mental movement" and thought as the manipulation of mental "objects."

The conference closed with a keynote address by Dr. M. Ito that covered levels of analysis from the subcellular to the cognitive and thus served as an excellent summary and conclusion for the symposium. Dr. Ito's many accomplishments were recognized at the symposium by awarding him the 1993 R. S. Dow Neuroscience Award.

This book is dedicated to the memory of Robert S. Dow, who died on 20 January 1995, approximately 1 year after the conference upon which this book is based. It is especially fitting that this book on motor learning and synaptic plasticity in the cerebellum be dedicated to Robert Dow, because the cerebellum was the focus of his scientific career from the 1930s until his death.

Robert Dow was born in 1908. He received his B.S. with honors from Linfield College in McMinnville, Oregon, in 1929, followed by an M.A. and M.D. from the University of Oregon Medical School in 1934 and a Ph.D. from the University of Oregon Medical School in 1935. While in medical school, he studied under Olof Larsell, a neuroanatomist, who made seminal contributions to the comparative anatomy of the cerebellum. It was through working with Larsell that Robert Dow developed his lifelong fascination with the cerebellum. Larsell encouraged Dow to pursue a career in research on the central nervous system. Dow then traveled to a number of laboratories around the world to pursue studies in the emerging field of brain research with John Fulton, Frederick Bremer, Gordon Holmes, and Herbert Gasser.

In 1939 Dow returned to Portland, where he became Associate Professor of Anatomy at the University of Oregon Medical School. He brought with him plans for the construction of a cathode ray oscilloscope, plans he shared with Howard Vollum and Jack Murdock, who later became founders of Tektronix Corporation. Dow popularized the use of the oscilloscope for studying brain waves in patients with traumatic head injuries sustained at the Portland shipyards during World War II. At the conclusion of World War II, Dow resigned his position as Professor of Anatomy and became Oregon's first neurologist in private practice.

In 1954 Dow received a Fulbright fellowship to collaborate with Giuseppe Moruzzi, the postwar leader of Italian physiology, with whom he wrote a book of mutual interest on the cerebellum. This book, published in 1958, is still regarded as the most thorough treatise on the disorders of the cerebellum.

After his year aboard, Dow continued to do research on the cerebellum, and in 1959, he established the Laboratory of Neurophysiology at Good Samaritan Hospital in Portland. Under his guidance, this laboratory grew into what is now known as the Robert S. Dow Neurological Sciences Institute, with 20 independent laboratories and a worldwide reputation of excellence in brain research.

Robert Dow maintained a dual career as scientist and clinician into his ninth decade. During his final years, he continued to serve as an advocate and fund raiser for the research institute that bears his name. He left a legacy of scientific and medical accomplishments that span the birth of modern brain research and clinical neurology.

Paul J. Cordo
Curtis C. Bell
Stevan Harnad

1

Cerebellar long-term depression as investigated in a cell culture preparation[1]

David J. Linden
Department of Neuroscience, The Johns Hopkins University School of Medicine, Baltimore, MD 21205.
Electronic mail: *dlinden@welchlink.welch.jhu.edu*

Abstract: Cerebellar long-term depression (LTD) is a form of synaptic plasticity, first described by Ito and co-workers, in which simultaneous activation of two excitatory inputs to a Purkinje neuron, parallel fibers (PFs) and climbing fibers (CFs), results in a sustained depression of PF synaptic drive. The purpose of this target article is not to assess the possible role of this synaptic alteration in motor learning, an issue which is addressed by other authors in this volume, nor is it to provide a detailed summary of the work on cerebellar LTD to this point (see Linden & Connor 1993; Crépel et al. 1993 for review) or to place cerebellar LTD within the context of other forms of persistent synaptic depression that occur within the mammalian brain (see Linden 1994b). Rather, it is to discuss results obtained using a very reduced preparation for the study of LTD, embryonic Purkinje neurons grown in culture and stimulated with exogenous excitatory amino acids, and to consider some advantages and limitations of this approach. Recent work using this preparation has suggested that three processes are necessary for the induction of cerebellar LTD: Ca influx through voltage-gated channels, Na influx through AMPA receptor-associated channels or voltage-gated Na channels, and protein kinase C activation – which is dependent upon activation of the metabotropic glutamate receptor mGluR1. In addition, input-specific induction of LTD has been demonstrated in this preparation under conditions where both spontaneous and evoked neurotransmitter release are reduced or eliminated, indicating that postsynaptic alterations are sufficient to confer this important computational property.

Keywords: climbing fiber; glutamate; motor learning; parallel fiber; Purkinje neuron; synaptic plasticity

1. Some properties of Purkinje neurons *in vivo*

Purkinje neurons, which function as a common inhibitory output stage for cerebellar cortical signals, receive two major excitatory inputs that are organized in very different ways. Single climbing fibers (CFs), originating in the inferior olive, make powerful one-to-one synaptic contacts with Purkinje neurons on the proximal portion of the dendritic arbor. In contrast, parallel fibers (PFs), which are the axons of cerebellar granule neurons, each make contact with many Purkinje neurons. Due to the vast numbers of granule neurons and their divergent input to Purkinje neurons (each Purkinje neuron receives ~150,000 PF contacts), this synapse is among the most abundant in the vertebrate central nervous system (CNS). The transmitter of the PFs is thought to be glutamate (Levi et al. 1985; Sandoval & Cotman 1978), and several types of excitatory amino acid receptor may be found at this synapse – including the AMPA (α-amino-3-hydroxy-5-methyl-4-isoxazole propionic acid) receptor and the metabotropic glutamate receptor. The transmitter of the CFs is still undetermined (Zhang et al. 1990; see Cuénod et al. 1989 for review), but is thought to be an excitatory amino acid (aspartate, homocysteate, and glutamate have all been proposed). The EPSCs evoked in Purkinje neurons by either CF or PF stimulation may be completely blocked by CNQX (6-cyano-nitroquinoxaline-2,3-dione), suggesting that the AMPA receptor is the primary mediator of ion flux at both synapses (Konnerth et al. 1990; Perkel et al. 1990). NMDA (N-methyl-D-aspartate) receptors are developmentally down-regulated in Purkinje neurons so that very few remain in the adult (Krupa & Crépel 1990; Rosenmund et al. 1992). The Purkinje neurons also receive considerable GABAergic inhibitory drive from local interneurons.

The Purkinje neuron is highly enriched in the components of several second messenger systems. Cerebellar Purkinje neurons contain metabotropic glutamate receptors, particularly mGluR1, a subtype linked to activation of the enzyme phospholipase C, in unusually high quantities (Masu et al. 1991), particularly in the distal dendritic spines where the parallel fiber synapses are received (Baude et al. 1993; Martin et al. 1992). The metabotropic receptor mGluR1 produces inositol-1,4,5-trisphosphate (IP_3) and 1,2-diacylglycerol (DAG), when activated. IP_3 binds IP_3 receptors and results in a mobilization of Ca from nonmitochondrial intracellular stores. IP_3 receptors are highly concentrated in the Purkinje neuron endoplasmic reticulum, including that present in distal dendrites and dendritic spines (Satoh et al. 1990). Also, DAG, together with Ca, activates protein kinase C (PKC), including the γ-PKC isoform present in Purkinje neurons (Hidaka et al. 1988).

Another second messenger system of interest is the nitric oxide/cGMP cascade (see Vincent, this issue, for review). Nitric oxide (NO), a gas that diffuses freely through cellular membranes, is formed by the action of NO synthase, which is activated by Ca/calmodulin, upon arginine to form citrulline and NO. NO in turn activates soluble guanylate cyclase, resulting in the production of cGMP, which then activates cGMP-dependent protein kinase. NO synthase immunoreactivity is found in granule cells and their PFs and in basket cells, but is absent in Purkinje neurons and CFs (Bredt et al. 1990; Vincent & Kimura 1992). Although Purkinje neurons contain guanylate cyclase (Ariano et al. 1982; Zwiller et al. 1981), and cGMP-dependent protein kinase (De Camilli et al. 1984; Lohmann et al. 1981), and have detectable resting levels of cGMP (Sakaue et al. 1988), they do not significantly accumulate additional cGMP in response to application of a NO donor (de Vente & Steinbusch 1992; de Vente et al. 1990; Southam et al. 1992). More recently, another gaseous messenger molecule that activates guanylate cyclase has been identified. Carbon monoxide (CO) is produced in the brain by the conversion of heme into biliverdin and CO by the enzyme heme oxygenase-2, which is present in both Purkinje neurons and granule neurons of the cerebellar cortex (Verma et al. 1993).

2. Some properties of Purkinje neurons in embryonic culture

Dispersed cultures of embryonic mouse cerebellum were grown in serum-free medium by the method of Schilling et al. (1991). Neurons used for LTD experiments were typically grown for 12–21 days in culture. At this time, Purkinje neurons in culture attained an elaborate dendritic morphology. While the dendritic arbor sometimes appeared similar to the seafan shape seen *in vivo* (Fig. 1, Top), other configurations, such as the bipolar morphology illustrated schematically in Figure 3 (Top), were also found. On a finer scale it may be seen that, similar to the case *in vivo*, most synaptic contacts from granule cells are received on dendritic spines (Dunn & Mugnaini 1993). Purkinje neurons in culture receive spontaneous excitatory and inhibitory synaptic input, as assessed by spontaneous postsynaptic currents when recorded in voltage-clamp mode (in the absence of tetrodotoxin), and spontaneous firing of simple spikes in current-clamp mode (Fig. 1, Bottom). They respond to stimulation of neighboring granule neurons and to exogenous excitatory amino acids with excitation. Likewise, they respond to stimulation of neighboring Purkinje neurons and exogenous $GABA_A$ agonists with inhibition. They express T, L, and P-type voltage-gated Ca currents as assessed by whole-cell voltage clamp recording (Linden, unpublished observations). One notable difference between cultured embryonic and adult Purkinje neurons is that the former express large NMDA-induced inward currents (Linden & Connor 1991), and the latter do not (Krupa & Crépel 1990; Rosenmund et al. 1992). It is interesting to note that over many weeks in culture, embryonic Purkinje neurons show a down-regulation of NMDA current (Linden, unpublished observations). In culture, Purkinje neurons display a similar pattern of reactivity for immunochemical markers to that seen *in vivo;* they are positive for calbindin-D_{28K}, PEP-19, γ-PKC, and cGMP-dependent protein kinase. They are negative for α-PKC and β-PKC, as well as for nitric oxide synthase activity as indicated by the diaphorase reaction (Schilling et al. 1994).

Further information is available about the baseline properties of Purkinje neurons in a *rat* cerebellar culture preparation that has also been successfully used for glutamate/depolarization-induced LTD. In a study that combined immunohistochemistry with electrophysiology of synaptic currents, it was determined that Purkinje neurons in culture received inhibitory contacts on the soma and proximal dendrites, whereas excitatory contacts from granule cells were made only on the dendrites (Hirano & Kasono 1993). This distribution is remarkably similar to that seen in the intact cerebellum.

In sum, Purkinje neurons in embryonic culture show many of the same basic properties as do those in the intact cerebellum (Table 1). A thorough parametric comparison is yet to be made, however. For example, even though both cultured and intact Purkinje cells express AMPA receptors, it is not clear that they express the same subtypes in the same combination(s). Furthermore, it is not clear whether there are pharmacological differences such as relative potencies of agonists and antagonists or microanatomical

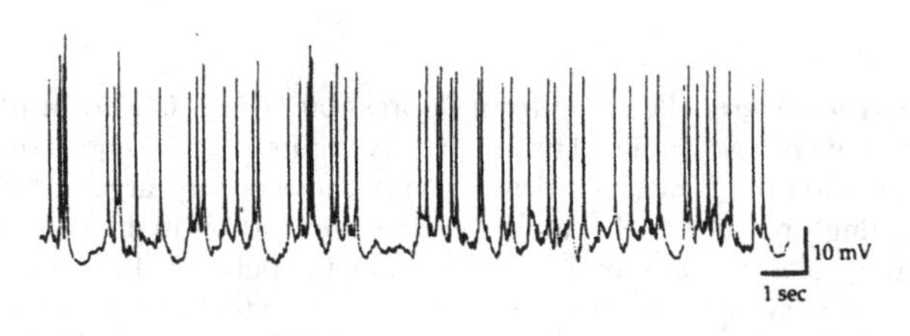

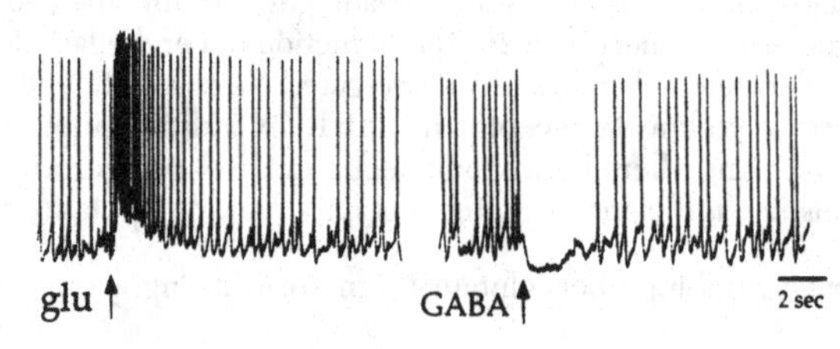

Figure 1. Embryonic mouse Purkinje neurons grown in dispersed culture. (Top) Purkinje neurons grown in culture maintain an elaborate dendritic morphology similar to that seen *in vivo* and are selectively immunoreactive for the marker calbindin-D_{28K}. This cell was from a culture of embryonic day 16 mouse cerebellum grown *in vitro* for 21 days. Photograph courtesy of Dr. Karl Schilling. (Bottom) Spontaneous activity and responses to iontophoretically applied GABA and glutamate recorded in current-clamp mode using a perforated-patch electrode attached to the cell soma.

Table 1. *A comparison of proteins potentially relevant to cerebellar synaptic plasticity: Cultured embryonic Purkinje neurons versus adult Purkinje neurons* in vivo[1]

	Embryonic cell culture	Adult *in vivo*
Receptors		
Metabotropic (mGluR1)	+	+
AMPA	+	+
NMDA	+	0
$GABA_A$	+	+
β-adrenergic	0	+
IP_3	+	+
ryanodine	+	+
Enzymes		
α-PKC	0	0
β-PKC	0	0
γ-PKC	+	+
Nitric oxide synthase (neuronal)	0	0
cGMP-dependent protein kinase	+	+
Ion channels		
I_{Ca} (T)	+	+
I_{Ca} (L)	+	+
I_{Ca} (P)	+	+
$I_{K(Ca)}$	+	+
I_{Na} (TTX-sensitive)		
Other proteins		
Na/Ca exchanger	+	+
calbindin D_{28K}	+	+

[1] Determinations in this table were made either by immunocytochemistry, electrical recording, or both.

differences such as distribution at synaptic versus extrasynaptic sites.

Finally, it should be noted that it is sometimes difficult to assess culture versus intact differences when considerable disagreement exists between different laboratories using similar preparations. As an illustrative case relevant to LTD, let us consider the electrophysiological effects of metabotropic glutamate receptor agonists on Purkinje neurons. Using the mouse embryonic culture protocol, we found that activation of Purkinje neuron metabotropic glutamate receptors by pressure-pulse application of the metabotropic agonist t-ACPD resulted in (1) intradendritic calcium mobilization that could be blocked by L-AP3 or pertussis toxin and (2) an inward current associated with an increase in membrane conductance that could be blocked by L-AP3 and was dependent upon external Na and internal Ca (Linden et al. 1994). The former result generally confirms previous findings that application of metabotropic receptor agonists produces an increase in internal Ca in Purkinje neurons. However, agonist effectiveness has varied considerably in different types of preparations. For example, in Purkinje neurons of rat cerebellar acute slices, both t-ACPD and 1S,3R-ACPD are relatively ineffective in stimulating Ca mobilization compared to quisqualate, which triggers large dendritic Ca increases (Llano et al. 1991; Takagi et al. 1992). In the second study, AP3 was shown to block the Ca increases. In a third study, done with normal external Ca, 1S,3R-ACPD caused increases in somatic Ca but these increases appeared to reflect influx via voltage-gated channels, as they were temporally associated with Ca spiking and were abolished when the membrane voltage was "manually" clamped at rest (Glaum et al. 1992). Dendritic levels were not measured. In contrast, studies on Purkinje neurons in organotypic culture have shown a large *somatic* Ca^{2+} mobilization in response to t-ACPD, with little change in the dendrites (Vranesic et al. 1991; Staub et al. 1992). Similarly, Purkinje neurons from an embryonic mouse studied in dispersed culture showed somatic Ca mobilization (Yuzaki & Mikoshiba 1992). Dendritic changes in this latter study were difficult to assess because of interference from the fluorescence of close neighboring neurons and glia. In our studies, Ca increases often spread to the soma in response to large stimuli, but the initial responses were clearly in the dendrites.

The tissue culture preparation used by Yuzaki and Mikoshiba (1992) is the technique most closely related to ours, but the characteristics of the Ca response differ from our results in that Ca mobilization was not blocked by either L-AP3 or pertussis toxin and the response, at least to a first approximation, appeared to be localized to the soma. While these differences might represent uncontrolled factors in tissue culture, it is quite possible that they arise from developmental regulation of different receptor subtypes. That is to say, the prominent soma response that is pertussis toxin-insensitive might reflect an early developmental stage (as also suggested by these authors), and the prominent dendritic response might reflect a later stage. The experiments of Yuzaki and Mikoshiba (1992) were done predominantly on cultures before 12 DIV (days *in vitro*), and these investigators showed a sharp fall off in the number of neurons displaying the mGluR response at times > 12 DIV. In our experiments, we concentrated on cultures that had been maintained for at least 18 DIV, in which the Purkinje neurons display mGluR-dependent LTD. The organotypic cultures, maintained as they are in very high serum, might also be held in an immature stage.

3. An incomplete review of mechanisms of cerebellar LTD as determined using intact and slice preparations

Some of the basic and less-controversial aspects of the phenomenology and mechanisms of cerebellar LTD are summarized below. Some of the more controversial findings using intact and slice preparations are discussed in later sections in specific reference to other studies conducted using cell culture techniques. LTD produced by coactivation of CFs and PFs may be detected as a depression of the PF-Purkinje neuron synapse, but not as a depression of the CF-Purkinje neuron synapse (Ito et al. 1982). Likewise, LTD is manifest as an attenuated response to test pulses of glutamate, the presumed transmitter of the PF-Purkinje neuron synapse, but not aspartate (Crépel & Krupa 1988; Ito et al. 1982). These results suggest that the synaptic modification that underlies cerebellar LTD is specific to the PF-Purkinje neuron synapse and is, at least in part, postsynaptic. The aspect of CF activation that contributes to LTD induction seems to be a prolonged depolarization of the Purkinje neuron that results in dendritic Ca entry (Konnerth et al. 1992). As such, induction of LTD is blocked when Purkinje neurons are electrically inhibited (Crépel & Jaillard 1991; Ekerot & Kano 1985) or loaded with a Ca chelator (Sakurai 1990). LTD may be induced when depolarization sufficient to produce Ca entry is substituted for CF activation (Crépel & Jaillard 1991; Crépel & Krupa 1988; Schreurs & Alkon 1993). Stimulation of PFs contributes to LTD induction by activating certain non-NMDA excitatory amino acid receptors. Stimulation of CFs together with iontophoretic application of glutamate or quisqualate to Purkinje neuron dendrites produces LTD of the PF-Purkinje neuron synapse (Kano & Kato 1987). LTD is not produced if aspartate or kainate is substituted for quisqualate or glutamate, nor is it produced by iontophoresis of quisqualate alone or by stimulation of PFs or CFs alone (Kano & Kato 1987; Sakurai 1987). Thus, non-NMDA excitatory amino acid receptor activation appears to be necessary, but not sufficient for LTD induction. NMDA receptors, which are not present on the Purkinje neurons of adult animals, do not seem to contribute to LTD induction (Kano & Kato 1987).

LTD is said to result from coactivation of PFs and CFs, but what are the precise timing constraints on this coactivation? A study using intracellular recording in rabbit cerebellar slices has indicated that LTD is optimally induced when CF stimulation precedes PF stimulation by 125–250 msecs (Ekerot & Kano 1989). Another study using a similar preparation has shown that LTD may be induced by CF–PF stimulation with an interval of 50 msecs, but claims that LTD induced by CF–PF pairing will not occur unless disynaptic inhibition is blocked by addition of a $GABA_A$ antagonist (Schreurs & Alkon 1993). Finally, a preliminary report using field potential recording in rabbit cerebellar slices has indicated that the optimal interval is in the opposite direction, with PF stimulation preceding CF stimulation by 250 msecs (Chen & Thompson 1992). The latter interval would require that some persistent signals from PFs, such as a consequence of metabotropic receptor activation, linger for at least 250 msecs to interact with the CF signal.

4. A cell culture preparation for the study of cerebellar LTD

The first studies of cerebellar LTD using a cell culture technique were conducted by Hirano (1990a; 1990b). These very difficult experiments used a preparation in which embryonic rat olivary and cerebellar neurons were cocultured, and Purkinje neurons were found that received input from both olivary and granule neurons. As seen *in vivo*, coactivation of granule neuron and olivary inputs resulted in a long-term depression of granule neuron to Purkinje neuron synaptic drive, and the latter stimulation could be substituted by direct depolarization of the Purkinje neuron sufficient to activate voltage-gated Ca influx.

Using cultured mouse Purkinje neurons, a preparation has been developed in which iontophoretic glutamate pulses and Purkinje neuron depolarization may be substituted for PF and CF stimulation, respectively (Linden et al. 1991). The depression so induced may be seen as a reduction of the glutamate-induced current as measured with a perforated-patch electrode attached to the Purkinje neuron soma. Recordings were typically made in tetrodotoxin/picrotoxin saline to prevent spontaneous synaptic transmission and the Purkinje neurons from firing Na spikes. An iontophoresis electrode was aimed at a large-caliber dendrite, at a point within one diameter-length of the soma, and glutamate pulses (30–80 msec duration) were applied at a frequency of 0.05 Hz. Following 15 min of stable recording, LTD was induced by pairing six successive glutamate test pulses with six, 4-second clamp depolarizations to −10 mV. The depolarization onset preceded the glutamate pulse by 500 msecs. The glutamate-induced current was almost always completely decayed by the end of the depolarization step. This treatment induced a reliable depression of the glutamate-induced inward current to 50%–60% of its baseline value, was usually evident in the first test pulses after conjunction, and persisted as long as the recording could be maintained. This preparation has several advantages that further the analysis of neuronal information storage. First, as LTD is induced and monitored without synaptic stimulation, alterations in Purkinje neuron responsiveness may be unambiguously attributed to postsynaptic processes. Second, the use of the perforated-patch recording technique (Horn & Marty 1988) allows for effective voltage-clamping of the Purkinje neuron while avoiding perfusion of the intracellular second messenger and calcium-regulating systems, as would occur with conventional whole-cell recording. Third, this preparation allows for simultaneous voltage-clamp recording and optical imaging using fluorescent dyes. These latter two advantages are, of course, not restricted to a cell culture preparation.

5. Mechanisms of LTD induction as determined using cultured Purkinje neurons

The experiments that have been undertaken with this technique have elucidated some principles of LTD induction and have confirmed and extended some of the previous observations detailed above. LTD induced by glutamate/depolarization conjunction was found to produce a depression specific to the AMPA subtype of excitatory amino acid receptor (Linden et al. 1991; Linden & Connor 1991). Further support for the notion that CF activation (or Purkinje neuron depolarization) exerts its effects via Ca influx was provided by the observation that LTD induced by glutamate/depolarization conjunction could be blocked in Ca-free external saline. By using selective agonists of excitatory amino acid receptors, it was shown that both AMPA and metabotropic receptors must be activated to induce LTD. Likewise, application of selective antagonists of either the AMPA or the metabotropic receptors were sufficient to completely block LTD induction (Linden et al. 1991; 1993). Recently, these findings have been confirmed and extended by the observation that application of an inactivating antibody directed against mGLUR1α blocks induction of LTD by glutamate/depolarization conjunction in cultured Purkinje neurons (Shigemoto et al. 1994). Therefore, three processes appear to be necessary and sufficient for LTD induction in voltage-clamped Purkinje neurons in culture: depolarization sufficient to cause Ca influx via voltage-gated channels, AMPA receptor activation, and metabotropic receptor activation.

As one consequence of metabotropic receptor activation is protein kinase C (PKC) activation via diacylglycerol liberation, PKC inhibitors and activators were applied to determine their effects on LTD induction in culture (Linden & Connor 1991). Inhibitors that act at both the catalytic site (RO-31-8220) and the regulatory sites (calphostin C or peptide pseudosubstrates) of PKC blocked LTD induced by glutamate/depolarization conjunction when applied during the conjunctive stimulus, but had no effect when applied 10 minutes after the conjunctive stimulus. These compounds did not exert their effects by attenuating voltage-gated Ca influx or the amplitude of AMPA-induced inward current. Application of phorbol-12,13-diacetate (PDA), a PKC activator, induced a depression of AMPA, but not NMDA test pulses. This selectivity was also seen when LTD was induced by quisqualate/depolarization conjunction. In addition, depression induced by PDA and quisqualate/depolarization conjunction were demonstrated to be nonadditive, suggesting that they share common mechanisms. These observations confirm and extend a previous report that LTD in slice may be induced by phorbol esters (Crépel & Krupa 1988). It has also been shown in trigeminal neurons that activation of PKC potentiates NMDA receptor-mediated current by relieving the Mg-dependent blockade (Chen & Huang 1992). It is not clear why PKC activation does not produce a similar effect in cerebellar Purkinje cells, but it is possible that it may result from a differential distribution of NMDA receptor subtypes.

These experiments indicate that activation of PKC is necessary for LTD induction. It is likely that PKC activation is not required for continued LTD expression. It is not clear if PKC activation is sufficient for LTD induction, or if other processes – possibly mediated by AMPA receptor activation – are required. In the simplest scenario, it might be imagined that PKC phosphorylates AMPA receptors or associated proteins at the PF synapse, resulting in receptor desensitization. At present, there is no evidence either to support or to eliminate consideration of this idea.

More recently, the role of AMPA-receptor activation in LTD induction has been addressed (Linden et al. 1993). In Purkinje neurons voltage-clamped to -80 mV in tetrodotoxin (TTX) saline, LTD of AMPA currents may be produced by Purkinje neuron depolarization together with pulses of glutamate or quisqualate but not t-ACPD or (quisqualate + CNQX), suggesting that AMPA receptor activation is necessary for LTD induction. The AMPA receptor in these cultured Purkinje neurons does not appear to exert its effects by directly gating Ca influx, as its associated channel is only weakly permeable to Ca as determined by reversal potential measurements ($P_{Ca}/P_{Na} = 0.17$ for AMPA as compared to $P_{Ca}/P_{Na} = 5.5$ for NMDA in these same cells). Replacement of external Na during quisqualate/depolarization conjunction, with either the impermeant ions NMG or TEA or with the permeant ions Li or Cs, caused a blockade of LTD induction, suggesting that Na influx through the AMPA associated channel is necessary for this process. Similarly, pairing quisqualate pulses with depolarizing steps near E_{Na} also failed to induce LTD. To determine whether activation of voltage-gated Na channels could substitute for AMPA receptor activation, responses to AMPA test pulses were measured in current-clamp mode following ACPD/depolarization conjunction in TTX-free saline. LTD was induced 3/16 times in normal (Na and Ca containing) medium and 7/16 times in medium containing the Na channel opener veratridine. This compares with an LTD induction frequency of 15/16 produced by quisqualate/depolarization conjunction. This finding is somewhat consistent with a recent report using the slice preparation that demonstrated LTD induction in 5/8 cells following direct depolarization (in TTX-free saline in current-clamp mode) together with bath application of ACPD (Daniel et al. 1992). Therefore, I suggest that, although Na influx via voltage-gated channels may suffice to induce LTD infrequently, activation of AMPA receptors is more effective.

Na influx through either channel might exert its effects by activating Na_i/Ca_o exchange, a process not stimulated by Li_i. (Baker & Dipolo 1984). Suggestively, antiserum to a bovine cardiac Na/Ca exchanger shows strong immunoreactivity in our cultured Purkinje neurons. Alternatively, Na might function to stimulate phospholipase C activity, as has been demonstrated in synaptosomal (Guiramand et al. 1991; Gusovsky et al. 1986) and slice (Benuck et al. 1989) preparations *in vitro*.

Some studies using the cerebellar slice technique have suggested that release of the gaseous second messenger, NO, in the cerebellar molecular layer and the consequent activation of soluble guanylate cyclase in the Purkinje neuron, is necessary for LTD produced by PF/CF conjunctive stimulation (Shibuki & Okada 1991) or PF/depolarization conjunctive stimulation (Crépel & Jaillard 1990; Daniel et al. 1993). These studies demonstrated that an LTD-like phenomenon could be induced when CF stimulation or direct depolarization of the Purkinje neuron was replaced by the application of NO (via donor molecules such as sodium nitroprusside) or membrane-permeable analogs of cGMP. Likewise, induction of LTD by more conventional means could be blocked by inhibitors of NO synthase (such as N^G-nitro-L-arginine) or agents that bind NO in the extracellular fluid (such as hemoglobin). Recently, it has been demonstrated that NO synthase inhibitors and NO scavengers fail to block LTD induction in slices treated with a gliotoxin, leading to the proposal that Bergmann glia exert a tonic negative regulation of LTD induction that is relaxed by NO (Shibuki 1993).

In contrast, it has been reported that LTD of glutamate currents produced without synaptic stimulation in cultured Purkinje neurons is unaffected by reagents that stimulate (sodium nitroprusside) or inhibit (hemoglobin, N^G-nitro-L-arginine) NO signaling (Linden & Connor 1992). What might underlie the difference between the lack of effect of NO-modulating reagents that we report and the actions found in previous studies? One possibility is that the difference lies in the type of stimulus used to monitor LTD. Both investigations from other laboratories used PF stimulation to monitor LTD, while our study used glutamate pulses. Therefore, if the target of NO action were a site other than the Purkinje neuron dendrite (such as the PF terminal), then the effect of NO-modulating reagents would be detected with PF stimulation but not with glutamate pulses. However, two observations suggest that this explanation might not be correct. First, preliminary data indicate that LTD induced in culture by stimulating a granule cell input to a Purkinje neuron together with direct depolarization is not blocked by hemoglobin or N^G-nitro-L-arginine (Linden, unpublished observations). Second, a recent report using the cerebellar slice preparation showed no effect of an NO donor on PF-evoked EPSPs recorded in Purkinje neurons (Glaum et al. 1992). It should be noted that similar confusion about the role of NO may be found in recent studies of hippocampal LTP, all of which have used the slice preparation (Lum-Ragan & Gribkoff 1993; Williams et al. 1993). This suggests that the divergent results using NO-modulating reagents in cerebellar LTD is probably not directly attributable to differences in slice vs. culture preparations.

Typically, one of the most frustrating aspects of experimentation with primary cultured neurons is cell-to-cell variation. The embryonic mouse cerebellar culture protocol has been refined in order to minimize variation in the electrophysiological properties of Purkinje neurons. While this strategy has been largely successful in the mouse, considerable variation remains in certain electrophysiological properties of Purkinje neurons recorded using a similar preparation of rat tissue (Linden et al. 1992). It should be noted that this does not represent any fundamental distinction between mouse and rat tissue, but rather that our preparation for rat is not optimal; other investigators have produced a better preparation of rat tissue (Shigemoto et al. 1994). Although this degree of variation precludes many types of experiments on cerebellar LTD, it does allow for a certain type of correlational study that exploits the variation inherent in some cell culture preparations. A form of variation present in our rat Purkinje neurons was the amplitude of voltage-gated Ca current. In addition, much to our annoyance, LTD could only be induced in about half of the rat Purkinje neurons tested. As previous work had indicated that activation of voltage-gated Ca channels was necessary for induction of LTD, we sought to determine if the amplitude of the voltage-gated Ca conductance and the probability of successful LTD induction were positively correlated.

Voltage-gated Ca current was recorded by applying depolarizing voltage steps from a holding potential of −90 mV. Recordings were made using the whole-cell patch-clamp technique with N-methylglucamine and tetraethylammonium-containing internal saline at t = −10 minutes. When Ca currents were present, they typically consisted of both a transient, low-threshold component and a larger, sustained, high-threshold component. Measures of peak Ca current largely reflected the amplitude of the latter component. Following establishment of a baseline response to glutamate test pulses, LTD induction by glutamate/depolarization conjunctive stimulation was attempted at t = 0 minutes. Measures of percentage of baseline current amplitude were made at t = 20 minutes. Analysis of these experiments, illustrated in Figure 2, indicates that LTD induction is more likely to occur in cells with larger voltage-gated Ca currents. However, the amplitude of LTD, once induced, is fairly constant and does not vary with Ca current amplitude. Ca current amplitude was not correlated with input resistance, degree of dendritic elaboration, or somatic diameter. Of course, a correlation such as this does not prove a causal link between Ca conductance and LTD induction. Ca current amplitude could merely co-vary with a third parameter, such as expression of an enzyme that is necessary for LTD induction. However, together with interventive evidence, correlational results exploiting cell culture variability can be an additional tool to evaluate the substrates of LTD.

Cerebellar LTD is attractive as a model system for the study of information storage not only because of its duration, but also because it demonstrates input specificity; LTD is confined to those PFs that are active at the time of CF stimulation (Chen & Thompson 1992; Ekerot & Kano 1985). This property confers enormous computational power upon the Purkinje neuron because it allows the strength of small groups of the ~150,000 PF inputs to be independently attenuated. Recently, it has been demonstrated that input specificity is preserved when LTD is

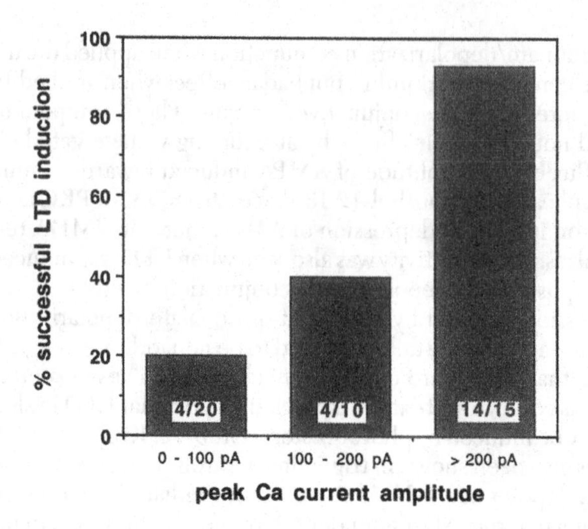

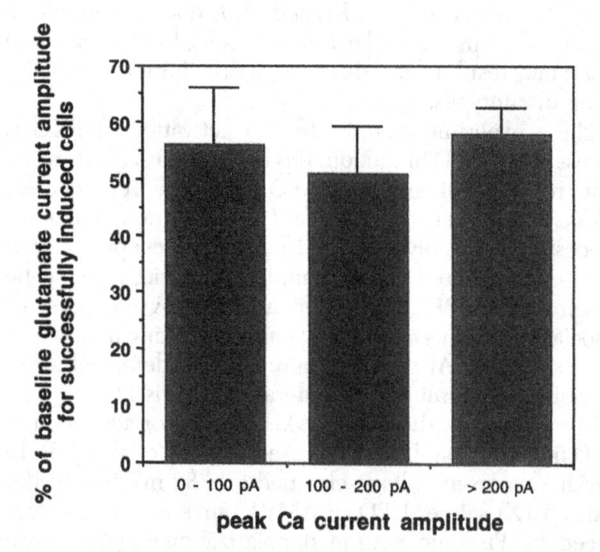

Figure 2. Amplitude of voltage-gated Ca conductance predicts induction frequency of LTD in cultured embryonic rat Purkinje neurons. (Top) Cells were divided into three groups based on peak Ca current amplitude (measured at t = −10 min), and the percentage of cells in which LTD was successfully induced was tallied. Successful LTD induction was defined as a reduction of glutamate-induced current to <80% of baseline. (Bottom) The degree of attenuation following glutamate/depolarization conjunctive stimulation was measured (at t = 20 min) *only* for those cells that showed successful LTD induction as defined by the criterion above.

induced by quisqualate/depolarization conjunction in cultured Purkinje neurons (Linden 1994a). When multiple, discrete quisqualate application sites are used, LTD is confined to those sites that are stimulated together with depolarization (Fig. 3). As these experiments are conducted in TTX/picrotoxin saline, it is unlikely that evoked release from presynaptic terminals is involved in determining specificity. Although this experiment suggests that presynaptic processes are not required for input-specific induction of LTD, it is not definitive. The Purkinje neurons grown in this culture system receive synaptic contacts and, although most synaptic transmission is abolished by the addition of tetrodotoxin/picrotoxin saline, some release persists under these conditions as indicated by the presence of spontaneous EPSCs that are blocked by kynurenate (data not

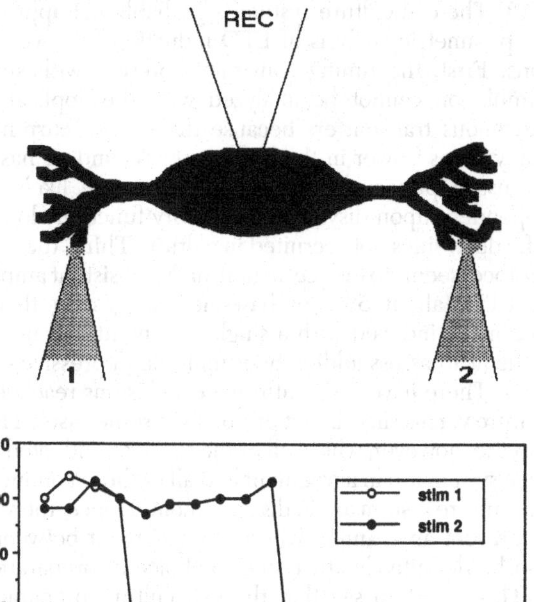

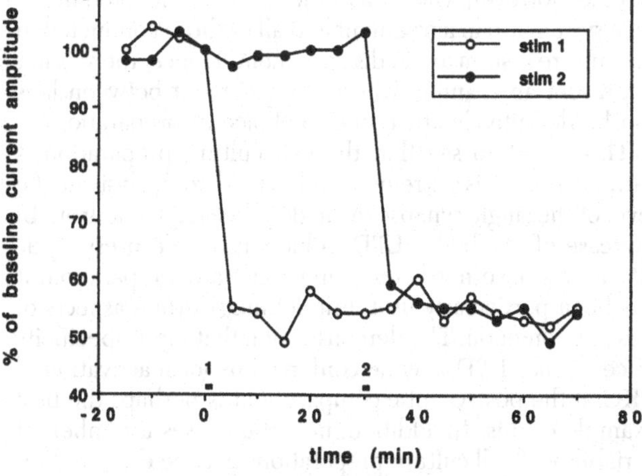

Figure 3. Input specificity of LTD may be seen in the absence of synaptic stimulation. (Top) A schematic diagram of the recording and stimulation protocol. Quisqualate iontophoretic electrodes (labeled 1 and 2 in diagram) are aimed at two widely separated sites on the Purkinje neuron dendritic arbor. Purkinje neurons with bipolar morphology are specifically chosen to facilitate nonoverlapping fields of stimulation. The responses to test pulses of quisqualate delivered alternately to the two stimulation sites are recorded with a perforated-patch electrode attached to the soma. This electrode also functions to deliver depolarizing voltage-clamp commands that may be paired with stimulation at either site to induce LTD. (Bottom) Pairing of quisqualate pulses delivered to site 1 with depolarization (t = 0 min) resulted in the induction of LTD as monitored with subsequent test pulses delivered to site 1 but not site 2. Site 2/depolarization conjunction (at t = 30 min) also resulted in LTD of that input with no further alteration of site 1 responsivity.

shown). It is possible that the level of spontaneous release could be increased either directly, as a result of a presynaptic action of quisqualate, or indirectly, by activation of the Purkinje neuron resulting in K-efflux or some other retrograde signal. Therefore two manipulations were performed to minimize the contribution of presynaptic processes. First, at the outset of the experiment, adenosine (100 μM) was included in the external saline to suppress spontaneous transmitter release (see Dunwiddie 1990 for review). In the cerebellum, large numbers of adenosine receptors are present on PF terminals (Goodman et al. 1983); application of adenosine has been shown to potently block evoked PF-mediated, but not CF-mediated, synaptic drive (Kocsis et al. 1984). Addition of adenosine (100 μM) to the external saline reduced the frequency of spontaneous synaptic currents to <10% of baseline values measured using either perforated-patch or conventional whole-cell recording, the latter using a Cs-containing recording electrode saline. Second, Purkinje neurons were physically isolated by scraping away adjacent cellular material in the culture dish, returning the dish to the incubator for 18–24 hours, and then conducting two-site LTD induction; see Figure 1. Following experiments with scraping, living cultures were stained with rhodamine 123, a vital stain for mitochondria sometimes used to image presynaptic terminals in neuronal culture (Yoshikami & Okun 1984), in order to confirm the absence of viable terminals on the isolated Purkinje cell. Neither application of adenosine, nor physical isolation, nor the two treatments in combination interfered with the induction of input-specific LTD. These findings strongly suggest that postsynaptic processes are sufficient to confer input-specificity upon cerebellar LTD in culture.

This finding suggests that a spatially restricted signal exists in the postsynaptic cell during the induction of LTD. The question remains as to which signal or signals are required to induce LTD in an input-specific manner. Because Ca influx through voltage-gated Ca channels causes an increase in internal Ca concentration that is distributed widely across the dendritic arbor (Konnerth et al. 1992), it is not a good candidate for such a process. The consequences of local glutamate release (or quisqualate application in the present experiments), namely activation of AMPA and metabotropic receptors and the consequent stimulation of PKC, seem most likely to fulfill this requirement.

To evaluate this possibility, the PKC-activating phorbol ester phorbol-12,13-dibutyrate (PDBu) was applied locally using the two-site stimulation protocol. Local application of PDBu (0.2 μM, at 0–5 min) to site 1 by microperfusion resulted in a specific depression of site 1 responses that was not altered by subsequent conjunctive stimulation of site 2. While phorbol esters such as PDBu are potent exogenous activators of PKC, it is not clear that they mimic endogenous routes of PKC activation. Whereas phorbol esters activate PKC by reducing the Ca concentration required for activation to intracellular resting levels, the natural activators of PKC (diacylglycerols, unsaturated fatty acids) are likely to act synergistically with intracellular Ca transients, and with each other, to cause PKC activation (Nishizuka 1992). In addition, phorbol esters have effects other than PKC activation. For example, they are activators of the enzyme phospholipase A_2 (Mallorga et al. 1980), which may or may not be activated as a consequence of metabotropic receptor activation in Purkinje neurons.

To address this issue, 1-oleoyl-2-acetylglycerol (OAG), a somewhat more physiological activator of PKC, was employed. Unlike PDBu, local application of OAG (10 μM) to site 1 together with AMPA test pulses did not result in a depression of Purkinje neuron responsivity monitored at either site. However, when OAG was applied to site 1 together with step depolarizations (10 steps to −10 mV, 4 seconds long, 20-second interstep interval) and AMPA test pulses, a depression specific to site 1 was induced. AMPA test pulses were used instead of quisqualate in this case because quisqualate/depolarization conjunction alone is sufficient to induce LTD (Linden et al. 1991).

Taken together, these experiments suggest that input-specificity of LTD is conferred, at least in part, by spatially constrained activation of PKC in the Purkinje neuron dendrite. It is likely that this spatially constrained activation results from the conjunction of a broad Ca signal contributed by direct depolarization of the Purkinje neuron (or by CF activation in the slice or intact cerebellum) and a constrained signal or set of signals contributed by AMPA and metabotropic receptor activation (or PF activation). The role of PFs seems limited to simple release of glutamate upon stimulation. Alteration in glutamate release as a consequence of LTD induction does not appear to be necessary, nor does release of any signal in addition to glutamate.

6. Conclusion: Some thoughts on method

There are obvious limitations to the study of a complex, multisynaptic phenomenon such as cerebellar LTD using a cell culture system. Even if the biochemistry of particular cell types, the pharmacology of particular synapses, and the gross morphology of the Purkinje cell itself remain similar to that *in vivo,* there are still significant problems of interpretation. Certainly, the cell culture system does not maintain the wiring diagram of the cerebellar cortex, and as a consequence, aspects of information processing and/or storage that are dependent on this level of structure cannot be examined using such a reduced preparation. There are several other limitations that should be explicitly considered:

(1) The monitoring of Purkinje neuron responsivity through the application of exogenous agonist is potentially problematic in that one will monitor the function of a certain number of extrasynaptic receptors that may not be relevant to alterations in synaptic strength. It should be noted, however, that the detection of LTD by test pulses of exogenous excitatory AMPA receptor agonists may be seen following LTD induction using either a glutamate/depolarization conjunction protocol (Linden et al. 1991; Shigemoto et al. 1994) or a PF activation/depolarization protocol (Crépel & Krupa 1988), suggesting that these test pulses activate a high proportion of synaptic (as opposed to extrasynaptic) receptors.

(2) The cell culture system may be less useful in evaluating the effects on LTD of extrinsic modulatory systems that project to the cerebellum, such as the noradrenergic (Bloom et al. 1971; Siggins et al. 1971) and serotonergic (Strahlendorf et al. 1979) systems. For example the β-adrenergic receptor, which is present on Purkinje neurons *in vivo,* is not strongly expressed in cultured embryonic Purkinje neurons as assessed electrophysiologically (Linden, unpublished observations).

(3) The cell culture system is probably not appropriate for parametric analysis of LTD induction for several reasons. First, the timing constraints present with synaptic stimulation cannot be analyzed with the application of exogenous transmitters because the evoked currents are many times slower in the latter case. Second, as has been discussed previously, these parameters are likely to be dependent upon disynaptic inhibitory function along specific delay lines not recruited in culture. Third, the present protocol seems to induce an unusually consistent amplitude of LTD (about 50% of baseline), suggesting that it is maximally induced with a single treatment and making it difficult to assess additivity of multiple depressive events.

(4) There have been differing conclusions reached with culture versus slice/intact protocols in some cases. There is no case, however, where all of the studies conducted in the slice system are in agreement and all of those conducted in the culture system are in disagreement. Hence, there is not at present an example where disagreement between labs can be definitively attributed to choice of preparation.

This is not to say that the cell culture preparation is without use. It is extremely well suited to analysis at the level of the single synapse or single postsynaptic neuron. In the case of cerebellar LTD induction, where presynaptic alterations have not been found using any preparation, it has been particularly useful in defining certain aspects of this phenomenon. The demonstration that input specificity of cerebellar LTD may be conferred by local activation of PKC in the postsynaptic compartment is perhaps the best example of this. In addition, in certain cases the inherent variability of cell culture preparations may be exploited for correlational studies (Fig. 2). In my view, analysis of cerebellar LTD in cell culture should be regarded as a useful tool in determining the biochemical and biophysical underpinnings of this plastic process.

ACKNOWLEDGMENT

Dorit Gurfel provided skillful technical assistance. Figure 1 (Top) was provided by Karl Schilling, and Figure 1 (Bottom) by Michael Dickinson. This work was supported in part by Public Health Service Grant MH51106, a Klingenstein Fellowship, an Alfred P. Sloan Research Fellowship, a McKnight Scholarship, a Young Investigator Award from The National Alliance for Research on Schizophrenia and Depression, and an award from the Develbiss Fund.

NOTE

1. This manuscript represents the state of the field in August, 1994, the time of its final revision.

2

Cellular mechanisms of long-term depression in the cerebellum

F. Crépel, N. Hemart, D. Jaillard, and H. Daniel
Laboratoire de Neurobiologie et Neuropharmacologie du Développement, CNRS URA 1121, Bat. 440, Université Paris-Sud, France
Electronic mail: *crepelf@dialup.francenet.fr*

Abstract: Long-term depression (LTD) of synaptic transmission at parallel fibre–Purkinje cell synapses is thought to be a cellular substrate of motor learning in the cerebellum. This use-dependent change in synaptic efficacy is induced by conjunctive stimulation of parallel fibres and climbing fibres. Researchers agree that the induction of LTD requires, as an initial step, a calcium influx via voltage-gated Ca^{2+} channels into a Purkinje cell, together with activation of ionotropic (AMPA) and probably metabotropic subtypes of glutamate receptors of this cell. Indeed, due to the lack of specific antagonist, the final demonstration of the contribution of metabotropic receptors in the LTD induction process, under functional conditions, remains unanswered. The debate is now focused on the second-messenger processes leading to LTD of synaptic transmission at parallel fibre–Purkinje cell synapses, after the calcium influx into the cell. All researchers agree that a calcium-dependent cascade of events including activation of protein kinase C is necessary for LTD induction. In contrast, the recruitment in the LTD induction of another cascade, also triggered by Ca^{2+}, that is, through synthesis of nitric oxide and cyclic GMP, remains controversial.

On the other hand, growing evidence suggests that these chains of reaction underlying LTD might ultimately lead to a genuine change in the functional characteristics of AMPA receptors at the parallel fibre–Purkinje cell synapses.

Keywords: cerebellum; cGMP; desensitization; excitatory amino-acid receptors; nitric oxide; protein kinase C; synaptic-plasticity

1. Introduction

Receptors to excitatory amino acids (EAA) can be broadly divided into two groups: N-methyl-D-aspartate (NMDA) and non-NMDA receptor types (Mayer & Westbrook 1987). Until recently, the non-NMDA group was thought to be constituted by two separate classes of receptors, that is, the Kainate (KA) and the Quisqualate (QA) receptors (Watkins 1981). The situation is now more complex for several reasons. First, QA receptors can be split into two subclasses: (1) the ionotropic QA receptors (Qi) coupled to a cationic ionic channel (also termed α-amino-3 hydroxy-5-methyl-isoxalone-4-propionate [AMPA] receptor), and the metabotropic QA receptors (mGlu) coupled to phospholipase C (PLC) (Recasens et al. 1987; Sladeczek et al. 1985; Sugiyama et al. 1987). (2), several subunits of the AMPA receptor have been cloned recently: These subunits – named GluR1, 2, 3, and 4 (GluR1–4) – exist in two versions (flip and flop) generated by an alternative splicing (Boulter et al. 1990; Hollman et al. 1989; Keinänen et al. 1990; Sommer et al. 1990) and are also characterized by a low affinity for KA (Nakanishi et al. 1990). A biochemical analysis of the AMPA receptor-channel purified from rat brain has shown that this receptor is a pentameric structure composed of combinations of the GluR1–4 subunits, the presence of additional subunits being very unlikely (Wenthold et al. 1992). Third, one knows now that high-affinity KA receptors are also likely to be heteromeric structures since five subunits (GluR5–7 and KA1–2) have been recently cloned (Sommer & Seeburg 1992; see also Gasic & Hollman 1992). Finally, five subunits of NMDA (NMDAR1 and NMDAR2A–D) and 6 mGlu receptors have also been recently identified (Nakanishi 1992), which makes the situation even more complex.

In most types of neurons studied so far, both NMDA and non-NMDA receptors to EAAs are well represented, for instance in hippocampal and neocortical cells (Artola & Singer 1987; 1990; Bindman et al. 1988; Collingridge et al. 1983; Hirsch & Crépel 1990). Moreover, they can be located on the same postsynaptic zones, as recently demonstrated in the hippocampus (Bekkers & Stevens 1989).

Another major reason for the current interest in receptors to EAAs is the discovery of their involvement in long-term changes in synaptic strength, which is thought to play a crucial role in learning and memory processes (Hebb 1949). It is now established that NMDA and non-NMDA receptors are responsible for the induction and the maintenance of long-term potentiation (LTP) of synaptic transmission in the hippocampus (Bliss & Lynch 1988; see also Collingridge et al. 1983).

The cerebellum of mammals is another interesting region of the brain where we can study the involvement EAA receptors in synaptic plasticity. First, the two main excitatory afferents to Purkinje cells (PCs), that is, parallel fibres (PFs) and climbing fibres (CFs, Eccles et al. 1967) are likely to use glutamate (Glu) as a neurotransmitter (Cuenod et al. 1989; Herdon & Coyle 1978; Hudson et al. 1976; Zhang et al. 1990). Second, and in marked contrast with most other neuronal cell types, mature PCs only bear non-NMDA receptors (Crépel et al. 1982; see also Crépel & Audinat 1991), which make them an interesting model for study of synaptic plasticity in the absence of this class of receptors. Finally, the participation of the cerebellum in motor learning was postulated by Brindley as early as 1964 (Ito 1984).

This hypothesis was formalized by Marr (1969), who proposed the so-called "external teacher" theory derived from the Perceptron, a cybernetic machine invented by Rosenblatt in 1962. According to the theory, CFs are able to modify the gain of synaptic transmission between PFs and CPs during motor learning in order to adjust the cerebellar output to the desired motor command. In this scheme, conjunctive activation of PCs by CFs and PFs lead to an LTP of synaptic transmission at PF-PC synapses. Later on, Albus revisited the theory and proposed instead that a long-term depression (LTD) of synaptic transmission occurs at PF-PC synapses following their coactivation with CFs (Albus 1971).

The present target article will deal with the main experimental data, gathered over the last 10 years, on the participation of EAA receptors of PCs in synaptic plasticity as a possible cellular basis of motor learning in the cerebellum.

1.1. Discovery of long-term depression. In keeping with the Marr-Albus theory of motor learning in the cerebellum, Ito and coworkers have been the first to demonstrate in *in vivo* experiments (Ito et al. 1982; see also Ito 1987; 1989) that, in rabbit cerebellum, conjunctive stimulation of PFs and CFs lead to LTD of synaptic transmission at PF-PC synapses. The fact that only those PF-PC synapses activated in conjunction with CFs are affected (Ito 1984) indicates that the changes in synaptic strength are restricted to the activated synapses. In the same paper (Ito et al. 1982), the authors also showed that coactivation of PCs by CFs and by direct application of Glu in their dendritic fields by ionophoretic electrodes leads to a persistent decrease of the responsiveness of PCs to this agonist, an observation that was later reproduced and extended (see below) and that suggests that part of the phenomenon occurs postsynaptically.

The demonstration by Kano and Kato (1987) that pairing CF input with Glu or QA application on PC dendrites also induces subsequent LTD of synaptic transmission between PFs and PCs, whereas KA, Asp, and NMDA are ineffective in this respect, is of importance because it strongly suggests that AMPA receptors of PCs are indeed involved in LTD.

1.2. Role of calcium in long-term depression of synaptic transmission. In *in vivo* experiments, involvement of Ca^{2+} in induction of LTD was initially suggested by the observation that stellate cell inhibition prevents LTD from occurring (Ekerot & Kano 1985), probably by blocking Ca^{2+}-dependent plateau potentials in PC dendrites following their activation by CFs (Ekerot & Oscarsson 1981). This observation led Ito to propose that the efficacy of PF-PC synapses is decreased as a result of both the activation of Glu receptors of PCs and the Ca^{2+} influx that occurs in these cells during their activation by CFs (Ito 1987; 1989).

Indeed, in more recent experiments in rat cerebellar slices, LTD of PF-mediated EPSPs (excitatory postsynaptic potentials) is consistently induced by pairing these synaptic responses with Ca^{2+} spikes directly induced in the postsynaptic cell by depolarizing current pulses (Fig. 1). In contrast, when only sodium spikes are induced in PCs during the pairing protocol, LTD is no longer observed and is replaced by LTP of PF-mediated EPSPs (Crépel & Jaillard 1991). Along the same line, in guinea pig cerebellar slices, LTD of PF-mediated EPSPs is not induced by their pairing with CF inputs when PCs are loaded with EGTA (Sakurai 1990). The involvement of Ca^{2+} in LTD induction

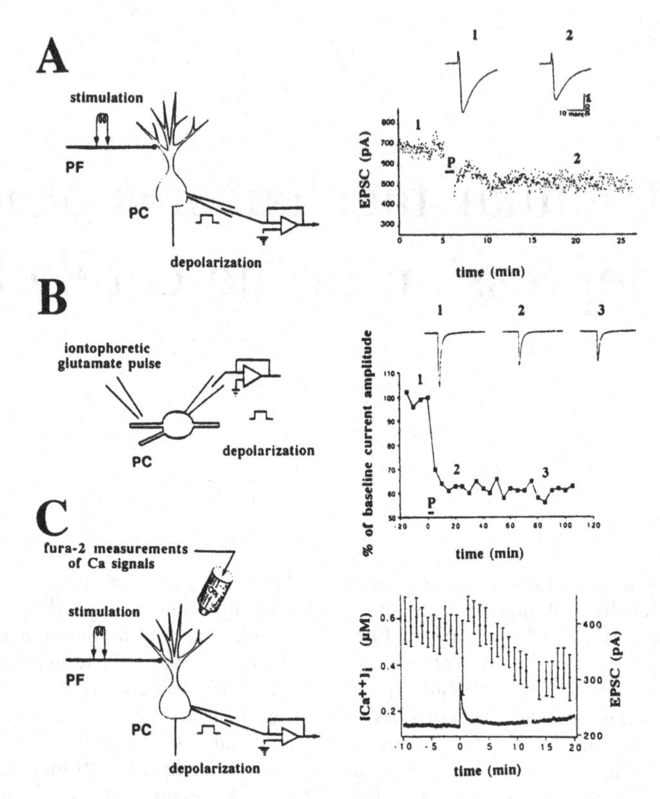

Figure 1. Role of calcium in long-term depression of PF-transmission (diagrams of experimental arrangements are shown on the left of each panel).

A: Example of LTD in a cerebellar slice, induced by pairing **(P)** PF-mediated EPSCs with depolarization of PCs giving rise to Ca^{2+} spike (duration = 1 min). Plot of EPSC amplitudes against time before and after the pairing. Insets display (1) control EPSC and (2) EPSC 15 min after the end of the pairing period. Adapted from Hemart et al. (1994).

B: Conjunctive depolarization of cultured PC to −20 mV (from t = 0 to 4 min, **P**) and glutamate application (at 0.05 Hz) produces LTD of glutamate currents. Insets display corresponding glutamate current traces before and after LTD induction. Adapted from Linden et al. (1991).

C: Time course of changes of PF-EPSC amplitudes and $[Ca]_i$ (measured with fura-2 fluorescence signal) in a PC of cerebellar slice. At time = 0 min, a series of eight depolarizing pulses (10 msec duration, from −60 to 0 mV) were paired with the PF stimulation. Adapted from Konnerth et al. (1992).

has been confirmed still more recently by two independent groups. First, in cultured PCs, Linden et al. (1991) have shown that LTD of AMPA-mediated currents is induced by conjunctive ionophoretic glutamate pulses and PC depolarization sufficient to produce Ca^{2+} entry through voltage-gated Ca^{2+} channels (Fig. 1). Second, by combining measurements of synaptic efficacy with fura-2 measurements of intracellular Ca^{2+} concentration in single patch-clamped PCs in thin slices, Konnerth and coworkers (1992) have shown that in pairing experiments, a transient rise in internal Ca^{2+} is sufficient to induce LTD (Fig. 1). It must be emphasized that the fact that Ca^{2+} concentration goes back to a normal level after LTD induction rules out the possibility that this form of synaptic plasticity is merely a pathological process. This is in keeping with the absence of LTD in pseudopairing experiments, that is, when PCs are depolarized to activate voltage-dependent Ca^{2+} channels in the absence of PF stimulation (Daniel et al. 1992), as well as with the fact that LTD was restricted to the activated synapses in the initial *in vivo* experiments of Ito and coworkers (see above).

1.3. Which glutamate receptors participate in LTD induction and expression? The role of AMPA receptors in induction and expression of LTD has been further investigated in whole-cell clamped PCs in acute slices, by using the same type of pairing protocol between PF-EPSPS and Ca^{2+} spike firing as before (see above). These experiments showed that when the pairing protocol was performed in the presence of a sufficient concentration of CNQX in the bath to block synaptic transmission, no LTD could be induced in any of the tested cells (Fig. 2). In contrast, when CNQX was bath applied after the induction of LTD, this did not prevent its maintenance after wash-out of the compound (Crepel et al., unpublished data). Similarly, in cultured PCs, Linden and coworkers (1991) also showed that no LTD occurs unless AMPA receptors are activated during its induction phase (Fig. 2). Of course, these experiments provide strong evidence of an involvement of AMPA receptors of PCs in LTD induction.

Now, in experiments performed with the grease-gap technique, Ito and Karachot (1990) have shown that coactivation of AMPA receptors of PCs by AMPA and of their mGlu receptors by trans-1-amino-cyclopentyl-1,3-dicarboxylate (trans-ACPD) is sufficient to induce a long-lasting desensitization of AMPA receptors of these cells (Fig. 3). According to these authors, this suggests that activation of mGlu receptors of PCs also plays an important role in LTD. This view was supported by more recent experiments by Linden and coworkers (1991) showing that in cultured PCs, induction of LTD by conjunctive depolarization and stimulation of Glu receptors requires both the activation of their AMPA and mGlu receptor subtypes.

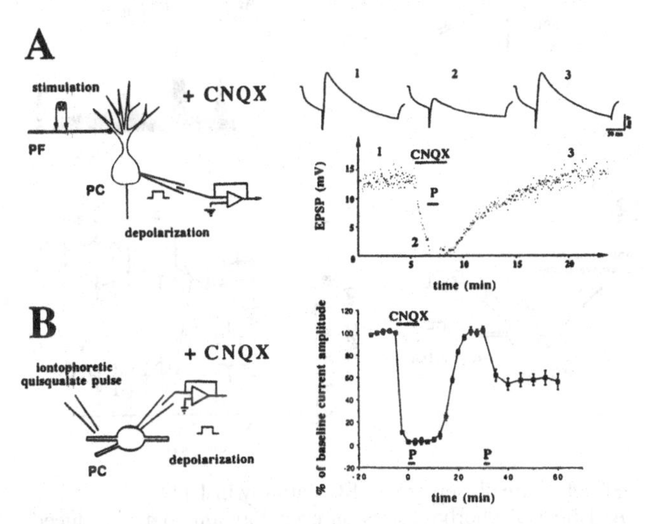

Figure 2. Involvement of AMPA receptor in LTD induction.

A: In cerebellar slices, pairing PF-mediated EPSCs with Ca spikes does not induce LTD in the presence of 4 μm CNQX in the bath. Plot of EPSC amplitudes against time before and after the pairing protocol (**P**). Insets: averaged EPSCs at the indicated times. Crépel et al., unpublished.

B: Conjunctive application of quisqualate with depolarization of cultured PC (**P**) in the presence of 20 μm CNQX failed to induce LTD of quisqualate currents after wash-out of CNQX. A second conjunctive stimulation (applied at t = 30 min) in the absence of CNQX-induced LTD. Adapted from Linden et al. (1991).

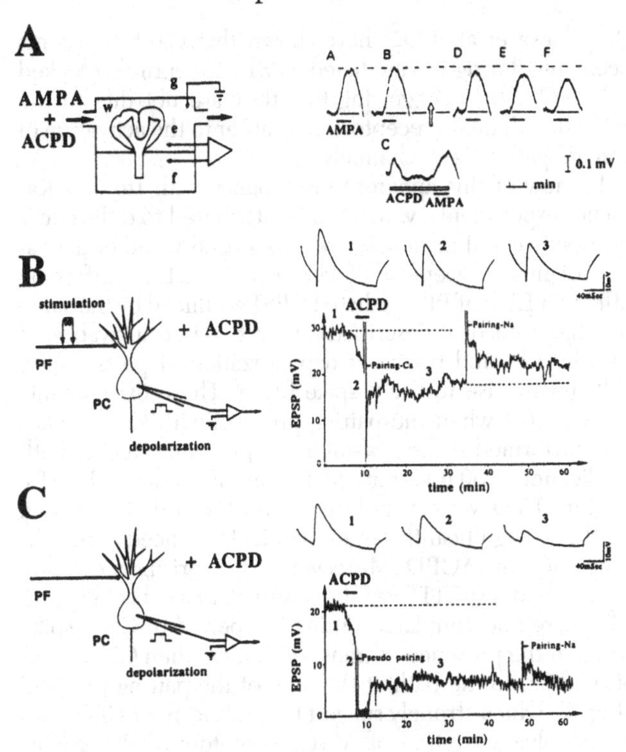

Figure 3. Involvement of metabotropic receptor in LTD induction.

A: Effects of conjunctive application of AMPA (10 μM) and t-ACPD (0.3 mM). Records A–F were taken at 10 min intervals and show AMPA-induced potentials. Upward arrow indicates that C lies between B and D. (Experimental arrangement: *w*–slice in a wedgeform; arrows indicate flow of perfusate; *e* and *f* are recording electrodes connected to d.c. amplifier; g–ground.) Adapted from Ito & Karachot (1990).

B: Effects of pairing of PF-mediated EPSPs with Ca^{2+} spikes performed in the presence of t-ACPD in the bath. Plot of EPSP amplitudes against time. Insets (1) display control EPSP, (2) EPSP under application of t-ACPD, and (3) EPSP 15 min after the pairing protocol performed in the presence of trans-ACPD.

C: Plot of EPSP amplitudes against time in another Purkinje cell before and after a pseudopairing protocol performed in the presence of trans-ACPD. Insets: EPSPs at the indicated time. (In this protocol, PF stimulation was omitted during the 1 min period of Ca^{2+} spike firing.) B and C adapted from Daniel et al. (1992).

We therefore decided to test this hypothesis differently in acute slices, specifically by coactivating AMPA and metabotropic receptors of intracellularly recorded PCs by PF stimulations and by bath application of trans-ACPD respectively. In all tested cells, trans-ACPD induced a marked decrease of PF-EPSPs. However, this effect was fully reversible after wash-out of the drug in all cells where no other effect of trans-ACPD was observed, and this, even in cells where PFs were stimulated at 1 Hz during trans-ACPD application. In a fraction of the recorded cells, trans-ACPD also induced a transient depolarization of PCs and a bursting firing of Ca^{2+} spikes. In this case only, an LTD of PF-mediated EPSPs was occasionally observed after wash-out of trans-ACPD. These results (Crépel et al. 1991), which have been recently confirmed by Glaum et al. (1992), do not support the hypothesis that mGlu receptors play a major role in LTD induction in acute slices, at least when voltage dependent Ca^{2+} channels are not activated at the same time. In keeping with this view, Ross and co-workers

(Miyakawa et al. 1992) have shown that Ca^{2+} transients occurring during PF-mediated EPSPs are entirely blocked by CNQX, thus suggesting that they are not due to the activation of mGlu receptors, but rather to the activation of voltage-gated Ca^{2+} channels.

In view of this apparent discrepancy with Ito and Karachot experiments, which can be attributed to differences in experimental protocols, we also looked whether activation of mGlu receptors of PCs has at least a reinforcing effect on LTD of PF-mediated EPSPs induced by the same pairing protocol as described earlier, that is, coactivation of PCs by PFs and by direct depolarization of postsynaptic cells giving rise to Ca^{2+} spike firing. These experiments showed that when the pairing protocol with Ca^{2+} spikes was performed at the peak of the depressant effect of bath application of 50 μM trans-ACPD on PF-mediated EPSPs, a robust LTD was now induced in most recorded cells (Fig. 3), that is, significantly larger than LTD induced in the absence of trans-ACPD. Moreover and surprisingly enough, the same robust LTD was also observed in most cells where PFs were not stimulated during the period of Ca^{2+} spike firing in the presence of trans-ACPD, or when CNQX was also added to the bath at the time of the pairing protocol (Fig. 3), which strongly suggest that this form of LTD does not involve activation of AMPA receptors of PCs for its induction (Daniel et al. 1992).

The present results therefore suggest that LTD can be induced by two different mechanisms: both involve activation of voltage gated Ca^{2+} channels of PCs, but one also depends primarily on the activation of ionotropic AMPA receptors (LTD_{AMPA}), whereas the other does not involve this class of receptors for its induction and depends instead on a strong activation of metabotropic Glu receptors (LTD_{mGlu}).

Interestingly enough, induction of the LTD_{mGlu} is blocked when slices are preincubated with 10 μM Thapsigargin, whereas LTD_{AMPA} is not (Hemart et al., unpublished observation). For the former, this suggests that Ca^{2+} release from intracellular stores plays an important role in LTD induction. For the latter, that is, LTD_{AMPA}, this mechanism does not seem to operate. Now, what is the origin of Ca^{2+} release from internal stores in LTD_{mGlu}? Fluorometric measurements show little evidence for Ca^{2+} release from intracellular stores in the presence of trans-ACPD) Lláno et al. 1991; Staub et al. 1992; Vranesic et al. 1991). Therefore, another hypothesis is that activation of metabotropic receptors at PF-PC synapses by trans-ACPD increases Ca^{2+} currents flowing through voltage-gated Ca^{2+} channels, which in turn would be now sufficient to activate a Ca^{2+} induced Ca^{2+} release in the dendrites. At this point, it must be emphasized that we still have no evidence that the AMPA receptor-dependent form of LTD in acute slices also depends on metabotropic receptors, in particular because no good antagonists of mGlu receptors are actually available.

1.4. Second messenger cascades involved in LTD. One knows that the calcium-dependent PKC I is very abundant in PCs (Hidaka et al. 1988; Nishizuka 1986). One also knows that mGlu receptors are located at PF-PC synapses (Martin et al. 1992) and are coupled with phospholipase C (PLC) (Nicoletti et al. 1986; Recasens et al. 1987; Sladeczek et al. 1985; Sugiyama et al. 1987). It was therefore tempting to postulate (Crépel & Krupa 1988) that the cascade of events leading to LTD involves a coactivation of PKC of PCs by Ca entry through voltage-gated ionic channels, and by diacylglycerol (DAG) produced by the activation of mGlu receptors by Glu released by PFs.

Indeed, a selective LTD of the responsiveness of PCs to Glu and QA (i.e., Asp-induced responses were unaffected) was obtained in about 40% of PCs in the presence of phorbol esters known to activate PKC (Fig. 4), whereas inactive analogs were without any effect (Crépel & Krupa 1988; 1990). Moreover, LTD of PF-mediated EPSPs following their pairing with Ca^{2+} spikes in rat cerebellar slices *in vitro* was nearly totally prevented by bath application of polymixin B, a potent blocker of PKC and, to a lesser extent, of calmodulin (CAM)-dependent kinase (Crépel & Jaillard 1990), and the same results were obtained when the PKC inhibitor peptide 19-36 was directly injected in the recorded cells (Hemart et al., unpublished data). Finally, these results were recently confirmed and extended in cultured PCs (Fig. 4; Linden & Connor 1991). It must be emphasized that they give indirect support to the view that mGlu receptors are involved in LTD induction.

There is now experimental evidence that LTD also depends on another cascade of events beyond Ca^{2+} entry into PCs. Indeed, it is known that Ca^{2+} can induce the formation of NO from arginine by activating a CAM-dependent NO-synthase (Garthwaite et al. 1988; 1989; see also, Ross et al. 1990). Nitric oxide (NO) is highly diffusible

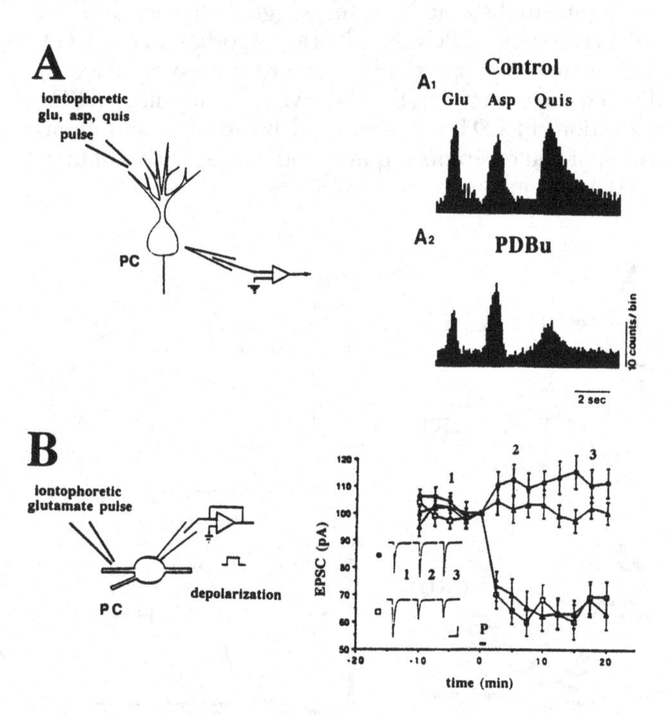

Figure 4. Involvement of PKC pathway in LTD.

A: Effect of phorbol esters on excitatory amino acid induced responses. Poststimulus time histograms of the responses of an extracellularly recorded PC to Glu, Asp, and Quis, before (A1) and after (A2) bath application of 400 μm of phorbol 12 13-dibutyrate (PDBu). Adapted from Crépel & Krupa (1990, pp. 323–329).

B: Conjunctive depolarization of cultured PC and glutamate application (**P**) produces LTD of glutamate currents with the following internal perfusates in the pipette: vehicle (Δ) and vehicle plus [glu27] PKC(19-36) (, 10 μm), but no LTD occurred with vehicle plus BAPTA (, 20 μm) and vehicle plus PKC (19-36) (, 10 μm). Insets: EPSCs at the indicated times. Scale bars = 100 pA, 2 sec. Adapted from Linden & Connor (1991).

and thus activates soluble guanylate cyclase (Tremblay et al. 1988) in cells where it is produced, as well as in surrounding cellular elements (Garthwaite et al. 1989 and in Ross et al. 1990). This cascade of events can therefore activate cGMP-dependent protein kinases in PCs where this enzyme is particularly abundant (Lohmann et al. 1981), as well as in PFs and glial cells.

It is therefore of prime importance that, in intracellularly recorded PCs with sharp electrodes in acute slices, N-monomethylarginine (L-NMMA), a potent inhibitor of NO synthesis (Knowles et al. 1989), nearly totally prevented LTD of PF-mediated EPSPs following their pairing with Ca^{2+} spikes; the same effect was obtained by methylene blue, which acts by blocking activation of soluble guanylate cyclase by NO (Crépel & Jaillard 1990). At the same time, Ito & Karachot (1990) also established (with the grease-gap method) that desensitization of AMPA receptors of PCs is likely to require the production of cyclic GMP via the production of NO. Later on, Shibuki & Okada (1991) showed that protocols known to induce LTD indeed lead to the production of NO in cerebellar slices. However, neither of these studies on the role of NO in LTD induction were performed at a cellular level (Ito & Karachot 1990; Shibuki & Okada 1991), and they only used blockers of NO to test its role in LTD (Crépel & Jaillard 1990). Furthermore, none of these studies dealt precisely with the site(s) of production and of action of NO. Finally, it has been recently claimed that LTD of glutamate currents in cultured PCs does not require NO signaling (Linden & Connor 1992). The role of NO in the induction of LTD at PF-PC synapses was therefore reinvestigated in whole-cell, patch-clamped PCs in thin slices *in vitro* (Daniel et al. 1993). It was thus confirmed that bath application of L-NMMA consistently partially prevents the induction of LTD of PF-mediated EPSPs induced by their pairing with Ca^{2+} spikes (Fig. 5), and that the effect of L-NMMA can be reversed by an excess of arginine. It was also shown that bath application of NO donors and of 8-bromoguanosine 3′:5′ cyclic monophosphate (8-bromo-c-GMP) are able to reproduce a LTD-like phenomenon. Finally, LTD of PF-mediated EPSPs was also induced when NO donors or guanosine 3′:5′ cyclic monophosphate (c-GMP) were directly dialyzed into PCs and these LTD-like effects partially occluded LTD induced by pairing protocols (Fig. 5). Therefore, these results show that NO indeed plays a role in LTD induction, and demonstrate for the first time that its site of action is probably the soluble guanylate cyclase of PCs. According to Ito and Karachot (1992), cGMP would in turn activate a cGMP-dependent protein kinase (PKG), thereby allowing phosphorylation of its specific substrate G-substrate, a potent inhibitor of phosphatases. In this scheme, LTD would involve both phosphorylation of AMPA receptors of PCs by PKC, and inhibition of their dephosphorylation by the NO route (Fig. 6).

However, in the cerebellum, NO-synthase has not been identified in PCs, but only in neighbouring elements (Bredt et al. 1990; Southam et al. 1992). Thus, one hypothesis is that NO might be produced by the NO-synthase located in PFs or in basket cells when these neurons are activated by both PF stimulations (Eccles et al. 1967) and by the large efflux of potassium that follows the entry of Ca^{2+} in PCs during pairing experiments. These possible paracrine interactions are illustrated in Figure 6. However, it is still possible that the NO-synthase involved in the induction of

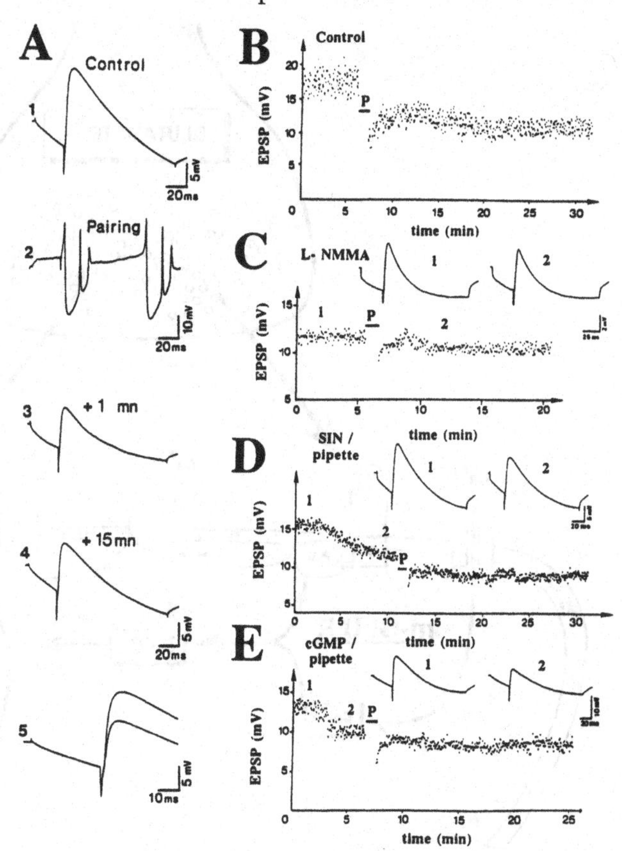

Figure 5. Involvement of NO pathway in LTD.
A: Effects of pairing on PF-mediated EPSPs in standard bathing medium. Averaged EPSP evoked in a Purkinje cell during the control period (A1), then 1 min (A2) and 15 min (A3) after the end of the pairing period with Ca spikes (A2). The EPSPs in A1 and A4 are superimposed (A5).
B: Plot of EPSP amplitudes against time for the cell illustrated in A, before and after the pairing protocol (**P**).
C to **E:** Plot of EPSP amplitudes against time in three other Purkinje cells before and after a pairing protocol (**P**) with respectively, 30 μm L-NMMA in the bath (C), 3 mM SIN-1 added in the recording pipette (D) and 0.5 mM cGMP added in the recording pipette (E). Insets: EPSPs at the indicated times. Adapted from Daniel et al. (1993).

LTD is located in PCs themselves, but is different from the already known forms of the enzyme, thus explaining why it has not been visualized so far.

On the other hand, in these experiments on the role of NO in LTD, the fact that the decrease in synaptic efficacy induced by a pairing protocol with Ca^{2+} spikes was not completely occluded by that induced by cGMP supports the view that, in acute slices, LTD involves two different routes, via PKC and via NO respectively (Fig. 6). Accordingly, we have shown recently that the induction of LTD is totally prevented when cells are bathed with L-NMMA and dialysed at the same time through the patch pipette with the PKC inhibitor peptide 19-36 (Crépel et al., unpublished data), whereas, as mentioned before, blockade of LTD induction with only one of the inhibitors is generally only partial. The fact that LTD may involve two different routes might also explain why no effect of NO blockers was seen in experiments on LTD performed in dissociated cell cultures since, in particular, putative NO donors as PFs and basket cells were possibly scarcely represented. It is also

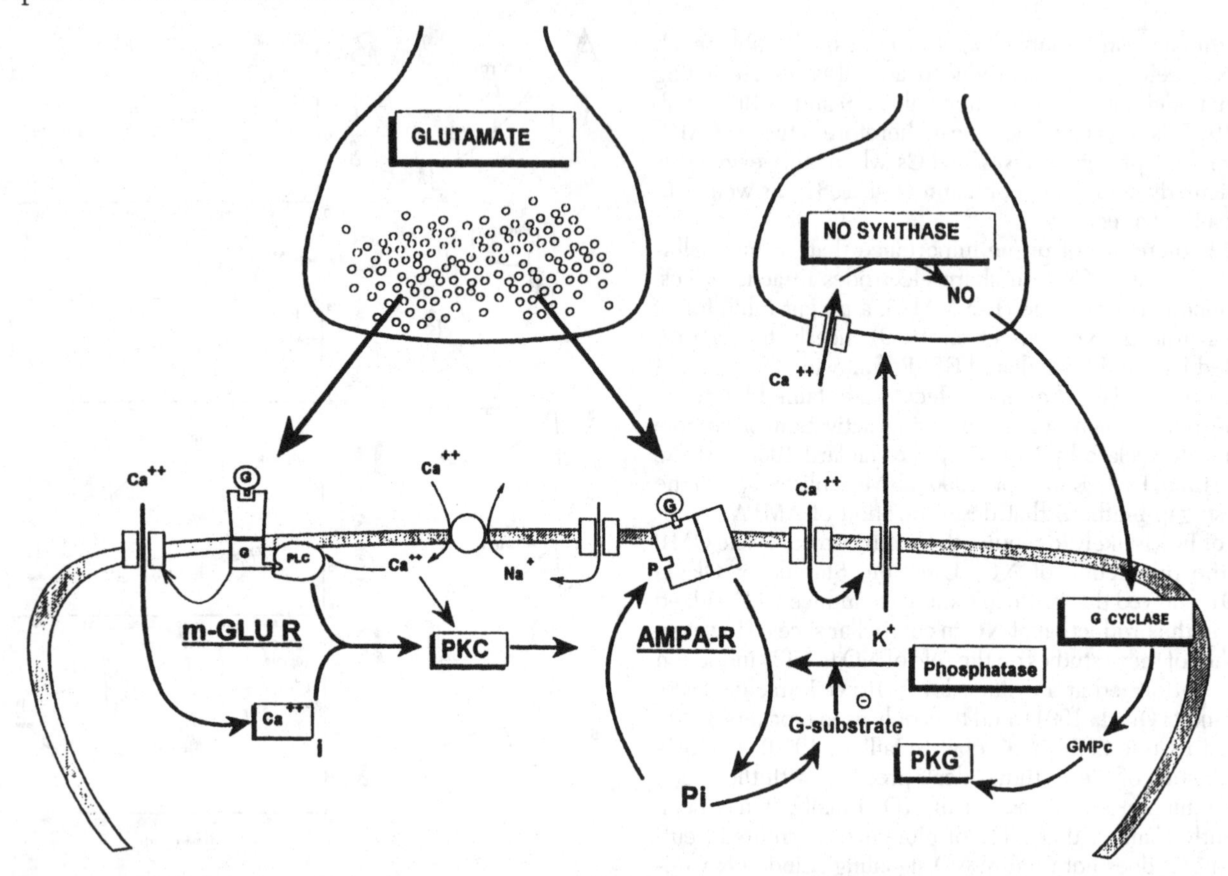

Figure 6. Schematic diagram of the signal transduction processes involving the PKC and NO-pathway that are presumed to underlie LTD. In this scheme, we have illustrated the hypothesis according to which NO-synthase involved in LTD is located outside PCs. Abbreviations: G – Glutamate; *m-GLU R* – Metabotropic receptor; *AMPA-R* – AMPA receptor; G – G protein; PLC – Phospholipase C; PKC – Protein kinase C; NO – Nitric oxide; cGMP – Cyclic guanosine monophosphate; PKG – Protein kinase G; Pi – Inorganic phosphate; P – Phosphorylation.

conceivable that cultured PCs only develop the PKC route and are thus insensitive to bath application of NO donors (Linden & Connor 1992).

Finally, an interesting observation of Ito and Karachot (1992) is that induction of LTD in slices by phorbol esters or by 8-bromo-cGMP requires concomitant activation of AMPA receptors. As we will see below, this fits well with recent experimental evidence showing that LTD induction requires, as an initial step, an agonist-dependent desensitization of AMPA receptors of PCs.

1.5. Does LTD involve desensitization of AMPA receptors of PCs? The fact that coactivation of PCs by CFs and by direct application of Glu in their dendritic fields leads to a persistent decrease of their responsiveness to this agonist (Ito et al. 1982), led Ito to propose that induction of LTD might ultimately lead to a long-term desensitization of ionotropic Glu receptors of PCs and thus to the observed decrease in synaptic efficacy. Accordingly, it was shown later that pairing ionophoretic application of Glu and Ca^{2+} spike firing of PCs induces LTD of their responsiveness to this agonist, both in acute slices (Crépel & Krupa 1988) and in dissociated cultures (Linden & Connor 1991). However, the observed decrease in efficacy of Glu in activating PCs might be due to causes other than a true desensitization of Glu receptors.

Recently, the nootropic compound Aniracetam has been shown to markedly reduce desensitization of AMPA receptors and/or to decrease the closing rate constant for ion channel gating (Isaacson & Nicoll 1991; Ito et al. 1990; Tang et al. 1991; Vyklicky et al. 1991). It was therefore an important finding to show that in whole-cell, clamped PCs in acute slices, Aniracetam has a larger potentiating effect on PF-mediated EPSCs during expression of LTD than is normal and that this compound also significantly blocks the induction of LTD. Indeed, these data strongly support the view that this change in synaptic efficacy involves a genuine change in the functional characteristics of these receptors. Furthermore, and in keeping with previous observations of Ito and Karachot (1992), these results also suggest that induction of LTD requires an initial agonist-dependent desensitization of AMPA receptors of PCs.

In contrast, the comparison of current-voltage curves of PF-mediated EPSCs in patch-clamped PCs in thin slices prior to and after LTD induction shows that this change in synaptic efficacy does not involve a change in reversal potential of the synaptic responses (Crépel et al., unpublished data).

1.6. Molecular composition of AMPA receptors of PCs. Until recently, the subunit composition of native AMPA receptors of PCs was unknown. In order to further characterize the AMPA receptors involved in the LTD, a new method was developed; it combines whole-cell recordings and a molecular analysis, based on the polymerase chain reaction (PCR), of the messenger RNAs harvested into the patch pipette at the end of each recording. We found (Lambolez et al. 1992) that each single Purkinje cell recorded in cerebellar cultures or in olivo-cerebellar cocultures expressed the messenger RNAs encoding the five following subunits of the AMPA receptor: the flip and flop versions of GluR1 and GluR2 as well as GluR3flip, GluR2 being the most abundant. In addition, GluR3flop and GluR4flip were scarcely expressed in half of these neurons and GluR4flop was never detected. These results strongly suggest that the AMPA receptors of Purkinje cells are heterogeneous with respect to their subunit composition. Whether this heterogeneity of the AMPA receptors within a single neuron could be the basis of a functional heterogeneity between synapses is still an open question.

In conclusion, LTD induction and expression can be tentatively explained in the following way, taking into account all experimental evidence mentioned above, as well as previously proposed schemes (Ito & Karachot 1990; 1992). When AMPA receptors of PCs are activated by Glu released by PFs, their agonist-dependent densensitization leads to a conformational change that exposes a phosphorylation site to PKC. Because this kinase has been activated by Ca^{2+} entry through voltage-gated Ca^{2+} channels and by DAG produced by PLC when mGlu receptors at PF-PC synapses are also activated by Glu released by PFs, the AMPA receptors of PCs can now be phosphorylated. On the other hand, the NO route ultimately inhibits phosphatases, thus allowing the phosphorylated state of AMPA receptors to be maintained. As a consequence of this phosphorylation, kinetics of opening and closing of AMPA receptor-coupled channels are likely to be affected, and/or a larger-than-normal fraction of Glu receptors is stabilized in a desensitized state at rest, thus explaining the maintenance of LTD (Fig. 6). In this scheme, the LTD_{mGlu} might be explained if one assumes that, in the absence of AMPA receptor activation by Glu, a phosphorylation of these receptors by PKC is still possible but requires a much larger activation of the kinase, which is the case with bath application of trans-ACPD. Finally, it remains to establish whether LTD is indeed involved in motor learning as suggested by recent *in vivo* experiments (Nagao & Ito 1991).

3

Long-lasting potentiation of GABAergic inhibitory synaptic transmission in cerebellar Purkinje cells: Its properties and possible mechanisms

Masanobu Kano
Laboratory for Neuronal Signal Transduction, Frontier Research Program, Riken, 2-1- Hirosawa, Wako-shi, Saitama 351-01, Japan
Electronic mail: *mkano@postman.riken.go.jp*

Abstract: The cellular basis of motor learning in the cerebellum has been attributed mostly to long-term depression (LTD) at excitatory parallel fiber (PF)-Purkinje cell (PC) synapses. LTD is induced when PFs are activated in conjunction with a climbing fiber (CF), the other excitatory input to PCs. Recently, by using whole-cell patch-clamp recording from PCs in cerebellar slices, a new form of synaptic plasticity was discovered. Stimulation of excitatory CFs induced a long-lasting (usually longer than 30 min) "rebound potentiation (RP)" of γ-amino-butyric acid A (GABAA)-receptor mediated inhibitory postsynaptic currents (IPSCs). As in LTD, induction of RP requires transient elevation of intracellular calcium concentration ($[Ca^{2+}]_i$) due to activation of voltage-gated Ca^{2+} channels. Activity of inhibitory synapses seems also to be necessary for RP to occur. RP is mainly due to up-regulation of postsynaptic GABAA receptor function, since PC response to bath-applied exogenous GABA is also potentiated with a time course similar to RP. The difference in the time scale between the Ca^{2+} transients (10–30 sec) and the durations of RP (>30 min) strongly suggests that some intracellular biochemical machinery is involved. Pharmacological evidence suggests that protein kinases are involved in RP of inhibitory synapses and LTD of excitatory PF synapses. Besides the well-described LTD, RP could be a cellular mechanism that plays an important role in motor learning.

Key words: Ca^{2+}; cerebellum; GABAA receptor; inhibitory synapse; long-lasting potentiation; protein kinase; Purkinje cell

1. Introduction

Activity-dependent long-term modification of transmission efficacy at synapses is thought to be a cellular basis of learning and memory (Kandel & Schwartz 1982). Since the discovery of long-term potentiation (LTP) in the hippocampus (Bliss & Lomo 1973), synapses that undergo plastic change have been described in various parts of the brain, for example, the visual cortex (Komatsu et al. 1981), motor cortex (Iriki et al. 1989), red nucleus (Tsukahara et al. 1975), superior colliculus (Okada & Miyamoto 1989), and cerebellum (Ito et al. 1982). In the cerebellar cortex, long-term depression (LTD) of excitatory parallel fiber (PF) synapse was first described by Ito et al. (1982), and this is thought to be a cellular basis of motor learning in the cerebellum (for review, see Ito 1989; Linden 1994).

It is somewhat surprising that modifiable synapses discovered and studied to date are mostly glutamatergic excitatory synapses. Despite the importance of inhibition in brain functions, plasticity at inhibitory synapses has not been demonstrated so far (except some related to epilepsy; Stelzer et al. 1987). Recently, it has been shown that GABAA receptor-mediated inhibitory synaptic transmission in Purkinje cells (PCs) of the cerebellum undergoes long-lasting potentiation (Kano 1992; Llano 1991b; Vincent et al. 1992). Llano et al. (1991b) and Vincent et al. (1992) reported that inhibitory synaptic transmission and responsiveness of PCs to exogenous GABA are potentiated after depolarization-induced Ca^{2+} entry to PCs. A robust and long-lasting potentiation of inhibitory synaptic transmission is induced following stimulation of excitatory climbing fibers (CFs) (Kano et al. 1992), which has been termed "rebound potentiation (RP)." This newly-found synaptic plasticity may play a role in the motor learning in concert with LTD at excitatory PF synapses. In this target article, I will describe the properties and discuss possible mechanisms of RP based on the data accumulated so far.

2. Properties of RP

RP is found by using whole-cell patch-clamp recording from visually identified PCs in cerebellar slices (Edwards et al. 1989; Kano & Konnerth 1992b; Konnerth 1990). This newly developed technique allows high-resolution recordings of membrane currents in neurons from slice preparations. Under whole-cell recording with the pipette solution containing CsCl (80 mM), Cs D-gluconate (80 mM), $MgCl_2$ (2 mM), EGTA (1 mM), Na-ATP (4 mM), Na-GTP (0.4 mM), and HEPES (10 mM) (pH 7.3), PCs have spontaneously occurring inward synaptic currents that are blocked in the presence of bicuculline (10 μM) (Konnerth et al. 1990; Llano et al. 1991c), which identifies them as GABAA-receptor mediated inhibitory postsynaptic currents (IPSCs). These IPSCs are recorded at a constant frequency. However, it is often observed that the IPSC amplitudes tend to decay slowly even with ATP and GTP in the patch-pipette (see prestimulus period in Figs. 1A and 2). Under whole cell recording conditions, stimulation of CFs induced large all-or-none excitatory postsynaptic currents (EPSCs) in PCs that are accompanied by characteristic regenerative responses (Llano et al. 1991c). CF stimulation (5 shocks at 0.5 Hz) readily induced a marked potentiation of IPSCs (Fig. 1A) in all cells tested (see Table 1). This phenomenon was termed "rebound potentiation" (RP) because it is the enhancement of inhibition triggered by strong excitation of PCs, and it counteracts the tendency of "run down" that is observed in some cells (see Figs. 1A and 2). The amplitude of RP was 139 ± 10.9% (mean ± S.E.M., n = 6, measured 10 min after the conditioning CF stimulation) of the control, and the time to half-recovery was 16.5 ± 3.6 min (n = 5) (Table 1). The RP reaches its maximum in 3–15 min after CF stimulation and a plateau-like phase follows (Fig. 1A).

3. Transient elevation of $[Ca^{2+}]_i$ is required for RP to occur

Stimulation of CFs is known to induce large increases in $[Ca^{2+}]_i$ in PCs (Knöpfel et al. 1991; Konnerth et al. 1992; Miyakawa et al. 1992; Ross & Werman 1987) that are spatially similar to depolarization-induced $[Ca^{2+}]_i$ transients (Lev-Ram et al. 1992; Ross et al. 1990; Ross & Werman 1987; Sugimori & Llinás 1990; Tank et al. 1988). LTD of excitatory PFs are shown to be triggered by this CF-induced elevation of $[Ca^{2+}]_i$ (Hirano 1990b; Konnerth et al. 1992; Sakurai 1990). The possibility that postsynaptic Ca^{2+} is required for RP is tested in the following experiments. When PCs were filled with a solution containing Ca^{2+} chelator BAPTA (30 mM), CF stimulation failed to induce RP (Fig. 1B). Furthermore, activation of voltage-gated Ca^{2+} channels by a single depolarizing pulse (from a

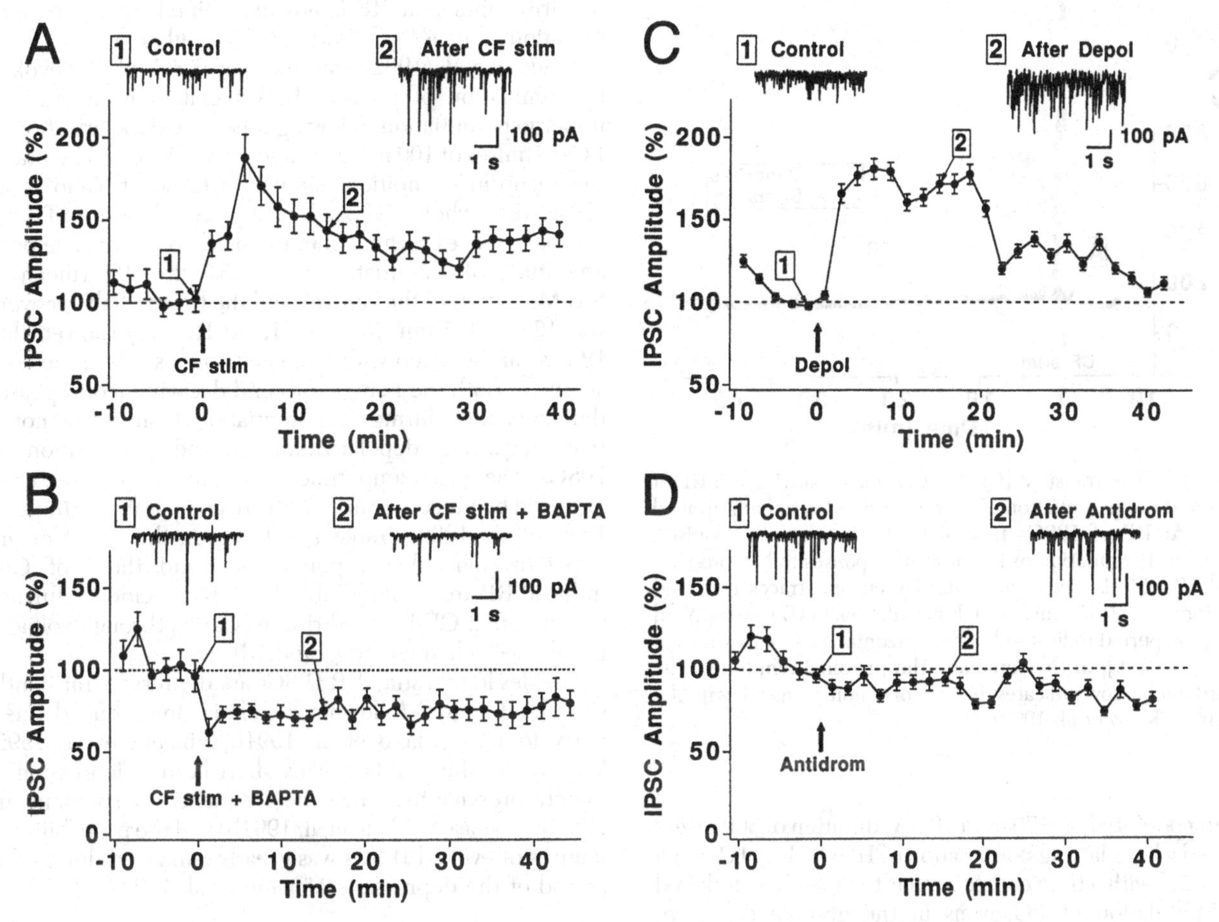

Figure 1. Rebound potentiation (RP) of inhibitory postsynaptic currents (IPSCs) in cerebellar Purkinje cells (PCs). **A:** Climbing fiber (CF) stimulation (5 pulses, 0.5 Hz) induces prolonged potentiation of IPSCs. **B:** CF stimulation (5 pulses, 0.5 Hz) fails to induce potentiation when the intracellular solution contained 30 mM BAPTA. **C:** Depolarization of PCs (single pulse from the holding potential of −70 mV to 0 mV, 500 msec) induced potentiation of IPSCs. **D:** Activation of antidromic spikes (5 spikes, 0.5 Hz) does not potentiate IPSC. Each point represents the mean ± S.E.M. of amplitudes of 100 to 200 consecutive IPSCs normalized to the values before conditioning. Since there is a tendency for the amplitudes of IPSCs to decay slowly during the control period, the last three values before conditioning were chosen for calculating the baseline (broken line at the 100% level). The threshold for detecting the IPSCs was set to −50 pA. Current traces are taken (1) before and (2) 14 min after conditioning. The traces in the insets were recorded at the time points indicated with thin arrows. The upward thick arrow indicates the time of conditioning (from Kano et al. 1992).

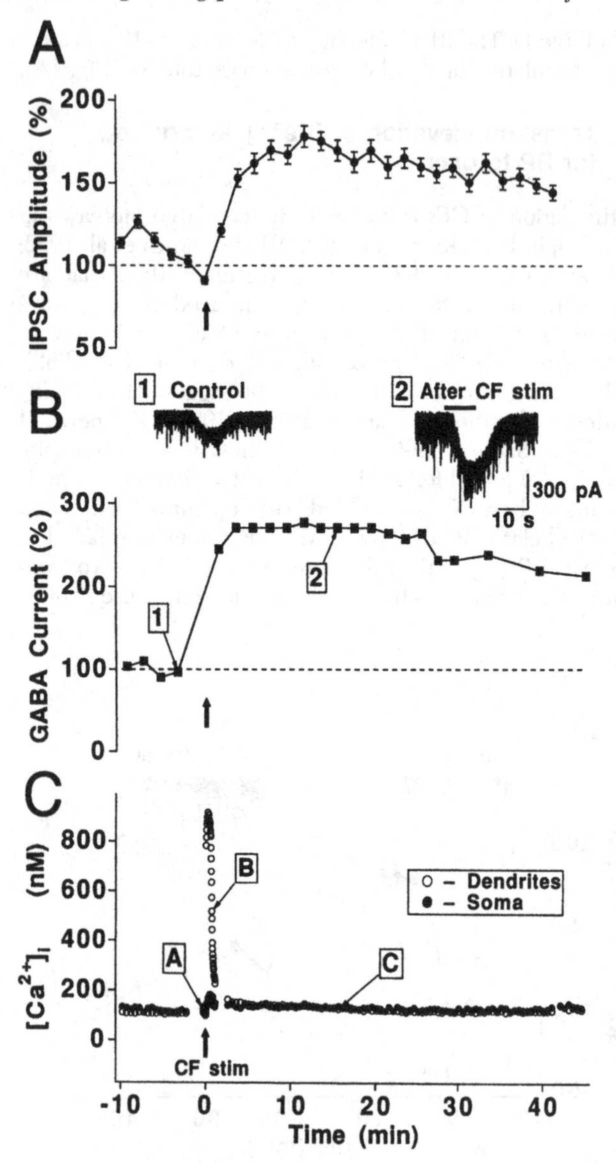

Figure 2. The transient $[Ca^{2+}]_i$ rises accompanying the RP of IPSCs and the potentiation of current responses to bath-applied GABA. **A:** RP of IPSCs following CF stimulation (5 pulses, 0.5 Hz). **B:** Potentiation of PC current responses to bath-applied GABA (2 μM, 10 sec). Insets display current traces recorded (1) before and (2) 15 min after CF stimulation. GABA was applied during the period indicated by the horizontal bars on top of each trace. **C:** $[Ca^{2+}]_i$ in the soma and the proximal dendrites. The upward thick arrow indicates the time of conditioning CF stimulation (from Kano et al. 1992).

holding potential of −70 mV to 0 mV, duration of 500 msec) induced a long-lasting potentiation of IPSCs (Fig. 1C) to an extent and with a time course similar to that of CF-induced RP. Stimulation of PC axons in the granule cell layer induced Na^+ action currents. These currents did not accompany characteristic long-lasting Cl^- tail currents that follow the Ca^{2+} entry through voltage-gated Ca^{2+} channels (Ca^{2+}-activated Cl^- currents, see Llano et al. 1991b). This indicates that Na^+ action currents did not cause significant elevation of $[Ca^{2+}]_i$. In fact, antidromic activation of Na^+ spikes with the same stimulation pattern (5 shocks, 0.5 Hz) as that used for CF stimulation did not induce any potentiation of IPSCs (Fig. 1D). Taken together, these findings strongly suggest that a CF-induced rise in $[Ca^{2+}]_i$ through voltage-gated Ca^{2+} channels triggered RP.

The correlation between the $[Ca^{2+}]_i$ elevation and RP was directly examined by combining whole-cell patch clamp recording with fluorometric $[Ca^{2+}]_i$ measurement (Fig. 2). PCs were filled with the Ca^{2+} indicator dye Fura-2 (200μM) through the patch-pipette and $[Ca^{2+}]_i$ was continuously monitored by digital fluorescence video ratio imaging technique (Llano et al. 1991a). Fig. 2C shows that the basal level of $[Ca^{2+}]_i$ in this cell had similar values in the soma and in the proximal dendrites of around 100 nM. On the conditioning CF stimulation, the dendritic $[Ca^{2+}]_i$ rose steeply to about 900 nM (Fig. 2C) and stayed elevated for about 40 sec, decaying then rapidly back to within about 30 nM of its initial baseline value. The $[Ca^{2+}]_i$ in the somatic region rose only slightly to less than 200 nM (Figs. 2C). A similar pattern of $[Ca^{2+}]_i$ change was observed in all of 5 PCs tested. After the transient rise, the $[Ca^{2+}]_i$ in both proximal dendrites and soma maintained a stable baseline value for more than 40 min (Fig. 2C). On the other hand, the simultaneously measured IPSCs were markedly potentiated (Fig. 2A). These results indicate that RP of IPSCs is triggered by a transient elevation of $[Ca^{2+}]_i$ in the PC dendrites, but that RP is not maintained by a persistent elevation of $[Ca^{2+}]_i$ following CF stimulation.

Vincent et al. (1992) reported that IPSCs of PCs evoked by stimulation of putative basket cells and their axons undergo potentiation following repetitive depolarization of PCs (8 pulses of 100 msec duration to 0 mV at 0.5 Hz) under the recording conditions similar to those of Kano et al. (1992) (i.e., whole-cell patch clamp recording from PCs in cerebellar slices with CsCl in the patch-pipette). The peak amplitude of potentiation was 146.7 ± 4.1% (mean ± S.E.M., n = 4) of the control, and the time to half-recovery was 12.9 ± 0.5 min (n = 3) (Table 1). They showed that IPSCs can be fitted with two exponentials. The time constants for both the rising (τon) and decaying (τoff) phases did not change during the potentiation. It should be noted that, regarding depolarization-induced potentiation of IPSCs, the peak amplitude and time to half recovery reported by Vincent et al. (1992) are comparable to those of Kano et al. (1992) (Table 1). Moreover, the values of the depolarization-induced potentiation and those of CF-induced RP are similar (Table 1). This coincidence further supports that CF-induced rise in $[Ca^{2+}]_i$ through voltage-gated Ca^{2+} channels triggered RP.

Besides long-lasting RP, IPSCs are depressed transiently (for about 1 min) following depolarization-induced Ca^{2+} entry to PCs (Llano et al. 1991b; Vincent et al. 1992; Vincent & Marty 1993). This short-lasting depression is mainly presynaptic, since the frequency of spontaneous IPSCs decreased (Llano et al. 1991b) and the probability of failures of evoked IPSCs was greatly enhanced during the period of the depression (Vincent et al. 1992).

4. RP involves up-regulation of postsynaptic GABAA receptor function

Several lines of evidence indicate that the inhibitory transmitter used by cerebellar cortical interneurons is GABA (Ito 1984, for review). The cause of RP may be either the enhancement of GABA release from the presynaptic terminals or the sensitivity increase at the postsynaptic receptors to GABA. The latter possibility was tested by measuring

Table 1. *Summary of potentiation of IPSCs and GABA-induced currents of cerebellar PCs*[1]

Measured currents (Method for induction)	Amplitude of potentiation (%)	Time to half recovery (min)	Reference
spontaneous IPSC (CF stimulation)	139.0±10.9 (n=6)[4]	16.5±3.6 (n=5)	Kano et al. (1992)
spontaneous IPSC (depolarization)[2]	134.9±14.8 (n=4)[4]	13.2±3.2 (n=4)	Kano et al. (1992)
GABA-induced current (CF stimulation)	179.7±24.3 (n=5)[4]	18.1±3.9 (n=5)	Kano et al. (1992)
evoked IPSC (depolarization)[3]	146.7±4.1 (n=4)	12.9±0.5 (n=3)	Vincent et al. (1992)
GABA-induced current (depolarization)[3]	144.0±7.2 (n=11)	3.7±0.5 (n=8)	Llano et al. (1991b)

Values are expressed as mean ± S.E.M.
[1]Comparison of the data from three papers.
[2]Single depolarizing pulse (500 msec, to 0 mV).
[3]Eight repetitive depolarizing pulses (100 msec, to 0 mV, at 0.5 Hz).
[4]Values were calculated 10 min after the conditioning.

membrane currents induced by bath-applied GABA (2 μM for 10sec) (Fig. 2B). At the same time, RP of IPSCs induced by CF stimulation was monitored (Fig. 2A). Following CF stimulation these currents induced by exogenous GABA were strongly enhanced. The time course of this potentiation was similar to that of RP of the IPSCs (Figs. 2A, 2B; see Table 1 for summary). Although the possible contribution of extrasynaptic GABAA receptors to this potentiation remains, this finding supports the notion that RP involves an up-regulation of postsynaptic GABAA receptor function. As in the control experiments for the RP of IPSCs, PC responses to bath-applied GABA were not potentiated following antidromic stimulation (5 pulses, 0.5 Hz). In contrast to the marked enhancement of postsynaptic sensitivity to GABA, no significant change in IPSC frequency could be detected during RP. The relative frequencies of IPSCs obtained 20 min after conditioning (in percentage of the control values, mean ± S.E.M.) were: 95 ± 6.7% (n = 5) for CF stimulation, 105 ± 9.5% (n = 4) for depolarization, 99 ± 4.5% (n = 6) for CF stimulation with intracellular BAPTA, and 100 ± 1.7% (n = 3) for the control experiments with antidromic stimulation. These results suggest that RP is mainly attributable to increased sensitivity of postsynaptic GABAA receptors on PCs.

Llano et al. (1991b) reported that GABA-induced currents are enhanced after repetitive depolarization of PCs (8 pulses of 100 msec duration to 0 mV at 0.5 Hz), a protocol that is also used to induce potentiation of evoked IPSCs (Vincent et al. 1992). However, the time to half recovery reported by Llano et al. (1991b) is significantly shorter than that of Kano et al. (1992) (see Table 1). One possibility for this difference would be due to the different methods for GABA application used in these two studies. Llano et al. (1991b) applied GABA presumably to the soma of PCs by "U-tube method." On the other hand, Kano et al. adopted simple bath-application and measured GABA-induced currents originating from both the dendrites and soma. Since large patch-pipettes (resistance, around 2MΩ) are used for recording from PCs, some intracellular factors necessary for a long-lasting up-regulation of GABAA receptors might have been "washed out" from the soma, but not from the dendrites. Llano et al. (1991b) may have measured GABA-mediated currents originating mainly from the soma that did not show long-lasting potentiation under whole-cell recording conditions. There is no other clear explanation for the difference, but this is an issue that should be addressed in future experiments.

5. Activity of inhibitory synapses is necessary for RP

RP can be induced by a few CF impulses or even a single depolarizing pulse. This is apparently quite in contrast to the stimulation parameter required to induce LTD of excitatory PF synapses. For LTD to occur, repetitive stimulation of CFs (50–960 pulses at 1 to 4 Hz) is necessary (Ekerot & Kano 1985; Hirano 1990a; Ito et al. 1982; Kano & Kato 1987; Sakurai 1987). Moreover, PFs have to be stimulated or postsynaptic quisqualate receptors should be activated in conjunction with CF stimulation (Ekerot & Kano 1985; Hirano 1990a; Ito et al. 1982; Kano & Kato 1987; Sakurai 1987) or with depolarization of PCs (Crépel & Jaillard 1991; Daniel et al. 1992; Hirano 1990b; Linden & Connor 1991; Linden et al. 1991). CF stimulation alone does not induce LTD.

Since CFs fire tonically at about 0.5 –1 Hz *in vivo*, it is puzzling that such a strong and long-lasting potentiation as RP can be produced by so few CF impulses. Measurement of $[Ca^{2+}]_i$ in PCs revealed that, under whole-cell patch-clamp recording with cesium as the major intracellular cation, Ca^{2+} transients induced by CF stimulation or depolarization are large in amplitude (up to 1 μM) and very long in duration (longer than 10 sec in many cases) (Kano et al. 1992; Konnerth et al. 1992; see Fig. 2C). By contrast, CF-induced Ca^{2+} transients have peak amplitudes of about 100nM and durations of 200–300 msec when measured under conventional intracellular recording with potassium in the recording electrode (Knöpfel et al. 1991). Therefore, a single CF stimulation with cesium inside might be equal to several tens of repetitive CF impulses under recording conditions with potassium inside. Moreover, since IPSCs occur spontaneously with rather high frequency (Konnerth et al. 1990; Llano et al. 1991b; see insets in Fig. 1), CF stimulation alone with cesium inside may result in the conjunction of inhibitory synaptic activity and elevation of postsynaptic Ca^{2+}. Therefore, the difference in stimulation parameters for the induction of LTD and RP cannot be compared directly.

In recent experiments in the author's laboratory, RP was induced using whole-cell patch-clamp recording with the pipette solution containing KCl (70 mM), K D-gluconate (70 mM), $MgCl_2$ (2 mM), Na-ATP (4 mM), Na-GTP (0.4 mM), EGTA (0.2 mM) and HEPES (30 mM) (pH 7.3). In comparison with the pipette solution used in the experiments shown in Figures 1 and 2, cesium was replaced by potassium. Because of the difficulty in stimulating inhibitory interneurons and their axons in isolation from excitatory PFs, CNQX (10 μM) and APV (50 μM) were used to block excitatory transmission mediated by both NMDA and non-NMDA receptors. IPSCs were recorded as inward synaptic currents at the stimulation frequency of 0.2 Hz (Fig. 3A). Following repetitive depolarization of PCs (240 pulses of 50 msec duration to −10 mV at 4 Hz, from the holding potential of −60 mV) in conjunction with stimulation of inhibitory interneurons, amplitudes of evoked IPSCs were potentiated (Fig. 3A). It is interesting to note that repetitive depolarization alone or stimulation of inhibitory interneurons alone at 4 Hz for 1 min failed to induce a long-lasting potentiation of IPSCs (Fig. 3B). Conjunctive repetitive depolarization with stimulation of interneurons potentiated the average amplitude of IPSCs to 129.3 ± 9.7% (mean ± S.E.M., n = 10) of control. By contrast, neither repetitive depolarization alone (94.7 ± 14.2%, n = 8) nor stimulation of inhibitory interneurons alone (97.6 ± 12.1%, n = 8) was effective in inducing potentiation (Fig. 3B). These results suggest that, under recording conditions with potassium inside, association of inhibitory synaptic activity with depolarization of postsynaptic membrane and accompanying Ca^{2+} elevation are required for triggering RP. It is still unclear whether potentiation of IPSCs can occur in physiological conditions with CF stimulation instead of depolarizing pulses. Ca^{2+}-imaging experiments show that both CF stimulations and depolarizing pulses induce similar pattern of $[Ca^{2+}]_i$ transients due to activation of voltage-gated Ca^{2+} channels (Llano et al. 1991a; Kano et al. 1992). This suggests that CF stimulation also can produce potentiation of IPSCs. It should also be noted that the parameter is comparable to that required for the induction of LTD of excitatory PF synapses. Thus, these lines of evidence suggest the possibility that RP and LTD work *in vivo* in a cooperative way to control PC excitability.

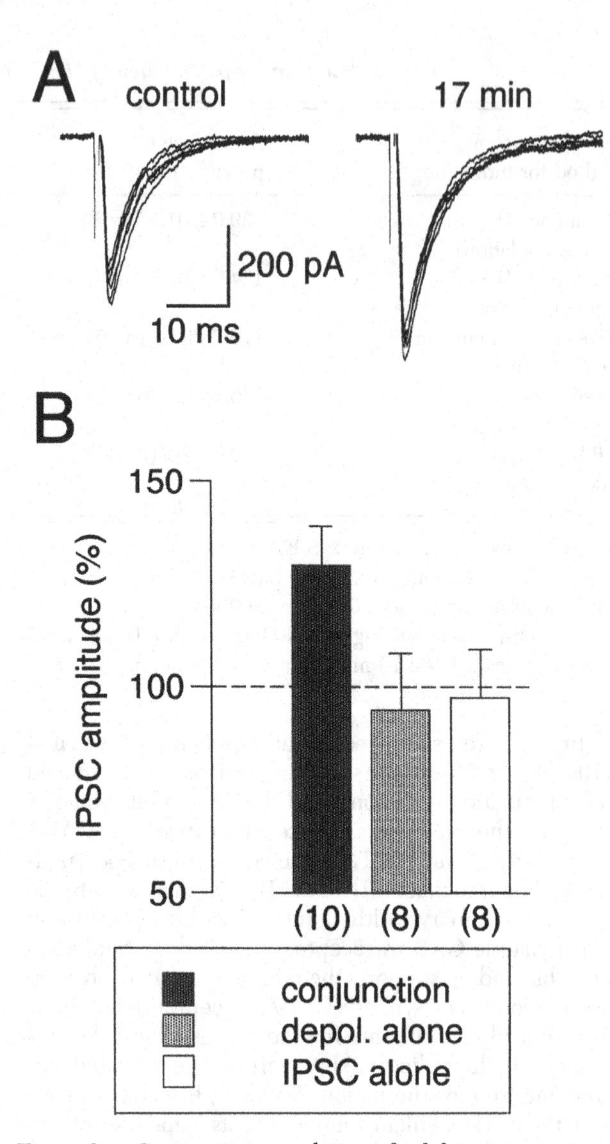

Figure 3. Conjunctive stimulation of inhibitory interneurons with depolarization of PCs is required for RP. **A:** Six consecutive traces of stimulus evoked IPSCs (0.2 Hz) are superimposed at time points before (control) and 17 min after repetitive stimulation of inhibitory interneurons (at 4 Hz for 1 min) in conjunction with depolarization (50 msec duration to −10 mV from the holding potential of −60 mV). Patch-pipette contained: KCl (70 mM), K D-gluconate (70 mM), $MgCl_2$ (2 mM), Na-ATP (4 mM), Na-GTP (0.4 mM), EGTA (0.2 mM), and HEPES (30 mM) (pH 7.3, adjusted with KOH). The external Ringer solution contained CNQX (10 μM) and APV (50 μM) to block excitatory transmission mediated by both NMDA and non-NMDA receptors. **B:** Average changes of IPSC amplitudes 17 min after the three types of conditioning: conjunctive stimulation of inhibitory interneurons with depolarization of PCs (conjunction), depolarization of PCs alone (depol. alone), stimulation of inhibitory interneurons alone (IPSC alone). Thirty-six consecutive trials of IPSCs were averaged in each cell to measure the amplitudes. Values represent percentage mean (±S.E.M.) change of average amplitudes of IPSCs of those taken before the conditioning. Number of recorded cells is shown in parentheses under each column.

6. Activation of protein kinase A induces up-regulation of GABAA receptor-functions

Experiments using extracellular recordings from PCs *in vivo* showed that 8-bromo-cAMP and forskoline reversibly potentiated the inhibitory effect of iontophoretically applied GABA on the firing of PCs (Parfitt et al. 1990; Sessler et al. 1989), suggesting a modulation of GABA receptors by cAMP-dependent protein kinase (PKA). Therefore, the effect of PKA on PCs was investigated directly by using the whole-cell patch-clamp recording from PCs in cerebellar slices (Kano & Konnerth 1992a).

Perfusing the slices with a saline containing 8-bromo-cAMP (500 μM, for 15 minutes), a membrane-permeable analogue of cAMP, significantly potentiated responses of PCs to bath-applied GABA (2–5 μM, 10 sec.) (Fig. 4A, upper panel). The magnitude of the potentiation was in average about 140% of the control value before application of 8-bromo-cAMP (Fig. 4B). The potentiation always per-

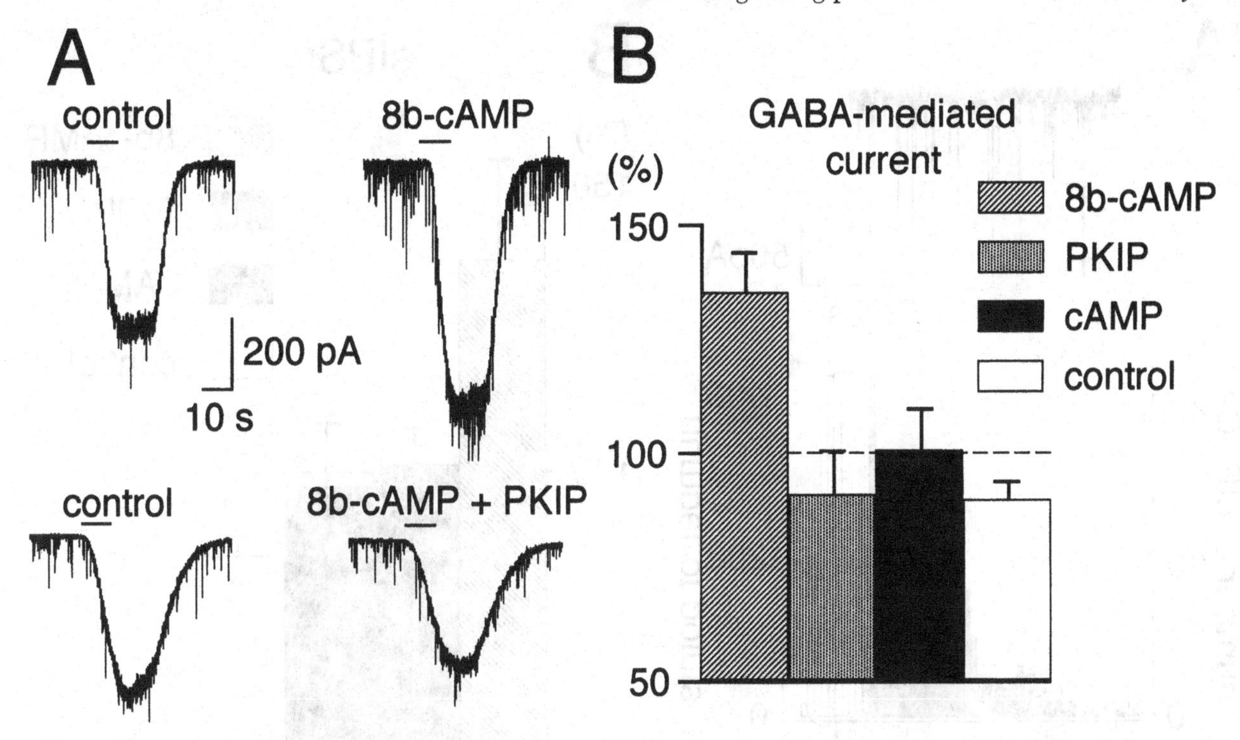

Figure 4. **A, upper panel:** Potentiation of GABA-mediated currents of PCs by 8-bromo-cyclic AMP (8b-cAMP). Records were taken in the presence of tetrodotoxin (TTX, 0.5 μM). Incubation of the slices with 8b-cAMP (500 μM, 15 min) produces a marked potentiation of the amplitudes of whole-cell current responses induced by bath-application of GABA (5 μM, 10 sec). Note that the amplitudes of spontaneously occurring IPSCs are also potentiated after 8b-cAMP application. **A, lower panel:** Same experimental protocol, but the protein kinase inhibitor peptide (PKIP 400 μg/ml) was added to the internal solution. Note that the GABA-induced currents are depressed rather than potentiated under these conditions. Traces displayed in the right column were taken at about 20 min after the end of 8 b-cAMP application. **B,** Average changes of GABA-induced whole-cell currents following different experimental manipulations: **8b-cAMP,** bath-application of 8b-cAMP (500 μM, 15 min) (n = 5); **PKIP,** bath-application of 8b-cAMP (500 μM, 15 min) and the addition of 400 μg/ml PKIP to the internal solution (n = 4); **cAMP,** bath-application of membrane impermeable cAMP (500 μM, 15 min) (n = 6); **control,** no experimental manipulations, records taken at about 30 min after the beginning of the recording (n = 5). Bars represent average values (± S.E.M.) of current changes normalized to control values taken in the same cells at about 10 min after the beginning of the whole-cell recording (modified from Kano & Konnerth 1992a).

sisted until the end of the experiment. By contrast, when the internal solution contained protein kinase inhibitor peptide (PKIP, 400 μg/ml), a specific inhibitor of PKA (Cheng et al. 1986), the same manipulation failed to induce potentiation of whole-cell GABA current (Fig. 4A, lower panel). Instead, a slight reduction of the amplitudes of the currents was observed. This reduction did not seem to be due to side effects of PKIP, since a similar "run-down" of the whole-cell current response to GABA was observed in control experiments without any treatment (Fig. 4B, control).

As well as the potentiation of currents induced by bath-applied GABA, 8-bromo-cAMP produced a similar potentiation of the amplitudes of spontaneous IPSCs measured in the presence of 0.5 μM TTX (sIPSC, Fig. 5A, upper panel for specimen record; Fig. 5B for summary). The threshold for detecting the sIPSCs (upward arrow in the lower panel of Fig. 5A) was set to −10 pA so that it was well over the background noise level. This potentiation lasted as long as the potentiation of whole-cell GABA current. The inclusion of PKIP in the patch-pipette prevented the potentiation of sIPSCs (Fig. 5B) similar to current responses induced by bath-applied GABA. It is important to note that bath-application of 500 μM cAMP (the membrane impermeable compound) affected neither GABA-mediated whole-cell current responses (Fig. 4B) nor sIPSCs (Fig. 5B), indicating that cAMP has only an intracellular site of action in PCs. The parallelism between the potentiation of the GABA-mediated currents and that of sIPSCs suggests that the same mechanism is mediating both effects (Figs. 4B, 5B). Moreover, the percentage changes of the GABA-mediated currents are strongly correlated with those of the sIPSCs (Fig. 6, $p < 0.01$ by Spearman's rank correlation test, $r = 0.77$), supporting the idea that both reflect the same effect, namely the change in the sensitivity of postsynaptic GABAA receptors.

The results presented above strongly suggest that GABAA receptors of cerebellar PCs are up-regulated by PKA through phosphorylation of the receptor protein itself or closely related protein(s). Thus, PKA is a strong candidate to play a role in RP. However, at present, a link between the elevation of Ca^{2+} and activation of PKA is unclear. To clarify whether PKA is involved in RP, blocking effects of PKA inhibitors (like PKIP) on Ca^{2+}-induced RP should be examined. Furthermore, it should also be tested whether PKA activation occludes the subsequent induction of RP.

GABAA receptor function is modulated in different ways in different regions of the nervous system. Note that Ca^{2+} elevation has opposite effect on GABAA receptors in other

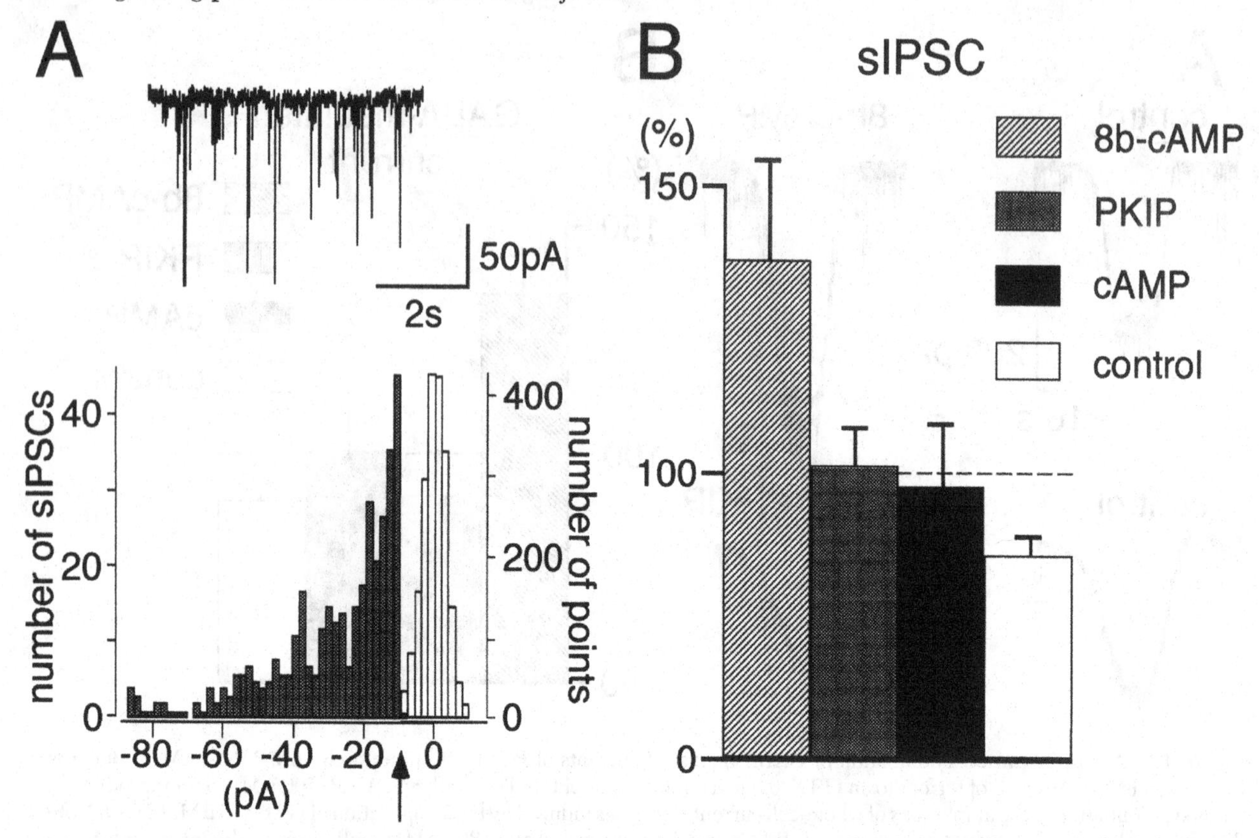

Figure 5. **A:** Record of spontaneous IPSCs (sIPSC, upper panel), amplitude distribution of the sIPSCs (hatched columns in the lower panel) and the background noise histogram (open columns in the lower panel). Records were taken in the presence of 0.5 μM TTX. The background noise histogram was constructed from the whole data points of the current traces where no IPSCs were present (total 1 sec, sampling frequency of 2 kHz). The threshold for detecting the sIPSCs (upward arrow in the lower panel) was set to −10 pA so that it was well over the background noise level. The amplitude distribution histogram of the IPSCs was constructed from the peak amplitudes of 400 consecutive sIPSCs. For demonstration purposes, a part of the whole sIPSC distribution with the current range from 0 to −90 pA is shown. The ordinate on the left is applied to the amplitude distribution of sIPSCs and that on the right is to the background noise. Bin width, 2 pA. **B:** Average changes of sIPSC amplitudes following different experimental manipulations similar to Figure 4B. Data from the same groups of cells shown in Figure 4B. Bars represent average values (± S.E.M.) of current changes normalized to control values taken in the same cells at about 10 min after the beginning of the whole-cell recording (modified from Kano & Konnerth 1992a).

cell types so far examined. In hippocampal (Chen et al. 1990), habenular (Mulle et al. 1992), and dorsal root ganglion neurons (Inoue et al. 1986), Ca^{2+} elevation depresses GABA responses. Effects of PKA are also opposite in hippocampal (Harrison & Lambert 1989), spinal cord (Porter et al. 1990) and sympathetic ganglion (Moss et al. 1992b) neurons. This suggests either that PCs possess GABAA receptors with a special molecular structure that undergo modulation in an opposite way to that of other cell types, or that GABAA receptor functions are regulated by yet unknown mechanisms that are influenced by Ca^{2+} or PKA in a way different from other cell types.

PKC and PKG are also rich in PCs and shown to be involved in LTD of excitatory PF synapses. (Crépel & Jaillard 1990; Crépel & Krupa 1988; Ito & Karachot 1990; 1992; Linden & Connor 1991; Shibuki & Okada 1991). However, no data are available to date that positively suggest the involvement of these kinase systems in RP of inhibitory synapses. In oocyte expression systems, PKC has been shown to depress GABAA receptors by directly phosphorylating the receptor protein (Moss et al. 1992a; Sigel et al. 1991; Whiting et al. 1990). Thus, it is possible that PKC is somehow involved in the regulation of GABAA receptor functions in PCs.

7. Plasticity of inhibitory synapses in other synapses

In the hippocampus, plasticity of inhibitory synapses was studied in relation to the mechanism of epilepsy, so that GABAergic inhibitory transmission was depressed during kindling (Stelzer et al. 1987). However, recent reports suggest that in some brain regions plasticity of inhibitory synapses plays a physiological role (Kano 1995, for review). Korn et al. (1992) reported that, in the goldfish, tetanic stimulation induces LTP at glycinergic inhibitory synapses on the Mauthner cells. Morishita and Sastry (1993) found the LTD of IPSCs of cells in the deep cerebellar nuclei induced by tetanic stimulation of PC axons. Komatsu and Iwakiri (1993) showed that, in the visual cortex of young rat, LTP of inhibitory transmission is induced by tetanus of inhibitory inputs, while NMDA receptor activation leads to LTD of the same synapses. Recently, Mitoma et al. (1994) reported that bath-application of serotonin causes en-

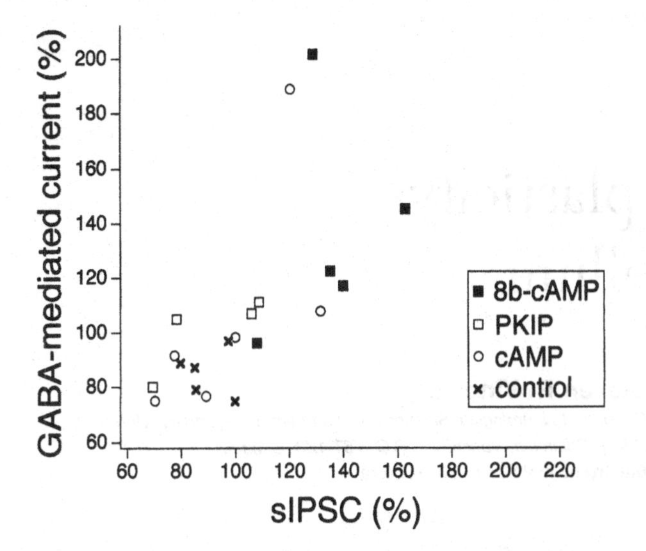

Figure 6. Scatter diagram showing the correlation between the percentage changes for the amplitudes of GABA-mediated currents (ordinate) and those of sIPSCs (abscissa) following the experimental manipulations indicated in Figures 4B and 5B.

hancement of inhibitory synaptic transmission in cerebellar Purkinje cells, which is presumably via presynaptic mechanisms. Thus, it appears that plasticity of inhibitory synapses exists in various parts of the brain, and that it may have important physiological functions such as learning and development.

8. Conclusion

In this target article, I have reviewed the properties and possible mechanisms of RP of inhibitory synapses of PCs, a newly found neural plasticity in the cerebellar cortex. The well-known form of synaptic plasticity in the cerebellum is LTD of excitatory PF-PC transmission. Accumulated evidence indicates that LTD is a cellular basis of motor learning in the cerebellum (for review, see Ito 1989; Linden 1994). LTD is induced by PF stimulation in conjunction with CF stimulation or depolarization of the postsynaptic membrane. It appears that RP also requires conjunction of inhibitory synaptic activity and postsynaptic depolarization. The parameters for the induction of RP (4 Hz, 240 conjunctive stimuli) are comparable to those of LTD (1–4 Hz, 100–960 conjunctive stimuli). It is therefore likely that, in concert with LTD of excitatory PF synapses, RP of GABAergic inhibitory synapses also play a role in motor learning.

The initial trigger for RP is CF activity that causes an elevation of $[Ca^{2+}]_i$ due to Ca^{2+} inflow through voltage-gated Ca^{2+} channels. RP is expressed mainly as an upregulation of postsynaptic $GABA_A$ receptors. PKA might play a role in the induction of RP, since activation of PKA by a cAMP analog-induced potentiation of $GABA_A$ receptor function similar to RP. On the other hand, no evidence is available so far that suggest the involvement of PKC and PKG in RP, although both of these kinases are shown to be involved in LTD of excitatory PF synapses (Crépel & Jaillard 1990; Crépel & Krupa 1988; Ito & Karachot 1990; 1992; Linden & Connor 1991; Shibuki & Okada 1991). Thus, it is likely that two distinct processes are present in PCs that lead to RP of inhibitory and LTD of excitatory PF synapses.

What are the implications of plasticity at inhibitory synapses for functions of the cerebellum? This newly found memory device will give the cerebellar cortical circuitry an additional computational capacity. In comparison with the sole site of plasticity at excitatory PF synapses, wider-range control of PC excitability by CFs becomes possible. Some 30% reduction of PF-mediated excitation occurs during LTD. If RP caused some 30% increase of inhibition at the same time, the sum would result in stronger depression of PC excitability than that achieved by LTD alone. Moreover, LTD and RP may not summate simply, because the sites of excitatory PF synapses and inhibitory basket/stellate cell synapses are spatially different. Some 100,000 PFs form synapses on dendritic spines of a single PC, while basket-cell and stellate-cell axons make relatively sparse contacts on the soma and proximal dendrites respectively (Ito 1984). This suggests that LTD of PF synapses is used for the fine control, whereas RP of inhibitory synapses is used for the robust and rough control of PC excitability.

The link between RP and motor learning is currently missing. Direct demonstration by single-unit analysis in behaving animals is not easy, because it is very difficult to distinguish between the reduction of excitation and the enhancement of inhibition. A pharmacological approach would be a possible alternative. This was taken to link LTD of excitatory PF synapses to motor learning. In cerebellar slice preparation, LTD is shown to be blocked by hemoglobin (a scavenger of nitric oxide). When hemoglobin was injected to the flocculus of monkeys, these animals showed no adaptation of the vestibuloocular reflex (Nagao & Ito 1991). Moreover, when hemoglobin was injected into the cerebellar vermis of walking decerebrate cats, these animals showed no adaptation to limb perturbation (Yangihara et al. 1994). By analogy with these works, if some pharmacological agents are found that specifically block RP and if the agents act from the extracellular space, they can be tested in behaving animals. Besides pharmacological agents, a recently developed gene-targeting technique will provide new experimental tools. Genetically created mutant mice that lack a specific molecule such as α-CaM-KII and PKC-γ have already been used to link hippocampal LTP and spatial learning (Abeliovich et al. 1993a; 1993b; Grant et al. 1992; Sakimura et al. 1994; Silva et al. 1992a; 1992b). Recently, this approach was also taken in linking LTD to cerebellar functions (Aiba et al. 1994; Conquet et al. 1994). Although it is not easy to assess the cerebellar functions in mice at the behavioral level, mutant mice would be a powerful new tool for studying the link between RP and motor learning.

ACKNOWLEDGMENTS

The author thanks Dr. N. Kawai for his kind support of this work. This work was also partly supported by grants from the Japanese Ministry of Education, Science, and Culture (05454677, 05260221, 05267242), and a grant from the Brain Science Foundation.

4

Nitric oxide and synaptic plasticity: NO news from the cerebellum

Steven R. Vincent
Division of Neurological Sciences, Department of Psychiatry, University of British Columbia, Vancouver, B.C. V6T 1Z3, Canada
Electronic mail: *sru@unixg.ubc.ca*

Abstract: Interest in the role of nitric oxide (NO) in the nervous system began with the demonstration that glutamate receptor activation in cerebellar slices causes the formation of a diffusible messenger with properties similar to those of the endothelium-derived relaxing factor. It is now clear that this is due to the Ca^{2+}/calmodulin-dependent activation of the enzyme NO synthase, which forms NO and citrulline from the amino acid L-arginine. The cerebellum has very high levels of NO synthase, and although it has low levels of guanylyl cyclase, cerebellar cyclic guanosine monophosphate (cGMP) levels are an order of magnitude higher than in other brain regions. A transcellular metabolic pathway is also present in the cerebellar cortex to recycle citrulline back to arginine. The NO formed binds to and activates soluble guanylyl cyclase to elevate cGMP levels in target cells. Studies employing NADPH-diaphorase, a selective histochemical marker for NO synthase, together with immunohistochemistry, *in situ* hybridization and biochemical studies have indicated that NO production occurs in granule and basket cells in the cerebellar cortex, whereas cGMP formation appears to occur largely in other cells, including Purkinje cells. Given that a long-term depression of AMPA currents can be seen in isolated Purkinje cells, this anatomical localization suggests that NO cannot play an essential role in the induction of this form of synaptic plasticity.

In vitro studies using cerebellar slices or cultures, as well as *in vivo* studies, have demonstrated that NO formation and release occur in the cerebellar cortex in response to NMDA receptor activation. Using intracerebellar microdialysis in awake-behaving animals, we have found that this is associated with a large increase in extracellular cGMP levels in the cerebellar cortex. A similar increase in cGMP efflux is seen in response to activation of AMPA or metabotropic glutamate receptors, or to activation of the climbing fiber input. The increase in extracellular cGMP required the Ca^{2+}-dependent activation of NO synthase, and was potentiated by inhibition of phosphodiesterases or organic anion transport. These results suggest a possible role for cGMP as an intercellular messenger in the cerebellar cortex.

In summary, stimuli that elevate Ca^{2+} levels in granule or in basket cells will activate NO synthase, which by binding to its receptor – soluble guanylyl cyclase – causes an increase in cGMP. cGMP may then act through protein kinases, phosphodiesterases, and ion channels or receptors to affect cerebellar function.

Keywords: cerebellar cortex; cyclic guanosine monophosphate; guanylyl cyclase; nitric oxide; Purkinje cells; synaptic plasticity

1. Introduction

The discovery that the increase in cGMP seen in cerebellar neurons in response to NMDA receptor activation resulted from the release of a diffusible messenger with properties similar to those of the endothelium-derived relaxing factor (EDRF), stimulated great interest in the role of this novel signaling pathway in the nervous system (Garthwaite et al. 1988). With the identification of EDRF as nitric oxide (NO), the enzyme responsible for its synthesis, nitric oxide synthase (NOS), was soon described in the brain (Bredt & Snyder 1990; Knowles et al. 1989; Mayer et al. 1990). Although the NO signaling pathway has now been described in many regions of the central and peripheral nervous systems, the cerebellum remains a key area for researchers interested in the functional significance of this novel messenger molecule.

In the rat, the cerebellum has about double the NOS activity found in other brain regions (Bredt et al. 1991; Förstermann et al. 1990). With the purification of neuronal NOS, immunohistochemical studies indicated that it was expressed by basket and granule cells and their processes but not by Purkinje cells or glial elements (Bredt et al. 1990). Similar observations were obtained using the NADPH-diaphorase reaction, which detects NOS histochemically (Hope et al. 1991; Southam et al. 1992; Vincent & Hope 1992; Vincent & Kimura 1992). Likewise, NOS messenger RNA (mRNA) is highly expressed in the granule cell layer, but is absent from the Purkinje cells (Bredt et al. 1991). The climbing fibers do not appear to contain NOS, since in numerous species the inferior olive is unstained with NADPH-diaphorase histochemistry (Kowall & Mueller 1988; Mizukawa et al. 1989; Vincent & Kimura 1992), and 3-acetylpyridine lesions of the inferior olive have no effect on cerebellar NOS activity (Ikeda et al. 1993). The deep cerebellar nuclei and their mossy fiber afferents to the cerebellar cortex also do not contain NOS-positive neurons.

There have been reports that Purkinje cells do stain weakly for NADPH-diaphorase during early postnatal development (Brüning 1993b; Yan et al. 1993). As noted by Brüning, this weak staining should be interpreted with caution. It may be the result of poorer fixation of these neonatal brains, since the relationship of NADPH-diaphorase to NO synthase is only apparent following formaldehyde fixation (Hope et al. 1991; Matsumoto et al. 1993). This might also explain a report of NADPH-diaphorase staining in the rat inferior olive (Southam & Garthwaite 1993). Indeed, we have seen weak staining of inferior olive or CA1 pyramidal neurons in material that has been poorly fixed (author, unpublished observations).

When NO is produced from arginine via NOS, the amino acid citrulline is a coproduct. Thus citrulline production would be predicted to occur in the NOS-containing basket and granule cells in the cerebellum. However, the enzymes needed to catalyze the conversion of citrulline back into the NO precursor arginine, argininosuccinate synthetase and lysase, are absent from these cells (Arnt-Ramos et al. 1992; Nakamura et al. 1990; 1991). Instead they are expressed in a discrete population of neurons just beneath the Purkinje cell layer, which might correspond with Lugaro cells or perhaps the recently described candelabrum cell (Lain & Axelrad 1994). This suggests that a transcellular metabolic pathway for citrulline and arginine metabolism occurs in the cerebellum. Indeed, a delayed release of arginine has been noted following stimulation of rat cerebellar slices (Hansel et al. 1992).

The cerebellum has relatively low levels of soluble guanylyl cyclase (Greenberg et al. 1978; Hofmann et al. 1977; Nakazawa & Sano 1974). A number of biochemical studies have sought to localize soluble guanylyl cyclase in the cerebellar cortex. Kainic acid lesions, which kill Purkinje, basket, stellate, and Golgi neurons but largely spare granule cells and glia, result in an almost complete loss of guanylyl cyclase activity and cGMP levels in the cerebellar cortex (Biggio et al. 1978). In contrast, lesions of the germinal layer of the developing cerebellum, which deplete the cerebellum of granule cells, result in an increase in guanylyl cyclase activity (Bunn et al. 1986). Also, the NO donor, sodium nitroprusside, dramatically elevates cGMP levels in granule cell-depleted mice (Wood et al. 1994). Thus granule cells do not appear to express significant soluble guanylyl cyclase activity *in vivo,* although NO-dependent cGMP formation has been well described in cultured granule cells (Kiedrowski et al. 1992; Novelli & Henneberry 1987).

Studies have also made use of various mutant mice exhibiting cerebellar degeneration to examine the localization of the NO/cGMP system. *Nervous,* mutant mice, which lack most Purkinje cells, were reported to have a corresponding loss of guanylyl cyclase and cGMP (Mao et al. 1975; Schmidt & Nadi 1977). *Nervous* and *Purkinje cell-degeneration* mutant mice also showed a loss of NOS activity in the cerebellar cortex (Ikeda et al. 1993). However, another group has found no change in cerebellar cGMP levels or in the response to harmaline in these mutant mice (Wood et al. 1994). Together, these confusing results indicate that secondary changes in enzyme expression occurring during development in these mutant mice might contribute to the observed neurochemical effects.

Immunohistochemical and *in situ* hybridization observations are largely consistent with the localization of guanylyl cyclase determined biochemically. The mRNAs for the $\alpha 1$ and $\beta 1$ subunits of soluble guanylyl cyclase are highly expressed in Purkinje cells, with moderate levels seen in stellate, basket, and Golgi cells and low levels of expression in granule cells (Furuyama et al. 1993; Matsuoka et al. 1992; Verma et al. 1993). Immunohistochemical studies with monoclonal antibodies to this enzyme have demonstrated strong staining of Purkinje cells, with weak staining of neurons in the molecular and granule cell layers (Ariano et al. 1982; Nakane et al. 1983), although earlier studies with a polyclonal serum that was not well characterized resulted in staining of all cerebellar cells (Zwiller et al. 1981).

In agreement with these studies, other biochemical and immunohistochemical observations have noted that cGMP-dependent protein kinase is highly expressed in Purkinje cells (Bandle & Guidotti 1977; DeCamilli et al. 1984; Dolphin et al. 1983; Lohmann et al. 1981; Nairn & Greengard 1983). Furthermore, a substrate for this kinase, termed G-substrate, is also concentrated in these cells (Detre et al. 1984; Schlichter et al. 1978; 1980).

Although a calmodulin-dependent cyclic nucleotide phosphodiesterase is present in Purkinje cells, where its expression is regulated by the climbing fiber input (Balaban et al. 1989), the activity of cyclic nucleotide phosphodiesterase, measured either in the presence or absence of calmodulin, is the lowest of any brain region (Greenberg et al. 1978; Hofmann et al. 1977). This might account for the fact that cGMP levels in the cerebellum are much higher than those in other brain regions and increase dramatically in response to activation.

Together, these anatomical studies suggest a model in which NO is produced by the calcium-dependent NOS present in basket and granule cells, and acts on its receptor, the soluble guanylyl cyclase, which is primarily localized in the Purkinje cells, as well as in inhibitory interneurons. However, pharmacological studies have provided results that make this relatively simple scenario controversial. In particular, immunohistochemical studies have been undertaken to localize those cells responding to NO with an increase in cGMP. A dramatic increase in cGMP-immunoreactivity in Bergmann glia, granule cells, and glomeruli (de Vente et al. 1989; 1990; de Vente & Steinbusch 1992; Southam et al. 1992) and moderate staining of Purkinje cell bodies (Southam & Garthwaite 1993) has been seen following sodium nitroprusside. Some cGMP staining of stellate and basket cells has also been described in unstimulated cerebellum, but Purkinje cells appeared to be unstained (Chan-Palay & Palay 1979). However, other immunohistochemical studies using glutaraldehyde fixation and high-affinity monoclonal antibodies have revealed intense cGMP-immunoreactivity in Purkinje cells (Sakaue et al. 1988). A possible explanation for these results might be provided by the discovery that NO induces a large increase in extracellular cGMP, which is then cleared from the extracellular space by a probenecid-sensitive mechanism (Luo et al. 1994; Tjörnhammer et al. 1986; Vallebuona & Raiteri 1993). Thus cGMP might be produced primarily in Purkinje cells but rapidly leaves these cells to be accumulated by the Bergmann glia.

2. NO and the regulation of cerebellar cGMP levels

Many studies using a wide variety of techniques are consistent with the presence of a tonic production of NO and cGMP in the cerebellar cortex. The levels of cGMP in the cerebellum are an order of magnitude greater than those in other brain regions. cGMP levels can be decreased by elevating cerebellar γ-aminobutyric acid (GABA) transmission (Dodson & Johnson 1980; Mao et al. 1974a; 1974b; 1975) and increased dramatically by blocking GABA action (Biggio et al. 1977a; Mailman et al. 1978; Mao et al. 1974a; 1974b). Conversely, activation of excitatory amino acid receptors increases cGMP levels in the cerebellar cortex (Briley et al. 1979; Danysz et al. 1989; Ferrendelli et al. 1974; Mao et al. 1974a; Wood et al. 1982). Furthermore, activation of the climbing fiber input using harmaline also produces large increases in cerebellar cGMP levels (Biggio & Guidotti 1976; Biggio et al. 1977b; Cross et al. 1993; Guidotti et al. 1975; Luo et al. 1994; Mao et al. 1974a).

How does climbing fiber activation regulate cerebellar NO and cGMP production? Climbing fibers activate Purkinje cells via an AMPA-type receptor, which can be blocked by CNQX, and mature Purkinje cells appear to lack NMDA receptors (Farrant & Cull-Candy 1991) and do not express NOS (Bredt et al. 1990; Southam et al. 1992; Vincent & Hope 1992; Vincent & Kimura 1992). Since the harmaline-induced increase in cerebellar cGMP can be blocked by NMDA receptor antagonists (Luo et al. 1994; Wood et al. 1982; 1990), the climbing fiber input to the Purkinje cells would not appear to be responsible. The NOS-containing granule and/or basket cells appear to be required for climbing fiber-induced increases in cerebellar cGMP (Wood et al. 1994), however, they do not appear to receive a substantial direct climbing fiber input (Leranth & Hamori 1981). This suggests that an indirect pathway may be involved. The inferior olive does innervate the deep cerebellar nuclei (Audinat et al. 1992; Llinas & Muhlethaler 1988), which in turn give rise to the major mossy fiber input to the granule cells, activating these neurons via both NMDA and AMPA receptors (Garthwaite & Brodbelt 1989). Thus, harmaline-induced stimulation of the inferior olive could result in excitation of the mossy fiber input to the granule cells from the deep cerebellar nuclei, and this might account for the activation of NOS seen following climbing fiber activation (Shibuki 1990).

The increase in cerebellar cGMP in response to excitatory amino acid receptor agonists or harmaline is mediated by activation of NOS, since it is effectively blocked by a variety of NOS inhibitors (Bansinath et al. 1993; Bredt & Snyder 1989; East & Garthwaite 1990; Garthwaite 1991; Garthwaite et al. 1989a; Luo et al. 1994; Southam et al. 1991; Wood et al. 1990). Various NO donors can themselves elevate cerebellar cGMP levels (Southam & Garthwaite 1991a). Furthermore, the increase in cGMP seen in cerebellar slices in response to glutamate receptor activation is inhibited by hemoglobin, suggesting that NO must travel between cells to affect guanylyl cyclase (Southam & Garthwaite 1991b).

Electrochemical methods have been used to demonstrate NO release from adult rat cerebellar slice preparations following electrical stimulation (Shibuki 1990; Shibuki & Okada 1991). Furthermore, NO release from cerebellar slices following depolarization or from NMDA receptor activation has been examined using a chemiluminescence assay (Dickie et al. 1990; 1992). NMDA receptor activation has also been shown to elevate NO release in the cerebellar cortex *in vivo* in awake, unrestrained animals (Luo et al. 1993). This study also provided evidence for a tonic NMDA-induced NO production in the intact cerebellar cortex.

In the cerebellar cortex, the NMDA-induced activation of NOS has been suggested to result largely from direct activation of NMDA receptors on the neurons containing NOS, since the increase in cGMP seen in response to NMDA is largely unaffected by tetrodotoxin (Luo et al. 1994; Southam et al. 1991). However, the recently characterized presynaptic NMDA receptor could also be involved (Smirnova et al. 1993). Indeed, some *in vivo* experiments imply a role for monoaminergic innervation in mediating the increase in cGMP seen in response to NMDA receptor activation. The increases in cerebellar cGMP levels seen in response to harmaline or local NMDA activation were blocked by prior reserpinization (Wood et al. 1992). Reserpine treatment also decreased basal cGMP levels by about 50% (Burkard et al. 1976; Ferrendelli et al. 1972; Wood et al. 1992). Pharmacological studies indicate that the stimulatory effect of harmaline or NMDA receptor activation was blocked by adrenergic antagonists of the α_{1A} type acting within the cerebellar cortex (Rao et al. 1991). Consistent with this are the observations that α_1 antagonists themselves can decrease, whereas α_1 agonists or noradrenaline itself can increase cerebellar cGMP levels (Haidamous et al. 1980). However, in some *in vitro* studies using slices from neonatal rat cerebellum, noradrenaline antagonized the cGMP increase induced by NMDA (Carter et al. 1988). Biochemical and electrophysiological experiments indicate that phencyclidine-sensitive NMDA receptors increase noradrenaline release in the cerebellum (Marwaha et al. 1980; Yi et al. 1988). This suggests that NMDA receptors on noradrenergic nerve endings in the cerebellum may, by stimulating noradrenaline release, affect cGMP levels via an α_1-receptor mechanism.

Although most attention has been paid to the role of NMDA receptors in the stimulation of NOS and the production of cGMP, activation of AMPA, kainate, or metabotropic glutamate receptors also produces an increase in cGMP, again, apparently via a direct action on NOS-containing cerebellar neurons (Garthwaite et al. 1989b; Luo et al. 1994; Okada 1992; Southam et al. 1991). The cGMP increase seen in response to AMPA is of similar magnitude to the NMDA-induced increase. In contrast, the cGMP increase seen in response to metabotropic receptor activation is considerably smaller. This suggests that releasing calcium from intracellular stores, which appears to mediate metabotropic receptor-induced NO production (Okada 1992), is much less effective in activating cerebellar NOS than is activation of the ionotropic glutamate receptors. Both the cerebellar basket cells (Gres et al. 1993) and the granule cells are known to possess metabotropic glutamate receptors (East & Garthwaite 1992; Fagni et al. 1991; Nicoletti et al. 1986), some of which may be present on the terminals making contact with Purkinje cells (Glaum et al. 1992; Takagi et al. 1992).

In addition to the excitatory amino acids, a variety of other neurotransmitter systems can affect cerebellar NO and cGMP production. Thus manipulations of cholinergic (Dinnendahl & Stock 1975; Dodson & Johnson 1979), dopaminergic (Biggio et al. 1977b; Breese et al. 1978; Burkard et al. 1976; Ferrendelli et al. 1972), and opioid (Biggio et al. 1977) systems outside the cerebellum are known to affect cerebellar cGMP levels. These actions appear to be mediated by activation of the mossy fiber afferents to the cerebellar cortex, which in turn leads to stimulation of the NOS-containing granule and/or basket cells (Wood 1991).

3. NO, cGMP, and cerebellar synaptic plasticity

When endothelium-derived relaxing factor (EDRF) was first suggested as an intercellular messenger in the cerebellum, its unique properties and potential significance for synaptic plasticity were noted (Garthwaite et al. 1988). The inhibition of vestibuloocular reflex adaptation by subdural hemoglobin suggested a possible role for NO in synaptic

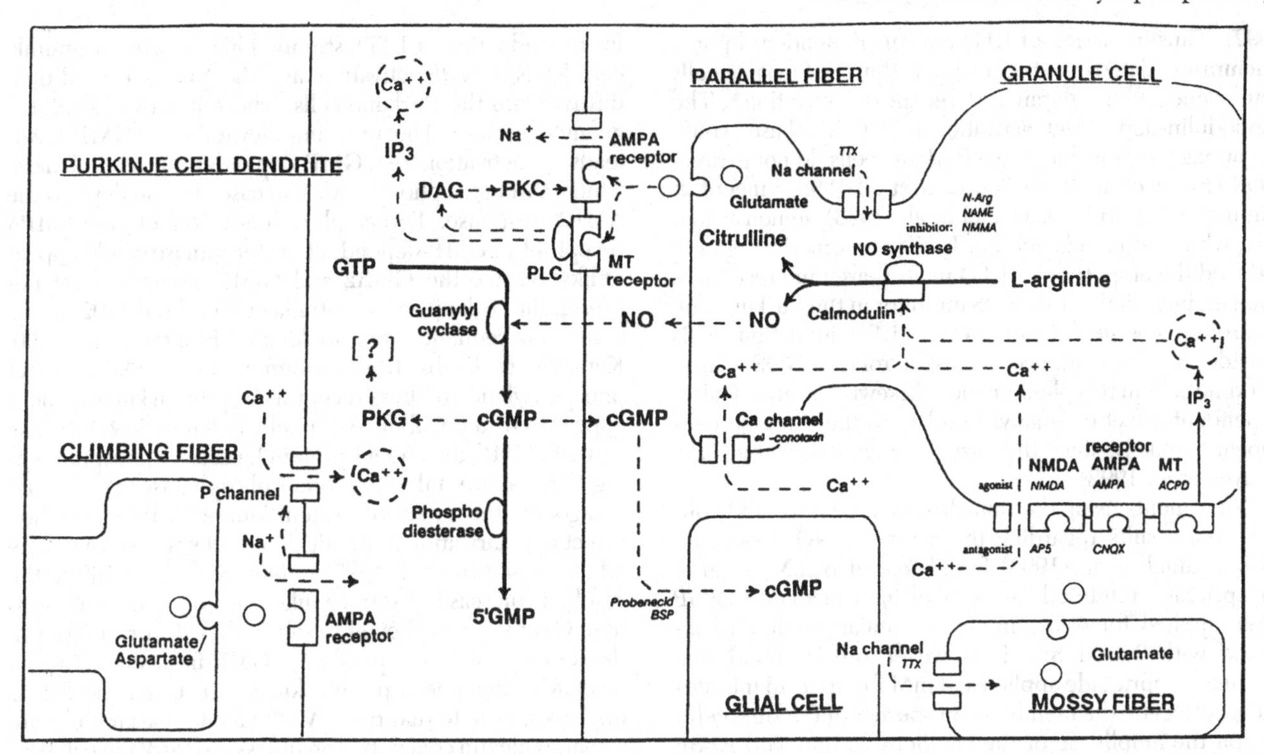

Figure 1. Schematic diagram illustrating the neuronal mechanisms involved in synaptic activity in the cerebellar cortex. Long-term depression (LTD) of the parallel fiber to Purkinje cell AMPA response requires coincident climbing fiber and parallel fiber activation.

The massive climbing fiber synapse, acting on AMPA receptors, can result in sufficient depolarization to activate P-type calcium channels in the Purkinje cell dendrites. This calcium increase may be further amplified through interaction with the calcium-dependent calcium release system (ryanodine receptor) present in Purkinje cell dendrites. The granule cells, which give rise to the parallel fibers, can be activated by mossy fibers via both AMPA and NMDA receptors. Activation of granule cells results in the calcium-dependent formation of nitric oxide (NO) within these neurons. Such NO formation in the parallel fiber nerve terminals could thus act on Purkinje cell dendritic spines receiving the parallel fiber input. The parallel fiber induced EPSP in Purkinje cells is mediated by an AMPA receptor, although a metabotropic glutamate receptor can also be activated, resulting in the generation of Ip_3 and diacyl glycerol (DAG) and activation of protein kinase C (PKC) within the Purkinje cell dendritic spines. Coactivation of both the AMPA and metabotropic receptors and subsequent PKC activation appear necessary for the induction of LTD of the AMPA response. Within the Purkinje cell, NO appears to activate soluble guanylyl cyclase (GC), which, via generation of cyclic guanosine monophosphate (cGMP), leads to activation of cGMP-dependent protein kinase (PKG). PKG could possibly influence LTD by various mechanisms, including actions on the two intracellular calcium release systems or phosphorylation of other targets such as protein phosphatase inhibitors.

plasticity in this area (Nagao & Ito 1991). There have been some reports of NO-catalyzed ADP-ribosylation of proteins being a possible action of NO in the brain (Brüne & Lapetina 1989). However, it now appears that NO does not catalyze such a reaction; rather, through S-nitrosylation, covalent binding of NAD to some proteins can occur (McDonald & Moss 1993; Zhang & Snyder 1992). S-nitrosylation of proteins may be of great importance in the neurotoxic actions of NO but is unlikely to be involved in the normal intercellular signaling by this molecule. Instead, NO appears to act by mediating agonist-induced increases in cGMP.

Data consistent with a role for the NO–cGMP signal transduction system in the long-term depression (LTD) of AMPA responses in Purkinje cells have been obtained by various groups. Ito and Karachot (1990) demonstrated that the desensitization of Purkinje cell responses to AMPA in rat cerebellar slices was abolished by Ca^{2+} chelators, hemoglobin, the nitric oxide synthase inhibitor N^G-nitroarginine, and an inhibitor of cGMP-dependent protein kinase. Experiments on 1-month-old rat cerebellar slices demonstrated an LTD of parallel fiber-mediated EPSPs when they were paired with depolarization-induced calcium spikes in Purkinje cells (Crépel & Jaillard 1990). This depression was prevented by bath application of polymyxin B, L-N^G-monomethylarginine or methylene blue, again suggesting a role for protein kinase C, NO, and cGMP. Furthermore, coapplication of sodium nitroprusside or 8-bromo-cGMP with AMPA or parallel fiber stimulation could induce LTD, suggesting that elevated cGMP could substitute for climbing fiber excitation of the Purkinje cells (Ito & Karachot 1990; Shibuki & Okada 1991; 1992). However, the parallel fibers contain NOS and would be an endogenous source of NO, whereas the climbing fibers lack NOS. Thus why NO donors should substitute for climbing fibers activation is not apparent.

In thin slices from 2-week-old rat cerebellum, whole-cell, patch clamp experiments demonstrated that climbing fiber activation or direct depolarization of Purkinje cells, delivered in conjunction with parallel fiber activation, induced an LTD of the parallel-fiber synapse (Daniel et al. 1993; Konnerth et al. 1992). The depolarization-induced rise in intracellular calcium in the Purkinje cell was both necessary and sufficient to initiate LTD (Konnerth et al.

1992). Thus induction of LTD appears dependent upon a calcium-mediated event occurring within the Purkinje cells coincident with glutamate receptor activation. The calmodulin-dependent activation of NO synthase would not appear responsible, since Purkinje cells do not express NOS (Bredt et al. 1990; Southam et al. 1992; Vincent & Kimura 1992). Indeed, Daniel et al. (1993) demonstrated that, while bath application of L-N^G-monomethylarginine reduced the amplitude of LTD in an L-arginine reversible manner, inclusion of this NOS inhibitor in the Purkinje cell recording pipette did not prevent LTD induction. This provides strong evidence against a role of NOS within Purkinje cells in this phenomenon. Likewise, a direct Ca^{2+}-dependent effect on guanylyl cyclase in the Purkinje cells appears unlikely, since the enzyme is unaffected by Ca^{2+} (Mayer et al. 1992).

Experiments using the thin-slice method also yield controversial results regarding the action of NO donors on LTD. Daniel et al. (1993) found that at 8 mM, sodium nitroprusside inhibited the parallel fiber-mediated EPSP when applied for 4–9 minutes. A similar result was observed with 3 mM SIN-1. However, others found that sodium nitroprusside applied at 3 mM for up to 15 minutes had no effect on the membrane properties of Purkinje cells, nor on the amplitude of the parallel fiber-induced EPSP (Glaum et al. 1992). Likewise, in perforated-patch recordings from cultured rat Purkinje cells, 3 mM sodium nitroprusside had no effect on the response to iontophoretic glutamate (Linden & Connor 1992). The slight differences in drug concentrations in these studies is unlikely to offer an explanation for these discrepancies, since the massive increase in cGMP levels produced would have been similar (Southam & Garthwaite 1991a). Indeed, these doses of sodium nitroprusside would likely be toxic to the neurons in these preparations (Dawson et al. 1991).

Recent studies have used NADPH-diaphorase to localize NO synthase in the chicken cerebellum. As in the rat, the granule cells and some cells in the molecular layer were stained, but the Purkinje cells and the neurons of the cerebellar nuclei were unstained (Brüning 1993). Studies undertaken in cultured chick Purkinje cells demonstrated that the response to iontophoretic AMPA, monitored by intracellular or whole-cell recording, was suppressed by trans-ACPD, a metabotropic glutamate receptor agonist, and this suppression could be blocked by hemoglobin, the NOS inhibitor L-N^G-monomethylarginine, or the cGMP-dependent protein kinase inhibitor KT5823, or mimicked by the NO donors sodium nitroprusside and 3-morpholinosydnonimine (Mori-Okamoto et al. 1993). However, potassium ferricyanide was also able to mimic the suppression, via a cGMP-independent mechanism. Furthermore, the trans-ACPD effect could also be prevented by blocking protein kinase C activity. Similar observations were made using slice preparations from adult rat cerebellum (Ito & Karachot 1990), although studies with cultured rat Purkinje cells found that preincubation with nitroarginine or hemoglobin for 25 minutes did not attenuate the depression of glutamate responses induced by glutamate/depolarization pairing (Linden & Connor 1992).

Most of these studies have made use of nonspecific NO donors, or NOS inhibitors that are nonselective and can affect numerous other cellular processes (Peterson et al. 1992; Schmidt et al. 1993). However, together these studies are consistent with the idea that although it is not necessary for the induction of LTD, stimulation of basket or granule cells leads to NOS activation, and the NO so formed then diffuses into the Purkinje cells, where it activates soluble guanylyl cyclase. The resulting elevation in cGMP levels leads to activation of cGMP-dependent protein kinase, which, through a phosphorylation cascade, can depress the AMPA response. Direct phosphorylation of the AMPA receptor by cGMP-dependent protein kinase would appear unlikely, since the GluR2 and GluR3 receptor subtypes present in Purkinje cells (Petralia & Wenthold 1992) lack a consensus sequence for this kinase (Boulter et al. 1990; Kennelly & Krebs 1991; Sommer et al. 1990). Direct phosphorylation of these receptors by protein kinase C does appear to be a possibility (Kennelly & Krebs 1991). It may be that cGMP-dependent protein kinase, by phosphorylating G-substrate, inhibits protein phosphatases that would antagonize this effect of protein kinase C. Perhaps when direct depolarization of the Purkinje cells is used to induce LTD (Konnerth et al. 1992; Linden & Connor 1992), the resultant increase in intracellular calcium and protein kinase C activity is sufficient to suppress AMPA responses in the absence of NO-dependent, cGMP-induced activation of cGMP-dependent protein kinase. In this regard, it is important to note that trans-ACPD alone does not elevate somatic calcium concentration unless the Purkinje cell fires calcium-dependent action potentials (Glaum et al. 1992).

4. Other roles for cerebellar NO and cGMP

cAMP-dependent protein kinase phosphorylation of CREB and related transcription factors allows the cAMP signal transduction pathway to regulate gene expression in target neurons. By analogy, NO-dependent cGMP production might influence gene expression in neurons such as the Purkinje cells, which express both soluble guanylyl cyclase and cGMP-dependent protein kinase. Recent work on PC12 cells indicates that NO can induce immediate early gene expression via the activation of cGMP-dependent protein kinase (Haby et al. 1994). NMDA receptor activation can induce *c-fos* expression in cultured cerebellar granule cells (Szekely et al. 1989). However, this appears to be mediated by protein kinase C, and dibutyryl cGMP did not increase *c-fos* expression. Likewise, evidence for a cGMP response element regulating gene expression in Purkinje cells or other neurons is lacking.

Other substrates for the cGMP-dependent protein kinase in the cerebellum might include the ryanodine receptor, which is highly expressed in Purkinje cells (Sharp et al. 1993) and has been shown to be phosphorylated by this kinase (Suko et al. 1993), causing an enhancement of its calcium-releasing activity (Herrmann-Frank & Varsanyi 1993). Furthermore, cGMP might enhance the production of the proposed endogenous ligand for this receptor, cyclic ADP-ribose (Galione et al. 1993). The IP_3 receptor, which is also highly expressed in Purkinje cells (Sharp et al. 1993), can also be phosphorylated *in vitro* by the cGMP-dependent protein kinase (Lincoln & Cornwell 1993).

A particularly exciting recent discovery is the demonstration that the gene for a cyclic nucleotide-gated channel is expressed in rabbit cerebellum (Biel et al. 1993). This channel has properties similar to those of the cation channel in the olfactory epithelium, opening in response to an increase in intracellular cGMP. It will be of great importance to determine which cells in the cerebellum express this channel, since one would predict that a NO-induced increase in intracellular cGMP might depolarize cells expressing such a channel.

Finally, given the extraordinarily high levels of cGMP in the cerebellar cortex and the demonstration of a NOS-dependent increase in extracellular cGMP following stimulation, a role for this cyclic nucleotide in intercellular communication might also be considered (Luo et al. 1994). Electrophysiological effects of extracellular cGMP have been reported in the cerebellum (Hoffer et al. 1971; Siggins et al. 1976). Thus cell-surface, G protein-linked receptors similar to those described for cAMP (Klein et al. 1988; Sorbera & Morad 1991) may exist for cGMP.

5. Conclusions

Intense interest has been focused on the role of NO since its discovery as a signal transduction molecule in the cerebellum. It is now apparent: stimuli that increase the intracellular free calcium within basket and granule cells can lead to the calmodulin-dependent activation of NOS. The NO produced will diffuse to reach and activate its target, soluble guanylyl cyclase, within Purkinje cells, and perhaps other cell types. It is clear from the biochemical neuroanatomy of this system that, although NO may be involved in synaptic plasticity, it does not appear to be essential for the induction of LTD. The function of NO is to activate soluble guanylyl cyclase. Thus, to understand the functional significance of this novel messenger molecule, it will be necessary to elucidate the roles of cGMP in cerebellar physiology.

ACKNOWLEDGMENT
The work from my laboratory on NO and cGMP was supported by a grant from the Medical Research Council of Canada.

5

Models of the cerebellum and motor learning

James C. Houk
Department of Physiology, Northwestern University Medical School, Chicago, IL 60611
Electronic mail: *houk@casbah.acns.nwu.edu*

Jay T. Buckingham
Department of Computer Science, University of Massachusetts, Amherst, MA 01003
Electronic mail: *buckingham@cs.umass.edu*

Andrew G. Barto
Department of Computer Science, University of Massachusetts, Amherst, MA 01003
Electronic mail: *barto@cs.umass.edu*

Abstract: This article reviews models of the cerebellum and motor learning, from the landmark papers by Marr and Albus through those of the present time. The unique architecture of the cerebellar cortex is ideally suited for pattern recognition, but how is pattern recognition incorporated into motor control and learning systems? The present analysis begins with a discussion of exactly what the cerebellar cortex needs to regulate through its anatomically defined projections to premotor networks. Next, we examine various models showing how the microcircuitry in the cerebellar cortex could be used to achieve its regulatory functions. Having thus defined what it is that Purkinje cells in the cerebellar cortex must learn, we then evaluate theories of motor learning. We examine current models of synaptic plasticity, credit assignment, and the generation of training information, indicating how they could function cooperatively to guide the processes of motor learning.

1. Introduction

Lesion studies carried out in the nineteenth century demonstrated that the cerebellum is important for coordinating movements (Florens 1824). Mechanistic models of the cerebellum, however, awaited an analysis of its histology (Braitenberg & Atwood 1958) and combined analyses of its histology and electrophysiology (Albus 1971; Marr 1969). The clear orthogonal relationships between parallel and climbing fibers and the dendritic trees of Purkinje cells (PCs) convinced Braitenberg that the cerebellum functions as a timing organ. He viewed the cerebellum's parallel fibers (PFs) as delay lines and its climbing fibers (CFs) as clock read-out mechanisms, with PCs firing only when there is a coincidence of a PF volley and a CF activation. In one of his examples, a PC innervating an antagonist muscle fires at a time delay appropriate to terminate a movement at its intended target. Even though PFs have small diameters and low conduction velocities, the time delays amount to only a few milliseconds, a rather short time for most problems in motor control. Another limitation of this theory is its requirement for foci of synchronized activation in the granular layer. For these and other reasons, Braitenberg's timing theory has not been as vigorously pursued as the learned pattern recognition theories developed subsequently by Marr (1969) and Albus (1971).

Like Braitenberg, Marr and Albus were impressed by the anisotropic structure of the cerebellar cortex. In constructing their information processing theories, they incorporated several additional anatomical features, for example, the marked differences in the convergence ratios of parallel (~100,000:1) and climbing (1:1) fibers onto PCs. Instead of serving as delay lines, PFs were thought to provide large vectors of potential input, transmitting a diverse array of information. The CFs, instead of serving as read-out devices, would function as training signals that adjust the synaptic weights of PF synapses, thus teaching PCs to recognize specific patterns signaled by their input vectors. Marr, Albus, and Braitenberg alike assumed that individual PCs control elemental movements that are evoked when the PC fires or pauses. Marr suggested that the cerebral cortex, by activating specific CF inputs, trains the cerebellum to recognize appropriate contexts for emitting the same movements in a more automatic fashion. Albus instead suggested that CFs signal errors, training the PCs to select movements that then reduce these errors. Many subsequent models have been based on extensions of these basic ideas.

In this target article, cerebellar models are assessed both on the basis of the cerebellar cortex's internal computations and in relation to their function in movement control. Figure 1 summarizes the position of the cerebellar cortex in the global scheme of movement control, as a regulator of premotor networks. While Florens (1824) realized that the cerebellar cortex only regulates, rather than controls, movements, it was more than a century later when Ito's

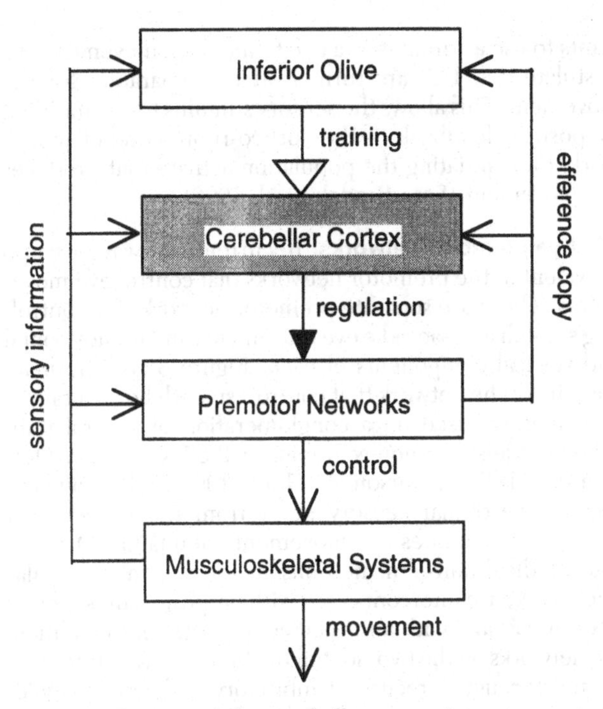

Figure 1. Position of the cerebellar cortex in motor control. Basic motor control actions are implemented by premotor networks in the brainstem, sensorimotor cortex, and spinal cord. The cerebellar cortex regulates premotor actions through inhibition and disinhibition. The inferior olive transmits the climbing fiber information that trains Purkinje cells in the cerebellar cortex how to perform their regulatory functions. In this and succeeding illustrations, regular arrow heads denote predominantly excitatory connections, closed arrows denote inhibition, and open arrows denote training influences.

(1969) electrophysiology explained why this is so. He demonstrated that inhibition is the sole action of the PCs that project out of the cerebellar cortex. The cerebellar cortex exerts its influence by inhibiting and disinhibiting motor control actions that are formulated elsewhere, in the premotor networks of the brainstem, sensorimotor cortex, and spinal cord. How does the cerebellar cortex perform these regulatory functions? We should answer this operational question prior to addressing the problem of motor learning, because a knowledge of operational principles will indicate, or at least provide insight about, what it is that PCs in the cerebellar cortex must actually learn. This places us in a better position to evaluate theories of how PCs are trained by CF signals originating in the inferior olive (IO).

This article deals generally with diverse models of the cerebellum and motor learning, although it emphasizes the adjustable pattern generator (APG) model that has been the focus of our research. We begin, in section 2, with a discussion of the properties of premotor networks, since this helps to specify what the cerebellar cortex may be attempting to regulate. In section 3 we discuss the operational features of several models of the cerebellum, analyzing the mechanisms whereby the cerebellar cortex is presumed to regulate different types of motor response. Then, in section 4, we discuss proposed roles of synaptic plasticity and the training signals sent from the IO in guiding processes of motor learning.

2. Premotor networks

What is the cerebellum regulating? This is one of the key questions that needs to be answered by a theorist attempting to model the cerebellum. Since the output actions of the cerebellar cortex are exclusively inhibitory, the cerebellar nuclear cells and vestibular neurons targeted by Purkinje inhibition must be regulated by a disinhibitory action. For this to work, nuclear cells need to have something to inhibit, either pacemaker activity or excitatory input from another source. In his original paper, Marr (1969) ignored this problem and simply assumed that the nervous system would somehow convert the inhibitory outputs of the PCs that had been selected through pattern recognition into an appropriate set of elemental movements. Albus (1971) proposed that PCs are trained to pause, rather than to fire, and that these pauses select elemental commands controlled in some unspecified manner by the individual nuclear cells. In both cases, the theory on the output side of cerebellum was not very well developed.

2.1. Limb premotor network. The problem of what is being regulated by the cerebellar cortex was briefly addressed in a follow-up article by Blomfield and Marr (1970). Their proposal was based on Tsukahara et al.'s finding (1968) of an excitatory recurrent pathway between the motor cortex and the cerebellar nucleus. Blomfield and Marr postulated that the motor cortex, through some unspecified mechanism, initiates limb movements by firing a large set of cortical neurons that tend to command many more elemental movements than are actually needed, and that the continuation of these commands is dependent on positive feedback circulating through the cerebellar nucleus. PCs were thought to eliminate, through inhibition of nuclear relays, those commands that are not needed. While the concept of positive feedback in limb premotor networks was pursued experimentally by Tsukahara et al. (1983), it did not receive much theoretical attention until recently (Eisenman et al. 1991; Houk 1989; Houk et al. 1993), the exception being Boylls's thesis (1975) and some commentary in reviews (Arbib et al. 1974; Ito 1970). Now, however, there is a substantial body of literature, reviewed by Houk, Keifer, and Barto (1993), that defines the anatomy and physiology of the limb premotor network and supports the concept that the spread of positive feedback through a recurrent network is crucial to limb command generation.

Figure 2 summarizes the known anatomy of the limb premotor network and its cerebellar cortical input. Since the interconnections between nuclei are topographically organized, we postulate that this diagram also applies at the level of microcircuitry and thus defines a modular architecture for generating the elemental movement commands envisioned by Marr and Albus. A band of PCs converges on a small cluster of nuclear cells (N) that participates in a topographically organized recurrent circuit that includes thalamic (T), motor cortical (M), rubral (R), pontine (P), and lateral reticular (L) neurons. Elemental commands are generated by positive feedback transmitted through this network and are then sent to the spinal cord in corticospinal and rubrospinal fibers. In contemplating the composite motor command transmitted by all the corticospinal and rubrospinal fibers, one needs to consider between a thousand and a million replications of this module. We envision that the individual elements in this large array of modules function in a partially autonomous manner, but that they

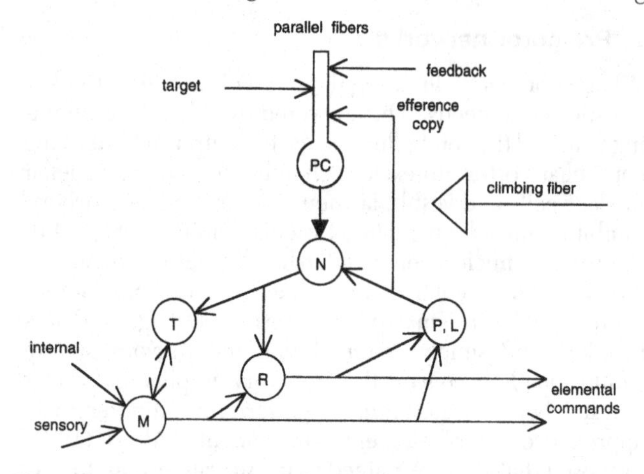

Figure 2. Organization of the limb premotor network and its regulation by cerebellar Purkinje cell (PC) inhibition. Neural stages in the limb premotor network are: N, cerebellar nuclear cells; T, thalamic relays; M, neurons in the primary motor cortex; R, neurons in the magnocellular red nucleus; P, pontine neurons; L, lateral reticular neurons.

also interact with each other through divergence in their loop connections (Houk et al. 1993). When this modular concept is combined with simple physiological assumptions, the resultant model provides plausible answers to several basic questions about the limb motor system.

For example, why are the movement-related discharges recorded from N, T, M, R, P, and L neurons so similar to each other? We believe that this is because elemental commands are generated as a collective computation in this cortico-rubro-cerebellar recurrent network (Houk et al. 1993). An elemental command is initiated when positive feedback causes activity to intensify nearly simultaneously at all stages in a given loop. The command is terminated when many of the module's PCs fire to inhibit N activity, whereupon activity at the other stages also dies out. In this manner an elemental command is expressed similarly at each of the nuclear stages.

What mechanisms are responsible for recruiting the large number of motor cortical and red nucleus neurons that participate in any given movement? We believe that commands become distributed to the population of cortical and rubral cells through divergence in modular loop connections. In this manner, positive feedback spreads from one module to another. This mechanism of progressive spread can explain the gradual rotations of population vectors in the motor cortex that occur when an animal is required to mentally rotate a visual target (Eisenman et al. 1991). A progressive spread of positive feedback also explains the common observation that reaction times are much longer than would be expected from the conduction delays seen as neural signals pass from sensory receptors, through the motor cortex, and back to motor neurons. The extra delay is explained by the need for positive feedback to intensify and spread through the limb premotor network (Houk 1989).

What is the function of the sensory response properties of cells in motor cortex and red nucleus? A sensory response to a discrete somatosensory stimulus is relatively weak and is confined to a relatively small population of neurons under conditions in which the subject ignores the stimulus, presumably because spontaneous PC discharge prevents the initiation of positive feedback. In contrast, when the subject wants to use a stimulus as a cue to initiate a movement, we postulate that PCs are turned off in preparation for the movement. This allows the sensory stimulus to be amplified by positive feedback in the cortico-rubro-cerebellar network, thus initiating the population activity that produces the movement (Sarrafizadeh et al. 1996).

2.2. Eye premotor networks. Recurrent pathways are also prevalent in the premotor networks that control eye movements. There are separate premotor networks for controlling smooth and saccadic eye movements and for horizontal and vertical components of each. Figure 3 is a summary diagram of the network that controls smooth horizontal eye movements, based on a conglomeration of several published models (Cannon & Robinson 1987; Galiana & Outerbridge 1984; Peterson & Houk 1991). This network receives vestibular sensory input from the semicircular canals and generates eye movement commands. On each side of the brainstem, neurons in the medial vestibular nucleus (V) are interconnected with prepositus hypoglosius neurons (P) and with intermediate types (PV). In addition, the networks on the two sides of the brainstem are interconnected through a recurrent inhibitory pathway. V signals are dominated by velocity coding, P signals by position coding, and many cells (e.g., PV) combine position and velocity coding. PCs in flocculus and posterior vermal regions of the cerebellar cortex inhibit some, though not all, of this family of neurons. The output commands from both sides of the brain converge upon ocular motor neurons to control the horizontal component of smooth pursuit and optokinetic eye movements and the vestibulo-ocular reflex.

Along with many obvious similarities, there are two important differences between this smooth eye network and the limb network discussed in the previous section. One is that the smooth eye network operates more or less continuously in its active state instead of making the many transitions from inactive to active states that are characteristic of the limb network. This fits with the need to control eye position continuously to stabilize the visual world, as opposed to the need to execute discrete limb movements to grasp and manipulate objects. A second important difference concerns coupling between corresponding networks on the two sides of the brain. Movements of the two eyes need to be coupled to insure fused

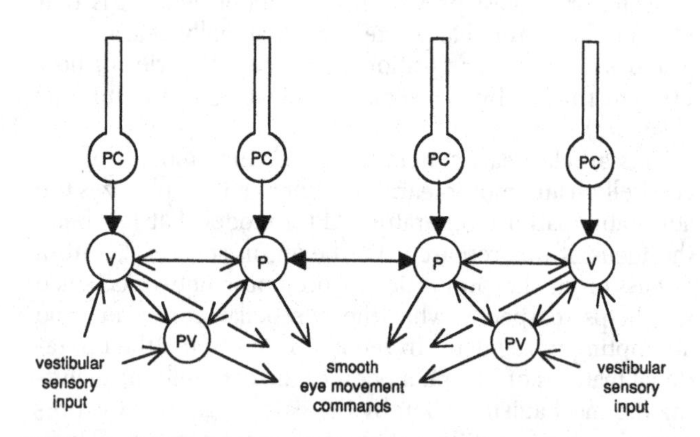

Figure 3. Organization of the premotor network controlling smooth eye movements in the horizontal plane. Neural stages are: PC, Purkinje cells; V, medial vestibular neurons; P, prepositus hypoglossius neurons; PV, intermediate types.

binocular images, whereas the left and right limbs can be operated independently in manipulation tasks. Coupling between the eye networks involves reciprocal inhibitory pathways that function as additional positive feedback loops. Modeling studies illustrate how this coupling could perform important operational and adaptive functions (Galiana 1985). The postulated operational function is to promote the conversion of vestibular head velocity sensations into eye position commands (an integration in the mathematical sense), and the adaptive function is to provide a sensitive site for regulating the gain of the vestibulo-ocular reflex. PCs in the cerebellar cortex are well situated to regulate the signals generated by this network, thus controlling integration and other dynamics. PCs also regulate the intensity of responses to vestibular input, thus modifying the effective gain of the vestibulo-ocular reflex.

Other examples of premotor networks could be highlighted, each having its characteristic processing mode and function. However, these two examples are adequate to illustrate the variety of collective computations that can be performed in a rapid and automatic fashion by the highly interconnected architecture of premotor networks. Now we turn to models of how these collective computations are operated upon by the regulatory actions of the cerebellar cortex.

3. Cerebellar cortex

How does the cerebellar cortex organize its internal computations so as to regulate the many recurrent pathways in premotor networks? Although the cerebellum has a remarkably uniform structure, the mechanisms of regulation may nevertheless differ, depending on the type of premotor network and movement that is being regulated. For example, the regulation of networks, such as the limb premotor network, that make transitions between passive and active states to control discrete limb movements may differ from the regulation of networks, such as the smooth eye network, that operate continuously to control smooth eye movements. We begin by considering models for regulating various types of discrete movements.

Elemental commands recorded from motor cortical and red nucleus neurons (M and R in Fig. 2) generally take the form of an intense burst of discharge, occurring at a frequency that corresponds to the velocity of the movement and having a duration that corresponds to the duration of the movement (cf. Houk et al. 1993). In addition to these phasic components, the neurons may also show tonic components. The elemental commands just described are seen in isotonic movement tasks, whereas analogous relations to force rate and force occur in isometric tasks. PCs recorded under similar conditions may show either bursts of discharge or pauses (cf. Thach et al. 1992). It is reasonable to assume that positive feedback is permitted to intensify in the loops that are regulated by pausing PCs, and that these loops generate commands to agonist muscles. Positive feedback would be inhibited from spreading to loops regulated by bursting PCs, and we assume that the latter loops are thus inhibited from generating commands to antagonist muscles. The question of how the cerebellar cortex exerts its control thus translates into a problem of understanding how appropriately timed bursts and pauses of PC discharge may be generated.

As might be anticipated, a variety of models of the cerebellar cortex are capable of explaining how these bursts and pauses could be generated. Therefore, it is helpful to have some constraints in addition to those already imposed by the basic anatomical and physiological properties of the network. In the following section we highlight a few particularly critical constraints that derive from studies of motor performance.

3.1. Additional constraints from experimental psychology. The concept that movement commands are centrally specified and are then executed in essentially an open-loop manner has evolved from a long line of studies in experimental psychology (cf. Schmidt 1988). According to this motor program concept, the system operates in a feedforward mode, as opposed to using sensory feedback from the periphery. Instead of providing feedback during the movement, sensory information is used to select the parameters of a motor program before it is initiated, to initiate the program, and to guide the subsequent adaptive process that mediates motor learning.

Although most investigators find sensory feedback to be ineffective in modifying ongoing motor programs, except in limited ways, Adams (1971; 1977) maintains that the endpoint of a movement is sensitive to proprioception. In support of Adams's conclusion, we found that the offsets of elemental motor commands recorded from the red nucleus can be prolonged appreciably by using a brake to prevent the animal from reaching the intended target (Houk & Gibson 1987). We also found clear support for the concept that the program operates mostly open loop, since application of the brake had relatively minor effects on discharge frequency during the burst.

Another key constraint from performance studies is the observation that the initiation and the programming of movement are separate asynchronous processes (Ghez et al. 1990; Gielen & van Gisbergen 1990). Programming can occur in preparation for a movement, at the time of a movement, or after a "default" movement has been initiated. We interpret this to mean that a decision to perform an action corresponds to the buildup of positive feedback in the limb premotor network, whereas the regulatory input to this network from the cerebellar cortex implements decisions about what movement to perform (Houk et al. 1993). Thus, in considering models of the cerebellar cortex, we will emphasize their capacity to regulate the metrics of elemental commands as opposed to the initiation of these commands. This fits well with the fact that cerebellar cortical lesions do not interfere with starting movements but instead result in dysmetria.

3.2. Reevaluation of the Marr and Albus models. It is instructive to reevaluate the Marr (1969) and Albus (1971) models in light of the above considerations. Marr's idea, or better the Blomfield and Marr (1970) revision of it, was that PCs fire in response to learned input patterns and this inhibits unneeded elemental movements. This model does not explain the pauses in PC discharge that release nuclear cell discharge. Pauses might be explained by assuming that input from inhibitory interneurons predominates over excitation to these particular PCs. Another issue is that both excitatory responses and pauses of PCs need to be more than a simple selection signal, since prolonged bursts and pauses are needed to inhibit and release the portions of the premotor network that control antagonist and agonist muscles. Dynamics that might produce extended bursts and extended pauses are not included in Marr's model. If one wishes to pursue the Marr model further, these various features need to be elaborated, and their performance implications need to be explored through simulation. Surprisingly, there was no actual simulation of Marr's model until quite recently (Tyrrell & Willshaw 1992). The latter authors found that much was unspecified in the original model. After providing these specifications, they were able to verify a number of Marr's predictions at an information-processing level. However, they have not yet attempted to use their simulation to drive a premotor network, let alone to control a movement.

The situation is somewhat different for Albus's (1971) theory, since early on it was developed into an executable model called "Cerebellar Model Articulation Controller," or CMAC (Albus 1975). CMAC has been developed to the point of using it to control actual robotic manipulators (Miller 1987). Because of these analyses, simulations, and applications, the CMAC architecture is now well understood. To summarize its operating principles, it essentially functions as a static associative memory that implements locally generalizing, nonlinear maps between mossy fiber inputs and PC outputs (Albus 1981). It does this by first treating the granule/Golgi cell network as an association layer that generates a sparse, expanded representation of mossy fiber input, and second by using adjustable weights to couple the large PF vector to PC output units with graded properties. It would not be difficult for the bursts and pauses observed in PC recordings to be simulated by this architecture.

However, it is not clear that either of the above models could satisfy the performance constraints summarized in section 3.1. For example, how might sensory information be used to select the parameters of a motor program and to trigger its initiation, but then be disconnected during the feedforward stage of execution? PC outputs must somehow be responsive to sensory input during a programming phase, but become unresponsive during execution. In the Marr and Albus models, responses to sensory input are mediated through PFs. There is no mechanism whereby these inputs would have actions during a programming phase, be disconnected during most of an execution phase, and then become active toward the end of execution in order to terminate the movement. Another issue concerns the independent control of the programming and initiation of a movement. While the Blomfield and Marr model could do this through an unspecified motor cortical process for initiating an excessive number of commands, in the Albus model there is no separation between programming and initiation.

3.3. Adjustable pattern generator model. The adjustable pattern generator (APG) model was developed by Houk, Barto, and colleagues specifically to address the limitations discussed in the previous paragraph (Berthier et al. 1993; Buckingham et al. 1994; Houk 1989; Houk & Wise 1995; Houk et al. 1990; Sinkjaer et al. 1990). The term "adjustable pattern generator" refers to the ability of an APG to generate an elemental burst command with an adjustable intensity and duration. The model has an anatomically based modular architecture that is summarized in Figure 4.

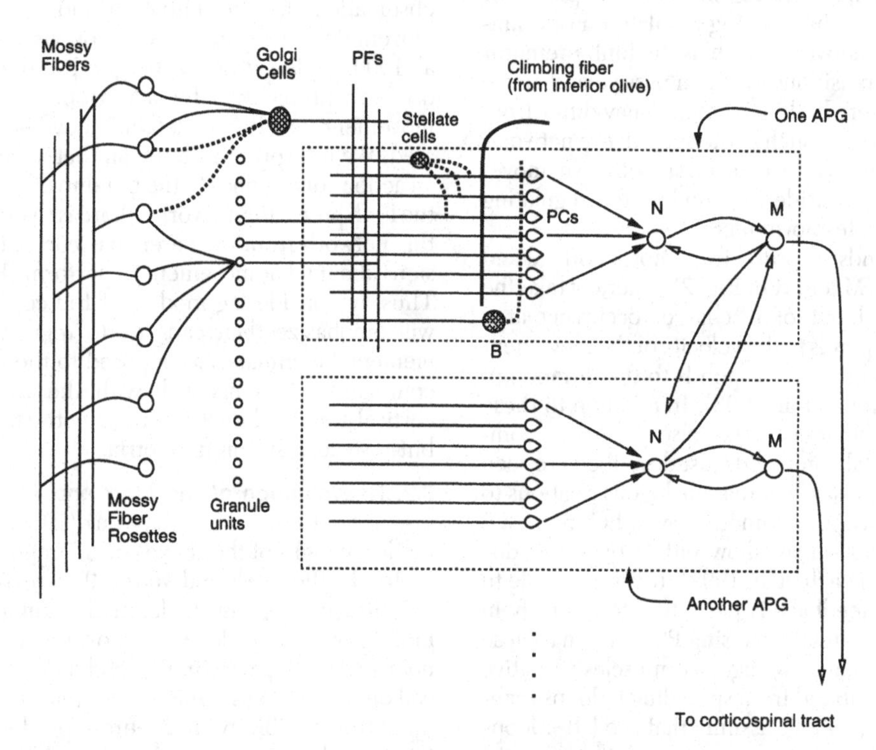

Figure 4. Modular architecture of the adjustable pattern generator (APG) array model of the cerebellum. PCs, Purkinje cells; B, basket cell; PFs, parallel fibers; N, cerebellar nuclear cells; M, neurons in primary motor cortex. Inhibitory interneurons in a module – the Golgi, basket, and stellate cells – are denoted with stippled cell bodies and axons.

Each module includes a positive feedback loop between a cerebellar nucleus cell (N) and a motor cortical cell (M), which provides an abstract representation of the cortico-rubral-cerebellar loops discussed in section 2.1. Each N receives inhibitory input from a private set of PCs. Each set of PCs receives a private CF training input, convergent input from an array of PFs, and inhibitory input from a basket cell (B) and from stellate cells (although the latter are not specifically simulated). In early versions of the model, the PFs were assumed to convey unprocessed mossy fiber input, whereas in our recent simulations (Barto et al. 1996) we have used a layer of granule units combined with Golgi cell feedback to create a sparsely coded representation of the mossy fiber input, along the lines assumed in the Marr and Albus theories.

Figure 5 illustrates how this model achieves independent control over the initiation and programming of a movement. In this example, programming occurs in a preparatory phase that is initiated by an instruction signal transmitted to Bs and PCs through mossy fiber input. The balance between direct PF excitation of PCs and their inhibition through a B unit causes some PCs to switch to a more intense on-state (stippled trace in Fig. 5, Left) and others to switch to an off-state (solid trace). In the APG model, motor commands are not initiated by these programming events in the cerebellar cortex. Instead, the initiation of command generation is triggered independently by sensory or internal inputs to the motor cortex or red nucleus. As discussed in section 2.1, this starts positive feedback and causes the limb premotor network to switch from an inactive to an active state, which then initiates elemental commands in a large number of APG modules. It is only after motor commands are initiated that the effects of programming events in the cerebellar cortex become expressed. Each elemental command intensifies to a level that is determined by the degree to which the module's PCs are switched off or on by the instruction signal. In this manner, the intensities of the different elemental commands can be preset to a variety of levels, thus offering an explanation of how population vectors in the motor cortex could be regulated (Eisenman et al. 1991). In agreement with the performance studies discussed earlier, this model allows the programming process to occur in a preparatory phase (the case illustrated in Fig. 5, Left), at the time of movement onset, or even after movement onset.

Once the PCs switch to a particular firing state, in response to the instruction stimulus, they are postulated to become refractory to further input until near the termination of the movement. As a consequence, simulated movements start to be executed in a feedforward manner. A key assumption is that CFs train PCs to recognize those particular patterns of PF activity that indicate when desired endpoints are about to be reached. Occurrences of these patterns cause PCs that were turned off in the programming phase to fire strongly, and this terminates positive feedback in the premotor network. We call this type of control "quasi-feedforward" (Houk et al. 1990) and have pointed out how it satisfies the motor program constraints from experimental psychology (Berthier et al. 1993). It is also advantageous in preventing delayed feedback from causing instability oscillations (Houk et al. 1990). The fact that mossy fibers display prominent sensory properties (Van Kan et al. 1993), whereas Purkinje and nuclear cells are relatively unresponsive to somatosensory stimulation (Harvey et al. 1977; 1979), is further explained by this feature.

In the original APG model, the quasi-feedforward characteristics of the network derived solely from the biophysical properties assumed for PC dendrites (Fig. 5, Right). Due to their high density of calcium channels, PC dendrites were assumed to behave like bistable binary elements possessing a zone of hysteresis, and, for simplicity, PCs were modeled as if they had only one dendrite. In a cell with many dendrites, each operating in a bistable manner, their summed effect on the soma would, of course, be multistable. An ionic model has been developed to demonstrate

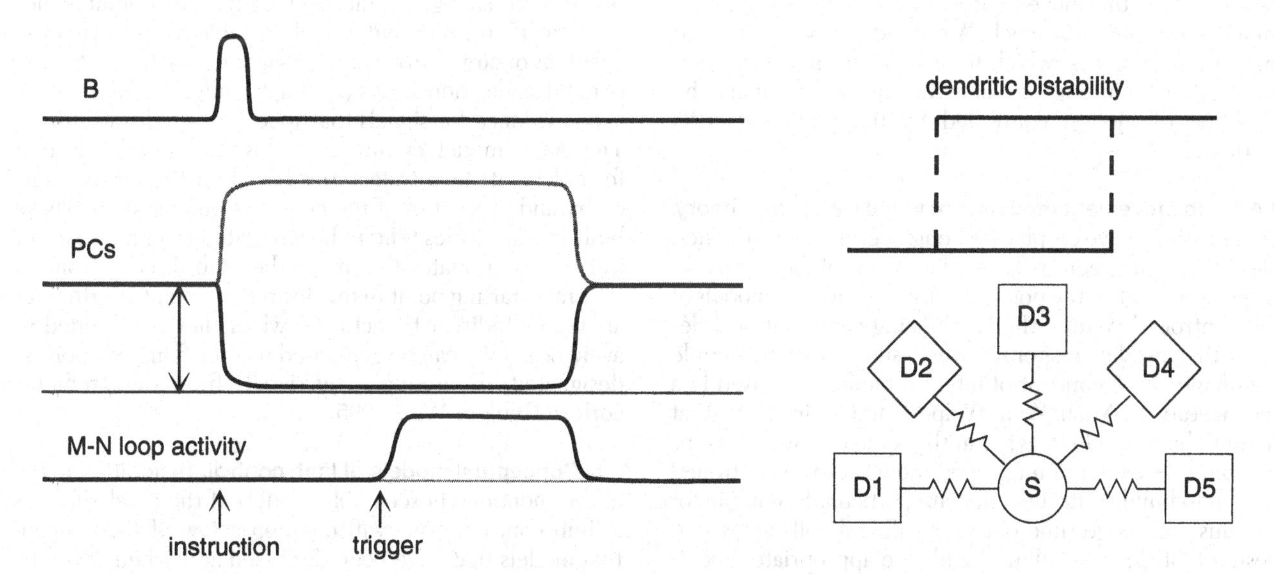

Figure 5. Signals utilized by the APG model in selecting and regulating the execution of a motor program. The instruction stimulus fires basket (B) cells and causes state transitions in Purkinje cells (PCs). The trigger stimulus initiates positive feedback in those M–N loops that are disinhibited. The inserts on the right illustrate the model of dendritic bistability in PCs and the concept that bistability in several dendrites (D1 through D5) gives rise to multistability in the soma (S) of a PC.

the feasibility of these ideas (Yuen et al. 1995). There are two other mechanisms that might contribute to the observed insensitivity to input, although both presume that PC dendrites behave in a binary (not necessarily bistable) fashion. If so, the recurrent inhibitory connections that Purkinje (and basket) cells make with each other might promote switching between states (Bell & Grimm 1969). Furthermore, we have recently found that insensitivity to input during the feedforward phase can result if PCs are trained to respond only to those specific patterns of PF input that occur at the ends of movements.

Motor programs might be stored in a lookup table as detailed lists of highly specific instructions, which is close to what Marr (1969) envisioned in his original theory. However, a literal application of this scheme would exceed the storage capacity of the cerebellum (Kawato & Gomi 1993). Instead, most investigators have favored the idea that memory may be used more frugally to store generalized motor programs that are then parameterized in order to control specific movements. In the APG model, the counterpart of a generalized motor program is a set of PF weights for proprioceptive and target inputs (Berthier et al. 1993). This is analogous to Adams's (1971) "perceptual trace," since a particular constellation of PF inputs, when processed by the set of PF weights, signifies that the desired endpoint of a movement is about to be reached, causing PC firing that terminates the movement command. Once weights are learned, the model's commanded velocity is then parameterized by B-cell firing in the selection phase of the model's operation (Houk et al. 1990). The velocity that is selected by turning PCs off is automatically scaled so as to depend on the distance between the initial position of the limb and the desired endpoint of the movement. Velocity can also be varied independently as a scaling factor controlled by diffuse neuromodulatory input, which can explain how velocity scaling can be applied simultaneously to all elements in a composite motor program (Schmidt 1988). Movement duration is parameterized in the execution phase of model operation. Duration turns out to be a dependent variable that evolves from the course of the movement as opposed to being determined by an internal clock. Amplitude is parameterized by the target inputs, which have synaptic influences that are graded according to where the target lies along the direction of motion controlled by the particular APG module.

3.4. Limb movement models motivated by control theory. Some authors have emphasized engineering control principles in designing cerebellar models. Control theorists have been fascinated by the potential utility of internal models of the controlled system, the latter being computational devices that predict responses when supplied with sample commands. The concept of internal models was used in a recent study by Miall, Weir, Wolpert, and Stein (1993) that treated the cerebellum as a "Smith predictor." Such systems combine delayed and undelayed models of the controlled system to build controllers that are particularly suitable for systems with large time delays. Miall and colleagues suggested that the cerebellum builds the appropriate models through a learning process. Once learned, these internal models become integral components of the controller. Since the motor system is characterized by large time delays, the cerebellum almost certainly needs to function as a predictive controller, but not necessarily as a Smith predictor. In another model motivated by engineering principles, Paulin (1989) suggested that the cerebellum computes like a Kalman filter, another form of predictor. The APG model has a much simpler control structure than either of these models, and it also functions as a predictive controller (Barto et al. 1996; Buckingham et al. 1995). Several of the assumptions that Miall et al. made in attempting to map their Smith predictor onto the gross anatomy of the nervous system and the microcircuitry of the cerebellum are tenuous. Paulin did not even attempt this exercise.

Kawato and Gomi (1992a) proposed that the lateral cerebellum learns to function as an "inverse model" of the limb's controlled system. Inverse models do the opposite of the forward models discussed in the previous paragraph; they predict commands when supplied with sample responses. If, instead of sample responses, they are supplied with signals representing desired trajectories, inverse models then generate the appropriate commands. Kawato and Gomi assumed that the motor cortex and cerebellum are simultaneously provided signals from the parietal cortex representing a desired trajectory of a limb movement. The motor cortex compares the desired trajectory with sensory feedback and issues a crude command while waiting for the cerebellum to use its inverse model to compute a precise command. After the latter is sent back to the motor cortex, it is used to compute an updated command. One problem with this model is that it ignores the time delays mentioned in the previous paragraph, which may result in serious problems with this control scheme. Although the authors map their controller onto the gross anatomy of the brain, no attempt is made to show how the microcircuitry of the cerebellum might be used in implementing an inverse model. These authors have also proposed models of the intermediate and medial cerebellum that are based on similar principles (Gomi & Kawato 1992).

The effective use of internal models presumes the existence of neural stages that compute desired trajectories for movements through space, and other stages for conversion into desired changes in muscle lengths and/or joint angles (Kawato 1990). Although signals capable of specifying the positions of targets in extrapersonal space are present in the parietal association cortex (Zipser & Andersen 1988), there is no evidence for signals that specify desired trajectories. The APG model circumvents this problem because it formulates its own trajectories based on the intrinsic circuitry and properties of the neuromuscular system. These built-in trajectories tend to be straight lines in the space of intrinsic coordinates that move the limb directly from an arbitrary starting point to the desired endpoint (Berthier et al. 1993). Indirect trajectories, when they are needed to avoid obstacles, can be generated by specifying "via points" designated, for example, by signals from the premotor cortex (Houk & Wise 1995).

3.5. Conceptual models of limb control. In addition to the above-mentioned executable models of the cerebellum as a limb controller, we will mention a few of the conceptual models that have been discussed and schematized by neurobiologists, but lack a formulation precise enough for simulation. In this vein, Thach and colleagues (1992) described why the PF architecture is ideal for presenting PCs with the information they need to coordinate multijoint movements. We accept this idea and tacitly include it in the APG model discussed earlier. Bloedel (1992) proposed that CFs selectively enhance microzones comprised of parasagittal rows of PCs rather than training PCs to recognize patterned input. This is compatible with the Llinás and Welsh (1993) scheme, whereby gap junctions in the IO synchronize clusters of olivary neurons that innervate parasagittal zones in the cerebellar cortex. Arshavsky, Gelfand, and Orlovsky (1986) offered some general guidelines about the role of the cerebellum in coordinating locomotion. Finally, Prochazka (1989) suggested that the role of the cerebellum is to control the gain of spinal and brainstem reflexes. The gain-control idea has also arisen in eye movement models that are discussed later.

3.6. Conditioned reflex models. Several models have been proposed to explain the role of the cerebellum in mediating conditioned reflex (CR) responses of the eyelid to tone and light conditioned stimuli (CS) (Buonomano & Mauk 1994; Gluck & Thompson 1990; Houk 1990; Moore et al. 1989; Thompson 1986). The circuit implicated in this response is similar in several respects to the limb premotor network discussed in previous sections (Fig. 2). However, the CR models have generally neglected pathways through the thalamus and motor cortex, since CRs can be learned and performed in decerebrate animals. Based on this simplification and other findings, Thompson (1986) proposed a conceptual model along the lines summarized in Figure 6A. Tone and light CSs project (via the pons) as mossy fibers into the cerebellum. Collaterals of the mossy fibers provide direct excitation of N cells and excitation of PCs via PFs. In this manner, N cells are both directly excited and indirectly restrained by PC inhibition to produce an N output. The combination generates a neural model of a CR that is then relayed via R to the abducens (VI) and other eye nuclei that output the CR. Unconditioned reflexes (URs) are mediated by air-puff unconditioned stimuli (US) to the cornea via the trigeminal nucleus (V). The US is also transmitted through CFs, and the latter train PCs and Ns to respond appropriately to their inputs.

This model is meant to explain how CRs are generated in delay conditioning tasks. In this paradigm, the CS is an extended period of stimulation that begins several hundred msecs in advance of the US and continues until the US is delivered. Thus, one can assume that the CS provides a constant excitatory drive to N that can readily be shaped by PC inhibition to form an appropriate CR. The control problem for PCs is to fire initially so as to cancel the excitatory drive to N and pause when it is time to initiate the CR; depending on the particular task, the PCs may also have to resume firing at the end to cancel the excitatory drive to N, thus terminating the CR. In several implementable models based on this concept (Buonomano & Mauk 1994; Gluck et al. 1990; Moore et al. 1989), PCs receive a set of PFs that transmit a diversity of temporal patterns related to the CS. The PC is taught to respond to those PFs that are active during periods when N needs to be inhibited and not to respond to PFs active when N needs to be disinhibited. In the Moore et al. (1989) model, the different timings of PF signals are generated in the pontine nuclei as a set of variously delayed responses to the CS input. In the Gluck & Thompson (1990) model, the different timings are again assumed to be generated outside of the cerebellum, but in this case are represented as spike trains that are modulated at several sinusoidal frequencies and phases. In the Buonomano and Mauk (1994) model, the different timings are assumed to be generated within the cerebellar cortex as dynamic interactions between granule and Golgi cells. This latter process is analogous to the mechanisms proposed by Fujita (1982) and by Chapeau-Blondeau and Chauvet (1991) in their dynamic filter models of cerebellar function. The model by Moore and colleagues is most readily accommodated by current single unit data that demonstrates that sources of mossy fiber input to the cerebellum have a distribution of latencies. The Moore et al. model is similar in this respect to a model proposed by Bell (1994) to explain the ability of the lateral line organ in electric fish to adaptively cancel the reafference from its own electrical discharge. The lateral line organ has phylogentic relations to the neuronal architecture of the cerebellum (Mugnaini & Maler 1993).

In contrast to the conditioning models described in the previous paragraphs, Houk (1989; 1990) suggested that the excitatory drive to N is produced not by direct sensory responses to the CS, but instead by positive feedback in the recurrent network that interconnects red nucleus (R), the lateral reticular nucleus (L), and the cerebellar nucleus (N); see Figure 6B. This hypothesis is similar to the APG model discussed earlier in relation to limb movement control (Figs. 2 and 4). Weak sensory (CS) inputs to R need to circulate and spread in order to start sufficient positive feedback in the R–L–N loop in order to initiate a CR command. The amplitude and duration of the CR command is then controlled by PC inhibition of loop activity. This model allows PCs to pause well in advance of the CR as a preparatory programming action, in agreement with observed firing patterns of PCs (Berthier & Moore 1986). This model is also in agreement with the diversity of signals that is recorded from R neurons during performance of CRs (Desmond & Moore 1991). In summary, the APG version of Thompson's (1986) conceptual model suggests that the CS–US association is mainly formed outside of the cerebellum (in R, L, or even the motor cortex), whereas the primary role of the cerebellum is to determine the topography of the CR (i.e., the motor program).

CR commands generated by APG modules are subsequently sent to a brainstem network that produces the final output. Activity-dependent labeling results led Keifer and Houk (1995) to suggest a model in which a trigeminal-reticulo-abducens network, illustrated schematically in Figure 6b, is used to output both CRs and URs. This brainstem network may be analogous to the network formed by segmental interneurons, propriospinal neurons, and motor neurons in the spinal cord, the latter being the target of the elemental commands illustrated in Figure 2.

3.7. Eye saccade models. Lesions of the cerebellum severely disrupt an animal's ability to adapt the accuracy of saccadic eye movements (Robinson & Optican 1981). Whereas there are many eye saccade models, only a few attempt to explain how the cerebellum might achieve this adaptive control (Dean et al. 1994; Grossberg & Kuperstein 1989; Houk et al. 1992). The anatomy indicates that cerebellar regulation of saccadic eye movements should occur at two levels, at the level of a tecto-cerebellar command network and at the level of a brainstem burst-generating network (cf. Houk et al. 1992). The model shown in Figure

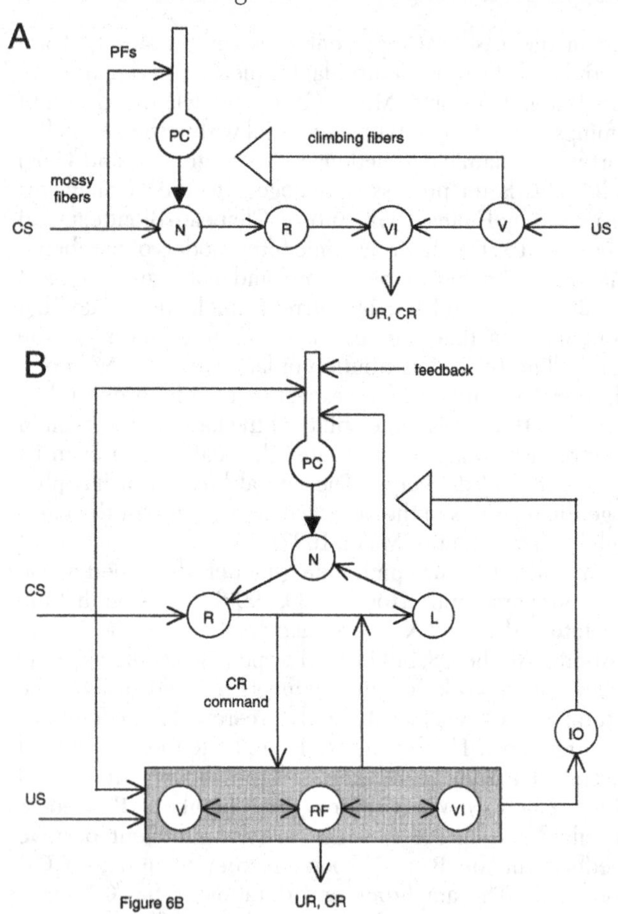

Figure 6. Cerebellar regulation of conditioned reflexes (CRs). Part A shows a simple model of CR generation: PC, Purkinje cells; N, cerebellar nuclear cells; R, red nucleus cells; VI, cells in the abducens nucleus; V, cells in the trigeminal nucleus. Part B shows a modified model of CR generation based on the APG theory. L, lateral reticular nucleus; RF, reticular formation; IO, inferior olive.

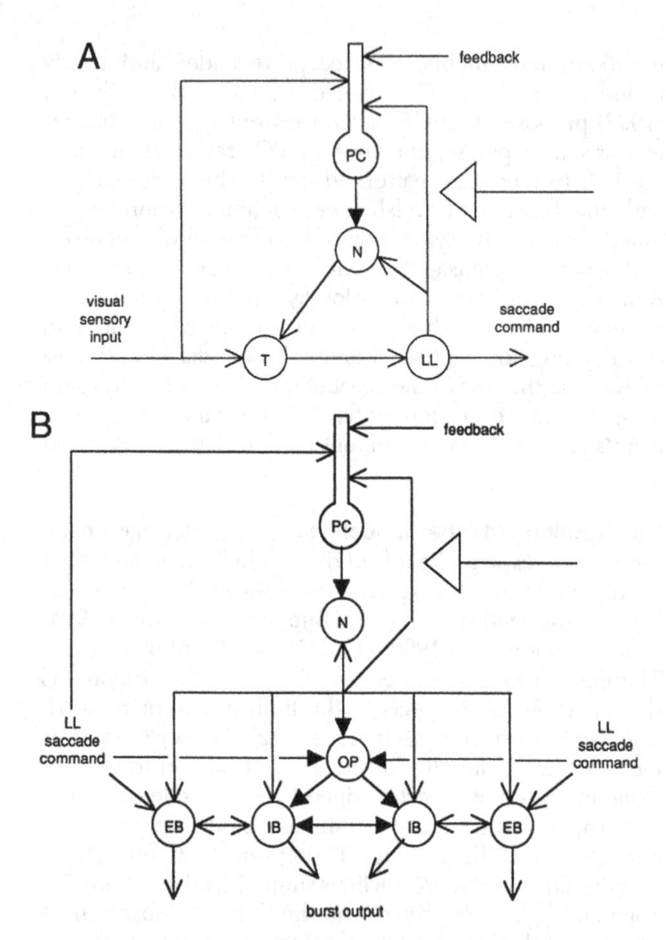

Figure 7. Cerebellar regulation of saccadic eye movements. Part A shows a model of the tecto-cerebellar network. PC, Purkinje cells; N, cerebellar nuclear cells; T, tectal neurons; LL, long-lead burst neurons. Part B shows a model of cerebellar regulation of the brainstem burst-generating network. PC, other Purkinje cells; N, other nuclear cells; OP, omnipause neurons; EB, excitatory burst neurons; IB, inhibitory burst neurons.

7A highlights the cerebellar control of the tecto-cerebellar command network, which we will consider first.

The intermediate layer of cells in the superior colliculus, or tectum (T), projects to a category of brainstem reticular neurons called long lead (LL) bursters. LLs are an important source of mossy fiber input to the cerebellum, with collaterals to the cerebellar nuclei (N). The N cells targeted by these collaterals project back onto T neurons, forming a recurrent network. Neurons in the T–LL–N premotor network generate bursts of discharge that typically precede saccadic eye movements by relatively long latencies and are assumed to function as saccade commands. The distributed nature of saccade commands is similar to the situation in the premotor networks controlling limb movements (M, R, P, L, N, and T neurons in Fig. 2). Pursuing this analogy further, Houk, Galiana, and Guitton (1992) hypothesized that positive feedback in the tecto-reticulo-cerebellar recurrent network functions as an important driving force for generating long-lead burst activity. The model explains the psychophysical observation (Gielen & van Gisbergen 1990) that the initiation of saccades and the specification of their kinematic parameters are controlled by separate processes. A saccadic burst command is initiated when weak visual sensory input to T neurons from the superficial layer of the model tectum initiates positive feedback in the T–LL–N loop, schematized in Figure 7A. PCs in the cerebellar cortex then regulate the intensity and duration of the bursts, thus specifying the motor program that controls the velocity, duration, and direction of saccadic eye movements.

Saccade commands are vectors comprised of many elemental commands, each specifying an elemental saccade in a particular direction. (This is in analogy with limb motor commands, except that elemental saccade commands are also specialized for particular movement amplitudes.) The model proposed by Houk et al. (1992) is an APG array analogous to the limb model discussed in section 3.3, in which each T–LL–N module is regulated by its own set of PCs. The vector sum of the set of elemental commands controls the direction of the saccade. Divergence in individual T–LL–N loops explains how a large population of T neurons can be recruited to form the composite command observed at the level of the tectum (McIlwain 1986), and the model equally explains the concurrent bursting seen in LL and N neurons. The pauses that occur in N neurons just before and just after bursts (Fuchs et al. 1993) appear to be expressions of PC inhibition in the process of regulating loop activity. These particular N neurons are clustered in the caudal fastigial nucleus along with other N neurons that are reciprocally connected with the brainstem burst-generating network.

The brainstem burst-generating network is shown in Figure 7B. It receives retinotopically organized saccade commands from LL neurons that participate in the tecto-cerebellar network of Figure 7A. Through a convergence and burst generating mechanism that has been modelled by several authors (Galiana & Guiton 1992; Krommenhoek et al. 1993; Robinson 1975; Scudder 1988; Van Gisbergen et al. 1981; 1989), these retinotopic inputs are converted into muscle-specific burst outputs that are sent to eye motor neurons to control saccadic movements. (Bursts are also sent to the smooth eye system to control fixations after saccades are completed.) Figure 7B shows the bilateral network for horizontal saccades, whereas only one side of the tecto-reticulo-cerebellar network was illustrated in Figure 7A (as another simplification, control of omnipause neurons by tectal fixation neurons was not discussed). The omnipause neurons (OPs) shown in the center maintain fixation between saccades by tonically inhibiting the burst neurons. Bursts are triggered by LL saccade commands; LL neurons on one side inhibit OPs while simultaneously exciting that side's excitatory burst (EB) neurons. Mutual inhibition between inhibitory burst (IB) neurons prevents both sides from firing simultaneously. N neurons in the fastigial nucleus project to EB, IB, and OP neurons and receive recurrent connections back from these and other sites, as collaterals of mossy fiber inputs to the appropriate regions of the cerebellar cortex.

The cerebellum regulates the burst-generating network through inhibition and disinhibition of N neurons, only one of which is shown in Figure 7B. A recent model of this system by Dean and colleagues (1994) gives the cerebellum the function of setting the gain of feedback to the burst generator circuit. They implemented gain control using a CMAC representation of the cerebellar cortex. Like the earlier model proposed by Grossberg and Kuperstein (1969), Dean's model computes gain factors that, when inserted into the brainstem saccade network, are capable of generating accurate saccades in the course of development. They also simulated the adaptive response that occurs after eye muscles are weakened or visual targets are displaced. As currently formulated, this is a trial-level model that does not predict the time course of neural signals. While some investigators see the cerebellum as exerting gain control, others have suggested that compensation is produced by subtracting an adjustable signal from a fixed-gain saccadic circuit (Optican & Robinson 1980). This subtraction scheme is the one favored by single unit data and also by the APG array model described in this article.

There has been considerable debate as to whether saccades are controlled as fixed vectors in retinotopic coordinates or as endpoint trajectories in head or body coordinates (Guitton et al. 1990; Robinson 1987; Scudder 1988; Sparks 1988). We have postulated that both strategies are used, but at different locations in the network (Houk et al. 1992). In the APG array model, the cerebellar cortex operates in head (or body if the head is free) coordinates. It exerts its executive regulation by preselecting an action (as in Fig. 5, Left), triggering it and then continuing it until the desired endpoint of a saccade is about to be reached, whereupon its PCs fire to terminate positive feedback in the loops that they regulate. PCs recognize desired endpoints on the basis of PF patterns derived from proprioceptive, efference copy and target signals in mossy fibers. Since different APGs regulate the tecto-cerebellar network (Fig. 7A) and the brainstem saccade generator (Fig. 7B), the PCs in the two systems probably utilize different vectors of PF input to perform their computation. We consider the computation to be a pattern recognition task that implements finite state control, as opposed to conventional feedback control. On the other hand, we accept the simpler feedback-controlled, fixed vector models as being valid for the computations that are performed more automatically within the tecto-reticulo-cerebellar network and within the brainstem saccade generator.

Since the saccade control system includes cerebellar connections with the final stage of motor output in the brainstem, as well as with the tectal-cerebellar network, it raises the possibility that a similar regulation may eventually be documented for limb and conditioned eyelid systems.

3.8. Smooth eye movement models. Starting with Ito (1970), many investigators have proposed that the cerebellum serves an important function in the regulation of smooth eye movements (Galiana 1986; Ito 1984; Kawato & Gomi, 1992b; Lisberger 1994; Peterson et al. 1991; Robinson 1976; Shidara et al. 1993). The basic circuit upon which these theories are based is shown in Figure 8. Neurons in the vestibular nucleus receive vestibular sensory input from the semicircular canals, visual sensory input from the retina, and an inhibitory input from PCs in floccular, paraflocular, and vermal regions of the cerebellar cortex. The sensory inputs to V neurons mediate basic vestibulo-ocular and optokinetic responses on a background of tonic PC inhibition, and these brainstem reflexes are then fine-tuned by modulations in PC discharge. The PC signals are computed from a variety of PF signals, including vestibular discharge related to the rotational velocity of the head, optokinetic discharge related to motion of the visual field, and efference copies of the eye movement commands computed by the network. Since motion of small visual targets has negligible input to V neurons other than via PCs, the models attribute pursuit movements entirely to transmission through the cerebellar cortex.

In normal animals, floccular PCs discharge steadily without modulation during vestibulo-ocular reflexes, even though powerful head velocity and efference copy inputs can be demonstrated (Miles & Lisberger 1981). This has led to the hypothesis that the basic vestibulo-ocular reflex through the brainstem is operating with an appropriate gain without the regulatory assistance of the cerebellum. These

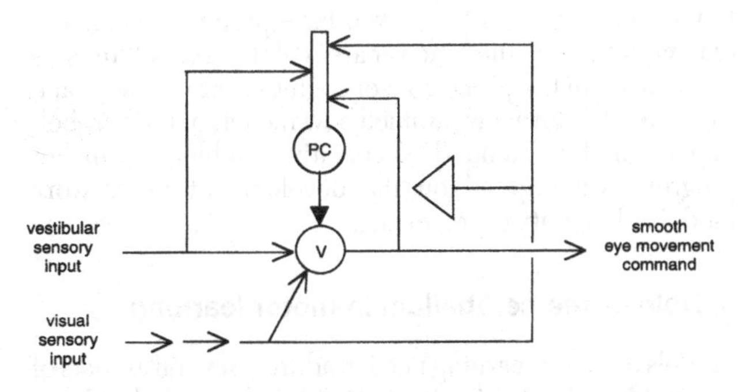

Figure 8. Cerebellar regulation of smooth eye movements in the horizontal plane. PC, Purkinje cells; V, medial vestibular neurons. Also see Figure 3.

PCs do fire, however, when the animal makes pursuit eye movements, compatible with the hypothesis that PC discharge in response to the visual PF input mediates pursuit responses through disinhibition (Lisberger 1994). Adaptation of the vestibulo-ocular reflex to visual distortion has been modeled in different ways. Ito (1989) attributes the adaptation to adjustments in PF synapses, whereas Lisberger (1994) argues that the requisite changes in gain are present in the brainstem part of this system. Other models have assigned gain change to both locations; rapid changes have been attributed to the cerebellar cortex and slower changes to the vestibular nucleus (Galiana 1986; Peterson et al. 1991). Tests of the latter hypotheses suggest that at least some of the rapid gain changes occur in the vestibular nucleus (Khater et al. 1993). A significant part of the compensatory response, however, is still attributable to the cerebellar cortex. On theoretical grounds it seems likely that adaptation would occur in both PCs and in the premotor network (Houk & Barto 1992) and this has recently been confirmed (Partsalis et al. 1995).

Models of smooth pursuit should address some of its unique properties. They should be capable of explaining the ability of pursuit to continue after visual input is interrupted. Young (1977) and others after him (cf. Lisberger 1994; Robinson 1987) attributed this to positive feedback through the efference copy loop (see Fig. 8; this assumes that the efference copy signals have a sign inversion in addition to the stage of PC inhibition.) The model in Figure 8 is oversimplified in the sense that it does not include the many recurrent loops in the smooth eye premotor network that were discussed in section 2.2 (Fig. 3). Positive feedback in these loops should also contribute to this storage property. Only the models proposed by Galliana (1986) and Peterson et al. (1991) include these features. A shortcoming of all of the above models of pursuit is their failure to confront predictive tracking, which is the ability to follow, without any significant time delay, targets that have certain deterministic properties. It is likely that much of our normal pursuit of moving targets involves at least some degree of predictive tracking. Mahamud et al. (1995) have recently shown that the APG model is capable of predictive tracking, and this feature is now being explored in more detail.

3.9. Cognitive processing models. The above models apply to the phylogenetically older parts of the cerebellum, which are clearly interconnected with the motor system. The newer parts of the cerebellum (much of the hemispheres and the dentate nuclei) are instead connected to the so-called association regions of the cerebral cortex. Leiner, Leiner, and Dow (1989) have proposed a conceptual model of how these newer areas of the cerebellum may be involved in the processing of cognitive information, and Ito (1993) has formally outlined how models of the cerebellum might be extended to cognitive problems. Further progress will depend on the development of network models of cognitive processing.

4. Role of the cerebellum in motor learning

Models of motor learning need to address several aspects of the problem. First, it is essential to identify precisely what is being learned in an information processing sense. The models of the cerebellum reviewed in the previous sections can serve this function. Second, they need to adopt a rule for modifying synaptic efficacy, hereafter referred to as a "learning rule." Preferably the learning rule should conform to, or at least be motivated by, the cellular mechanisms that underlie neuronal plasticity in the region (or regions) of brain that is (are) being modeled. Third, they need to confront the credit assignment problem, which concerns the difficulty of directing training signals to appropriate sites in the network and at appropriate moments in the training process, in order for learning to be adaptive. Fourth, they must define the training information that is provided to the model, and this should be justified in terms of the information that is likely to be available for guiding the learning process in the organism. In this section we attempt to address each of these issues in relation to the role of the cerebellum in motor learning.

4.1. Cellular mechanisms defining learning rules. The publication of pattern recognition models of the cerebellum by Marr (1969) and Albus (1971) encouraged experimentalists to search for a cellular mechanism of synaptic plasticity in PF synapses that might implement one of the postulated learning rules. Marr hypothesized that the PF synaptic weight would be increased if the PC fired at about the same time that the PF was active. This is analogous to long-term potentiation (LTP) as expressed in the hippocampus (Bliss & Collingridge 1993) and amounts to a Hebbian rule in which a synapse is strengthened whenever the presynaptic ending and postsynaptic cell are simultaneously active. The role of the CF input was to fire the PC unconditionally, thus reinforcing PF synapses active at the time of climbing fiber discharge. No provision was made for weight decreases.

Albus envisioned a learning rule with an opposite sign and more complex properties. He postulated that synaptic weight would be decreased, that is, a long-term depression (LTD) instead of an LTP would occur, and only in the presence of a three-way coincidence between a CF input (training signal), PC firing (postsynaptic factor), and PF synaptic activity (presynaptic factor). This amounts to a unidirectional version of the training rule used in research on perceptrons. He also postulated synaptic decrements on the spiny synapses of basket and stellate cells, thus providing a mechanism for countering the generally reduced excitatory input to PCs that would occur with training experience.

The experimental search for cellular mechanisms defining a cerebellar learning rule is discussed in other articles in this issue. To summarize these observations, there now seems to be good agreement that CF activity, when coupled with other factors, produces an LTD as opposed to an LTP, in agreement with Albus's model of the learning rule (Ito 1989). Although there may be more than one mechanism for LTD, the one that is best supported by current data involves the intracellular activation of protein kinase C in the spine (Linden 1994). This mechanism appears to be mediated by a combination of intracellular steps, as outlined in Figure 9.

When a PF is active, it releases glutamate neurotransmitter, which produces both depolarization of the spine via AMPA receptors and an activation of mGluR1 metabotropic receptors (left side of Fig. 9). The activation of metabotropic receptors should be localized to those synapses that are activated by presynaptic transmitter release;

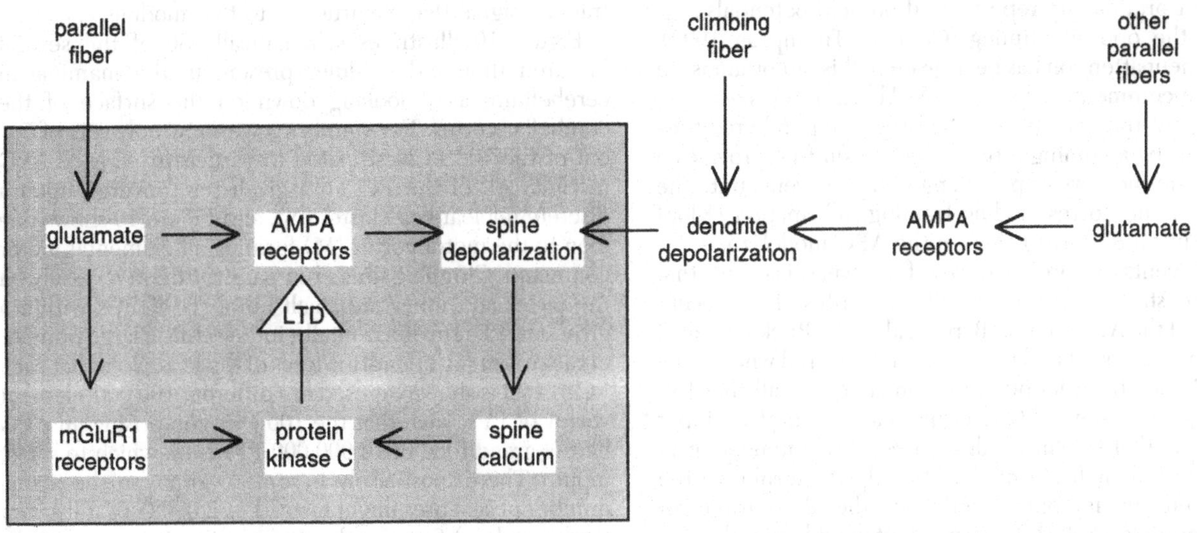

Figure 9. Cellular mechanisms in a learning rule for long-term depression (LTD). Stippled box defines boundaries of a spine. According to this model, LTD requires three factors: parallel fiber input to the spine (a presynaptic factor), dendritic depolarization produced by responses to other parallel fibers (a postsynaptic factor), and dendritic depolarization produced by climbing fiber input (a training factor).

this step has been interpreted to represent the presynaptic component in a three-factor learning rule (Houk & Barto 1992; Houk et al. 1990). The other two factors relate to dendrite and spine depolarization, which are required to open calcium channels and elevate spine calcium. According to the model, accumulation of enough spine calcium to mediate LTD requires a higher degree of spine depolarization than can be produced by the spine's own synapse. Additional spine depolarization is caused by the depolarization of the adjacent dendrite, the latter being mediated partly by CF input (the training signal) and partly by postsynaptic responses to other PF inputs (the postsynaptic factor). If both of these influences are present, the spine becomes sufficiently depolarized to elevate spine calcium appreciably. Then protein kinase C can be activated by its cofactors, calcium and the diacylglycerol produced by mGluR1 activation. Activated protein kinase C produces LTD through an action on AMPA receptors (Crepel, this issue). Thus, presynaptic, postsynaptic, and training factors are required in combination to produce LTD. Additional support for the above model of LTD has recently come from studies of mice that lack mGluR1 (Aiba et al. 1994; Conquet et al. 1994). These animals display both impaired Purkinje cell LTD and severe cerebellar ataxia.

Less is known about the learning rule for weight increases, although one seems to exist. LTP has been induced in brain slices by stimulating PFs without CFs (Sakurai 1987). This led Houk and Barto (1992) to postulate that the presence of presynaptic input without either postsynaptic depolarization or CF discharge mediates LTP. However, LTP can be produced in tissue culture by combining AMPA-receptor activation with depolarization, provided mGluR1 activation is omitted (Linden et al. 1991). This result suggests a quite different rule for LTP, namely, the presence of CF and postsynaptic activity coupled with the absence of presynaptic activity. The latter rule, which would have the effect of normalizing input to PCs, has not yet been tested computationally. It receives support from the recent finding that gain decreases of the vestibulo-ocular reflex (presumably mediated by LTP) are resistant to metabotropic antagonists, whereas gain increases (presumably mediated by LTD) are blocked (Carter & McElligott 1994).

Simulation studies using the learning rule for LTD outlined in Figure 9 and the first mechanism for LTP mentioned above demonstrate the capability of finding correct PF weights, but only in a simple learning task under highly restricted conditions (Berthier et al. 1993). The model's learning process was not robust enough to learn arbitrary movements in different parts of the work space, nor was it capable of training the weights of sets of PF synapses onto individual PCs. Most cerebellar modeling studies have not attempted to conform to the mechanisms of synaptic plasticity to this degree. They have simply used variants of the well-known perceptron or LMS rules and training strategies (discussed by Houk & Barto 1992). Such studies can be looked upon as testing the operational features of a model adequately, and in some cases they may test some of the organizational issues of learning discussed later, but they fall short of addressing the basic neurobiology of motor learning, which has been an important goal in our research.

In analyzing the shortcomings of the above cellular learning rule, we found that a major problem is its failure to adequately address the temporal credit assignment problem. This is the problem of delivering appropriately timed training information to the network's neurons, to insure that learning is adaptive. The actions produced by PCs are completed before they are detected by sensory feedback to generate the training information in CFs. To compensate for this problem of delayed feedback, the cerebellar learning rule needs to modify synaptic actions that occurred prior to a CF's discharge. Most synaptic physiologists have not addressed this problem of temporal credit assignment.

However, the one full study that is available reported that CF stimulation must actually precede PF stimulation by 125–250 msec for an optimal LTD (Ekerot & Kano 1989). A recent preliminary report based on field potentials suggested the opposite timing (Chen & Thompson 1992). Insufficient attention has been given to this important issue (but see commentary by Houk & Alford, this issue).

Network theorists typically address temporal credit assignment by assuming a trace mechanism that provides a short-term memory of preceding synaptic events until the arrival of the corresponding training information (Klopf 1982; Sutton & Barto 1981). In the APG model, the most critical events to store are traces of the synaptic events that promote state transitions in PC dendrites. In a recent version of the APG model (Barto et al. 1996; Buckingham et al. 1995), we postulated that a trace is triggered whenever a PC switches from its off- to its on-state. Recall that this should occur when a PC recognizes a pattern of PF input predicting that the limb will soon reach its intended goal. The subsequent firing of the PC helps to terminate the movement, and it is only after this that the CF returns error information (sect. 4.3). The important point here is that CF firing, in trials when it occurs, arrives several hundred msec after the PC response that needs to be evaluated. The postulated mechanism saves a trace in those spines that helped to switch the PC into its depolarized state. With this learning rule, PCs very effectively learn to respond to complex patterns of PF input to terminate movements at an intended goal (Buckingham et al. 1995). Cellular studies need to explore the possibility that spines receiving PF input concurrent with the onsets of plateau potentials undergo LTD in response to subsequent CF input.

The cellular mechanisms of LTD and LTP discussed above may not be adequate or appropriate for forming more permanent, very long-term memories of motor programs. Gilbert (1975) postulated that noradrenaline-containing cells in the locus ceruleus function to evaluate motor performance on a slower time scale than do CFs. Their firing might signal, for example, that a succession of movements performed over the course of several minutes was sufficiently successful to warrant conversion of the LTDs and/or LTPs that occurred during their execution into a more permanent memory. The latter might, for example, take the form of changes in spine density as seen by Greenough and colleagues (Black et al. 1990).

4.2. Structural credit assignment. The learning rules discussed in the previous section apply to each PF synapse, there being on the order of 10^{15} of these synapses. If training information was conveyed by climbing fibers indiscriminately to all of these synapses, learning would be quite impractical. Since each PC is innervated by only one CF, and since CFs transmit diverse training signals (sect. 4.3), there is an opportunity for learning to be guided in an efficient manner. For this to be achieved, however, requires the routing of each training signal to appropriate PCs in the network. The problem of doing this, called the structural credit assignment problem, is potentially very difficult. Houk and Barto (1992) hypothesized that the unique modular organization of the cerebellar cortex is an evolutionary adaptation that helps alleviate structural credit assignment. Small clusters of IO neurons with similar receptive fields innervate parasagittally oriented strips of PCs called microzones (Ekerot et al. 1991), and the PCs comprising this set project to a common cluster of nuclear cells (Gibson et al. 1987). This anatomical organization occurs early in development and insures that each APG module will receive a training signal that is particular to that module.

Figure 10 illustrates schematically 30 of the several hundred thousand modules present in the mammalian cerebellum, as if looking down on the surface of the cerebellar cortex. For simplicity, we show only sets of 5 – out of the 100 PCs estimated to participate in each APG module. All of the PCs in a given set converge upon a discrete nuclear cell cluster (N), and Figure 10 shows one loop to the motor cortex (M) forming an elemental motor command. Climbing fibers run parasagittally (vertically on the page) and innervate small numbers of PCs within a given set. PFs run horizontally, intersecting a large number of the rectangular dendritic trees of the PCs. Note that each PC in a set is shown exposed to a different 100,000-element vector of PFs, such that the 100 PCs comprising an APG have a grand total of 10,000,000 PFs from which to select input. (There most likely is some overlap, so the actual number of synapses might instead be 2,000,000, still a very large number.) Since each APG receives its own semiprivate training signal, the storage potential of the network is indeed quite exceptional (Gilbert 1974).

While the modular organization shown in Figure 10 has great potential for appropriate credit assignment, due to precision of connectivity in the parasagittal plane, the realization of this potential requires a precise alignment between the elemental commands generated by a given module and the training information conveyed by its CF input. The elemental command has to be one capable of diminishing the firing probability of the CF (Houk & Barto 1992). In the case of CFs with tactile receptive fields, the appropriate alignment is one that would mimic the spinal withdrawal reflex elicited by a noxious stimulus applied to its receptive field (Ekerot et al. 1991; Houk & Barto 1992). If it is assumed that the premotor network is capable of Hebbian-like, NMDA-mediated synaptic plasticity, one can outline a plausible developmental sequence that could automatically perform this alignment (Guzmán-Lara 1993). Once alignment is completed, the module would be ideally structured to elaborate conditioned withdrawal responses, and the same organization would be useful in developing motor programs for guiding the limb in a workspace containing obstacles. Given our hypothesis that CFs with proprioceptive receptive fields detect when movements are too small (Berthier et al. 1993; sect. 3.3), we postulated that the elemental command produced by the corresponding APG should move the limb in the direction that maximally activates the CF. With proper alignment, the feedback-error learning scheme discussed in section 4.3 (Kawato & Gomi 1993) could also use this mechanism to achieve excellent structural credit assignment.

Cerebellar models of conditioning and of eye movement control have not yet confronted the structural credit assignment problem. This is because they have generally dealt with single control modules (typically single PCs) that produce net commands as opposed to sets of modules regulating sets of elemental commands. The tacit assumption here is that all PCs in, say, the horizontal zone of the flocculus have identical output connections and thus can be represented by a single equivalent PC. If there is only one PC and only one CF in the model, there is no need to design an alignment scheme. However, this means that the model

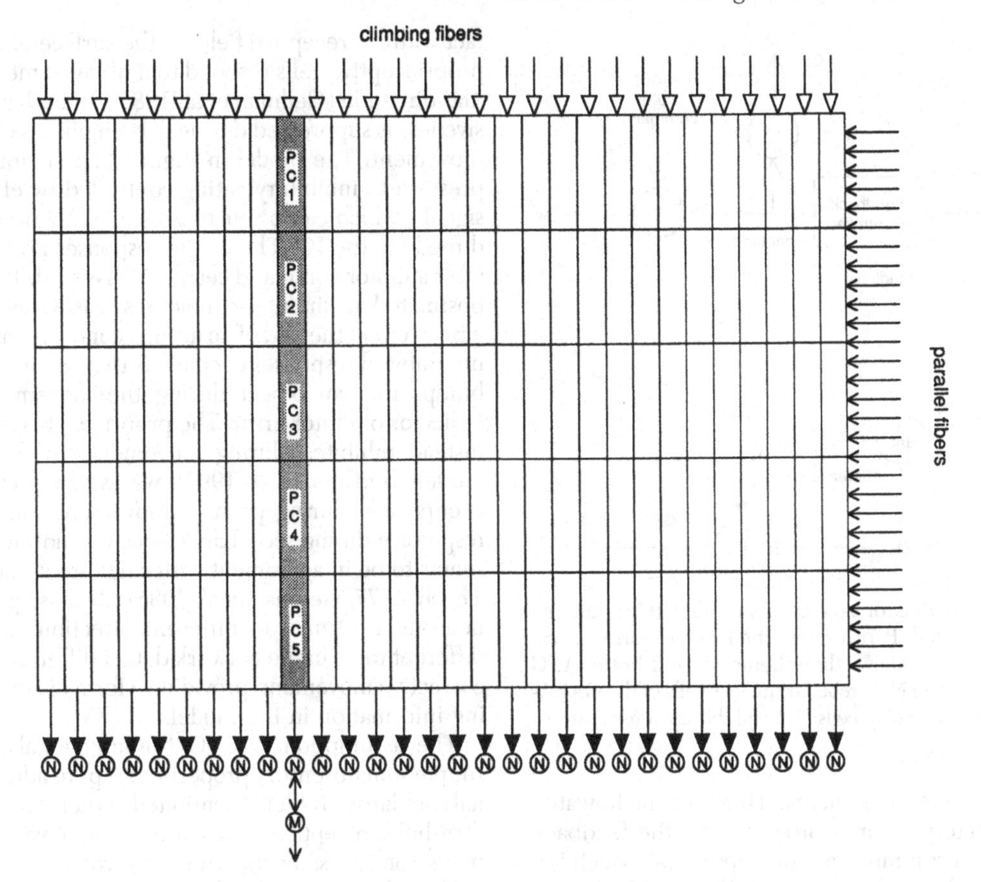

Figure 10. Structural credit assignment in cerebellar modules. Open arrows at the top are climbing fiber inputs oriented in a parasagittal plane. Parallel fiber inputs instead are oriented in the horizontal plane. The rectangles schematize the dendritic trees of individual Purkinje cells (PC1–PC5). Longitudinal sets of Purkinje cells provide focused inhibitory input to cerebellar nuclear cells (N). Stippling highlights 5 PCs that participate in one APG module, with loop connections to the motor cortex (M) shown forming an elemental motor command.

is not suitable for investigating the mechanisms that normally coordinate the cooperative actions of many modules working in parallel.

4.3. Training signals. Most theories of the cerebellum assume that CFs from the IO transmit the essential training information that guides motor learning. However, different theories are based on different assumptions regarding the specific nature of these signals. Marr (1969; for an elaboration, see Blomfield & Marr 1970) assumed that the IO transmits specific instructions from the motor cortex designating which elemental movements need to be executed. This begs the question of how the motor cortex acquires such elaborate knowledge about a large set of required movements. Albus (1971) assumed that the IO compares sensory feedback with desired trajectories to signal errors in performance. His desired trajectories are less demanding, but inherently similar to Marr's instructed movements. Neither theory confronts the problem of how internal standards might be acquired.

Recordings from IO neurons or from their CF axons provide useful constraints on the type of training information that is available. The different regions of the IO receive various combinations of sensory fibers and collaterals of motor fibers (Bloedel & Courville 1981) signaling efference copies. The electrophysiology of the ascending sensory pathways from the spinal cord originally suggested a very limited responsiveness to low-threshold somatosensory signals, leading Oscarsson (1980) to postulate that the IO computes error signals that are dominated by efference copy inputs. However, in awake animals somatosensory responsiveness is marked, both to tactile and to proprioceptive stimuli; furthermore, the motor responses stressed by Oscarsson are difficult to demonstrate except as inhibitory influences that gate off somatosensory responsiveness during certain phases of movement (Gellman et al. 1985; Weiss et al. 1990). The regions of cerebellum that regulate smooth eye movements receive CFs that are directionally selective to very low velocities of visual motion across the retina; some are sensitive to retinal slip of large, optokinetic images (Simpson 1984), and others are sensitive to slip of small, visual pursuit targets (Stone & Lisberger 1990). Retinal slip is a natural error signal since it designates a failure of the smooth eye control system to stabilize visual images on the retina. It is a sensory signal as opposed to an efference copy.

Kawato and Gomi (1992b; 1993) refined and developed Oscarsson's error hypothesis into a theory of feedback-error learning, summarized in Figure 11A. According to this theory, the IO transmits a motor error signal that is generated by a simple feedback controller. The difference between a desired trajectory and sensory feedback reporting on the actual trajectory forms a trajectory error analogous to

A

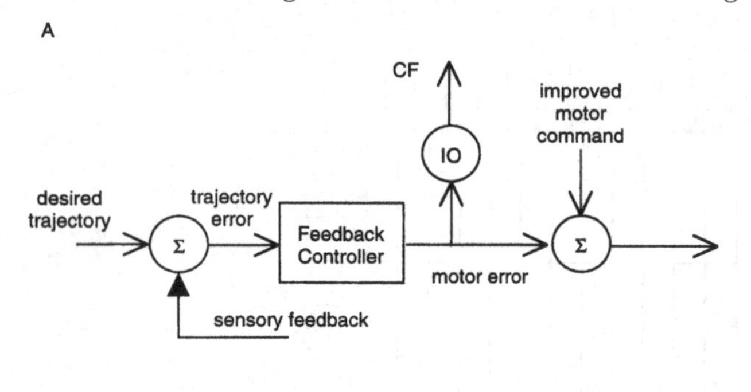

B

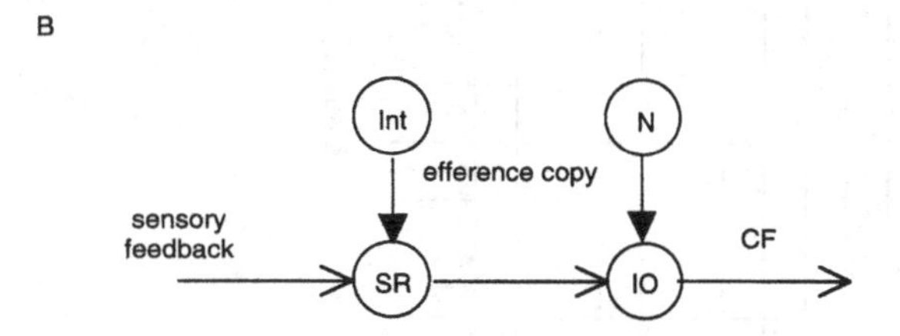

Figure 11. Two models of how climbing fiber (CF) training information is generated. Part A shows the feedback-error learning scheme, and part B shows the scheme utilized in the APG theory. IO, inferior olive; N, cerebellar nuclear cells of the GABA-ergic type; SR, sensory relay cells; Int, inhibitory interneurons.

the error signal in Albus's theory. However, in Kawato's theory the trajectory error is processed by the feedback controller to convert it into a motor error signal, which is a vector with quite desirable training capabilities. The motor error also serves as a crude motor command that is eventually replaced by an improved motor command, after the cerebellar cortex has learned the inverse model that was described in section 3.4. While this theory functions well in robot manipulation tasks, its consistency with the anatomy and physiology of different cerebellar control systems needs to be examined.

In the case of limb movements, Kawato and Gomi (1992a) assumed that the motor cortex functions as the simple feedback controller and that it also contains the two summing junctions. The motor cortex receives desired trajectory signals from the association cortex and actual trajectory information via sensory feedback. It computes the difference to form a trajectory error, which is processed further to generate the motor error signal that is sent through the IO to provide training information to the cerebellum. The motor cortex is also the summation site where the improved command generated in the cerebellum is added to the motor error to produce the net command sent to the spinal cord. A basic problem with this model may be its assumption that IO activity reflects motor error. As pointed out earlier in this section, IO neurons are highly responsive to sensory input and are actually suppressed by motor signals. Another disadvantage is that feedback-error learning requires a higher authority, the association cortex, to produce a desired trajectory signal. Kawato's model suggests that the cerebellar cortex progressively takes over control from extracerebellar mechanisms in the course of learning, which is opposite to the shift that has been postulated by others (Galiana 1986; Houk & Barto 1992; Houk et al. 1992; Peterson et al. 1991).

In the APG model, Houk and Barto (1992) accepted the sensory nature of IO signaling as a basis for the generation of training information. Tactile cells respond to light contact within a receptive field on the surface of the limb, and proprioceptive cells respond to limb movement in a particular direction (Gellman et al. 1985). In both cases, responsiveness is suppressed during certain phases of the animal's movement. The model in Figure 11B attributes the suppression to inhibitory gating controlled by efference copy signals, which can occur in sensory relay neurons (SR) or directly in the IO. The tactile responses are inhibited just after a motor command ceases (Weiss et al. 1990), which is postulated to eliminate contact responses that would otherwise occur at the end of an accurate movement. This leaves uninhibited responses to contacts that occur when the limb bumps into an object during the movement, a simple indicator of motor error. The proprioceptive responses are instead inhibited during movement. In the APG array model (Berthier et al. 1993), we assumed that inhibition occurs only during primary movements, leaving the IO responsive during secondary corrective movements – which seems to be in agreement with single-unit data (Gilbert & Thach, 1977; Gellman et al. 1985). Because proprioceptive neurons are tuned to different directions of movement, different units in the network detect different directions of corrective movement, providing a form of supervised training information in the model.

The assumption that IO training signals derive from simple somatosensory properties helps to address the internal standards problem mentioned earlier. The fact that low-threshold receptive fields are aligned with nociceptive fields for the same neurons (Ekerot et al. 1991) would suggest that IO neurons learn to respond to low-threshold predictors of nociceptive stimuli in the course of normal development (Houk & Barto 1992). According to this theory, nociception as a punishment signal provides the ultimate training information for the cerebellar network.

The origin of internal standards may not be an issue for smooth eye movements, since the negative image of a head movement, as sensed by vestibular receptors, can be thought of as the desired trajectory for stabilizing visual images on the retina. Furthermore, motion detectors in the retina directly transduce the trajectory errors of the optokinetic system. The accessory optic system routes crude commands directly to vestibular neurons and efference copies to the appropriate region of the IO (Kawato & Gomi 1992b). If smooth pursuit is treated as a reflex, a similar though somewhat more complex argument can be made regarding this system. The movement of a visual target, as analyzed in the parietal cortex, could serve as the desired trajectory.

In applying feedback-error learning to saccadic eye movements, Dean and colleagues (1994) assumed that the retinal projection to the superficial layer of the tectum in effect computes an endpoint error in visual coordinates, as assumed earlier by Grossberg and Kuperstein (1989). The internal standard is thus replaced by the assumption that a target in the peripheral visual field needs to be foveated. Note that a trajectory error is not computed in this case, only an endpoint error. This endpoint error is then transformed by the intermediate layer of the tectum into a kind of motor error signal that was found to be suitable for driving feedback-error learning in a CMAC model of the cerebellum (Dean et al. 1994). Again, we note that single-unit recordings from CFs do not support this theory. The CFs that innervate the saccadic region of the cerebellum do not encode motor signals, although they do respond to proprioceptive inputs. In reviewing this issue, Houk, Galiana, and Guitton (1992) suggested that these CFs might be sensitive to their proprioceptive input only during corrective saccades. This should be relatively easy to test experimentally.

Models of classical conditioning have assumed that the IO transmits the US, for example, a strong puff of air to the cornea, as a training signal. These CFs should then mediate the associative learning of a CS (Thompson 1986). In analogy with IO neurons involved in limb control, olivary neurons projecting to the eye blink system have exquisitely sensitive tactile receptive fields around the eye. This suggested to Weiss and colleagues (1993) that these IO neurons might be detecting eyelid closure, as opposed to the US, in which case they could provide training signals for adjusting the amplitude of the eye blink motor program. This would fit with the APG version of the conditioning model of the cerebellum discussed earlier (Fig. 6B). It also fits with conditioning studies of mutant mice lacking both mGluR1 receptors and cerebellar LTD (Aiba et al. 1994). These animals retain an ability to initiate CRs, but are impaired in their ability to regulate the amplitude of the response.

In summary, one can make a substantial case that the training information transmitted by the IO is based on relatively simple sensory responsiveness.

4.4. Distributed learning. The cerebellar cortex is of course not the only CNS structure that mediates motor learning. While not a primary topic in the present article, perspective may be served by offering a brief statement of our views about how plasticity in the inferior olive, premotor networks, basal ganglia, and cerebral cortex might interact with the cerebellar cortex to create an advantageous environment for overall motor learning (cf. Barto 1995; Houk & Barto 1992; Houk et al. 1992; 1995; Houk & Wise 1995).

The IO may be a site of learning that can improve its ability to generate training information. This seems likely in the course of development, during which IO neurons may acquire their ability to respond to low-threshold predictors of the nociceptive stimuli that also activate them (sect. 4.3). Learning at the level of the IO may also operate in the adult, which could offer explanations for the puzzling CF responses that have been observed during some motor tasks (Mano et al. 1986; Ojakangas & Ebner 1992). In the Mano experiment, CFs responded in the interval between the random transitions in a visual target and the onsets of the movements that the animals made in attempting to follow the targets. Ito (1989) pointed out that the observed CF responses do, in fact, detect errors, namely the discrepancies between target motion and the animals' responses, which are delayed by a reaction time. Due to the random occurrences of target transitions, the observed responses might represent the earliest possible predictors of these errors. In a similar vein, the CF responses observed by Ojakangas and Ebner (1992) may represent acquired predictors of the corrective movements their animals used to compensate for gain changes in the visual target display. In both of these cases, we would postulate that IO neurons have somehow acquired responses to signals that predict the proprioceptive or tactile signals that activate these neurons at the onsets of the corrective movements. The IO system might also learn how to better use inhibitory gating to improve the quality of the training information that is supplied by CFs (Houk & Barto 1992). Learned gating patterns could be transmitted by the inhibitory projections to the olive from the GABAergic cerebellar nuclear cells (N in Fig. 11B), since these neurons are probably adaptively controlled by input from PCs in the cerebellar cortex. Future microelectrode studies, if appropriately designed, might be able to detect adaptive alterations in IO responsiveness.

The cerebellar cortex is probably capable of transferring some of its motor program knowledge to the premotor network. Elsewhere we discussed how this could come about for limb and saccadic eye movements through Hebbian mechanisms supervised by PC forcing functions generated in the cerebellar cortex (Houk & Barto 1992; Houk et al. 1992). It has also been suggested that the cerebellum provides the information that is used to train the brainstem vestibulo-ocular reflex (Galiana 1986; Miles & Lisberger 1981; Peterson et al. 1991). The cerebellar cortex might acquire this information during task rehearsal, or it might already have adequate information and need only export it to the premotor network. Formerly it was predicted that premotor networks would learn relatively slowly as compared with the cerebellar cortex (Galiana 1986; Houk & Barto 1992; Houk et al. 1992; Petersen et al. 1991). However, several recent studies suggest that rapid learning can occur at brainstem (Luebke & Robinson 1992; Khater et al. 1993) and cerebral (Raichle et al. 1994; Sanes et al. 1988) sites. In the Luebke and Robinson study, a process of deadaptation, which normally required only 30 minutes, was suspended by inactivation of the flocculus, supporting the hypothesis that the cerebellar cortex guides the learning. In the Raichle study, PET scans showed metabolic activity associated with a cognitive process moving from the lateral cerebellum to a sylvian site in the cerebral cortex following less than 15 minutes of practice. These cases are consistent with the hypothesis that the cerebellar cortex can export knowledge that it has previously acquired to neurons that are the targets of its regulation. This should be a fruitful area for future investigation.

Learning how to perform complex behavioral acts clearly requires the cerebral cortex and basal ganglia. Nevertheless, the cerebellum would need to learn how to use the information sent through cortico-ponto-cerebellar pathways to coordinate and program its own participation in these actions (Houk & Wise 1995).

6

On climbing fiber signals and their consequence(s)

J. I. Simpson, D. R. Wylie, and C. I. De Zeeuw
Department of Physiology and Neuroscience, New York University Medical Center, New York, NY 10016
Electronic mail: *simpsj01@popmail.med.nyu.edu*

Abstract: The persistence of many contrasting notions of climbing fiber function after years of investigation testifies that the issue of climbing fiber contributions to cerebellar transactions is still unresolved. The proposed capabilities of the climbing fibers cover an impressive spectrum. For many researchers, the climbing fibers signal errors in motor performance, either in the conventional manner of frequency modulation or as a single announcement of an "unexpected event." More controversial is the effect of these signals on the simple spike modulation of Purkinje cells. In some hands, they lead to a long-term depression of the strength of parallel fiber synapses, while, in other hands, they lead to a short-lasting enhancement of the responsiveness of Purkinje cells to mossy fiber inputs or contribute to the often-seen reciprocal relation between complex and simple spike modulation. For still other investigators, the climbing fibers serve internal timing functions through their capacity for synchronous and rhythmic firing. The above viewpoints are presented in the spirit of trying to reach some consensus about climbing fiber function. Each point of view is introduced by summarizing first the key observations made by the respective proponents; then the issues of short-lasting enhancement, reciprocity between complex and simple spikes, and synchrony and rhythmicity are addressed in the context of the visual climbing fiber system of the vestibulocerebellum.

Keywords: cerebellum; complex spikes; eye movements; flocculus; inferior olive; mossy fibers; movement; nodulus; posture; Purkinje cells; simple spikes; synchrony

1. Introduction

Despite nearly three decades of investigation, the issue of climbing fiber contributions to cerebellar transactions is still unresolved. Although opposing and controversial points of view exist about the roles of climbing fibers, some aspects of climbing fiber anatomy are agreed upon. It is accepted that all climbing fibers originate from the contralateral inferior olive (Desclin 1974; Szentágothai & Rajkovits 1959; see also Dow 1942) and that the inferior olive is composed of subdivisions whose climbing fibers terminate in the cerebellar cortex in sagittally oriented zones (Groenewegen & Voogd 1977; Groenewegen et al. 1979; Jansen & Brodal 1954; Voogd 1964), Figure 1. It is further agreed that in the adult each Purkinje cell receives only one climbing fiber (Eccles et al. 1966) and that, in general, Purkinje cells within a given zone project to only one cerebellar nucleus, which receives collaterals from the inferior olivary axons that terminate as climbing fibers in that zone (De Zeeuw et al. 1994b; Eccles et al. 1967; Van der Want et al. 1989; Wylie et al. 1994). Furthermore, the cerebellar nuclei contain GABAergic neurons that project contralaterally to those parts of the inferior olive (De Zeeuw et al. 1989; Nelson & Mugnaini 1989) that provide the collaterals to that particular cerebellar nucleus (Dom et al. 1973; Graybiel et al. 1973). The anatomical unit comprised of a particular sagittal zone with its climbing fibers and their collaterals (along with the corresponding Purkinje cells and the associated cerebellar nucleus) has been called a module (Voogd & Bigaré 1980) and is considered to be the functional unit of the cerebellum (Andersson & Oscarsson 1978; Oscarsson 1969; 1979). Thus it is agreed that the climbing fibers are central to parcelling the cerebellum into anatomical modules. In some parts of the cerebellar cortex, each anatomical zone is mirrored by the uniformity of its climbing fiber responses to natural stimulation (e.g., De Zeeuw et al. 1994b; Wylie et al. 1994), but in other parts, the zones break up into patches of climbing fiber responses with natural stimulation (e.g., Robertson 1984; Robertson et al. 1982).

The olivocerebellar system has several remarkable physiological characteristics that make it something of a curiosity in the sensorimotor operations of the brain. One conspicuous feature is the extremely powerful excitatory action of the climbing fiber on the Purkinje cell that results in the burst discharge of the Purkinje cell (its complex spike, Eccles et al. 1966; Ito & Simpson 1971; Mano et al. 1989; Thach 1967). Another unusual feature is the extremely low firing frequency of olivary neurons. In both anesthetized and awake animals, olivary neurons discharge either a single spike or a burst of spikes (2–5 spikes with an interspike interval of 2–3 msec) about once or twice per second (Armstrong 1974; Armstrong & Rawson 1979; Crill 1970). Even with application of chemical "excitants" such as harmaline, only 8–10 discharges per second are observed (de Montigny & Lamarre 1973; Lamarre et al. 1971; Llinás & Volkind 1973). The limited frequency range of olivary discharge stands in contrast to the range of several hundred spikes per second for the Purkinje cell simple spikes (Thach 1967), whose excitatory drive is conveyed by the mossy fiber–parallel fiber input (Eccles et al. 1967).

Despite the well-documented anatomy of a seemingly

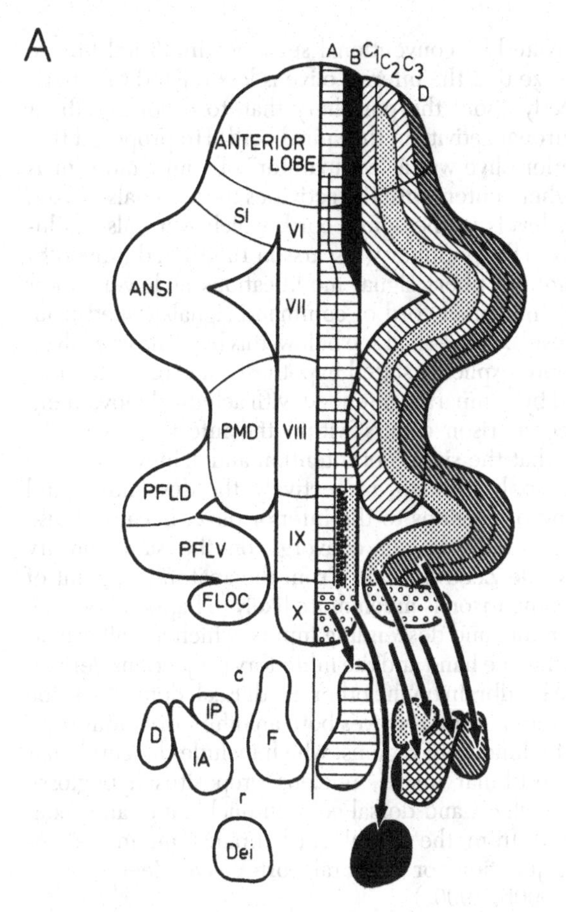

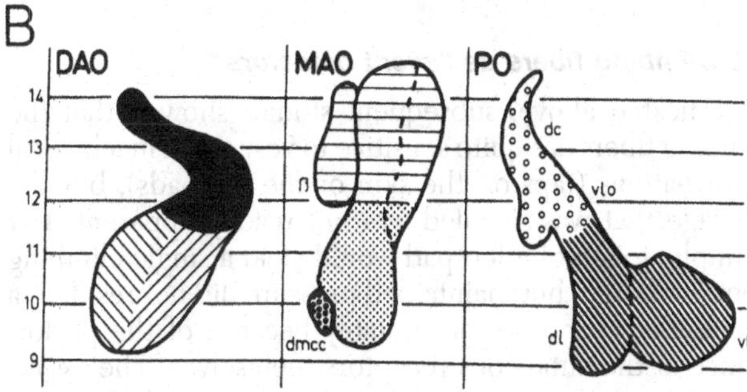

Figure 1. The cerebellar modules of the cat. **A** shows a schematic diagram of the longitudinal zonal organization of the inferior olivary projection to the cerebellar cortex (top) and a representation of a horizontal section through the deep cerebellar nuclei (fastigial, anterior and posterior interposed, and lateral) and the lateral vestibular (Deiters') nucleus (bottom). **B** shows a horizontal reconstruction of the inferior olive. Stereotactic levels are indicated to the left (level 14 is caudal; level 9 is rostral). Corresponding symbols in **A** and **B** indicate reciprocal connections between the inferior olive and the cerebellar nuclei. Likewise, each cortical zone receives from the correspondingly marked subdivision of the inferior olive. The corticonuclear relations are indicated with arrows (i.e., Purkinje cells of zone A project to the fastigial nucleus; Purkinje cells of zone B project to Deiters' nucleus, etc.). Some of the finer subdivisions of the olivocerebellar projection, such as that to the flocculus (see Fig. 7), are not illustrated. Abbreviations: ANSI–ansiform lobule; β–subnucleus beta; D–dentate (lateral cerebellar) nucleus; DAO–dorsal accessory olive; dc–dorsal cap of Kooy; Dei–Deiters' nucleus (lateral vestibular nucleus); dl–dorsal lamella of the principal olive; dmcc–dorsomedial cell column; F–fastigial nucleus; FLOC–flocculus; IA–anterior interposed nucleus; IP–posterior interposed nucleus; MAO–medial accessory olive; PFLD–dorsal paraflocculus; PFLV–ventral paraflocculus; PMD–paramedial lobule; PO–principal olive; SI–simple lobule; vl–ventral lamella of the principal olive; vlo–ventrolateral outgrowth; A, B, C_1, C_2, C_3, and D–designations of the longitudinal cortical zones; c–caudal, r–rostral (from Groenewegen et al. 1979).

simple, repetitive circuit, there is little agreement about the meaning or consequences of climbing fiber activity. In trying to synthesize the various hypotheses on the function of the climbing fibers, one has the sense of looking at a drawing by Escher. Each point of view seems to account for a certain collection of findings, but when one attempts to put the different views together, a coherent picture of what the climbing fibers are doing does not appear. For the majority of researchers, the climbing fibers signal errors in motor performance, either in the usual manner of discharge frequency modulation or as a single announcement of an "unexpected event." For other investigators, the message lies in the degree of ensemble synchrony and rhythmicity among a population of climbing fibers. Beyond the matter of what messages are being carried by the climbing fiber signals (sect. 2) is the more controversial issue of their consequence(s) (sects. 3 and 4). The central questions are whether climbing fiber activity (1) leads to long-term depression of the strength of parallel fiber synapses, (2) results in short-lasting enhancement of Purkinje cell responsiveness to mossy fiber inputs, or (3) constitutes a largely independent channel with little influence on the simple spikes. In the following pages these views of the possible roles of the climbing fibers are presented in the spirit of trying to reach some consensus about their function. Each point of view is introduced by summarizing the basic observations made by the respective proponents. Subsequently, the issues of short-lasting enhancement and synchrony and rhythmicity are addressed (sect. 5) in the context of the vestibulocerebellar visual climbing fiber system, whose modulation by retinal image motion provides a palpable example of climbing fibers signaling an error in performance.

2. Climbing fiber messages

2.1. Climbing fiber modulation

The initial attempts to find a message in climbing fiber activity involved the spino-olivary pathways to the anterior lobe of the vermis (Oscarsson 1969; 1973; Thach 1967). With natural stimulation, the responses of the spino-olivary systems to this part of the cerebellum were weak or absent and the receptive fields were large and vaguely delimited. These findings were perplexing because they appeared at variance with the high degree of specificity typifying the connectivity from the inferior olive to the cerebellum. Presumably Oscarsson's initial findings with natural stimulation reflected the extensive partial spinal cord sectioning done to study separately the several spino-olivary pathways. In contrast to these early observations, subsequent investigators have been able to activate climbing fiber inputs to the vermis and pars intermedia of the anterior lobe with the use of natural stimulation, particularly cutaneous stimulation (Armstrong & Edgley 1984; Armstrong & Rawson 1979; Eccles et al. 1972; Gellman et al. 1983; 1985; Ishikawa et al. 1972; Leicht et al. 1977; Robertson et al. 1982; Rushmer et al. 1976). Also in contrast to the earlier findings on the inability of climbing fibers to signal peripheral events transmitted through the spinal cord, Rubia & Kolb (1978; Kolb & Rubia 1980) found that the kinematics of passive movement of the cat's forepaw were, in fact, reflected in the modulation of the climbing fiber responses of Purkinje cells recorded in the anterior lobe. To reveal the relation between complex spike activity and the amplitude, velocity, and acceleration of the movement, a number of trials had to be summed, but such an averaging may occur during a single movement when ensembles of Purkinje cells are considered. The potential capability of ensembles of Purkinje cells to signal kinematic variables through their conjoint climbing fiber responses has been described theoretically (McCollum 1992; Robertson & McCollum 1989) by applying set theory to data obtained by Robertson and colleagues (Robertson 1984; Robertson & Laxer 1981; Robertson et al. 1982) on the somatotopy of cutaneously evoked complex spike responses in the anterior lobe of the decerebrate cat.

Because of their low discharge frequency, climbing fibers were for some time held to be unable to encode messages in the customary way, that is, with frequency coding. While it was realized that complex spikes did respond to cutaneous and muscle stimulation, the constraint on firing frequency suggested to some that climbing fibers are "phasic" (Llinás 1970; 1974) and that their responses encode an event rather than an on-going process (Armstrong 1974; Rushmer et al. 1976). However, the capability of climbing fiber modulation to encode signals in the conventional manner of frequency modulation is clearly apparent in the responses of visually activated climbing fibers in the flocculonodular lobe (Fig. 2) (De Zeeuw et al. 1994b; Fushiki et al. 1994; Graf et al. 1988; Kano et al. 1990a; 1990b; Kusunoki et al. 1990; Maekawa & Simpson 1973; Simpson & Alley 1974; Stone & Lisberger 1990; Wylie & Frost 1993). These "visual" climbing fibers originate in the dorsal cap of Kooy and the ventolateral outgrowth, and signal the direction and speed of movement of large parts of the visual world across the retina. In the rabbit, they are optimally responsive to low speeds (about 1°/s) (Alley et al. 1975; Barmack & Hess 1980; Simpson & Alley 1974). The signals of retinal image motion provide a measure of the "shortcomings" of the compensatory eye movement system in stabilizing the retinal image. For the low speeds of retinal image motion to which the climbing fibers are best responsive, their low firing rate does not prohibit signaling of speed and direction in the conventional manner of frequency encoding, and the modulation has both phasic and tonic components. In contrast to the complex spike modulation produced by passive paw movement (Kolb & Rubia 1980; Rubia & Kolb 1978), the floccular complex spike modulation produced by visual stimulation can be seen without resorting to averaging (Simpson & Alley 1974). Similarly, Barmack et al. (1989; 1993a) found that neurons of the beta nucleus of the inferior olive, which projects to parts of the nodulus and uvula, modulate robustly in response to natural stimulation of the otoliths and vertical semicircular canals (Fig. 3). For some beta nucleus neurons the response is maintained for maintained tilt, thus providing another example of "tonic" signaling by the climbing fiber system.

2.2. Climbing fibers as "comparators"

Oscarsson's (1969; 1973) original findings that the climbing fibers transmitting input from the spinal cord could not be well-activated by conventional sensory stimuli led him to hypothesize that the inferior olive is less related to reporting directly about the periphery than to reporting about interneuronal activity in the spinal cord. He proposed that the inferior olive was a "comparator" of command signals from higher centers with the activities these signals evoked at lower levels in the spinal cord, which were also influenced from the periphery (Oscarsson 1980). In this hypothesis, climbing fibers signal modifications at lower motor levels in the spinal cord of command signals issued from higher levels. As we shall see below, this hypothesis evolved into a more explicitly stated hypothesis of error detection, obtained by comparing intended with achieved movement. If the comparison occurs within the inferior olive, it is unlikely that the signals of intention and achievement are conveyed only through, respectively, the descending and ascending projections to the inferior olive, because these inputs generally do not converge on the same olivary neurons (De Zeeuw 1990). From the anatomical point of view, a comparison is much more likely to happen between the ascending and descending inputs, which are all excitatory, on the one hand and the inhibitory projections derived from the hindbrain on the other hand. Each dendritic spine of an olivary neuron receives both an inhibitory input from one of the hindbrain regions, which include the cerebellar nuclei, vestibular nuclei, nucleus prepositus hypoglossi, solitary nucleus, and dorsal column nuclei, and an excitatory input from the spinal cord, brainstem, mesodiencephalic junction, or cerebral cortex (De Zeeuw et al. 1990a; 1990b; 1990c).

2.3. Climbing fibers as "event detectors"

As indicated above, subsequent studies showed that the climbing fibers are quite sensitive to very small mechanical perturbations (taps on the skin or the footpads), but the message that was signaled was not readily apparent. For example, touch of a footpad would yield a single climbing fiber discharge, but maintained pressure did not result in a maintained response, presumably because of the performance of the cutaneous receptors themselves. The exquisite sensitivity of the climbing fibers to pressure on the footpad led Rushmer et al. (1976; see also Armstrong 1974) to conclude that they would provide no information during locomotion other than the fact that the footpad had contacted an object because the forces and displacements encountered during locomotion would insure that the probability of firing during each touchdown or lift-off would be virtually one. Therefore, they would not be able to discriminate variations in the normal range of forces and displacements. As a consequence of the response characteristics of this climbing fiber system to passive touch of the paw, Rushmer et al. proposed that it serves as an "event detector," signaling simply that the foot has made or lost contact with a weight-bearing surface. The questions then arose as to (1) whether such a stereotyped response signaled only the time of occurrence of footfall or lift-off and no other aspects of performance, and (2) whether such sensory responses occurred during voluntary movements or only during passive stimulation.

2.4. Climbing fibers as "unexpected event detectors"

The notion of the climbing fiber as an "event detector" was unsatisfying to many investigators. For example, as stated

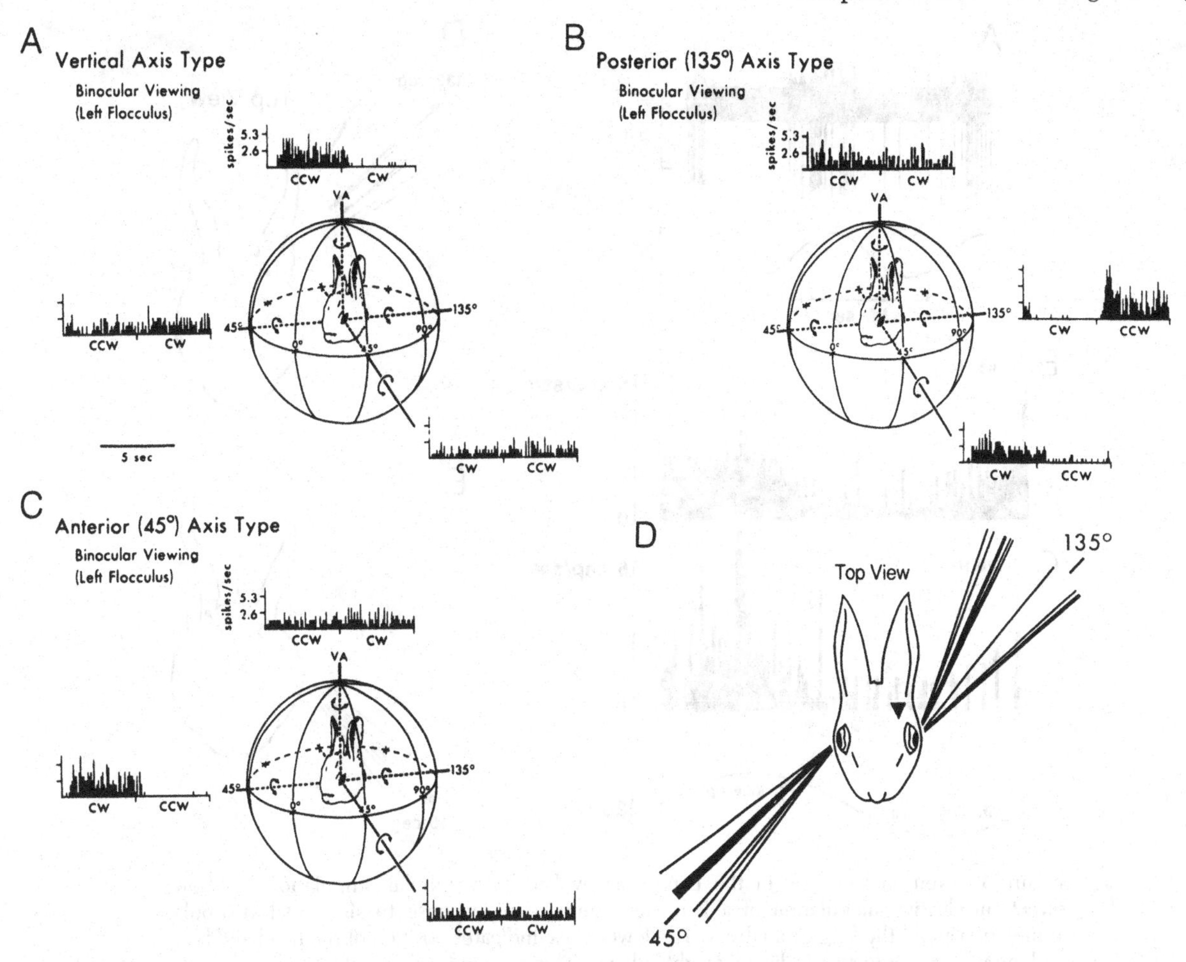

Figure 2. Responses of the complex spikes of floccular Purkinje cells to optokinetic stimulation. These panels provide an overview of the spatial organization of the preferred and null axes characterizing the rotation selectivity of the visual climbing fiber responses for each of the three principal classes (**A, B,** and **C**) of Purkinje cells in the rabbit's cerebellar flocculus. A "planetarium" projected stimulus was rotated about each of three orthogonal axes. The peristimulus histogram (15 repetitions, 100-msec binwidth) next to each axis depicts the firing rate during clockwise (CW) and counterclockwise (CCW) constant speed (0.5°/sec rotation. CW and CCW indicate the sense of visual world rotation when viewed from the rabbit along the axis that is directed toward the respective histogram. The ordinate scale on the vertical axis histogram applies to all histograms. The Purkinje cells were classified according to which one of the three axes of rotation produced the deepest modulation and according to which eye was dominant. The numbers on the equator of the reference sphere enclosing the rabbit's head indicate the ipsilateral and contralateral azimuth angles. The neuron type shown in **A** responds best to rotation about the vertical axis (VA neurons). The neuron types shown in **B** and **C** both respond best to rotation about a horizontal axis oriented at 45° contralateral azimuth/135° ipsilateral azimuth. They differ with respect to ocular dominance: the neuron in **B** is ipsilateral dominant (posterior [135°] axis or ipsi-135° neuron) whereas the neuron in **C** is contralateral dominant (anterior [45°] axis or contra-45° neuron). **D** shows the distribution of the preferred axes for dominant eye stimulation for 10 anterior (45°) axis Purkinje cells and 10 posterior (135°) axis Purkinje cells. For each cell the orientation of the preferred axis was determined by fitting azimuthal tuning data with a sine curve. The line indicating the preferred axis orientation is drawn from the dominant eye. The solid triangle indicates that the recordings were made from the left flocculus, as was also the case for **A, B,** and **C.** The spatial organization of climbing fiber responses to rotation is similar to that of the vestibular semicircular canals and eye muscles. That is, the VA neurons respond best to rotation about an axis that is approximately perpendicular to the horizontal vestibular canals and the plane of the horizontal recti. Likewise, the ipsi-135° and contra-45° neurons respond best to rotation about an axis that is approximately perpendicular to the ipsilateral anterior canal (and contralateral posterior canal) and the plane of the ipsilateral vertical recti (and contralateral obliques) (adapted from Graf et al. 1988).

above, Kolb and Rubia (see also Ojakangas & Ebner 1992; 1994) found that the climbing fiber system can represent movement kinematics, which is more than just the occurrence of a peripheral event. Moreover, a number of investigators showed that when animals are allowed to make voluntary movements at their own pace the complex spike activity does not occur in close correlation with specific aspects of the movement (Andersson & Armstrong 1985; 1987; Armstrong et al. 1982; 1988; Boylls 1980; Gellman et al. 1985; Mano et al. 1986; 1989; Thach 1968; 1970). For example, Boylls (1980) noted that in the decerebrate, walking cat some olivary cells tended to discharge at specific times during the step cycle, yet there was an absence of a strong time-locked relation between any aspect of the locomotor cycle and the firing of the olivary neurons. Such findings required modification and qualification of the "event detector" proposal. The changes in view harkened back to the proposal of Oscarsson that the inferior olive acts

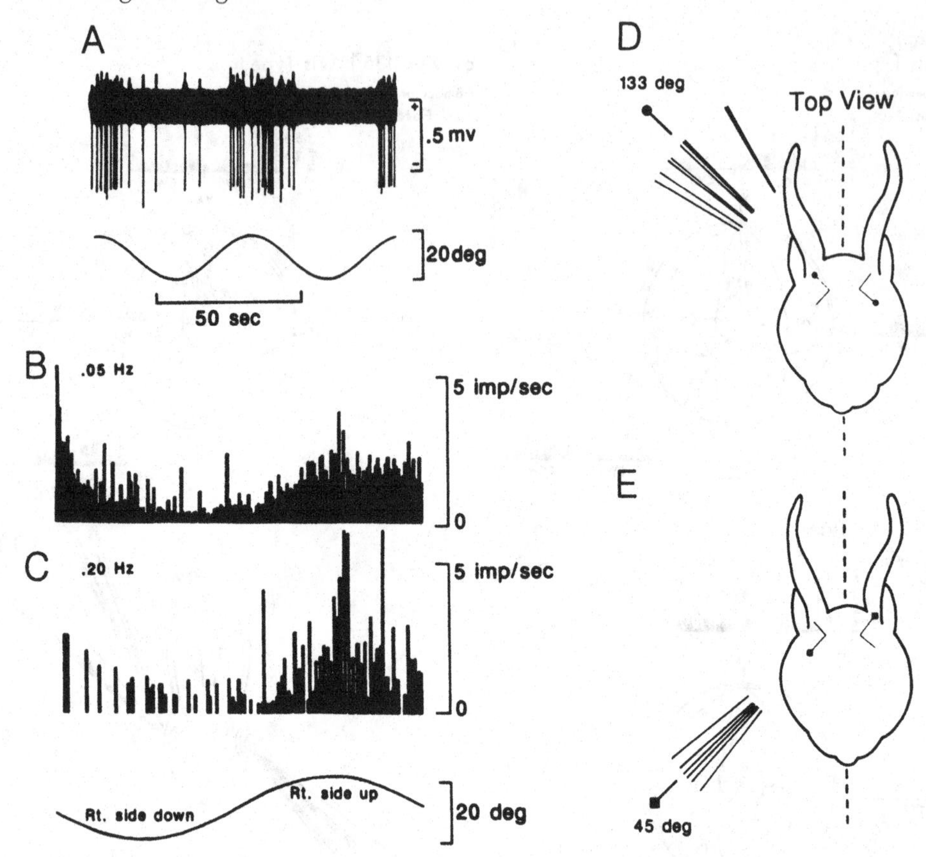

Figure 3. Neuronal activity in the β-nucleus evoked by vestibular stimulation. **A** shows direction-selective modulation of a β-nucleus neuron in response to sinusoidal vestibular stimulation about the longitudinal axis. The lower trace indicates position of the head and body with an upward deflection indicating right side up. This neuron increased its activity when the rabbit was rotated onto its left side. **B** and **C** show peristimulus time histograms (compiled from 40 cycles) of the activity of the same olivary neuron in response to rotation about the longitudinal axis (see lower trace) at two frequencies, 0.05 Hz (**B**) and 0.20 Hz (**C**). Each bin represents 2°. **D** and **E** show the distribution of the "null axes" of β-nucleus neurons. These axes were determined by sinusoidally rotating the animal about a horizontal axis and shifting the head about the vertical axis, either clockwise or counterclockwise, until the neuron was no longer modulated. The "null axes" clustered around two loci that correspond to optimal orientations of pairs of vertical semicircular canals. The null axes in **D** are consistent with vestibular modulation originating from neurons with right-posterior/left-anterior semicircular canal sensitivity. These neurons were clustered in the rostral β-nucleus on the right side. The null axes in **E** are consistent with vestibular modulation originating from neurons with right-anterior/left-posterior semicircular canal sensitivity. These neurons were clustered in the caudal β-nucleus on the right side (adapted from Barmack et al. 1993a).

as a comparator of intended with achieved movement, and what was an "event detector" became an "unexpected event detector." To illustrate, Gellman et al. (1985) found in the awake cat that complex spike responses to a passively applied stimulus generally failed to occur when a similar stimulus was produced by a voluntary movement, unless the receptive field of the complex spike was "unexpectedly" brought into contact with an object during active movement. Similarly, Andersson & Armstrong (1985; 1987) showed that when an awake cat walking along a circular horizontal ladder encountered a rung that unexpectedly gave way, an increase in complex spike discharge occurred prior to the unlatched rung hitting its end stop. These two sets of observations are in line with a proposal by Oscarsson (1980) that built upon his earlier comparator hypothesis. The more explicit hypothesis is that the olive would detect perturbations of two types: perturbations in the commands introduced into lower spinal cord motor centers as a consequence of reflex activity, and perturbations in the evolving movement due to unexpected changes in load or resistance. Andersson and Armstrong recast this proposal specifically to apply to the cat walking on the horizontal ladder by pointing out that when the cat stepped on the unlatched rung a mismatch occurred between the intended and achieved trajectory of the stepping limb. The many other reports of a relation between complex spike activity and movement perturbations produced by changes in load or contact are also under the umbrella of the comparator hypothesis of error detection (e.g., Gilbert & Thach 1977; Kim et al. 1987; 1988; Lou & Bloedel 1992a; 1992b; Ojakangas & Ebner 1992; 1994; Wang et al. 1987).

In the decerebrate cat performing treadmill locomotion, Kim et al. (1987) found that about one-half of the Purkinje cells showed complex spike modulation in relation to the unperturbed step cycle. The presence and absence of complex spike modulation during unperturbed locomotion in the decerebrate and intact cat, respectively, can both be accommodated by the comparator hypothesis because the decerebration altered the usual complement of signals to be compared. The difference between the findings of Boylls (see above) and Kim et al. may have its basis in the difference in the population of recorded neurons; Boylls recorded in the caudal medial accessory olive, which projects to the medial A-zone (Voogd & Bigaré 1980), whereas Kim et al. recorded complex spikes in the intermediate zones (C1 and C2).

2.5. Climbing fibers as "error detectors"

The word "unexpected" is unfelicitous because we can only presume what the animal expects. Moreover, in Andersson & Armstrong's (1985; 1987) studies the complex spike discharges were in some instances related to perturbations that the animal quite likely expected, but chose not to adjust for in the locomotor cycle. After the forelimb of the cat encountered an unlatched rung, the hindlimb also encountered the unlatched rung and this perturbation was also signaled to the cerebellum. After this sequence has occurred several times, one can reasonably presume that the cat is expecting that the hindlimb will be perturbed. Consequently, expectation or no expectation is not the issue, but rather whether a mismatch occurs between the intended and achieved movement. Gellman et al. (1985) acknowledged that "unexpected" contact detection could in some circumstances be considered error detection, but they had difficulty in reconciling the responses of climbing fibers to passive displacement in the absence of movement with error detection. Perhaps in such a situation a climbing fiber response would be indicative of an error in posture as opposed to an error in movement. If a cat in a posture of repose is disturbed by a footpad touch, then a climbing fiber response can be a signal that the achieved posture is not satisfactory. Since the cerebellum is involved in controlling both posture and movement, errors in postural performance must also be reported.

2.6. Summary

We now know that the climbing fibers can be modulated both tonically and phasically and that they can be exquisitely sensitive to sensory stimulation. Although different investigators have used different phrases to describe what the climbing fibers are signaling, for the majority the climbing fibers are signaling an error in motor performance, which includes posture as well as overt movement. The visual climbing fiber signals of error are reliable and robust in comparison to those of most other climbing fiber systems. This difference is probably a consequence of the fact that the error in performance, the retinal image motion, is not computed by a neural comparator dealing with intended and achieved movement, but is computed mechanically by the movement of the eye relative to the external world. The relative movement represents the difference between the intended and achieved movement, which is detected directly at the sensory surface by specialized speed- and direction-selective retinal ganglion cells.

3. Interactions between complex and simple spikes

The low firing frequency of the climbing fibers not only prompted thinking about them in terms of encoding phasic events, but also encouraged the idea that their low frequency should be compensated for by an interactive influence on the Purkinje cell simple spikes. Proposals as to how complex spikes may magnify their influence through interaction with simple spikes run the gamut from transient to tonic and from short-lasting to long-term.

3.1. The climbing fiber pause and beyond

A transient climbing fiber effect on simple spike activity can be seen by recording spontaneous activity from Purkinje cells. After the onset of a complex spike, the simple spikes pause for a variable period of time, from 10 to several hundred milliseconds depending on the state of the animal (Bell & Grimm 1969; Bloedel & Roberts 1971; Granit & Phillips 1956; Mano et al. 1986; McDevitt et al. 1982; Murphy & Sabah 1970; Sato et al. 1992; 1993). The pause in simple spike firing, which is called the climbing fiber pause (e.g., Armstrong 1974), may be due to a combination of inactivation of the Purkinje cell membrane and the influence of climbing fiber collaterals on cerebellar cortical neurons that could decrease (via Golgi cells) transmission in the mossy fiber – granule cell pathway and could also directly inhibit (via basket cells) the Purkinje cells (Bloedel & Roberts 1971).

After the pause, the simple spike activity has one of three general patterns (Fig. 4) – a return to baseline within a few milliseconds (pure pause), a gradual return to baseline over several tens of milliseconds (pause-reduction), or a rapid increase in activity to a level that is greater than baseline for a period of several tens of milliseconds (pause-facilitation). These observations of McDevitt et al. (1982) were recently confirmed and given a statistical appraisal by Sato et al. (1992). Like McDevitt et al., Sato et al. used the decerebrate, unanesthetized cat and recorded spontaneous activity of individual Purkinje cells. They also found the same types of temporal patterns in the simple spike activity subsequent to the complex spike. Of these patterns, the most common by far was pause-facilitation (71%), followed by pure pause (25%), and then pause-reduction (4%). Sato et al. also quantified the parameters of the pause and concluded that it is unlikely that the pause found in the pure pause and pause-facilitation types of Purkinje cells contributes any substantial decrease to their transient simple spike output. With regard to action potentials transmitted down the Purkinje cell axon, the pause, in fact, need not even exist because in approximately half of the Purkinje cells a burst of 2–5 action potentials was found to occur with climbing fiber activation (Ito & Simpson 1971; see also Mano et al. 1989).

The complex spike influence on the ensuing simple spike activity is highly variable and includes the possibility that the complex spike does not effectively influence the simple spike activity, so that under physiological conditions the two systems may be considered independent, as advocated by Llinás (1970; 1974). Sato et al. (1993) ascribed the variability to competition between mechanisms that facilitated and depressed the simple spike activity. A proposed mechanism for depression coupled the projection of climbing fiber collaterals to inhibitory interneurons, with the obser-

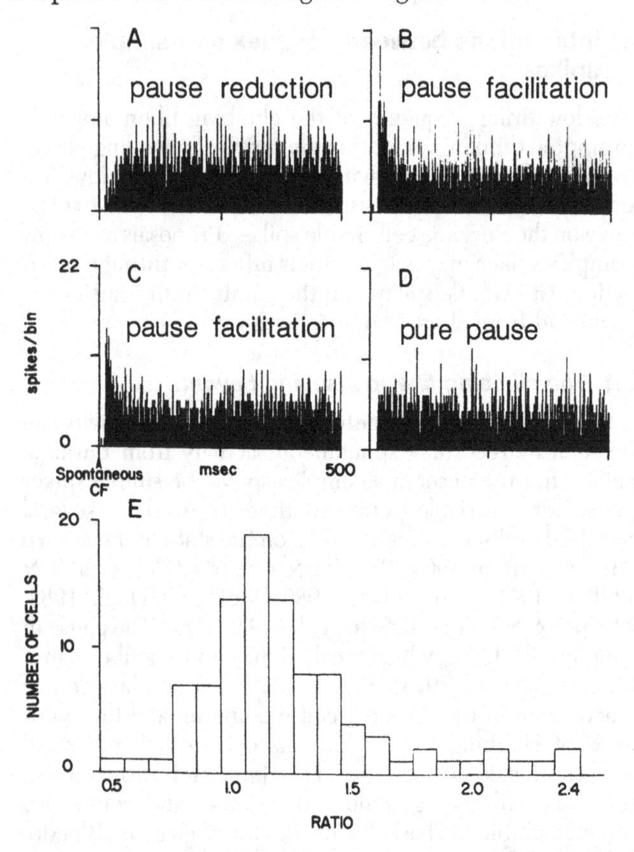

Figure 4. Changes in simple spike discharge rates following spontaneous climbing fiber (CF) discharges of Purkinje cells in unanesthetized, decerebrate cats. **A–D** show poststimulus time histograms (PSTH) of simple spike (SS) activity triggered by 100 spontaneous climbing fiber inputs, time zero (t = 0) indicating the time at which the triggering complex spike occurred (bin width = 1 msec). Each histogram was obtained from a different neuron. **E** shows a plot of the number of Purkinje cells with the indicated ratio of the average simple spike rate in the period from 10 to 50 msec following the complex spike trigger to the average simple spike rate in the period from 400 to 440 msec (from McDevitt et al. 1982).

vation by Kano et al. (1992) that $GABA_A$ receptor-mediated inhibitory effects on Purkinje cells *in vitro* can be potentiated by electrical stimulation of climbing fibers. Kano et al. also suggested that a high complex spike frequency *in vivo* would maintain the $GABA_A$ receptor sensitivity of Purkinje cells at a high level, whereas a low frequency would maintain it at a low level, thereby controlling the tonic simple spike firing rate (see below). For post-climbing fiber facilitation of simple spike activity, Sato et al. (1993) offered three tentative mechanisms: (1) inhibition of inhibitory interneurons by Purkinje axon collaterals, (2) inhibition of Golgi cell activity after climbing fiber activation (Schulmann & Bloom 1981; Yamamoto et al. 1978) leading to disinhibition of the mossy fiber-granule cell relay, and (3) prolonged depolarization of the Purkinje cells due to a voltage-dependent increase in calcium conductance following a complex spike (Ekerot & Oscarsson 1981; Llinás & Sugimori 1980).

A pause in the extreme can be produced by increasing the level of inferior olive activity to 8–10 Hz, either with the use of tremogenic drugs such as harmaline or with electrical stimulation, which results in elimination of simple spike activity (Demer et al. 1985; Llinás & Volkind 1973; Rawson & Tilokskulchai 1981; 1982). Conversely, silencing the inferior olive, whether achieved by lesion, lidocaine, or cooling, results in a gradual increase in the tonic background level of simple spike activity (Benedetti et al. 1984; Colin et al. 1980; Demer et al. 1985; Leonard & Simpson 1986; Montarolo et al. 1982). This relation between inferior olivary activity and tonic simple spike activity has been found in all parts of the cerebellum so far studied, but its physiological meaning remains unclear (Ojakangas & Ebner 1992; 1994). The tonic relationship between complex and simple spike activity should not be confused with the reciprocal relationship that is typically seen in the modulation of complex and simple spikes in the flocculus (Demer et al. 1985; Ghelarducci et al. 1975; Graf et al. 1988; Stone & Lisberger 1990). Leonard & Simpson (1986) recorded the modulation of the simple spikes in the flocculus before and after silencing the climbing fibers with submicroliter injections of lidocaine into the dorsal cap of the inferior olive. They found that while the simple spike background activity increased on average 35%, the absolute depth of modulation surprisingly remained unchanged (Fig. 5).

3.2. Gain change hypothesis

One interactive view of the consequence of climbing fiber signals is that for a short period (about 200 msec) after a complex spike the responsiveness of the Purkinje cell to its mossy fiber input is enhanced. This view of climbing fiber function, called the gain change hypothesis, evolved from a series of studies conducted by Ebner & Bloedel (1981a; 1981b; 1984; Ebner et al. 1983). In these studies, both spontaneous and naturally evoked Purkinje cell activity recorded from decerebrate, unanesthetized cats was used to compile cross-correlograms and poststimulus time histograms to examine the relation between a complex spike and the pattern of the ensuing simple spike activity. The findings made with each of a variety of paradigms led to the same conclusion – the climbing fiber input to a Purkinje cell frequently results in a short-lasting enhancement of its responsiveness to mossy fiber inputs. The change in responsiveness occurred during a period of up to several hundred milliseconds following a complex spike, but it did not persist in the absence of the complex spike. Increased responsiveness was manifest as an increase in the amplitude of both the excitatory and the inhibitory components of the simple spike response.

More recently, Bloedel and colleagues (Bloedel & Kelly 1992; Lou & Bloedel 1986; 1992a; 1992b) have tied the gain change hypothesis to the modular organization of synchronously active climbing fibers to propose the "dynamic selection hypothesis," which bears some resemblance to the "synchronizing pulse hypothesis" of Mano et al. (1986; 1989). Under the "dynamic selection hypothesis" the climbing fibers are seen as a means to spatially focus modulation of Purkinje cells and, in turn, nuclear cell activity. According to the hypothesis, Purkinje cells that have a synchronous activation of their climbing fiber input would, through enhancement of their responsiveness to the mossy fiber input, be much more dramatically modulated than Purkinje cells that are activated by a comparable mossy fiber input, but that are not activated by their climbing fiber input. Therefore, the climbing fiber input acts to select or emphasize those mossy fiber inputs that will

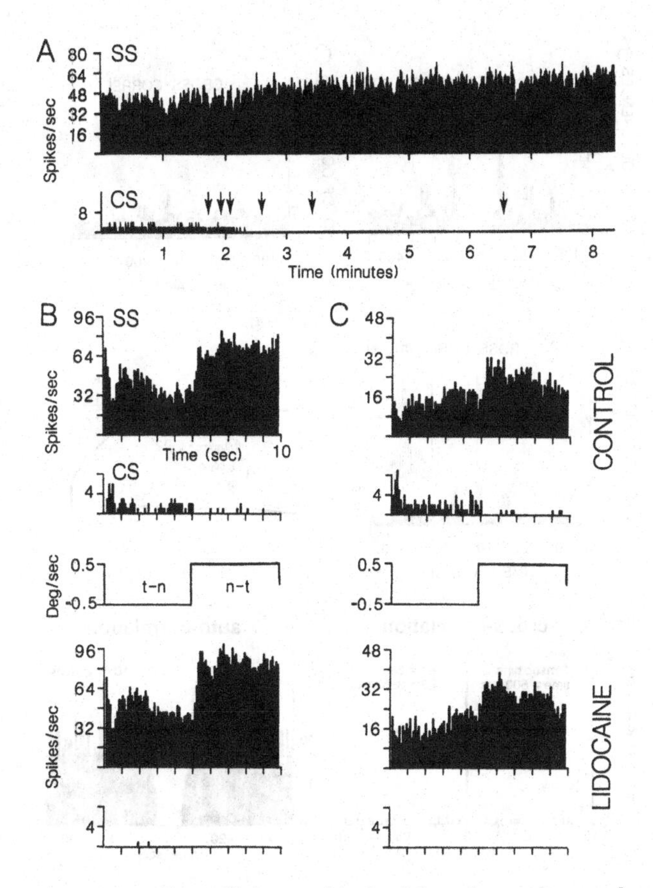

Figure 5. Effect of lidocaine block of the inferior olive on the simple spike activity of floccular Purkinje cells in the rabbit. **A** shows the effect on spontaneous simple spike activity produced by blocking the complex spike. The arrows indicate the times of the lidocaine injections into the dorsal cap. The simple spike firing rate gradually increased by 55% following blockade of the complex spikes. **B** and **C** show two examples of visually modulated VA Purkinje cells in the presence and absence of complex spike modulation. The optokinetic stimulation was produced by rear-projecting a random dot pattern (70° × 70°) onto a tangent screen centered on the optic axis of the ipsilateral eye. The CONTROL histograms were compiled before lidocaine injections, and the LIDOCAINE histograms were compiled after the complex spike firing rate was suppressed. Both cells have stereotypical simple spike modulation (simple spikes and complex spikes are reciprocal) that is unaffected by silencing the complex spike modulation, even though the spontaneous rate increased. Abbreviations: SS–simple spikes; CS–complex spikes; t-n–temporal to nasal (in reference to the ipsilateral eye); n-t–nasal to temporal (from Leonard 1986; see also Leonard & Simpson 1986).

produce the greatest modulation of the simple spikes. Evidence supporting the dynamic selection hypothesis has been obtained by Lou & Bloedel (1992a; 1992b) using a perturbed locomotion paradigm in decerebrate, locomoting ferrets. The responses of 3–5 sagittally aligned Purkinje cells were recorded simultaneously in response to an intermittent perturbation of the forelimb during the swing phase, and the amplitude of the combined simple spike responses across the population of Purkinje cells were correlated with the extent to which their climbing fiber inputs were synchronously activated. (The word "synchronous" as used by Bloedel and colleagues refers to a much less constrained timing relation than when used by Llinás and colleagues. This difference will be elaborated below.)

3.3. Climbing fibers as "teachers"

Marr (1969) and later Albus (1971) hypothesized that the cerebellum learns to perform motor skills through climbing fiber induction of long-term changes in the strength of parallel fiber synapses on Purkinje cells. Ito (1970) put this idea into a specific context by proposing that the flocculus adjusts the performance of compensatory eye movements in response to climbing fiber signals of "visual blur" functioning as a teaching input. The hypothesis that climbing fibers function as teachers is an interactive one, but unlike the interactive gain change hypothesis, the focus is on the permanency rather than the transientness of the climbing fiber influence on the simple spike activity. A further distinction between these two interactive views lies in the polarity of the influence of the complex spike on the simple spike modulation. Ito and colleagues (Ito 1989; Ito et al. 1982a; Kano & Kato 1988) emphasize the reciprocal relation between complex and simple spike activity in the flocculus because that relation is the one that would be established by a long-term depressive action of climbing fibers on simple spike activity. On the other hand, recordings advanced in support of the gain change hypothesis show that complex and simple spikes often increase together (Ebner & Bloedel 1981b; Ebner et al. 1983; Lou & Bloedel 1992b).

Because the issue of cerebellar learning is explored in detail in the other articles in this volume, and because our studies of the visual climbing fiber input to the flocculus have been directed toward considerations other than cerebellar plasticity, we will not recycle the conflicting views of the proponents and opponents of a teaching function of the climbing fibers (Bloedel 1992; Gilbert & Thach 1977; Ito 1982; 1984; Kawato & Gomi 1992; Lisberger 1988; Lisberger & Sejnowski 1992; Llinás & Welsh 1993; Miles & Lisberger 1981; Ojakangas & Ebner 1992; 1994; Schreurs & Alkon 1993; Tempia et al. 1991; Thach et al. 1992; Thompson 1986; Welsh & Harvey 1989). Even so, we would like to add several observations. First, it is interesting to note that the version of the teaching hypothesis presented by Marr (1969) stated that if complex spike activity reflected a response of sensory receptors, then the movement produced by discharge of the respective Purkinje cells should act to move the receptor so that the complex spike activity decreased. If otherwise, then a positive feedback situation would exist, because Marr envisioned that the climbing fibers acted to potentiate the simple spike activity. In the version of the teaching hypothesis advocated initially by Albus and later by Ito, the opposite relationship should hold, because the parallel fiber input that is paired with climbing fiber input is postulated to become less effective. That is, reduction of the simple spike activity of a Purkinje cell that receives from a given climbing fiber should lead to a sensory input that opposes the occurrence of that climbing fiber's discharge. Conversely, stimulation of that Purkinje cell would be expected to evoke a motor response that, in turn, leads to a climbing fiber response. Such an association between complex and simple spikes is present for the floccular visual climbing fiber system of the rabbit (Van der Steen et al. 1994). While this reciprocal relation is consistent with the Albus–Ito version of the teaching hypothesis, its presence does not demonstrate the validity of that hypothesis. For instance, visual modulation of the floccular complex spikes cannot be held to carve out a reciprocal relation from a monolith of tonic simple spike activity because the reciprocal relation between visually induced complex and simple spike modulation is present in the flocculus of dark-reared rabbits (Soodak et al. 1988). Furthermore, concomitant activation of complex and simple spikes can occur in relation to movements that have already been well-learned (Mano et al. 1986; 1989; Ojakangas & Ebner 1994). Also, as described below, some of the rabbit floccular Purkinje cells that showed reciprocal simple and complex spike modulation during rotation in the presence of vision, showed concomitant simple and complex spike activation during rotation in the absence of vision (De Zeeuw et al. 1995b).

As a second observation, we note that when the rabbit's compensatory eye movement gain is adapted to a new, stable level, the modulation of the visual climbing fibers would continue to be substantial because of the retinal image motion still remaining (e.g., Nagao 1983; 1988). Therefore, even though the signals of retinal image motion on the visual climbing fiber input to the flocculus are those proposed to be necessary for the climbing fibers to function as "teachers," those signals are apparently not sufficient. Complex spike modulation in conjunction with parallel fiber modulation is not by itself guaranteed to induce changes in the parallel fiber synapses. This conclusion was reached by another route by Ekerot & Kano (1985), who proposed that the original Marr and Albus hypotheses required revision to include an additional condition in order for the conjunction of complex and simple spikes to change the weight of the parallel fiber synapses. Conjunction alone is not enough. The level of inhibitory input to the Purkinje cell and its influence on the degree to which the complex spike results in calcium entry are also critical (Callaway et al. 1995; Campbell et al. 1983; Eckerot & Oscarsson 1981; Llinás & Sugimori 1980; Sakurai 1987; 1990). The extent to which these conditions are satisfied in the normal cerebellum is a matter of fervent debate.

4. The olivocerebellar system as a "timing device"

The above notions about the consequences of climbing fiber signals have in common the view that the climbing fibers evoke their effects through a change in the simple spike activity. Standing apart from this view is the proposal by Llinás and colleagues that the climbing fiber input acts on its own, forcing the Purkinje cells to fire in a burst-like manner to evoke an effect in the cerebellar nuclei neurons that is distinguishable from the one induced by the simple spikes (Jahnsen 1986; Llinás 1974; 1985; Llinás & Muhlethaler 1988a; 1988b). In this proposal, the climbing fibers serve as a "timing device" for movement execution (Llinás 1985; 1991; Llinás & Welsh, 1993; Llinás & Yarom 1986). The timing hypothesis has at least three components: (1) the climbing fibers convey motor commands, (2) these commands are for phasic motor acts, and (3) the inferior olive controls the timing among the different components of a motor act.

4.1. Synchrony and rhythm of olivary neurons

The "timing device" proposal is based on two observations (see Fig. 6). First, because of particular membrane conductances, olivary neurons (and consequently Purkinje cell complex spikes) tend to fire rhythmically, typically at a characteristic frequency of about 8–10 Hz (Bloedel & Ebner 1984; Llinás & Yarom 1981a; 1981b; 1986; but see Keating & Thach 1993). Second, olivary neurons show a tendency to fire synchronously, that is within a few milliseconds of each other (Bell & Grimm 1969; Bell & Kawasaki 1972; Llinás & Yarom 1981a), due to the fact that they are electrotonically coupled by dendrodendritic gap junctions (De Zeeuw et al. 1989; Llinás 1974; Llinás et al. 1974; Sotelo et al. 1974). Multiple electrode recording has shown that complex spike synchrony is most prevalent among

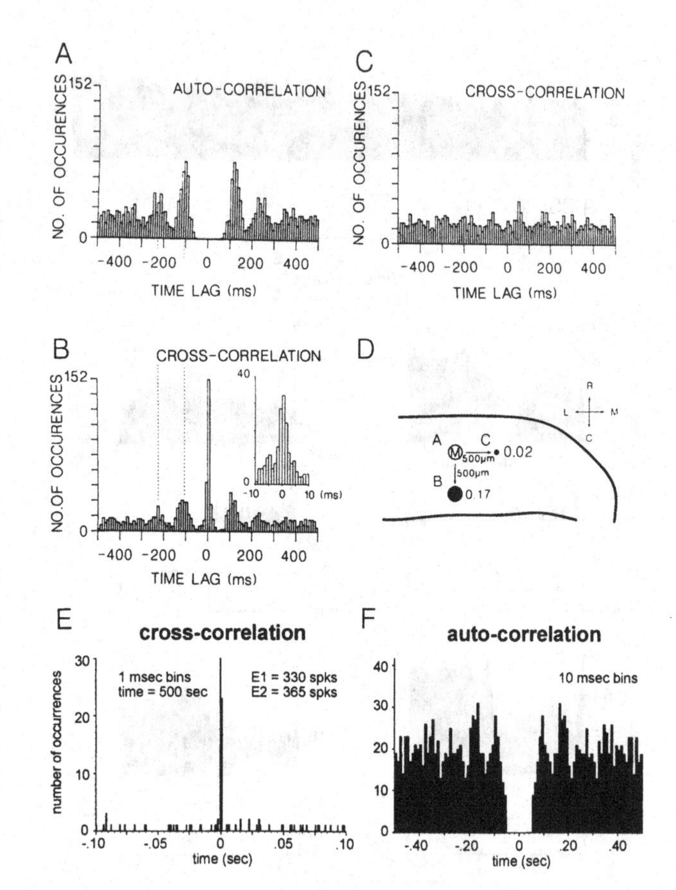

Figure 6. Synchronous and rhythmic complex spike activity of Purkinje cells. **A–D** are adapted from Sasaki et al. 1989. **A–C** show the auto- and cross-correlograms of complex spike activity of three cells recorded simultaneously from the superficial vermis of the rat. The relative location of these cells is shown in **D.** The cell at position A is the "master cell" (M), which is located 500 μm rostral to the neuron at position B and 500 μm lateral to the neuron at position C. The area of the dots at positions B and C represents the degree of cross-correlation between each cell and the master cell. **A** shows the auto-correlogram of the complex spike activity recorded at position A. As indicated by the two peaks occurring at about 120 and 240 msec, this cell had a characteristic frequency of about 8 Hz. **B** shows the cross-correlogram of the two sagittally aligned cells, located at positions A and B. The presence of the time-zero peak indicates that these two cells tended to fire within the same msec (see inset in **B**). In contrast, the cross-correlogram for the cells at positions A and C, shown in **C,** does not have a peak. **E** and **F** show examples of synchrony and rhythmicity in the ventral nodulus of the anesthetized rabbit (Wylie et al. 1995). **E** shows the cross-correlogram between two VA1 zone neurons that have a synchronous temporal relationship. **F** shows an auto-correlogram of a third VA1 neuron with 3 to 4 peaks spaced about 85 msec apart, indicating a characteristic frequency of about 12 Hz.

Purkinje cells in the same rostrocaudal band (Bell & Kawasaki 1972; Llinás & Sasaki 1989; Sasaki et al. 1989; but see Welsh et al. 1995).

It is important to recognize that the use of the term "synchronous" by Llinás and colleagues differs from that of Bloedel and colleagues (Bloedel & Kelly 1992; Lou & Bloedel 1992a; 1992b), who use "synchronous" to describe the clustering of complex spikes of a set of Purkinje cells within a much broader period of time, typically tens of milliseconds. With such a temporal dispersion, synchrony does not necessarily mean that the timing relations among the complex spikes are due to coupling via gap junctions. At the same time, it should be noted that synchrony in the millisecond range can be achieved with chemical synapses if a strong afferent input is shared by a set of neurons. However, the importance of electrotonic coupling for the olivocerebellar system is supported by the distribution of dendritic lamellar bodies that are associated with dendrodendritic gap junctions (De Zeeuw et al. 1995a); these dendritic lamellar bodies are distributed in all olivary subdivisions and their density in the inferior olive is higher than in any other area of the brain.

Lang et al. (1989; 1990; Lang 1995) have shown that complex spike synchrony is enhanced by administration of GABA-antagonists or by lesioning the GABAergic neurons in the cerebellar nuclei that innervate the inferior olive. Many of these cerebellar nuclei GABAergic neurons terminate apposed to the dendrodendritic gap junctions inside the olivary glomeruli (De Zeeuw et al. 1989; Nelson & Mugnaini 1989; Llinás 1974; Llinás et al. 1974; Sotelo et al. 1974; 1986). These GABAergic neurons can vary the coupling among olivary neuronal clusters representing different muscles. In that way, particular sets of muscles may be called into play at appropriate times during a movement (Lang 1995; Welsh et al. 1995). Indeed, dynamic repatterning of complex spike synchrony was found by Welsh et al. (1995) for a population of rat Purkinje cells recorded during rhythmic tongue movements. Lang (1995) noted that the degree of synchrony alters computational ability. Low levels of synchrony within the inferior olive reflect many small ensembles of neurons whose activity is relatively independent of each other, permitting many independent computational tasks to be performed. In contrast, a highly synchronized inferior olive has much less computational ability, but that would have advantages for tasks requiring simultaneous contraction of many different muscles. Llinás (1991) has noted that human reaction-time movements have been reported to be paced by a normal 10 Hz physiological tremor (Goodman & Kelso 1983), which may be mediated by the inferior olive. Moreover, harmaline increases the occurrence of synchrony and rhythmicity of olivary neurons and can, thereby, drive a 10-Hz body tremor (Llinás & Volkind 1973).

4.2. Climbing fibers and motor commands

Many investigators of the climbing fiber system have found that complex spike activity can be influenced by sensory stimuli (see above). But what is the relation of complex spike activity to voluntary motor behavior and to motor commands in particular? This question has been addressed by focusing on complex spike activity that occurs in relation to the onset of movement. (Fukuda et al. 1987; Lang 1995; Lang et al. 1992; Mano et al. 1986; 1989; Welsh et al. 1992; 1993; 1995).

Mano et al. (1986; 1989) recorded the responses of single Purkinje cells in the intermediate and lateral parts of the cerebellar hemispheres of trained monkeys performing visually guided wrist-tracking movements. For those Purkinje cells whose complex spike activity was response-locked to the movement, about 60% showed a phasic increase of complex spike firing rate at the onset of the movement. The complex spike activity increased with both rapid and slow tracking movements, but the increase was larger with faster step-tracking movements than with slower ramp-tracking movements. In most of the Purkinje cells that showed complex spikes locked to movement, the increase in complex spike firing rate occurred during "motor time," which is the period from the onset of the EMG change in the prime mover muscles to the beginning of the movement. Interestingly, the complex spikes were not response-locked when the monkey returned the manipulandum to center position after completing the tracking task, even in those Purkinje cells that showed a significant increase of complex spike activity during visually guided tracking. This difference, which was present even when differences in direction, starting point, and speed were controlled for (Mano et al. 1989), may be due to the fact that the initial movement was triggered by an external stimulus and the timing constraints on performance were greater than for the return movement.

Consideration of the effect of the complex spike on the subsequent simple spike activity led Mano et al. (1986; 1989) to propose the "synchronizing pulse hypothesis." Under this hypothesis, the role of the climbing fibers may be to "assist the effective onset or cessation of simple spike frequency modulation when animals are required to perform voluntary movement with precise timing as in externally triggered movement." The improvement in effectiveness is held to be due to short-term actions of the complex spike on the simple spike firing pattern in combination with electrical coupling among inferior olivary neurons.

Several studies have used multiple electrode recording to examine complex spike activity in relation to spontaneous vibrissae movement (Fukuda et al. 1987; Lang 1995; Llinás 1991) and in relation to vibrissae movement evoked by electrical stimulation of the motor cortex (Lang 1995; Lang et al. 1992). With spontaneous movements in animals anesthetized with ketamine and in awake animals, complex spikes often occurred within 20 milliseconds prior to onset of vibrissae movement. Lang (1995) pointed out that the ability to observe complex spike timing in relation to vibrissae movement may have been aided by the fact that each vibrissa is moved by a single muscle. This simple arrangement may, in part, explain why some previous investigations (e.g., Andersson & Armstrong 1987; Thach 1968) found only a weak or nonexistent relationship between complex spikes and the more complicated multijoint limb movements. While complex spikes mostly preceded onset of the vibrissae movement, they also occurred soon after movement onset and thus were coincident with the movement. Lang et al. (1992) found that the probability of vibrissae movement increased when the complex spikes were more synchronous. Lang (1995) noted that the inferior olive may not only be determining the times at which the movements occur, but also may be playing a permissive role in allowing movements to occur because the movements do not necessarily have to occur at every opportunity. The permissive aspect could be a reflection of the subthreshold oscillation of olivary membrane potential, while the occurrence of a movement may be a reflection of the synchronous discharge of the coupled ensemble of olivary neurons.

In another study revealing time-locking of complex spikes to movement, Welsh et al. (1995) used multiple electrodes to record from Purkinje cells in crus 2a of alert rats trained to tongue lick in response to an acoustic tone. The complex spike population response revealed a gradual increase in activity preceding a sharp peak at the time of maximal tongue protrusion. Complex spike activity occurring after that time was suggested to be associated with tongue movements that would be required to hold the water drop on the tip of the tongue. A less pronounced second peak in climbing fiber activity occurred coincident with the closing of the mouth.

The three studies described above have in common the finding that the complex spike activity can increase in relation to rapid, precisely initiated motor behavior, but the exact relationship was different in each case. In one case the increase occurred mostly between EMG onset and movement onset, in another the increase often occurred within 20 milliseconds prior to movement onset, while in the third the increase was centered on a particular facet of the movement. Thus, in some instances the complex spikes of some Purkinje cells precede movement, while in other instances the complex spikes occur after movement onset, but still in close temporal relationship to movement onset. Those complex spikes occurring prior to the movement can be associated with the beginning of the movement, while those occurring later may be related to subtle features of the movement that occurred immediately after onset, or they may be associated with stopping the movement.

4.3. Floccular climbing fibers and motor signals

The signals of retinal image motion on dorsal cap and ventrolateral outgrowth neurons arise from midbrain projections that are non-GABAergic and presumably excitatory (Horn & Hoffmann 1987; Nuñes-Cardozo & Van der Want 1990). In addition, the dorsal cap and the ventrolateral outgrowth are innervated by neurons in the nucleus prepositus hypoglossi (Gerrits et al. 1985; McCrea & Baker 1985) that are partly cholinergic (Barmack et al. 1993b) and partly GABAergic (De Zeeuw et al. 1993). The GABAergic input from the prepositus hypoglossi together with the GABAergic input from the dorsal group *y* and the ventral dentate nucleus (De Zeeuw et al. 1994a) account for most, if not all, of the inhibitory input to the dorsal cap and ventrolateral outgrowth. As in other regions of the inferior olive (De Zeeuw et al. 1989), the inhibitory synapses account for approximately half of the synaptic input. At least one-fifth of the inhibitory synapses in the dorsal cap and ventrolateral outgrowth are located in the glomeruli and act to control the strength of coupling among olivary neurons, as discussed above. The other inhibitory synapses are located mainly elsewhere on the dendrites. Their contribution to floccular climbing fiber signaling remains to be determined, but some indication of what these inhibitory inputs may be doing is available from recordings in the awake rabbit of floccular complex spike activity during vestibular stimulation in the absence of vision (De Zeeuw et al. 1995b). The complex spikes of a minority of Purkinje cells that were modulated by retinal image motion were also modulated during sinusoidal rotation about the vertical axis in the dark. This modulation was not as strong as that for visual stimulation. The great majority of these Purkinje cells showed increased complex spike activity with contralateral head rotation in the dark, which is opposite to the complex spike behavior with rotation in the light. Since the polarity of the complex spike modulation during rotation in the dark was generally opposite to that in the light, the complex spike modulation in the dark was not due to an incomplete darkness. The most likely candidate for the input underlying the modulation during rotation in the dark is the nucleus prepositus hypoglossi, which contains neurons that project to the caudal dorsal cap (De Zeeuw et al. 1993), as well as neurons that have eye velocity and eye position signals associated with the horizontal vestibulo-ocular reflex (Escudero et al. 1992; Lopez-Barneo et al. 1982).

In further consideration of the possibility that motor signals are present on some floccular climbing fibers, we note that one of the five floccular zones delineated with acetylcholinesterase histochemistry (Tan et al. 1995a) – the C2 zone – does not receive signals of retinal image motion (De Zeeuw et al. 1994b). The source of its climbing fibers is the rostral pole of the medial accessory olive (Tan et al. 1995b), which receives a major input from the nucleus of Darkschewitsch (De Zeeuw & Ruigrok 1994). The part of the nucleus of Darkshewitsch that innervates the rostral pole of the medial accessory olive (Porter et al. 1993) receives input from the frontal eye fields of the cerebral cortex (Miyashita & Tamai 1989). This connectivity suggests that the climbing fiber input to the C2 zone of the flocculus may carry motor signals that are related to voluntary gaze shifts involving both the head and eyes.

5. Visual climbing fibers in the rabbit vestibulocerebellum

5.1. Modules of the rabbit flocculus

In attempting to reach some consensus about several of the prominent views of climbing fiber function, we have performed experiments in the rabbit flocculus, where the visual climbing fiber input is known to carry a reliable and robust error signal of retinal image motion. A functional overview of the modular input–output relations of the visual climbing fibers in the rabbit's flocculus is shown in Figure 7.

The flocculus of the rabbit consists of five zones (1, 2, 3, 4, and C2) whose borders can be delineated in the floccular white matter by using acetylcholinesterase (AChE) staining (De Zeeuw et al. 1994b; Tan et al. 1995a; 1995b; Van der Steen et al. 1994). The climbing fiber projections to zones 1 and 3 are derived from the rostral dorsal cap and ventrolateral outgrowth; the climbing fiber projections to zones 2 and 4 are derived from the caudal dorsal cap; the climbing fiber projection to zone C2 is derived from the rostral pole of the medial accessory olive (Tan et al. 1995b). In zones 1–4 the complex spikes are optimally modulated by rotational optokinetic stimulation about either the vertical axis or about a horizontal axis approximately perpendicular to the ipsilateral anterior semicircular canal (Graf et al. 1988; Kano et al. 1990b; Kusunoki et al. 1990; Simpson et al. 1981; Wylie & Frost 1993). These optimal axes thus have a geometry similar to that of the best-response axes of

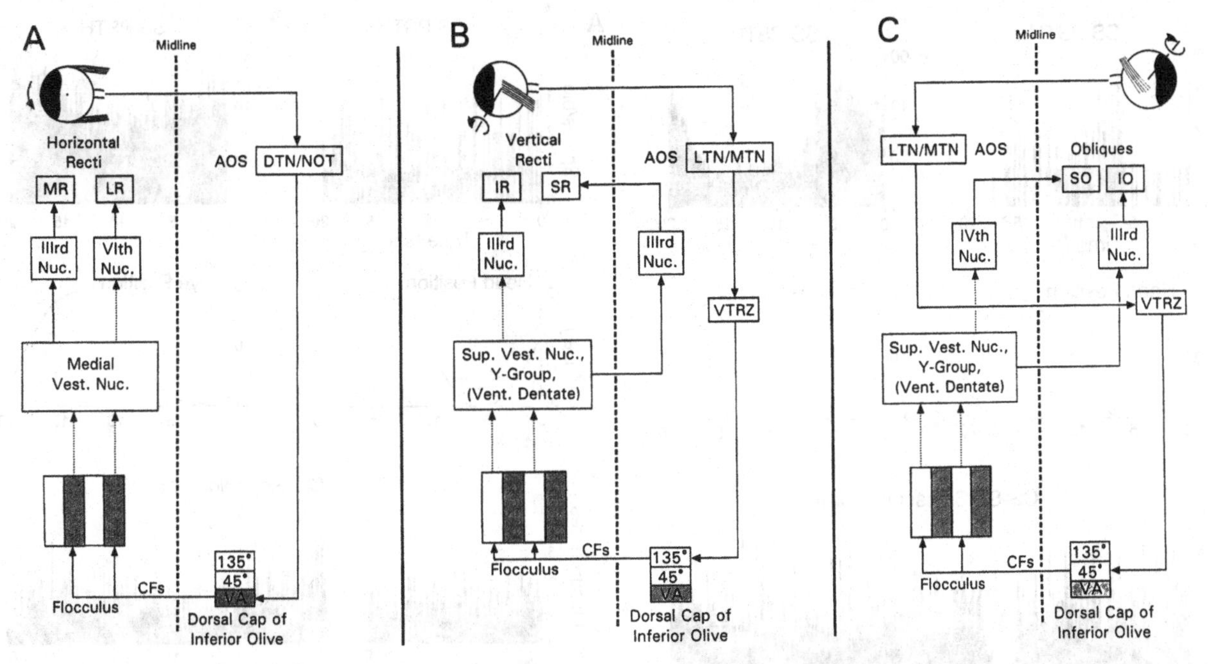

Figure 7. Input–output relations between retinal image motion and the floccular modular control of extraocular muscles in the rabbit. Signals from retinal ganglion cells are processed in various parts of the accessory optic system before being differentially distributed to the inferior olive. Each panel (**A, B,** and **C**) relates to one of the three classes of olivary neurons that are distinguished on the basis of the orientation of their preferred axis for responding to visual world rotation and their ocular dominance (see Fig. 2). Each class is located within a particular part of the dorsal cap and ventrolateral outgrowth and gives rise to a specific group of climbing fibers that traverse particular acetylcholinesterase-delineated, white-matter compartments to innervate the Purkinje cells of particular zones. Two zones receive their input from the caudal dorsal cap, and two other zones receive their input from the rostral dorsal cap and ventrolateral outgrowth. Note that the 135° and 45° classes of climbing fibers both project to the same two floccular zones. The Purkinje cells in the floccular zones, in turn, have a differential projection to the vestibular complex from which particular sets of extraocular muscles are controlled. Activation of these sets of muscles causes eye rotation about axes that are spatially close to the preferred axes of the visual climbing fibers. Abbreviations: AOS–accessory optic system; DTN, LTN, MTN–dorsal, lateral, and medial terminal nucleus; MR–medical rectus; LR–lateral rectus; IR–inferior rectus; SR–superior rectus; SO–superior oblique; IO–inferior oblique; VTRZ–visual tegmental relay zone (based on De Zeeuw et al. 1994b; Graf et al., 1988; Ito et al. 1977; Leonard et al. 1988; Simpson et al. 1988; Soodak & Simpson 1988; Tan et al. 1995a; 1995b; Van der Steen et al. 1994; Yamamoto 1979).

the semicircular canals and to the axes about which the three pairs of extraocular muscles rotate the eye (see Fig. 2). The horizontal axis neurons in the rabbit flocculus can be divided into two classes based on ocular dominance. Purkinje cells dominated by the contralateral eye respond best to rotation about an axis at about 45° contralateral azimuth; neurons dominated by the ipsilateral eye respond best to rotation about an axis at about 135° ipsilateral azimuth. We refer to these neurons as contra-45° and ipsi-135° neurons, respectively. The vertical axis (VA) neurons are located in zones 2 and 4, while the contra-45° and ipsi-135° neurons are located in zones 1 and 3. The climbing fibers of zone C2 do not respond to optokinetic stimulation (De Zeeuw et al. 1994b).

5.2. Simple spike transients in the rabbit flocculus

The gain change hypothesis (Ebner & Bloedel 1981a; 1981b; 1984; Ebner et al. 1983) asserts that the climbing fiber input to a Purkinje cell can produce a short-lasting enhancement of the responsiveness to mossy fiber inputs. We investigated the patterns of simple spike transients of Purkinje cells in the flocculus of awake rabbits presented with vestibular stimulation in the light and dark, and with optokinetic stimulation. Vestibular stimulation was provided by rotating the restrained rabbit about the vertical axis with the use of a servo-controlled turntable; optokinetic stimulation was provided by a planetarium projector rotating about the vertical axis. Eye movements were recorded using the scleral search coil technique (Robinson 1963), and Purkinje cells, identified by the presence of a brief pause in simple spike activity following a complex spike, were recorded extracellularly. Complex spike and simple spike peristimulus time histograms (PSTH) were computed along with complex spike–simple spike cross-correlograms.

A statistical method similar to that of Sato et al. (1992) was used to assign the Purkinje cells to one of three categories – pure pause, pause-facilitation, or pause-reduction – on the basis of the cross-correlogram patterns obtained with at least 100 sequential complex spikes. To be considered a pause-facilitation cell, the average simple spike activity during at least one 20-millisecond period within the first 50 milliseconds following the end of the brief pause had to be significantly greater ($p < 0.01$) than the average simple spike activity during the 100 milliseconds prior to the onset of the complex spike. This criterion is less restrictive than that of Sato et al. since the 20-millisecond period used to test for significance was not constrained to begin immediately following the end of the pause in the cross-correlogram. A similar test was used to assign cells to the pause-reduction category. From our sam-

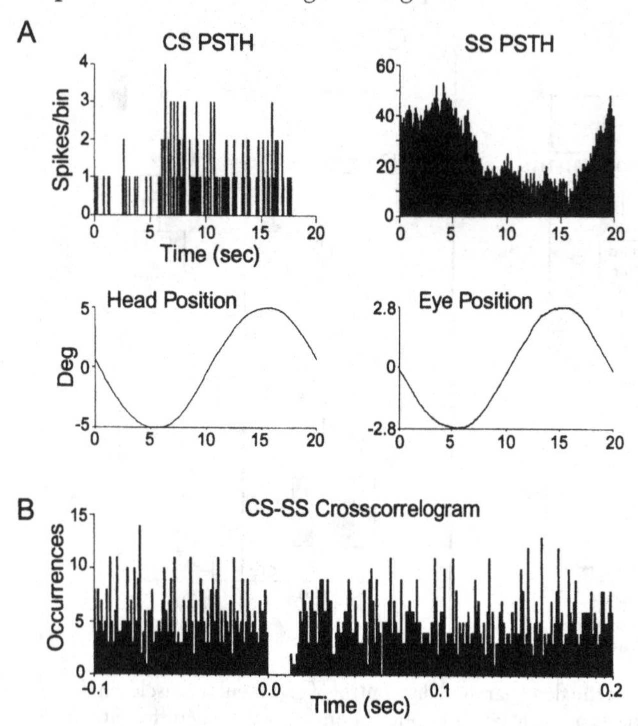

Figure 8. Neural activity of a "pure pause" floccular Purkinje cell recorded in the awake rabbit during sinusoidal vestibular stimulation in the light at 0.05 Hz. **A** shows the complex and simple spike poststimulus time histograms (PSTHs) compiled over 6 cycles (100-msec binwidth) and averages of the turntable and inverted eye positions. Note the reciprocity in the complex and simple spike modulation. The complex spike activity increased during rotation toward the side of recording. **B** shows the complex spike–simple spike cross-correlogram (1-msec bins) compiled from the same 6 cycles. After a brief pause, the simple spike activity rapidly returned to an average rate that was not significantly different from the average rate during the 100-msec pre-complex spike period.

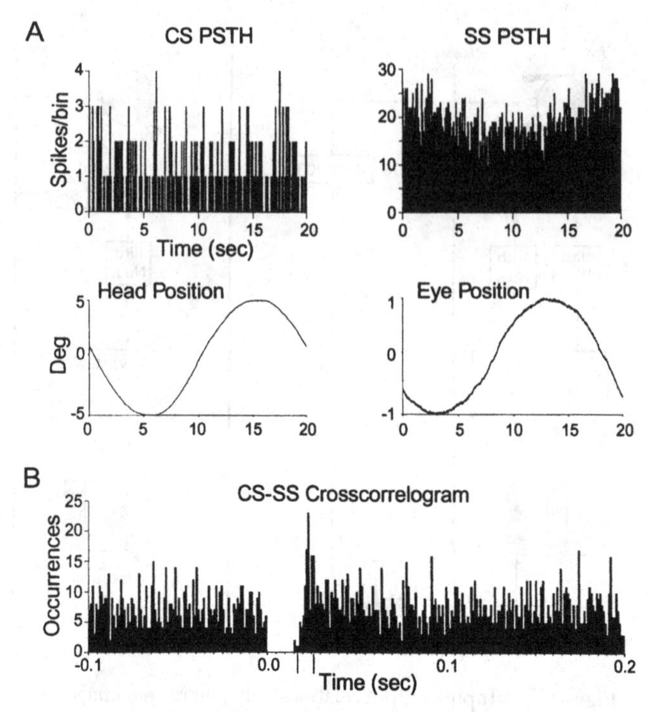

Figure 9. Neural activity of a "pause-facilitation" floccular Purkinje cell recorded in the awake rabbit during sinusoidal vestibular stimulation in the dark at 0.05 Hz. **A** shows the complex and simple spike poststimulus time histograms (PSTHs) compiled over 6 cycles (100-msec binwidth) and averages of the turntable and inverted eye positions. The simple spike (SS) activity was modulated, but the complex spike (CS) activity was not. **B** shows the complex spike–simple spike cross-correlogram (1-msec bins) compiled from the same 6 cycles. After a brief pause, the simple spike activity rapidly rose to a level that for a short time was higher than the remainder of the analysis period. For each 20-msec period that began at the times between the two lines on the time axis of the correlogram, the average simple spike rate was significantly greater ($p < 0.01$) than the average rate during the 100-msec pre-complex spike period.

ple of 43 Purkinje cells, 26 were pure pause (Fig. 8), 12 were pause-facilitation (Fig. 9), and 5 were pause-reduction. Even with our more liberal statistical criterion, the percentage of pause-facilitation cells was less than half of the 60%–70% we had anticipated from the findings of Ebner & Bloedel (1981a), McDevitt et al. (1982), and Sato et al. (1992). Also, the duration of the facilitation was substantially shorter than anticipated. Part of these differences may be attributed to the decerebration employed by these investigators, because in the awake monkey Mano et al. (1986) found (as we did in the awake rabbit) that the pure pause category was the predominant one (66%), with the remaining Purkinje cells in the pause-facilitation category. In this context, it is noteworthy that in another study in the awake monkey Ojakangas and Ebner (1994) found that about 75% of the Purkinje cells were in the pure pause category.

The significance of the pause and any ensuing simple spike transients for the control of slowly changing compensatory eye movements eludes us. The duration of the pause (10–20 msec) of the pure pause and pause-facilitation Purkinje cells is very brief in comparison to the time course of the compensatory eye movements produced in response to the stimuli employed. The reciprocity often seen between simple and complex spikes in the flocculus is not due to the brief pause, because, as also found by several other investigators (Demer et al. 1985; Leonard 1986; Sato et al. 1992), it is too brief and too infrequent to make a substantial contribution to the reciprocal relationship. The duration, size, and infrequent occurrence of the transient facilitations and reductions indicate that they too make little contribution to the overall simple spike modulation. On the other hand, transients in simple spike activity following complex spikes do occur in the flocculus. Perhaps their contributions will become apparent in relation to other aspects of eye movements such as voluntary gaze shifts, which in the rabbit typically occur as a combined eye and head movement.

5.3. Synchrony and rhythmicity in the rabbit vestibulocerebellum

Complex spike synchrony was investigated by recording from pairs of Purkinje cells in the ventral nodulus of ketamine-anesthetized rabbits (Wylie et al. 1995). Like the flocculus, the ventral nodulus is comprised of zones. Three of the four zones receive visual climbing fibers from the dorsal cap and ventrolateral outgrowth. The climbing fibers of the most medial zone are not responsive to retinal image motion, but receive a vestibular climbing fiber input from the beta subnucleus. Two zones (VA1 and VA2) receive their climbing fiber input from the caudal dorsal cap, which contains VA neurons. The zone between the two VA zones contains contra-45° and ipsi-135° neurons, which receive their climbing fibers from the rostral dorsal cap and the ventrolateral outgrowth (zone HA) (Balaban & Henry 1988; Barmack et al. 1989; 1993a; Barmack & Shojaku 1992; Kano et al. 1990a; 1990b; Katayama & Nisimaru 1988; Shojaku et al. 1991; Wylie et al. 1994).

In confirmation of the studies of Llinás and colleagues described above, we found that pairs of Purkinje cells from the same zone often showed complex spike synchrony, which we defined as the tendency for the complex spikes of the two Purkinje cells to discharge within at most 2 milliseconds of each other. To assess the temporal relationship of a cell pair, cross-correlograms were constructed using 1-, 2-, 5-, 10-, and 20-millisecond binwidths for 100 bins on either side of time zero. To quantify the temporal relationship of a cell pair, the cross-correlation coefficient was used as a synchrony index (Sasaki et al. 1989; Sugihara et al. 1993; Wylie et al. 1995). The tendency of a neuron pair to fire within a given time period was determined as significant if one of the two time-zero bins was denoted as a peak. A time-zero bin was denoted as a peak if, (1) it had a value of at least 5, (2) it was the highest bin, and (3) it was at least 3 standard deviations above the mean. The activity was considered to be synchronous when one of the two time-zero peaks in the cross-correlograms with 1-, or 2-millisecond bins was significant. The cross-correlogram for two Purkinje cells from the VA1 zone is shown in Figure 6E. During a 500-second period one cell had 330 spontaneous complex spikes and the other had 365. On 53 occasions, the complex spikes discharged within a millisecond of each other. Of 82 pairs consisting of two Purkinje cells in the same zone, 33 (40%) showed a tendency to fire within 2 milliseconds. Included in these 82 are 7 pairs comprised of a contra-45° Purkinje cell and an ipsi-135° Purkinje cell located in the HA zone. Five of these pairs showed a synchronous relationship. The occurrence of complex spike synchrony has recently been confirmed in the flocculus of the awake rabbit (Fig. 10).

We also recorded from 16 pairs consisting of one Purkinje cell in each of the VA1 and VA2 zones, which are spatially separated by the 1 mm-wide HA zone. Nonetheless, 6 pairs showed a synchronous relationship. Moreover, 3 of 14 pairs consisting of one floccular VA Purkinje cell and one nodular VA Purkinje cell showed a synchronous relationship, which is notable when one considers that the climbing fibers to these zones are probably of different lengths. Sugihara et al. (1993) have shown that despite the fact that some olivocerebellar branches in crus 2a are considerably longer than others, their conduction times are quite uniform (but see Aggelopoulous et al. 1994).

The value of the synchrony index varied depending upon the nature of the response of the particular complex spike pair to optokinetic stimulation. For cross-correlograms computed at the 2-millisecond binwidth, the largest average synchrony index was 0.039 for 39 pairs of complex spikes that were recorded in the same zone and responded best to optokinetic stimulation about the vertical axis (Wylie et al. 1995). While this value may appear small, it is about one order of magnitude larger than the value expected for two random independent spike trains having a mean firing rate in the usual range of complex spike firing rates (Sugihara et al. 1993). Even so, some may still question the importance of this degree of synchrony. As two tentative answers, it is suggested that the effect of synchrony on the cerebellar nuclei may not be directly proportional to the strength of the synchrony (Welsh, personal communication), and that changes in the pattern of the synchrony across the cortex could be more important than changes in the value of the synchrony index (Welsh et al. 1995).

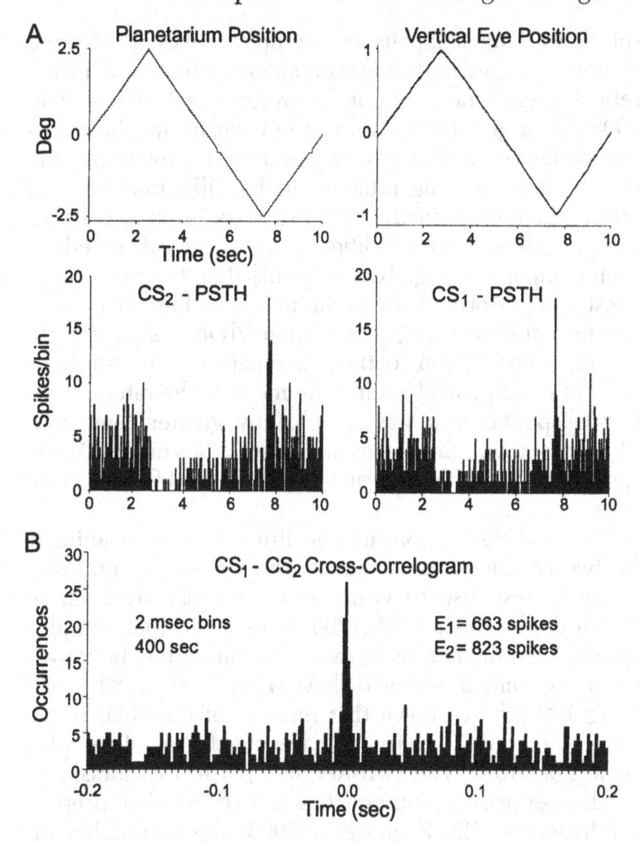

Figure 10. Synchronous complex spike activity recorded from a pair of floccular Purkinje cells in the awake rabbit. Both of these Purkinje cells received visual climbing fibers of the contra-45° axis class. Panel **A** shows the average planetarium position, the average vertical position component of the compensatory eye movement, and the peristimulus time histograms (PSTHs; 50-msec bins) for the complex spike activity of the two Purkinje cells cumulated over 40 cycles. A clear synchronous relationship between the firing of the complex spikes of these two Purkinje cells is seen in **B** as a peak in their cross-correlogram.

We do not know the functional significance of synchrony between different floccular and nodular modules that receive similar climbing fiber inputs. In broad terms, the nodulus and ventral uvula are involved in control of the velocity storage mechanism, whereas the flocculus is important for controlling the gain of the vestibulo-ocular and optokinetic reflexes (Cohen et al. 1992; Waespe et al. 1983; 1985). In addition, both the nodulus (Ito et al. 1982b; Nagao 1983) and the flocculus (Stahl & Simpson 1995; De Zeeuw et al. 1995b; Ito et al. 1982b; Nagao 1983) influence the phase of compensatory eye movements, but in opposite directions. Thus, the modules in the nodulus and flocculus may be synchronized to coherently combine different aspects of compensatory eye movements.

Because retinal image motion effectively modulates the complex spike activity of nodular and floccular Purkinje cells, we investigated whether that stimulus would affect the strength of the temporal relation of these complex spikes. Purkinje cell pairs with complex spikes of the same class were recorded during spontaneous activity and during retinal image rotation about the preferred axis (Wylie et al. 1995). As a quantitative measure of synchrony, the cross-correlation coefficient was calculated during rotation in the On direction, during rotation in the Off direction, and during spontaneous activity for 53 pairs of Purkinje cells. To compare these three conditions, ratios were calculated for each comparison (e.g., [On − spont]/[On + spont], and a t-test was performed versus the null hypothesis that there was no difference (i.e., [On − spont]/[On + spont] = 0). For cross-correlation coefficients obtained from 2-millisecond binwidth correlograms, the mean of the ratio for On versus spontaneous was significantly greater than zero (0.18, $p < 0.02$); the means of the two other ratios (Off vs. spont; On vs. Off) were not significantly different from zero.

The level of synchrony may be different in awake animals for two reasons. First, compensatory eye movements will occur in response to visual and vestibular stimulation; Welsh et al. (1992; 1993; 1995) have noted that complex spike synchrony in crus 2a increases during tongue movements. Second, as outlined above, Lang et al. (1989; 1990; Lang 1995) have shown that removal of the GABAergic input to the inferior olive increases the degree of complex spike synchrony. The GABAergic input to the dorsal cap and the ventrolateral outgrowth is from the nucleus prepositus hypoglossi (De Zeeuw et al. 1993), the ventral dentate nucleus, and the dorsal group y (De Zeeuw et al. 1994a). Neurons in prepositus hypoglossi and dorsal group y are modulated during visual-vestibular stimulation in alert preparations (Chubb et al. 1984; Lopez-Barneo et al. 1982; McFarland & Fuchs 1992; Partsalis et al. 1993), and this modulation may result in stronger synchrony.

Many nodular Purkinje cells showed rhythmic complex spike activity as revealed in auto-correlograms (Fig. 6F; Wylie et al. 1995). Thirty-two percent (36 of 114) displayed one or more peaks in their complex spike auto-correlogram. The average rhythm was 8 Hz (range 2.5–12.5 Hz), in agreement with the findings of Llinás & Yarom (1981a; 1981b; 1986). For the 36 rhythmic cells, one to six peaks on each side of the correlogram were apparent (mean = 2.4 peaks). While most cells (25) had two or more peaks, 11 had only one peak. (Cells showing a single peak in the auto-correlogram would accurately be described as producing pairs of action potentials separated by a characteristic time interval.) Thirty-five of the 114 Purkinje cells were recorded for sufficient time to obtain more than 600 complex spikes. A 10-millisecond binwidth auto-correlogram with 100 bins on either side of time-zero was constructed for these cells. On the basis of visual inspection, the initial part of these auto-correlograms was judged to be rhythmic for 9 of the 35 cells, as exemplified in Figure 6F. The rhythm strength for these 9 cells was quantified by first determining the best-fit sine wave for the initial 2 or 3 periodicities of the auto-correlogram and then calculating the ratio of the amplitude to the mean of the sine wave. For the cell shown in Figure 6F, the rhythm strength was 0.38 for the first 2 periodicities of the auto-correlogram. The average rhythm strength for the 9 cells was 0.35 (range 0.26–0.53); the average frequency of the best-fit sine wave was 8.0 Hz (range 4.8–11.5 Hz).

We found that the occurrences of synchronous firing and rhythmic oscillation were independent. Synchrony occurred with or without the rhythmic firing of one or both neurons in the pair. These observations, made in the anesthetized rabbit, are in contrast to those made in the slice preparation where spontaneous oscillations of olivary neurons occurred synchronously in all cells examined (Llinás & Yarom 1986).

6. Conclusion

We have attempted to put some of the prominent hypotheses of climbing fiber function into the context of the visual climbing fiber input to the rabbit's vestibulocerebellum. With regard to the gain change hypothesis, the majority of floccular Purkinje cells in the awake, behaving rabbit showed only brief pauses in the simple spike activity that followed a complex spike. Although transient increases and decreases in simple spike activity did occur following the climbing fiber pause in some Purkinje cells, their functional meaning for the control of slowly changing compensatory eye movements escapes us. Perhaps their contribution is to be found in relation to rapid, voluntary eye movements associated with changes in gaze. With regard to climbing fibers and cerebellar learning, we can say from our studies in the flocculus that the error message is right, but apparently it alone is not sufficient to ensure plasticity. With regard to synchrony and rhythmicity of complex spikes, both were observed in the vestibulocerebellum. Even though the synchrony changed in relation to visual stimulation, its contribution to control of compensatory eye movements awaits elucidation in the awake animal.

In sum, the function of the climbing fiber input to the cerebellum remains an intriguing question that must be answered prior to understanding cerebellar performance. Although many of the electrophysiological characteristics of this afferent system are known, and a number of hypotheses have been developed as to its function, a consensus has not been reached and the function of the climbing fibers remains an enigma. This situation reflects, in part, our limited understanding of the various ways in which cerebellar cortical inhibitory interneurons influence the consequences of the climbing fiber signals.

ACKNOWLEDGMENTS

This research was supported by grants from NWO (R 95-260), KNAW, and NIH (NS-13742). D. R. Wylie was supported by a postdoctoral fellowship from NSERC (Canada). We thank Evgeney Buharin and Ilan Kerman for technical assistance.

7

Does the cerebellum learn strategies for the optimal time-varying control of joint stiffness?

Allan M. Smith
Département de physiologie, Centre de recherche en sciences neurologiques, Université de Montréal, Montréal, Québec, H 3C 3J7, Canada
Electronic mail: *smitha@ere.umontreal.ca*

Abstract: Although there is increasing agreement that the cerebellum plays an important role in motor learning, the basic substance of what constitutes motor learning has been difficult to define. Unless motor learning is somehow radically different from other forms of learning, it must involve relatively simple stimulus-stimulus and stimulus-response associations. All forms of learning, including purely sensory associations and cognitive learning as well as motor learning, effect changes in behavior. However, a singular characteristic of motor learning is that it adjusts joint and limb mechanics by altering the neural input to muscles through practice and mental rehearsal. The hypothesis proposed here is that the cerebellum plays an important role in motor learning by forming and storing associated muscle activation patterns for the time-varying control of limb mechanics. By modulating the cocontraction of agonist-antagonist muscles through adjustments in the timing and amplitude of muscle activity, the viscoelastic properties of joints can be appropriately regulated throughout movement and adapted for transitions between postures and movements. Optimal control of joint viscoelastic properties cannot be achieved by online corrections initiated by reflex feedback because of the delays and consequent instabilities incurred. Instead, strategies for optimizing muscle activation patterns or synergies must be learned from the temporal association of proprioceptive stimuli signaling muscle lengths and forces and the rates of changes in these parameters, with reinforcement occurring when the movement achieves its objective. Such strategies would involve varying degrees of cocontraction or reciprocal inhibition of agonist-antagonist muscles that ultimately contribute to joint and limb stiffness. Evidence from neural recordings and clinical and experimental lesion studies are presented, suggesting that the cerebellum uses teleceptive and proprioceptive feedback as feedforward conditioned stimuli for specific muscle activation patterns contributing to joint stiffness (i.e., agonist-antagonist muscle synergies) for particular tasks and postures. A wide variety of observations are thought to be consistent with such a role for the cerebellum, but ultimately additional experiments could confirm or disconfirm this hypothesis.

Keywords: agonist-antagonist cocontraction; cerebellum; joint stiffness; motor learning; muscle activation strategies; muscle synergies; multijoint limb control

In the twenty-five years since the publication of the *Cerebellum as a Neuronal Machine* (Eccles et al. 1967), the notion that the cerebellum plays an important role in motor learning has become commonly accepted. The important question facing neurophysiologists is no longer *whether* the cerebellum participates in skill acquisition, but rather *how* it operates. In view of the current controversy over whether the cerebellum is the single exclusive storage site for motor learning, I propose to assume (but will rely on the evidence provided by others in this issue) that some significant plastic changes in synapses do occur within the cerebellar cortex during motor learning. Learning in general is thought to be a distributed process, however, and hence it is not reasonable to assume that all motor learning is limited or otherwise exclusive to the cerebellum. Also missing from this review because of space limitations are considerations of the regional specialization within the cerebellum itself and the sensory afferent systems that converge there. Instead, I propose to expand upon and modernize an idea first elaborated by Babinski (1899), who drew upon the animal lesion studies of Luciani as well as his own clinical experience to formulate his concept of cerebellar function.

Ataxia, which is associated with large cerebellar lesions, represents an inability to coordinate synergies, or according to Babinski (1899), an "asynergia," which he felt was pathonomonic of cerebellar dysfunction. It was this definition that drew the comment from Holmes (1939) that it was "unnecessary as it would include symptoms of different origins." Retrospectively, this seems unnecessarily parsimonious considering that many medical terms are used to refer to collective groups of symptoms of different origin. Nevertheless, the critique has found its way into many textbooks of neurology with the explanation that cerebellar asynergia can be explained by increased reaction time, dysmetria, muscle weakness, and hypotonia. Yet how accurate is this assertion in the light of modern experimental neurology? For example, Mai et al. (1988) found muscular power to be normal in patients with a wide range of cerebellar syndromes, although these same patients had a variety of deficits in controlling isometric finger forces. Although numerous studies have confirmed an increased reaction time and dysmetria after cerebellar lesions (Brooks et al. 1973; Conrad & Brooks 1974; Diener et al. 1993; Spidalieri et al. 1983), muscle weakness remains a doubtful explanation for the deficient braking action of antagonist muscles. Flament & Hore (1986) demonstrated that cooling the cerebellar nuclei produced hypermetric movements accompanied by delayed recruitment of antagonist muscles, and this observation has now been confirmed in patients with cerebellar lesions (Diener et al. 1993). However, both these studies showed that although the onset of the antagonist EMG was delayed, the amplitude was increased, which argues strongly against the suggestion that muscles are weakened.

In 1902, Babinski published another very short paper reporting that patients with cerebellar lesions show difficulty in performing a rapid succession of alternating movements. He introduced the term "adiodocokinesia" to describe a deficit in the alternate activation and relaxation of antagonist muscles. Babinski considered that asynergia and adiodocokinesia together represented a breakdown in the spatial and temporal organization of movement, which I will argue is a fundamental deficit in reciprocally activating the agonist-antagonist muscles regulating joint stiffness.

Simply put, the hypothesis advanced here has three postulates. The first suggests that the cerebellum, by repeated associations through practice, forms and stores rewarded muscle activation patterns contributing to the time-varying joint or limb stiffness values for particular movements or postures. This limb or joint stiffness is controlled both by individual muscle activation levels as well as by the degree of cocontraction between mechanically opposing muscles. The second postulate is that inhibition from separate microzones of Purkinje cells in the cerebellar cortex can relax the activity in single muscles or groups of muscles, and that multiple microzones acting together optimize time-varying changes in joint stiffness. Finally, the third postulate is that plastic changes in the cerebellar cortex result from repeated temporal associations of proprioceptive and teleceptive stimuli that act as conditioned stimuli to evoke the optimal time-varying changes in joint stiffness.

The following presentation is organized in two sections. The first reviews the evidence that muscle, joint, and limb stiffness are controlled by synergies involving mechanically opposing muscles, and the second evaluates the data suggesting a cerebellar involvement in the command strategies that control muscle, joint, and limb stiffness.

1. Controlling the mechanical properties of muscles, joints, and limbs

1.1. Clarifying the concept of synergy

Unfortunately the term synergy has become somewhat confusing due to inconsistent usage and the absence of an accepted meaning, which results in at least two different and potentially contradictory definitions (see reviews by Lee 1984; Macpherson 1991; McCrea 1992; Windhorst et al. 1991). Originally, Babinski (1899) called synergy an "association of movements" and, in a later paper, "the ability to simultaneously accomplish diverse movements which constitute a single act" (Babinski & Tournay 1913). A different use of the term, frequently attributed to Bernstein (1967), suggested that synergies were centrally organized motor programs for fixed actions. In most tasks requiring movement, the number of parameters that can be controlled by the nervous system exceeds the number needed to achieve the goal, thereby creating excess degrees-of-freedom and a problem of response selection. Bernstein (1967) hypothesized that the nervous system might simplify motor control by grouping the parameters together under the coordination of a fixed single directive that might be considered a synergic command. According to this view, synergy was the command algorithm rather than the group of muscles that implement the algorithm. As Macpherson (1988a; 1988b; 1991) has previously noted, there is little empirical evidence in favor of fixed single command synergies. Instead, synergies appear to be flexible muscle activation patterns in which any combination of linkages is potentially available.

Macpherson (1991) also suggested that the nervous system uses motor *strategies* to achieve task-relevant motor objectives requiring the control of position, force, or velocity and that are subsequently implemented as muscle synergies. One way in which the notions of fixed and flexible synergies might be brought closer together, although not entirely reconciled, would be to consider that strategies form the command algorithm employed by the nervous system to optimize muscle synergies by a process of continuous adaptive learning. Consequently, synergies would become progressively more efficient and stereotyped with the proper anticipation of ensuing teleceptive and proprioceptive stimuli. In any purposeful movement, achieving the task objective (i.e., reinforcement) is the primary criterion for optimization, although the reduction of effort (Hasan 1986), optimizing the speed-accuracy tradeoff (Fitts 1954), maximizing speed or endurance (Alexander 1989), or minimizing jerk (Hogan & Flash 1987) might be other important influences shaping the behavior. Once optimized, the synergies would remain fixed as long as the objective, the strategy, and the feedback signals remain constant, but alteration in any one of these three would initiate a new process of optimization. This paper deals with the contribution of the cerebellum to the development of motor strategies controlling muscle synergies that alter the compliance or stiffness of joints or limbs.

After reviewing the problem of synergy at considerable length, Macpherson (1991) concluded that since no natural movement involves only a single muscle, synergy could be defined as simply "a group of muscles acting together." Moreover, since mechanically opposing muscles must act together, their cooperation might be considered as a fundamental unit of muscle synergy. That is, a synergistic group should include not only the prime-mover agonist muscles, but also the antagonists that must be either inhibited, disfacilitated, or cocontracted. However, as Macpherson (1991) noted, this definition of muscle synergy is extremely broad and implies some sort of control mechanism for motor coordination without providing any insight as to what this mechanism might be. This question of how the nervous system specifies strategies of muscular activation and inhibition is currently of great interest to neurophysiologists.

The combination of biomechanical analysis and EMG recording techniques has also revealed the presence of hidden, and heretofore unsuspected, agonist-antagonist cocontraction synergies. For example, van Zuylen et al. (1988) discovered that the triceps muscle is activated during forearm supination to compensate for the flexor action of the biceps muscle. Although the triceps makes no contribution to the forearm supination, it insures that the net flexor-extensor torque at the elbow remains zero. The presence of covert synergies in addition to the heterogeneous recruitment of motoneurons during the cocontraction of agonist-antagonist muscles has led Gielen and his colleagues (Jongen et al. 1989; ter Haar Romeny et al. 1984; van Zuylen et al. 1989) to propose the existence of a centrally-driven mechanism for controlling reciprocal inhibition or coactivation of parts of the flexor and extensor motoneuron pools. Taken together, these findings suggest that although some muscle synergies may be extremely covert, they nevertheless perform important functions related to both task requirements and biomechanical constraints. In the light of these data, Macpherson's (1991) definition of muscle synergies might be further qualified to the following: A muscle synergy is *a group of muscles acting together whose actions contribute to movement efficiency and postural stability.*

1.2. The mechanical properties of a multijointed limb

Motor control neurophysiologists have been turning increasingly to the study of mechanics for clues as to how the brain controls complex limb movements. The mechanical impedance (i.e., the resistance to movement) of a single joint can be characterized by scalar values of inertia, viscosity, and stiffness. This mechanical impedance determines how much the joint will resist a perturbing force. Although the inertial component of the mechanical impedance is approximately constant, the viscous and elastic components depend on reflex gain and the amount of prior muscle activation. The mechanical properties of a multisegment limb are more complex and require a matrix of coefficients of inertia, viscosity, and stiffness.

One of the first great insights into motor control was Sherrington's (1909; 1947) discovery of the spinal reflexes controlling reciprocal inhibition. Today, both reciprocal inhibition and cocontraction are recognized as muscle activation patterns controlled by the brain, which would seem to include all skeletal muscles and even the extraocular motor system (Robinson 1981). Reciprocal inhibition and agonist-antagonist cocontraction reflect strategies of motor control that directly affect joint stiffness. Although the purpose of this strategic control is to optimize the limb's mechanical interaction with its environment, the specific objectives to be achieved may vary widely. For example, activation of a muscle as it is being stretched increases the elastic energy of the muscle, as in crouching preparatory to jumping in cats (Walmsley et al. 1978; Zomlefer et al. 1977). Cocontraction can also briefly increase the joint stiffness to optimally absorb the momentum of a visually predictable perturbation such as catching a ball (Lacquaniti & Maioli 1989; Lacquaniti et al. 1993) or absorbing the impact of ground reaction forces in the foot-contact, E_2 phase of quadrupedal locomotion (Goslow 1973). Alternatively, when the task is to stabilize unpredictable and unstable loads, the subject may resort, with some effort and risk of fatigue, to a strategy of cocontraction (e.g., Akazawa et al. 1983; Hasan 1986; Humphrey & Reed 1983; Milner & Cloutier 1993). Nevertheless, with practice, and because cocontraction is fatiguing, even unstable loads are usually transferred to one of the muscle groups whose actions are mutually opposing rather than alternating or cocontracting. For example, Clement and Rezette (1985) found that better-trained gymnasts performing handstands prefer a strategy of transferring the load to the forearm extensor muscles alone, whereas cocontraction or rapid sequential activation of the antagonists produced significantly more body sway (and probably more fatigue) in less experienced gymnasts.

A recent examination of muscle activity in bicycle pedaling has pointed out that the direction of joint rotation of individual joints is in inherent conflict with the required joint torque necessary to generate a directed force on the pedal (van Ingen Schenau et al. 1992). According to these authors, the monoarticular hip extensor muscles cocontract with the biarticular hip flexors to produce the correct balance of joint flexor torque needed to transfer force in the appropriate direction at the distal joints. This pattern of muscle activation is reminiscent of a particular type of cocontraction that occurs during the opposition of the fingers in pinching (Smith 1981). In this instance, cocontraction of the forearm flexors and extensors of the wrist and fingers increases the stiffness at the carpal and metacarpal-phalangeal joints in order to transmit forces to the tips of the fingers more effectively.

It is also important to recognize that cocontraction and reciprocal inhibition are not mutually exclusive. Agonist-antagonist activity can overlap even during reciprocal activation (e.g., the triphasic muscle activation pattern of rapid voluntary movements). In fact, because of the relatively lengthy electromechanical delay between the arrival of the action potential, the consequent change in muscle tension, and the duration of muscle twitches, cocontraction, and stiffness increases can occur even when the EMG bursts show no temporal coincidence whatsoever. An analysis of EMGs during rapid alternating movements similar to those studied by Babinski have shown a modulated reciprocal activity superimposed on a background of cocontraction to control joint stiffness (Feldman 1980a; 1980b; Humphrey & Reed 1983; Levin et al. 1992).

1.3. The control of muscle stiffness

Variations in the stiffness of individual muscles is an important factor contributing to joint or whole limb stiffness. Muscle stiffness is determined, in part, by inherent properties such as the degree of overlap between actin and myosin filaments limiting the number of cross-bridges that can be formed at a given muscle length (Gordon et al. 1966). In addition, the stiffness of an active, fully innervated muscle can be further modulated through reflex gain changes (e.g., Hoffer & Andreassen 1981; Houk & Rymer 1981; Nichols & Houk 1976; Rack & Westbury 1974), or supraspinal commands (e.g., Lacquaniti & Maioli 1989; Lacquaniti et al. 1993; Milner & Cloutier 1993). A full discussion of the reflex control of muscle is beyond the scope of this article. The major concern of the present paper is not whether the brain or the cerebellum in particular can contribute to the reflex changes in the stiffness of individual muscles, but whether a more efficient control of joint and limb stiffness can be achieved by an anticipatory command strategy dictating the degree of cocontraction in posture and movement. As will be discussed later, there is good reason to believe that feedback mechanisms (i.e., reflexes) are by themselves inadequate to control the overall stiffness of multiarticular limbs.

1.4. The control of joint viscoelastic properties

Lacquaniti and Maioli (1989) were among the first to point out that combined voluntary and reflex changes in joint stiffness and viscosity significantly reduced both the oscillation, amplitude, and damping time induced by perturbations. Postural stability results partly from the combined activities of agonist-antagonist muscles about a particular joint (Bizzi et al. 1982; Feldman 1980a; 1980b; Mussa-Ivaldi et al. 1985; Rack & Westbury 1969; 1974). This stability of a fixed position in a plane has been represented as a two-dimensional postural force field. The postural forces are springlike and can be characterized as the product of a displacement vector and a stiffness matrix. The stiffness matrix is often represented graphically as an ellipse in which the length and direction of the major and minor axes of the ellipse represent the magnitude and direction of the eigenvectors of the stiffness matrix. Limb stiffness was measured for the entire upper limb and represented both as an endpoint stiffness and a joint stiffness (Flash & Mussa-Ivaldi 1990; Mussa-Ivaldi et al. 1985; Shadmehr et al. 1993). In general, the shape and the orientation of the stiffness fields were determined by biomechanical properties such as muscle moment arms and limb postures, and were similar between subjects over time. In contrast, only small changes in the magnitude of the stiffness ellipse occurred when subjects voluntarily increased the stiffness to resist sinusoidal force-pulse perturbations (Mussa-Ivaldi et al. 1985). From a subsequent study it appeared that, to some degree, the changes in the magnitude of the single-joint stiffness ellipses were correlated with changes in agonist-antagonist EMG activities (Flash & Mussa-Ivaldi 1990). Furthermore, Shadmehr et al. (1993) found that the strength of the postural force fields decreased as the amplitude of the displacement from the original position increased.

To date, none of the studies have proven that limb or joint stiffness per se is a parameter controlled by the nervous system. Conceivably, these fields might be an incidental by-product of the activation of individual muscles. However, from a theoretical standpoint, Hasan (1986) has suggested that optimizing joint stiffness may have certain advantages in executing unperturbed displacements of an inertial load, particularly as far as reducing movement "effort" (defined as the nonreflex drive to motoneurons) is concerned. Also Hogan (1990) has pointed out the advantage of making a limb more compliant to avoid "contact instability" when a multiarticular arm of springlike actuators encounters an object in the environment. These theoretical analyses indicate that because of the delay, mechanical control cannot be adequately regulated by feedback compensation alone. Evidence supporting the notion that myotatic reflexes may be inadequate to maintain unstable loads has been provided by Milner and Cloutier (1993). They found that when particular joint forces and rotation frequencies combined to produce reflexes that were 180° out of phase with angular velocity, the mechanical instability was greatly aggravated. Apart from reflexes, the nervous system has three additional preparatory strategies for modulating limb stiffness: changing individual muscle stiffness, changing limb position, or changing the degree of cocontraction in agonist-antagonist muscles.

1.5. The control of time-varying stiffness

Measures of time-varying stiffness and viscosity depend on the amplitude, duration, and frequency of the perturbation used as well as the joint position. Nevertheless, measures under static conditions indicate that significant modulation of the viscoelastic property of joints can occur under some conditions (Hunter & Kearney 1982; Lacquaniti & Maioli 1989). Estimates of viscoelastic properties during movement have been more difficult to obtain. Indirect estimates made by Latash & Gottlieb (1991) and direct measurement by Milner (1993) suggested that dynamic stiffness was greater for fast movements than for slow movements. One of the few studies to measure joint stiffness directly both at rest and during movement was conducted by Bennett et al. (1992). Applying pseudorandom force-pulses through a wrist-mounted air jet, Bennett et al. (1992) showed that for the range of planar elbow movements and velocities, the static (i.e., postural) stiffness was greater than the dynamic (i.e., movement) stiffness. Figure 1 illustrates the time-varying changes in stiffness throughout the movement. The stiffness was lowest at the midway point and rose at the turning points from flexion to extension, from extension to flexion. These data also suggest that limb stiffness is higher in static postures, which is consistent with what is known about short-range muscle stiffness. Recently both Bennett (1993b) and Milner (1993) found time-varying changes in elbow stiffness during flexion–extension movements that increased with movement velocity. In addition Milner (1993) found that elbow stiffness was greater for viscous loads than for no-load movements of equal velocity. One unresolved difference between these two studies relates to whether stiffness was greater during posture or during movement. However, both studies agree that tuning joint stiffness to movement speed greatly facilitates the execution of movement. Whether these time-varying changes in stiffness reflect a changing control signal or merely the fortuitous sum of mechanical properties in individual muscle still remains to be determined.

2. Cerebellar function in the time-varying control of joint stiffness

2.1. Ataxia results from a feedforward not a feedback deficit

The cerebellum exerts a significant modulatory action over gamma motoneurons (Gilman 1969a; 1969b; Schieber &

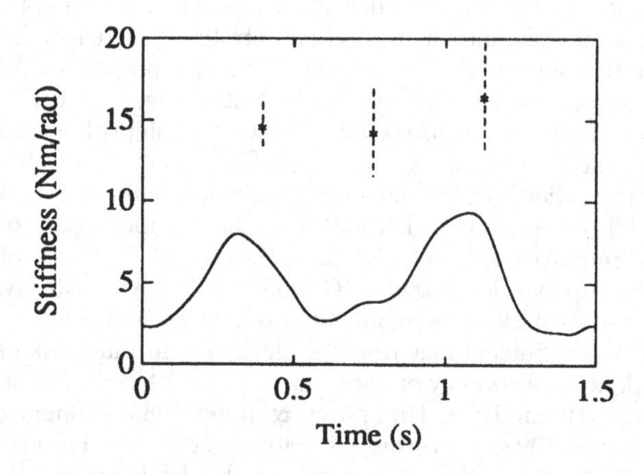

Figure 1. The mean time-varying stiffness for 300 oscillations of the elbow joint of fixed duration and amplitude. The * indicates the mean postural stiffness averaged over four trials as the subject pointed to targets located at the two extremes and midway between. The vertical bar shows one standard deviation. From Bennett et al. 1992.

Thach 1985), and the notion of cerebellar control over stretch reflex gain has been widely accepted (MacKay & Murphy 1977). However, the adequacy of a breakdown in this mechanism to explain cerebellar ataxia has recently been seriously challenged (Gorassini et al. 1993). According to this view, cerebellar ataxia is not primarily caused by deficient proprioceptive reflexes but results instead from an inability to program muscle synergies. In spite of Holmes's (1939) opinion that with respect to cocontraction of agonist-antagonists "no such disturbance exists in sufficient degree to play a part in cerebellar ataxia," many contemporary neurologists have shown that cerebellar lesions are associated with disturbances of the agonist-antagonist muscle relations. For example, Rondot et al. (1979) recorded from a variety of limb muscles and noted that the initiation of fast movements was accompanied by short EMG bursts appearing frequently and simultaneously in agonist and antagonist muscles of the shoulder and elbow in patients with cerebellar lesions. Hallett et al. (1975) found that patients with cerebellar lesions, when asked to perform ballistic elbow flexion against a tonic triceps activity, showed cocontraction of both biceps and triceps that resulted in hypometric movements. Conversely, if a tonically active biceps is suddenly released, the delayed activation of the antagonist triceps causes a hypermetric flexion (Terzuolo et al. 1973). Similarly, the poorly coordinated sequential activation of agonist-antagonist muscles in multijoint movements severely impaired the ability of cerebellar-damaged patients to throw a ball at a target accurately (Fig. 8 in Becker et al. 1990). Cocontraction also disrupts and slows rapidly alternating movements of the wrist (Diener et al. 1993). Figure 2, taken from Diener et al. (1993), compares the EMG activity of antagonist muscles during rapidly alternating wrist movements in a patient with a unilateral cerebellar lesion. Figure 2B shows that the movements are slower and more irregular, and the temporal relationship between EMG onset and the movement turning points is lost compared to the contralateral hand shown in 2A.

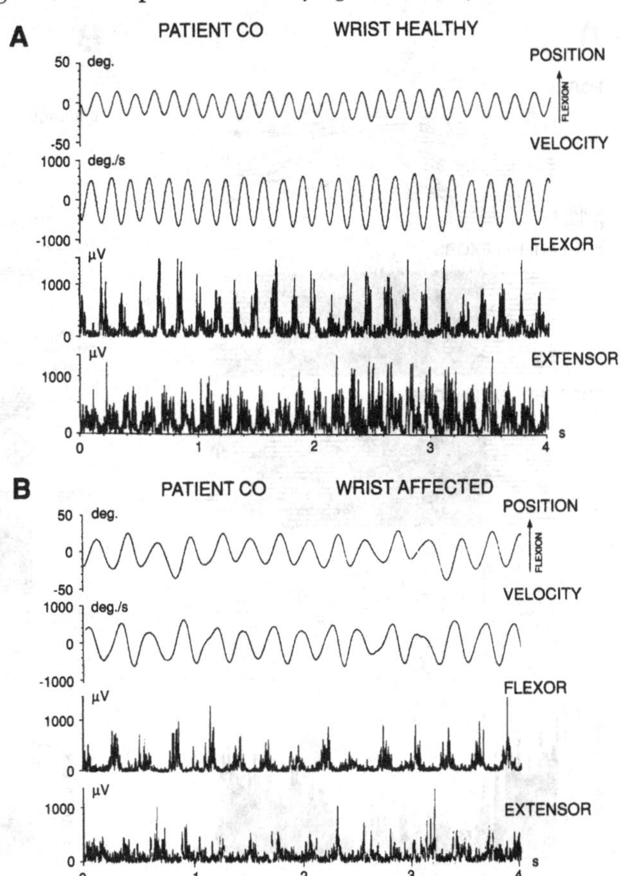

Figure 2. EMGs from wrist flexor and extensor muscles in a patient with a unilateral cerebellar lesion asked to perform rapid alternating movements. The hand ipsilateral to the lesion (B) shows slowing and irregularity of movement as well as a desynchronization between EMG onset and the movement turning points. Redrawn from Diener et al. 1993.

It is also important to emphasize that the cocontraction seen in the EMG activity of patients with cerebellar damage is quite unlike the cocontraction seen in Parkinsonian patients. Parkinsonian rigidity is essentially an increase in tonic EMG activity, whereas the cocontraction deficit of cerebellar patients is most clearly revealed in the dysfunctional time-varying modulation of activity between agonist-antagonist muscles during movement.

2.2. Cerebellar activity related to agonist-antagonist muscles

It has been suggested that the increased discharge of some Purkinje cells may be related to muscle relaxation (Smith 1981). If this were true, then one would expect the majority of Purkinje cells to decrease firing frequency during the cocontraction of agonist-antagonist muscles. To test this hypothesis, several monkeys were trained to perform two types of hand movements. In the first task, the animals performed flexion and extension movements of the wrist. In the second task, the monkeys maintained an isometric pinch of the thumb and index finger for one second. The former task was achieved by a reciprocal inhibition of antagonists, whereas the latter was accomplished by a cocontraction of both flexor and extensor muscles of the hand. About two-thirds of the hand-related Purkinje cells decreased firing in one direction of the reciprocal wrist-movement task as well as during the cocontraction associated with pinching (Frysinger et al. 1984). Figure 3 shows the decreased activity of a Purkinje cell during cocontraction and the reciprocal activity of the same cell in alternating wrist flexion and extension. The decreased Purkinje cell activity shown in Figure 3A is coincident with the cocontraction of forearm flexors and extensors. Also during extension movements, the decreased Purkinje cell activity shown in Figure 3B (right) was time-locked to flexor relaxation, whereas the increase in Purkinje cell activity lagged the onset of flexor muscle activity during flexion by approximately 250 milliseconds (Fig. 3B, left). Moreover, an analysis of the activity of 22 individual forearm muscles failed to reveal any flexor muscles with an activity profile that fit with this Purkinje cell discharge pattern (Smith et al. 1983).

In the same two tasks, over 90% of cells in the dentate and interposed nuclei increased discharge frequency during pinching, and of these 70% showed reciprocal activity during reciprocal wrist movements (Wetts et al. 1985). Figure 4 shows an example of a dentate neuron that showed increased discharge frequency during extensor muscle activation and reduced discharge during flexor muscle activity. During cocontraction however, the activity of most nuclear

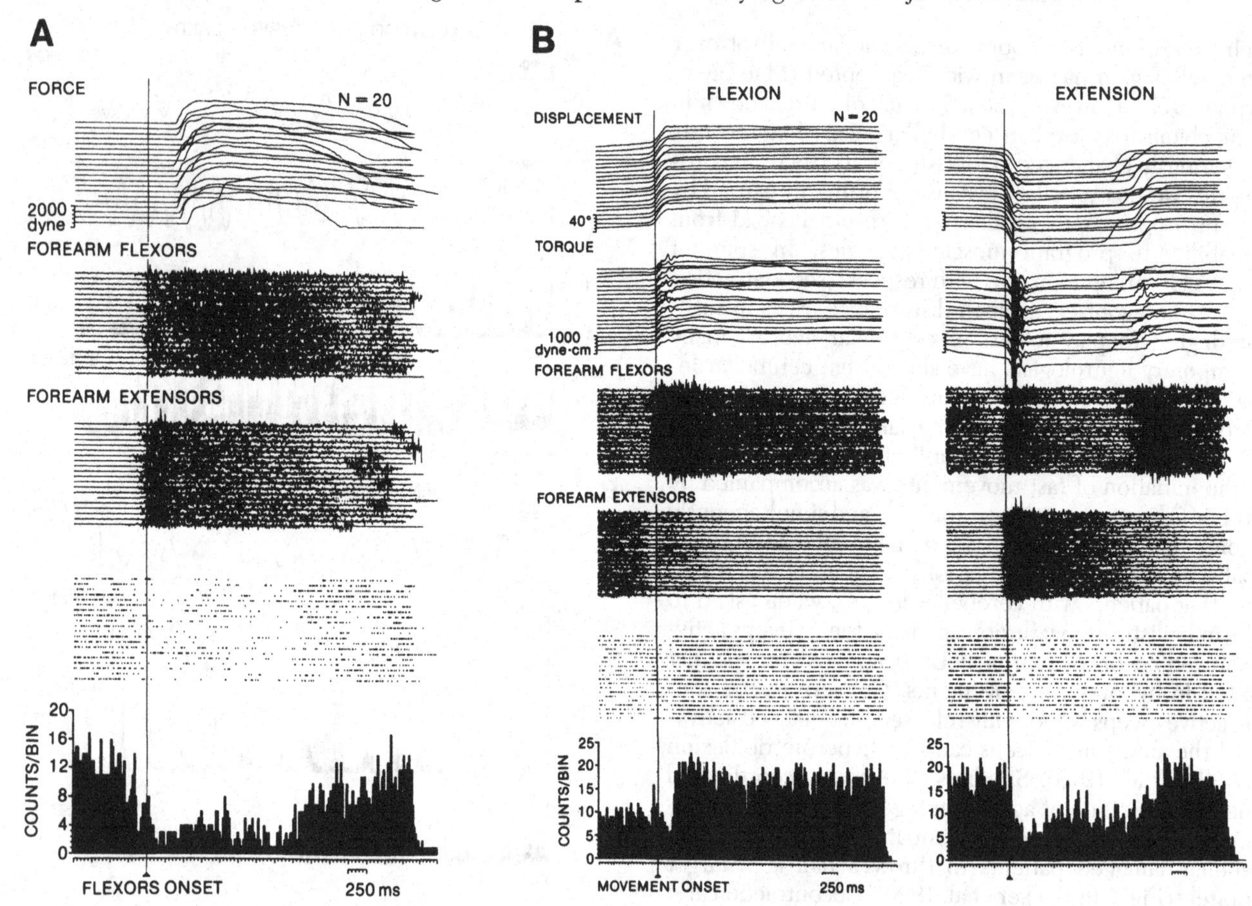

Figure 3. Activity of a single Purkinje cell recorded in a monkey during a maintained pinch accomplished by a cocontraction of the forearm flexors of the wrist and fingers shown on the left. On the right, the same Purkinje cell demonstrated reciprocal discharge during performance of a wrist flexion and extension executed by reciprocal activation of wrist muscles. From Frysinger et al. 1984.

cells – including this neuron – increased activity, in contrast to Purkinje cells, which decreased their activity during cocontraction. This cerebellar nuclear cell behavior is approximately what one would expect from neurons having a firing frequency proportional to muscle activity. That is, the activity of the nuclear cell shown in Figure 4 is probably related to forearm extensors regardless of whether the functional context is one of cocontraction or reciprocal inhibition. These two studies suggest that the output from cerebellar cortical Purkinje cells depends on whether antagonist muscles are reciprocally inhibited or cocontracted, whereas the activity of cerebellar nuclear cells does not.

In spite of these differences, the cerebellum may not be the only supraspinal structure involved in reciprocal inhibition and agonist-antagonist cocontraction. In a similar study of monkeys performing two tasks, one involving reciprocal inhibition and the other agonist-antagonist cocontraction, Humphrey and Reed (1983) found two spatially separated populations of motor cortical neurons that were preferentially activated with either reciprocal muscle control or with agonist-antagonist cocontraction respectively. These data are an important indication that other, perhaps all, motor system structures may play a role in setting joint stiffness.

A second series of experiments was undertaken to examine the effect of the loss of Purkinje cell inhibition on the simple reciprocal organization of ankle muscle antagonists during treadmill locomotion. The mutant mouse Lurcher has a progressive cerebellar cortical atrophy such that no Purkinje cells survive beyond early adulthood. Nonetheless, the Lurcher mouse has a nuclear cell density and volume that is essentially normal (Caddy & Biscoe 1979). Additional degeneration of granule cells and cells of the inferior olive have been shown to be secondary to the loss of Purkinje cells (Wetts & Herrup 1982a; 1982b).

Using a shuttered video camera, the locomotion of Lurcher mice walking on a treadmill was studied and EMGs from two antagonist muscles of the ankle, the anterior tibial and the triceps surae, were recorded (Fortier et al. 1987). The video analysis demonstrated a significant uncoupling of the movements between the fore and hind limbs on the same side of the body. A similar aberrant phase relation was found between the two hind limbs, which normally should have been 180° out of phase since the mouse, rarely, if ever, gallops. Figure 5 shows that the start of flexor activity with respect to the time the foot contacted the ground was highly irregular, and the reciprocal behavior, particularly of the extensor muscles, appeared to have been replaced by a modulated cocontraction. The locomotion was very ataxic and the frequent interruptions from a loss of equilibrium accounted for the absence of modulation in the contralateral limb, also seen in Figure 5. In general, mice without Purkinje cells show deficits in both the ability to simultaneously (e.g., asynergia) and sequentially (e.g., dysdiadocokinesia) command the desired muscle synergies. As Babinski (1899) once remarked "It is in locomotion that the cerebellar asynergia is most evident."

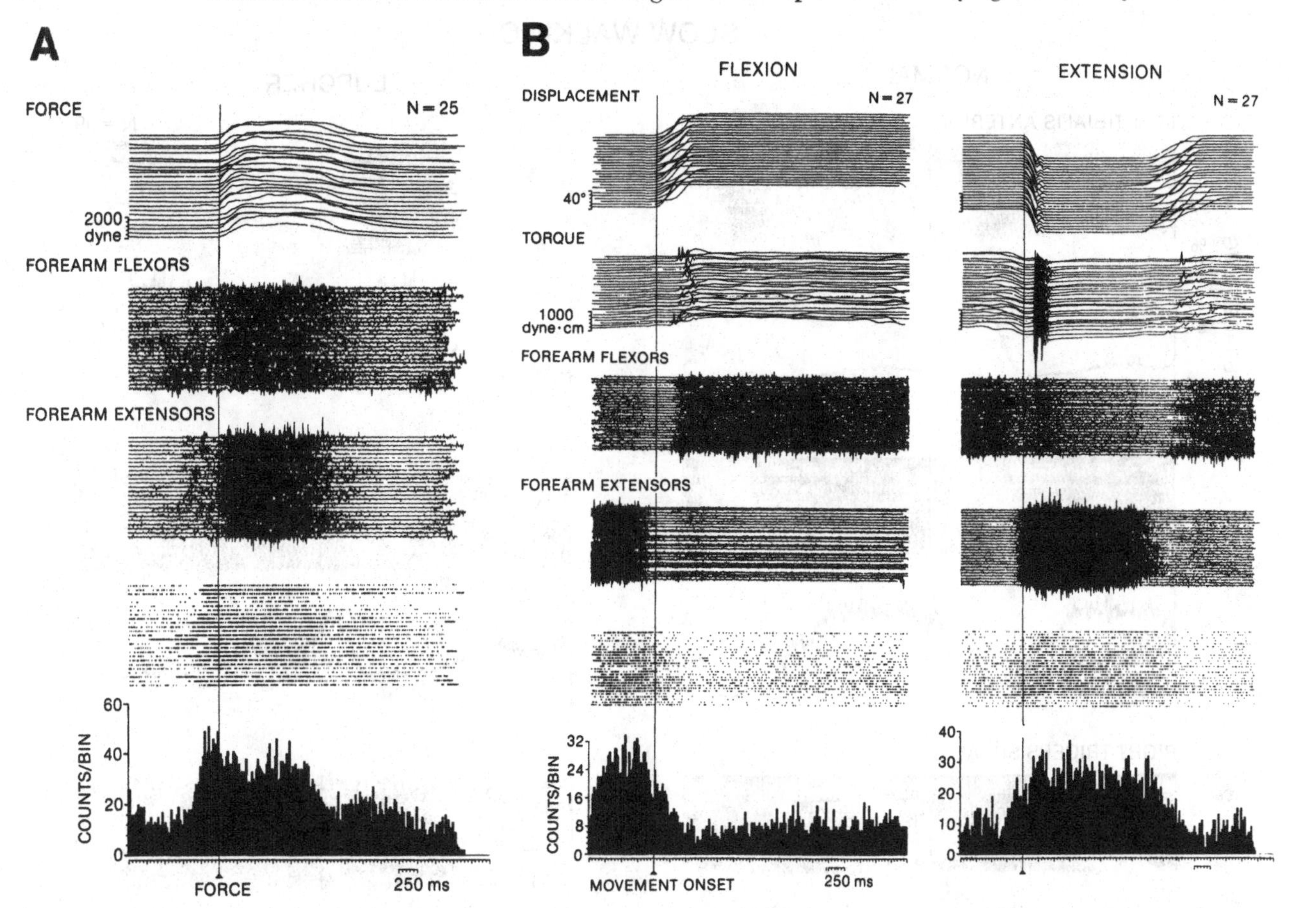

Figure 4. Activity of a single dentate cell recorded in a monkey performing a maintained pinch accompanied by a cocontraction of the forearm flexors of the wrist and fingers shown on the left. On the right, the same dentate cell demonstrated a reciprocal discharge during performance of a wrist flexion and extension executed by reciprocal activation of wrist muscles. From Wetts et al. 1985.

The activity of simple spikes recorded in Purkinje cells during locomotion are also consistent with (but do not prove) a role for the cerebellum in the time-varying control of limb stiffness. For example Armstrong and Edgley (1984) found that Purkinje cells were more active during locomotion than at rest, and that the Purkinje cell population as a whole was more active during the swing phase compared to the stance phase. Although most antagonist muscles are reciprocally active in locomotion, there are occasional brief exceptions. For example, Udo et al. (1981) found that Purkinje cell activity related to extensor muscles peaked during the E_1 phase to prevent excessive extensor stiffness (i.e., to increase limb compliance) at the moment of foot contact with the ground.

In spite of the circumstantial evidence favoring a reciprocal behavior between cerebellar and nuclear neurons, this has never been adequately proven by simultaneous recording between connected pairs of Purkinje cells and their target nuclear neurons. It is quite possible that parallel modulation can also be present under some circumstances, although such activity would moderate the net effect on joint stiffness.

2.3. Cerebellar activity related to reaching movements

The cerebellum is thought to play a more important role in compound, multiarticular movements than in simple, single-joint movements. Single cell recording studies during limb movements in primates have yielded somewhat controversial results. Some investigators failed to find any consistent discharge related to the direction of movement (MacKay 1988; Mano & Yamamoto 1980; Schieber & Thach 1985), whereas others (Gibson et al. 1990) have stressed that hand movements produced stronger modulation than whole-arm reaching. In spite of these reports, ataxic reaching movements and deficits in visuomotor tracking are consistent findings with either degenerative cerebellar disease or experimental inactivation (e.g., Becker et al. 1991; Miall et al. 1987).

Our own studies were intended to determine whether the discharge of cerebellar neurons in awake monkeys trained to execute visually triggered whole-arm pointing movements is direction-related or not. A monkey seated before a two-dimensional work surface was trained to move a pendulum from a central start position to one of eight radially arrayed targets in response to a visual cue. Movements to each target was randomized and repeated five times. Most shoulder area neurons in both the cerebellar cortex and cerebellar interpositus and dentate nuclei had a single peak of activity grouped about movements in a particular direction, despite that fact that cerebellar neurons discharged with movements over a wide range of directions (Fortier et al. 1989).

We also compared cerebellar and motor cortical unit activity in monkeys trained to perform the same multijoint pointing task (Fortier et al. 1993). The most salient differ-

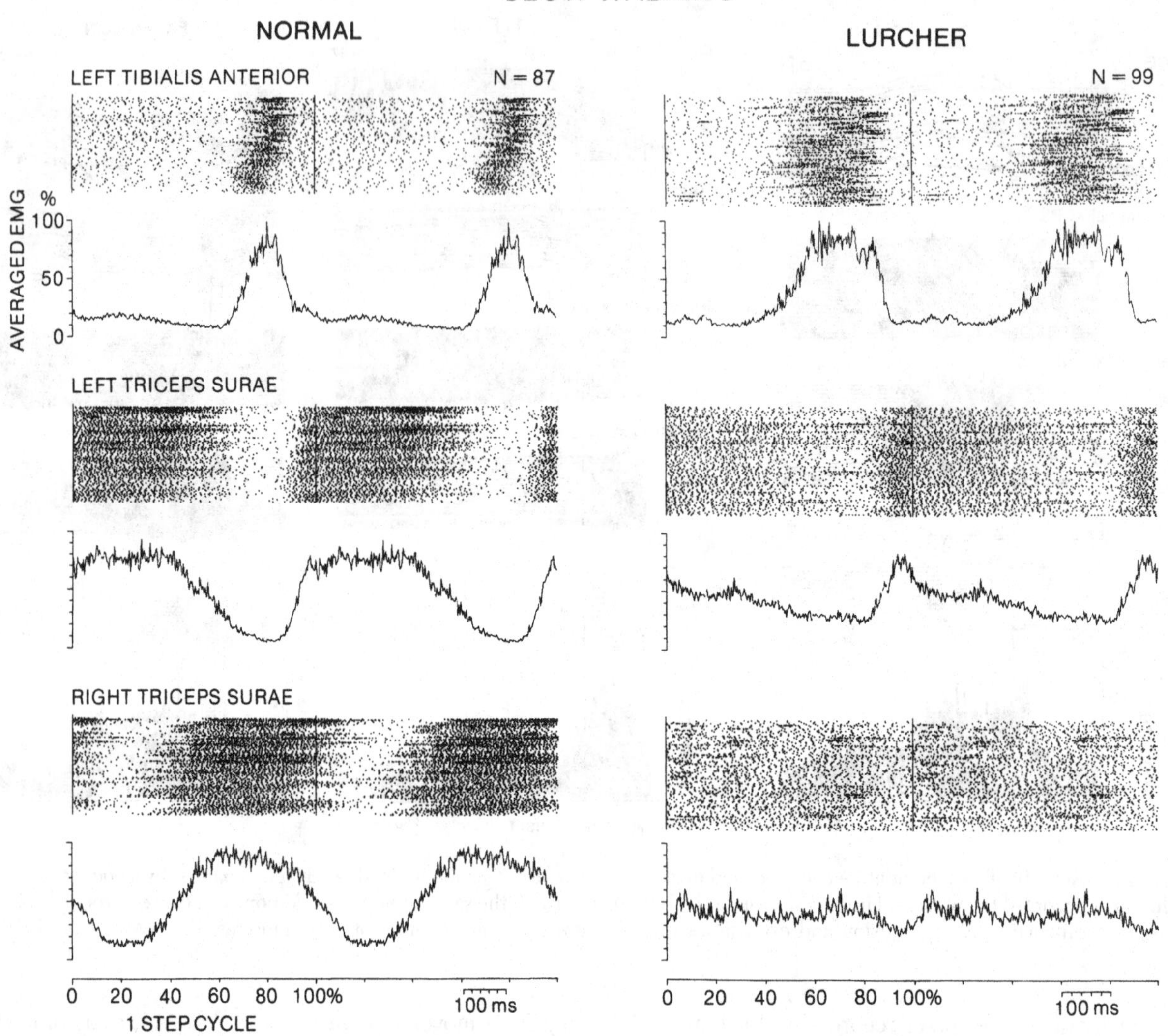

Figure 5. Activity of the ankle muscles during walking in normal and Lurcher mice. The EMGs are displayed as both rasters and summed activity profiles. Each step cycle was normalized to 100% from foot contact to foot contact. In the Lurcher, the left triceps surae shows little reciprocal inhibition and the activity of the right triceps surae is desynchronized with respect to the left footfall. From Fortier et al. 1987.

ence between the activities of motor cortex and cerebellum lay in the greater percentage of "graded" discharge patterns found in the cerebellum compared with a greater number of "reciprocal" patterns found in the motor cortex. A majority (58%) of motor cortical cells had reciprocal discharge patterns, with the greatest activity for movements in the preferred direction and significant inhibition for movements in the opposite direction. In contrast, about 70% of cerebellar neurons had graded increases in activity for all movement directions, but with a single peak direction. In particular, Purkinje cells showed a step-like increase in activity for all eight movement directions (compared to the stationary posture prior to the go stimulus) in addition to a modulated increase in activity for movements approximating the preferred direction (see Fig. 9B in Fortier et al. 1989).

To what parameters of motor control might this Purkinje cell activity be related? Although the Purkinje cell excitability co-varies with movement direction, it seems unlikely that this activity is correlated with kinematic parameters such as position, velocity, or acceleration. A direct relationship to either dynamic parameters, such as muscle force or rate of force change, or to the underlying EMG amplitude, seems equally unlikely because there is no adequate explanation for the step-like increase in Purkinje cell activity for all directions of reaching. However, the inhibitory action of Purkinje cells on the cerebellar deep nuclei, which are closely related to muscle tone, would be consistent with reducing limb stiffness during movements as opposed to active holding (Bennett et al. 1992).

2.4. The synchronization of movement and posture

As mentioned earlier, initial posture has a significant impact on the forces required for movement because of muscle movement arms and the length-tension properties of muscle. Therefore, both Hogan (1990) and Massion (1992) have postulated the need for a control system to maintain equilibrium and compensate for reaction torques in multiarticular movements. Massion (1992) further pointed out that a semi-independent postural control system can also be used to preserve certain body parts in relation to each other while allowing other segments to move, as with maintaining head-to-trunk position or eye-to-hand positions. Although Massion enumerated various reasons for believing that posture and movement may be controlled by partially independent systems, normally these two must be very closely integrated to produce well-coordinated movements. In general, postural adjustments and movements are triggered either simultaneously by external perturbations of body equilibrium, or in the case of self-initiated movements, postural adjustments occur in anticipation of equilibrium changes (Cordo & Nashner 1982; Horak et al. 1984; Lee et al. 1987).

Although the basal ganglia (Viallet et al. 1987) and premotor areas (Viallet et al. 1992), also play a role in postural adjustments, lesions of the cerebellum produce significant postural deficits (Diener et al. 1990; Viallet et al. 1994). Horak and Diener (1994) described normal EMG latencies and normal spatial patterns of muscle activation accompanying the postural responses to an imposed rotation of the support surface in cerebellar patients. The response amplitude of these reactive postural adjustments were excessively large but nevertheless appropriately scaled to the rotational velocity of the support surface. The other postural deficit described by these authors was an inability of cerebellar patients to use prior experience in order to scale their postural responses to the amplitude of the platform displacement.

2.5. Sensory afferents to the cerebellum serve as conditioned stimuli

The cerebellum receives a wide variety of sensory inputs including proprioceptive, cutaneous, vestibular, visual, and auditory stimuli arising from both the multiple sources of mossy fiber afferents and the single source climbing fiber inputs from the inferior olive. The literature on these sensory inputs to the cerebellum has been thoroughly reviewed recently by Stein and Glickstein (1992). All movements occur within a sensory context, and movements that produce unexpected results or encounter perturbations give rise to error signals that are conveyed back to the cerebellum by fast feedback pathways triggering both error awareness and long-loop corrective reflexes. However, rapid, well-practiced voluntary movements are generally too fast for online error correction (Lashley 1951), and the phase lag of reflex feedback would be inappropriate to stabilize some limb oscillations (Milner & Cloutier 1993). Consequently, a feedforward control system is preferable to a control system based on feedback involving detrimental delay. In such a model, feedback drives an adaptive, inverse-dynamics control strategy that models the limb and its interaction with the environment with increasing accuracy. Learning theory suggests that two or more simultaneously active neural activities will tend to become associated with one another and when one activity is elicited, the associated activities will also be evoked (Hebb 1949).

In the cerebellum, the sensory afferents could serve as conditioned stimuli for cueing associated and rewarded muscle synergies, which would set critical values of joint stiffness by the same principals of learning theory (stimulus-to-stimulus or stimulus-to-reward associations) as those involved in classical or instrumental conditioning.

A visual stimulus, such as the sight of the ball dropping in the Lacquaniti and Maioli (1989) catching experiment, is sufficient (with practice) to trigger anticipatory EMG activity. Similarly, auditory stimuli have been known to trigger anticipatory EMG activity since the classical conditioning experiments of Pavlov. However, the demonstration by Rossignol and Melvill Jones (1976) that movement precedes the musical beat in dancing is perhaps a better example not only of auditory triggered EMG activity, but also of time-varying changes in joint stiffness as well. One would predict that cerebellar-damaged patients would perform equally poorly at both ball catching or dancing because of the inability to associate a muscle control strategy with the requisite teleceptive stimuli.

Evidence is now increasing that the cerebellum plays a role in functions affecting motor control indirectly as well as directly. For example, patients with cerebellar lesions have difficulty in making accurate duration estimates of auditory stimuli (Ivry & Keele 1989). Although the reason for these deficits is unclear, a difficulty in predicting the duration of teleceptive stimuli is more likely to be the cause of motor dysfunction rather than the result of it. This is shown, for example, by the impaired velocity perception of patients with cerebellar lesions (Ivry & Diener 1991). Because the rapid execution of movement sequences requires the precise timing of EMG activity based on the anticipation of teleceptive and proprioceptive cues, the inability to anticipate these stimuli accurately could lead to serious movement timing errors (Ivry et al. 1988) and deficits in organizing sequential movements (Inhoff et al. 1989).

The inability of cerebellar patients to use prior experience in order to scale their postural responses reported by Horak and Diener (1994) is a particularly relevant example of how the cerebellum uses proprioceptive cues as conditioned stimuli. In these experiments, the perturbation amplitude could not be fully appreciated by the subjects until the termination of the perturbation. The subjects were therefore required to use prior experience in order to scale their responses, and cerebellar damaged patients were particularly deficient.

Another example of this function is illustrated by single-cell recordings of Purkinje cell discharge in alert monkeys subjected to a series of predictable force-pulse perturbations while trying to maintain a stable hand position. Many Purkinje cells in the anterior lobe have proprioceptive or cutaneous receptive fields on the hand, and their activity is strongly modulated during grasping and holding movements of the hand as well as perturbation of the hand position. Figure 6 illustrates the average discharge from five-trial blocks during a preperturbation control period (A), the first five perturbed trials (B), and five trials after adaptation to the 100-millisecond downward force-pulse (C). The reflex responses appeared immediately upon the first perturbation (B), whereas the anticipatory discharge emerged progressively with repetition. The anticipatory discharge and preparatory increase in grip force preceding the perturbation were visible in Figure 6B even during the first five trials. However, both the anticipatory cellular discharge and the grip-force increase commenced earlier after repeated perturbations. In addition, the perturbation produced a smaller change in position after 10 trials (Fig. 6C) and the reflex response, which probably functioned as an error signal, decreased as the adaptive responses became more efficient. During extinction, the reverse was true. The

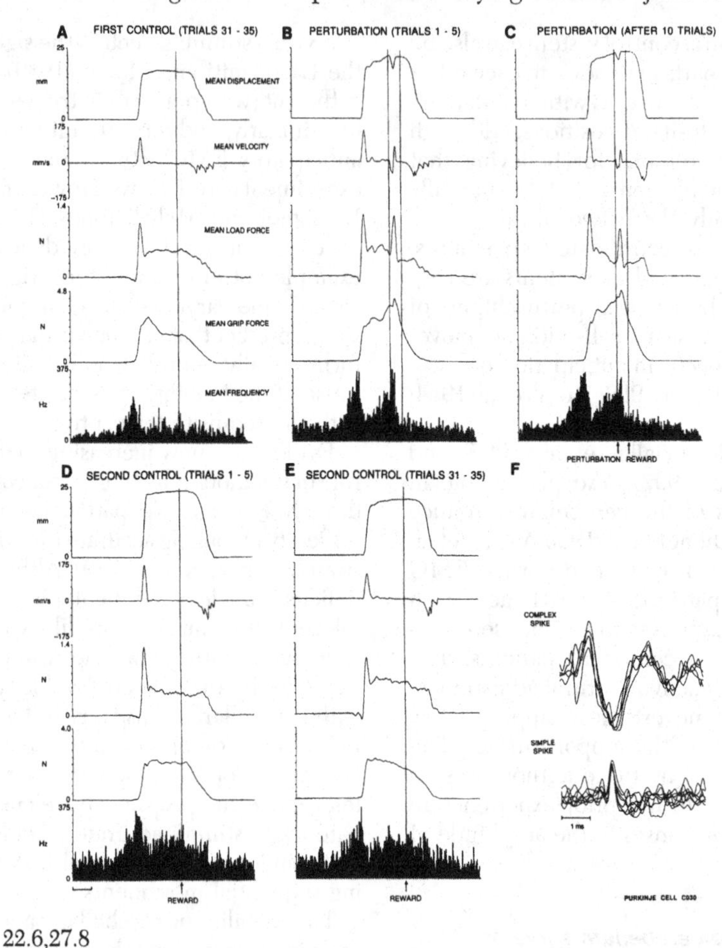

22.6,27.8

Figure 6. Acquisition and extinction of an anticipatory response in a Purkinje cell associated with a predictable force-pulse perturbation applied to the hand during position stabilization. Both the behavior and the associated discharge pattern changed gradually during the acquisition and extinction of the anticipatory response. From Dugas & Smith 1992.

reflex response ceased immediately once the perturbation had been withdrawn (Fig. 6D), whereas the anticipatory activity diminished progressively for many trials and had not completely disappeared after 35 extinction trials (Fig. 6E). The anticipatory activity shown in Figure 6 clearly paralleled the preparatory EMG activity, but the relation between Purkinje cell discharge and stiffness of the wrist and fingers joints is admittedly moot in this example. However, increased stiffness was the primary objective of the task since it allowed the animal to compensate for the perturbation and obtain its reward.

2.6. What does the cerebellum learn?

It is suggested the cerebellum learns by forming and storing associated muscle activation patterns for the time-varying control of limb or joint stiffness (i.e., muscle strategies) that contribute to movement efficiency and postural stability. A review of the evidence supporting cerebellar involvement in the formation and control of muscles synergies was recently presented by Thach et al. (1992; 1993) and will therefore not be restated except to reiterate the potential importance of the long parallel fiber system in establishing facilitatory associative connections between different mediolateral regions of the cerebellum that control muscle groups of different parts of the body. Similarly, evidence that the cerebellum associates teleceptive, particularly visual, information as conditioned stimuli for the sequential organization of movement as been reviewed recently by Stein and Glickstein (1992). The arguments from both these reviews appears to be compatible with the notion that the cerebellum plays an important role in learning muscle synergies.

However, the second part of the thesis – that the optimal time-varying modulation of joint stiffness described above is controlled in part by the cerebellum – requires further justification. First of all, it is important to stress that this hypothesis would include the cerebellar contribution to segmental reflexes causing changes in the stiffness of individual muscles. Moreover, it is the reflex pathways that provide the proprioceptive feedback necessary to developing an anticipatory strategy as a more efficient means to control multi-articular limbs with viscoelastic muscles. It is hypothesized that the cerebellum provides the muscle coactivation strategies for optimizing joint or limb stiffness values for particular movements or postures. The degree of reciprocal inhibition of antagonist muscles could be increased or decreased by greater or lesser Purkinje cell activity. It is well known that functionally related microzones of Purkinje cells impinge on nuclear cells that excite the motoneurons of functionally agonist muscles over the long descending motor pathways (Andersson & Oscarsson

1978; Ito 1984). Increasing Purkinje cell inhibition of tonic nuclear excitation to the descending motor pathways could decrease motoneuron excitability through the known segmental interneuronal circuits to provide reciprocal relaxation of antagonist muscles. Basket cell inhibition of "off-beam" Purkinje cells (Eccles et al. 1967) could accomplish this function. During agonist-antagonist cocontraction the "on-beam" Purkinje cells would also be inhibited, thus initiating a chain of disinhibition culminating in cocontraction. Similar but more mathematical models incorporating these features have been described recently by Gomi and Kawato (1992), Houk and Barto (1992), and Kawato and Gomi (1992).

This hypothesized time-varying control of limb or joint stiffness has two additional implications. First, an important function of the cerebellum would be to provide an essential postural stabilization by appropriately anticipating the required dynamics of incipient movements. Second, the cerebellum could learn to use available teleceptive and proprioceptive stimuli as conditioned triggering cues (Stein & Glickstein 1992), developing the necessary feedforward commands needed to control the mechanical impedance of multijoint limbs (Kawato & Gomi 1992). Limb or joint stiffness must be optimized in a time-varying, moment-to-moment manner in order to achieve smooth coordinated movements, and any failure to achieve adequate stiffness will result in ataxia.

2.7. How the cerebellum might control joint stiffness

The inhibitory action of Purkinje cells and therefore the entire output of the cerebellar cortex on deep cerebellar nuclei is an important element of cerebellar physiology (Ito & Yoshida 1966; Ito et al. 1966; 1968). Furthermore, with the significant exception of the nuclei-olivary neurons (Nelson & Mugnaini 1989), the entire cerebellar nuclear outflow exerts a tonic excitatory action on descending brain stem motor pathways and the ascending thalamocortical system (Massion 1973). Although some earlier studies of electrical stimulation of the interpositus nuclei had suggested a preferential action on flexor limb muscles, a more recent study in awake monkeys found that low-intensity stimulation from a single point in the dentate nucleus could evoke responses in multiple muscles in both fore and hindlimbs; these responses were thought to represent muscle synergies (Rispal-Padel et al. 1982). Although not commented upon by the authors, cocontraction of agonist-antagonist muscles is also clearly evident in their illustrative EMGs (Rispal-Padel et al. 1982). In contrast, stimulation of the cerebellar cortex inhibits nuclear cell activity and produces a disfacilitation of spinal motoneurons (Llinás 1964). However, the effect of cerebellar cortical or nuclear stimulation on joint or limb stiffness has yet to be determined.

Some years ago it was suggested that Purkinje cells might function like supraspinal Ia inhibitory interneurons (Smith 1981). By inhibiting deep cerebellar nuclear cells, groups of Purkinje cells (e.g., microzones) could evoke a cascade of effects culminating in the disfacilitation of motoneuron pools, resulting in antagonist muscle relaxation. It is further speculated that the Purkinje cells themselves must be inhibited during the voluntary cocontraction of agonist-antagonist muscles. According to the sequence of actions proposed schematically in Figure 7, two populations of Purkinje cells acting together would be needed to execute

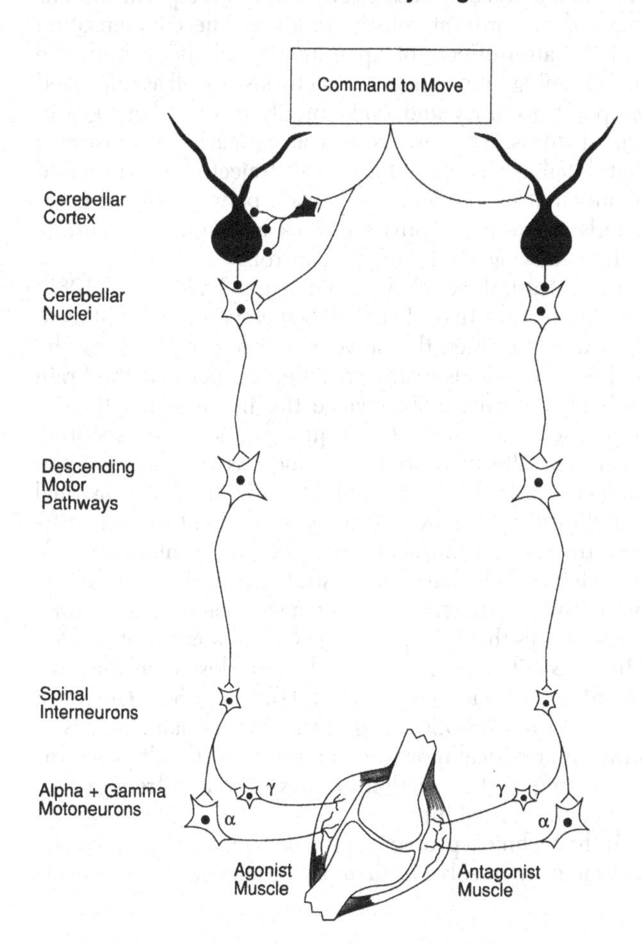

Figure 7. Two populations of Purkinje cells acting together upon agonist-antagonist muscle groups could facilitate either reciprocal inhibition or antagonist cocontraction depending on the command to move originating outside the cerebellum.

the reciprocal switching of excitation from one agonist group to another, or alternatively, to allow agonists and antagonists to cocontract, depending on the desired goal of the task. Nevertheless, Purkinje cells respond to a much greater variety of afferents and it would be overly simplistic to consider cerebellar Purkinje cells as merely ectopic spinal Ia inhibitory interneurons.

2.8. Measuring stiffness and testing the hypotheses

The speculation that the cerebellum controls the time-varying modulation of joint stiffness would be a more readily testable hypothesis were it not for the fact that stiffness is a difficult parameter to measure. To make matters worse, the force-pulse perturbations that are used to cause changes in position are almost certain to evoke reflexes and compensatory strategies. That is, the act of measurement is disruptive to the ongoing motor strategy. The application of either small pseudorandom perturbations during movement such as employed by Bennett et al. (1992) or, alternatively, tonic bias forces (Bennett 1993a; 1993b) would seem to have the best chance of success. A more elaborate example of the application of tonic bias forces was explored by Lackner and DiZio (1994), who found that subjects could adapt visually triggered pointing movements to compensate for coriolis force perturbations induced by constant velocity rotation. The compensation and the after-effects of opposite sign included both the straightening of movement trajectories as well as improved endpoint accuracy and undoubtedly involved changes in arm stiffness. This experiment also clearly demonstrates that detailed aspects of movement trajectories are revised by moment-to-moment comparison between motor commands and proprioceptive feedback and supports the utility of time-varying joint stiffness control.

In contrast, the evidence from Mussa-Ivaldi et al. (1985) and Flash and Mussa-Ivaldi (1990) casts some doubt as to the extent to which the nervous system can influence the limb stiffness fields to any great degree, because the brain and spinal cord cannot change the inherent mechanical properties of the limb. Consequently, the nervous system must find alternate strategies, such as by changing the initial or starting posture of a limb or the level of individual muscle stiffness or by regulating whole joint or limb stiffness through cocontraction and reciprocal inhibition. Although it is likely that all three strategies are used at various times, the fatiguing nature of agonist-antagonist cocontraction requires that it be brief (e.g., Clement & Rézette 1985; Milner & Cloutier 1993) and demand less than maximal voluntary contractions (Tyler & Hutton 1986). However, lesson 3 from robotics suggest that even small changes in stiffness at critical moments can resolve difficult problems of intersegmental reaction torques, resonant frequencies, and contact instability.

If the technical problem of stiffness measurement can be satisfactorily resolved, then a number of experiments should be able to establish whether the cerebellum plays any role. The displacement of loads made unstable by supplying positive position or velocity feedback, which evokes voluntary cocontraction as a compensatory strategy, offers one promising approach (DeSerres & Milner 1991; Milner & Cloutier 1993); the addition of tonic bias forces (Bennett 1993a; 1993b; Lackner & DiZio 1994) offers another. Recording single Purkinje cell activity during the displacement of stable and unstable loads should be adequate to demonstrate differences related to changes in stiffness. Increasing movement velocity should also increase stiffness as well as the frequency of rapid alternating movements. Alternatively, patients with cerebellar damage or animals subjected to reversible cerebellar inactivation should show deficits in stiffness regulation.

In an earlier BBS target article on cerebellar function, Bloedel (1992) suggested that, because of its homogeneous structure, the cerebellar cortex should perform a unitary neuronal operation applicable to the entire musculature. The hypothesis that the cerebellum implements real-time muscle control strategies over joint and limb stiffness would appear to satisfy at least this one particular requirement.

ACKNOWLEDGMENTS

The research conducted by the author was supported by a grant from the Medical Research Council of Canada to the Groupe de recherche en sciences neurologiques of the Université de Montréal. I am grateful for the honest criticisms of H. A. Buchtel, T. Drew, J. F. Kalaska, T. E. Milner, and Stephen Scott as well as comments from several helpful BBS reviewers.

8

On the specific role of the cerebellum in motor learning and cognition: Clues from PET activation and lesion studies in man

W. T. Thach
Department of Anatomy and Neurobiology, Washington University School of Medicine, St. Louis, MO 63110
Electronic mail: *thachw@thalamus.wustl.edu*

Abstract: Brindley proposed that we initially generate movements "consciously," under higher cerebral control. As the movement is practiced, the cerebellum learns to link within itself the context in which the movement is made to the lower level movement generators. Marr and Albus proposed that the linkage is established by a special input from the inferior olive, which plays upon an input-output element within the cerebellum during the period of the learning. When the linkage is complete, the occurrence of the context (represented by a certain input to the cerebellum) will trigger (through the cerebellum) the appropriate motor response. The "learned" movement is distinguished from the "unlearned" conscious movement by its now being automatic, rapid, and stereotyped. The idea is still controversial, but has been supported by a variety of animal studies and, as reviewed here, is consistent with the results of a number of human PET and ablation studies. I have added to the idea of context-response linkage what I think is another important variable: novel combinations of downstream elements. With regard to the motor system and the muscles, this could explain how varied combinations of muscles may become active in precise time-amplitude specifications so as to produce coordinated movements appropriate to specific contexts. In this target article, I have further extended this idea to the premotor parts of the brain and their role in cognition. These areas receive influences from the cerebellum; they are active both in planning movements that are to be executed and in thinking about movements that are not to be executed. From recent evidence, the cerebellar output extends even to what has been characterized as the ultimate frontal planning area, the "prefrontal" cortex, area 46. The cerebellum thus may be involved in context-response linkage, and response combination even at these higher levels. The implication would be that, through practice, an experiential context would automatically evoke a certain mental action plan. The plan would be in the realm of thought, and could – but need not – lead to execution. The specific cerebellar contribution would be one of the context linkage and the shaping of the response, through trial and error learning. The prefrontal and premotor areas could still plan without the help of the cerebellum, but not so automatically, rapidly, stereotypically, so precisely linked to context, or so free of error. Nor would their activities improve optimally with mental practice.

Keywords: cerebellum; cognition; mental movement imagery; motor learning; planning; sequence; timing

1. Introduction

1.1. The traditional position of the cerebellum in the hierarchic organization of the central nervous system

The functions of the cerebellum have been obscure, and, accordingly, speculation about them rather free. But the last century has seen the development of a systematic approach to the understanding of brain structure and function. The steps in the approach include: (1) connectivity – could the known circuitry support the functions proposed? (2) ablation – is the function abolished or impaired by removal of the part in question? (3) stimulation – can the function be evoked by stimulating the part in question? (4) natural activation – does naturally increased activity of the part in question correlate with the behavior in question?

The fundamental organizational plan of the central nervous system in general and the motor system in particular is hierarchical. Hierarchic structuring is evident in its anatomy, phylogeny, ontogeny, and in the effect of discharging and destroying lesions (Jackson 1870). This principle assumes fresh and crucial significance in understanding cerebellar functions.

In Jackson's scheme, the lowest level in the hierarchy is the spinal cord (Fig. 1); the motor neuron is the final common pathway for all motor commands to muscle. Nowadays, spinal interneurons are thought to control muscle force, length, stiffness, and viscosity and provide a compliant interface with the environment. Withdrawal and crossed extension reflexes provide protective responses to noxious stimuli, and also the substrate for locomotion. When set to oscillate, the reflex circuits generate the rudiments of walking and running behavior.

The middle level consists of a number of motor pattern generators, each projecting down to the spinal cord mechanisms to provide movements with a specific goal. Vestibulo-, reticulo-spinal, and tonic neck reflexes provide the

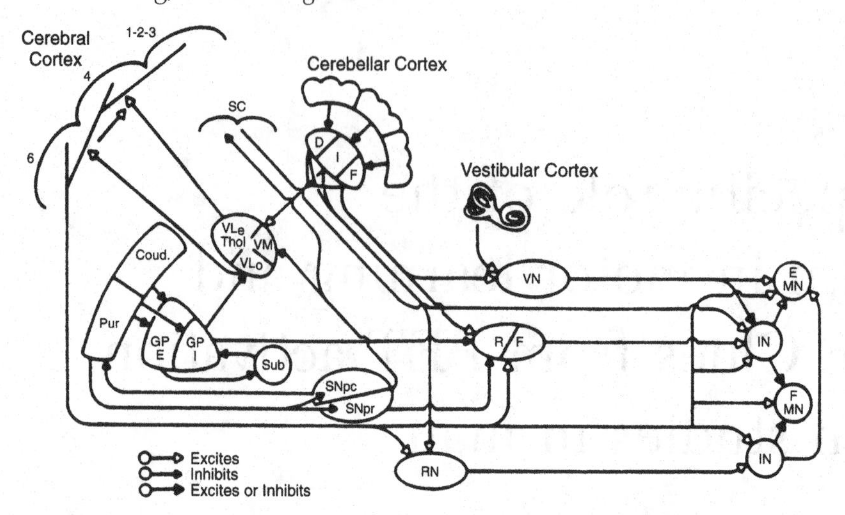

Figure 1. Descending pathways to the spinal cord and their origins in brainstem, cerebellum and cerebrum. Caud, caudate; Put, putamen; GPe, globus pallidus, external segment; GPi, globus pallidus, internal segment; Thal, thalamus; VLc, caudal ventrolateral nucleus; VLo, oral ventrolateral nucleus; VM, ventromedial nucleus; Sub, subthalamic nucleus; D, dentate nucleus; I, interposed nucleus; F, fastigial nucleus; SNpc, substantia nigra pars compacta; SNpr, substantia nigra pars reticulata; RN, red nucleus; RF, reticular formation (e.g., reticular nucleus of the pontine tegmentum); VN, vestibular nuclei; IN, interneuron; E MN, extensor motoneuron; F MN, flexor motoneuron (from Thach & Montgomery 1990, p. 170).

automatic anti-gravity component of upright stance and locomotion. The superior colliculus provides automatic orienting movements of eyes, head, neck, and possibly limbs to visual, auditory, and somatosensory stimuli. Cortico- and rubrospinal neurons provide volitional movement of individual muscles, particularly of face, fingers, and toes.

The highest level of motor control in this scheme consists of those parts of cerebral cortex associated with thinking and knowing. The back half of the brain – the occipital, temporal and parietal lobes – are devoted to higher order processing of vision, audition, and somesthesis. The front half – the frontal lobe – is devoted to movement and motivation. In the human brain, the left hemisphere specializes in language analysis, the right, spatial relationships within self and extrapersonal space. The upper portions of the parietal lobes partake in localizing objects as potential targets for movement, the lower portions and posterior temporal lobes in recognizing the objects. The upper portions of the temporal lobe partake in language and spatial relations on the left and right sides, respectively; the lower portions, declarative memory. Cognition is the product of associations between the secondary receiving and sending areas.

1.2. The traditional role of the cerebellum

In animals, cerebellar ablation by Rolando (1809; 1823), Fluorens (1824), and especially Luciani (1891; 1915) clearly showed that what the cerebellum *was* involved in was the control of posture and movement. The studies of human cerebellar ablation by Babinski (1899; 1906), and especially Holmes (1917; 1922a,b,c,d; 1939) confirmed these observations, and appeared also to determine what the cerebellum *was not* involved in: sensation (except weight discrimination, which requires movement), perception, attention, learning, memory, mood, apperception, language – that is to say, all forms of cognition (Holmes 1917; 1922a,b,c,d; 1939). As to the specific nature of the cerebellar contribution to movement, "fine control" was suggested because large focal cerebellar lesions often caused only the slightest (recognized) deficits in behavior; "coordination," because (to some observers) compound movements appeared to be more affected than simple (Babinski 1899; Fluorens 1824; but see Holmes 1939). Since then, the traditional teaching in neurology has been that the cerebellum facilitates the fine control and coordination of movement.

Physiological thinking about what the cerebellum does has also been greatly influenced by what comes into it. Sherrington called it "the head ganglion of the proprioceptive system." The cerebellum receives every sensory modality that has been looked for – muscle spindle, Golgi tendon organ, skin and joint receptors, vestibular, acoustic, visual, and lateral line. And the "sensory" input is only a small part of the overall input to the cerebellum. In man, motor, sensory, cognitive, and associational cerebral cortex contributes the greatest input via the large ponto-cerebellar projection. Yet even the "sensory" input is in all cases from second order (or greater) relay neurons, and is therefore pre-processed. Further, the ventral and rostral spino-cerebellar paths report on inter-neuronal locomotion generators that are premotor and distinctly *not* sensory (Arshavsky et al. 1972a,b). Nonetheless, some modern physiologists regard the cerebellum more as a sensory analyzer than a movement controller (Nelson & Bower 1990).

The cerebellum was thus supposed to sit off the main line ("metasystemic," MacKay & Murphy 1979), to receive information from virtually all parts of the nervous system, and to funnel down to the motor generators of only the middle level. The purpose was to provide for their fine control and coordination. The cerebellum was seen as being *both* motor and sensory, a "middleman." The middleman used sensory information to facilitate or optimize the motor operation (Bloedel 1992).

2. Cerebellar motor learning theories

2.1. The need: What would the cerebellum learn about movement?

We do not and cannot think about all of the muscle actions in compound and sequential movements. There are too many of them, their transitions too fast, their temporal and magnitude relationships too precise. They occur through some process that is automatic, subconscious. One is aware of individual muscles and joints only when one is beginning to learn new compound movements. Then the movement is altogether different: it is slow, irregular, and "uncoordinated"; all the muscles that are to be involved are not working properly together in amplitude and time.

2.2. The theories: Automation through trial-and-error learning of context-response linkage

Several theories and a number of lines of evidence have pointed to the crucial role of the cerebellum in the adaptation and learning of movement. Brindley first suggested (1964) that the acquisition of skilled movements, such as playing the piano, begins as a conscious act mostly under the control of the cerebral cortex, without help from the cerebellum (Fig. 1). But the cerebellum itself can also initiate the performance, and it immediately begins to acquire control of the task. It recognizes the contexts in which each "piece" of consciously initiated movement occurs. After repeated tries, it links that context within itself to the movement generators so that the occurrence of the context automatically triggers the movement. Thus, with time and practice, the cerebellum largely controls the process, with little or no help from the cerebrum. The cerebrum and the conscious mind are free to do and think about other things. Control of the task has been shifted from a conscious cerebral cortical process to a subconscious one mostly under the control of the cerebellum.

Marr (1969), Albus (1971), and Ito (1972) independently modeled the process using the cerebellar circuit design and function as sketched by Ramon y Cajal and updated by Eccles, Llinás, and Sasaki (Eccles et al. 1967; Llinás 1981). Gilbert (1974) added that synapses were not simply turned "on" (Marr) or "off" (Albus) but were adjusted to give the continuum of Purkinje cell firing frequencies which are actually observed in awake behaving animals (cf. Thach et al. 1992). The circuit models were based on the great differences between the two main cerebellar input systems. The highly convergent mossy fiber-parallel fiber-Purkinje cell system brought information from most parts of the nervous system, and the information was represented as modulations in high-frequency firing of 0–500/sec. The relatively one-to-one climbing fiber Purkinje cell system arose exclusively from neurons of the inferior olive; the synaptic contact was very powerful, but the firing rates were so low (around 1/sec) as to raise questions about the information content.

Mugnaini (1983) showed that the parallel fiber is 6 to 10 times longer than had been supposed, and contacts Purkinje cells along a beam spanning $\frac{1}{3}$ to $\frac{1}{2}$ the width of the cerebellar cortex. Recent animal studies have shown at least one more or less complete body map in *each* of N. fastigius, interpositus, and dentatus (Asanuma et al. 1983, Fig. 2). Further, each nucleus appears to control a different aspect, or mode of movement for the entire body it maps (Thach et al. 1992): dentate, synergist muscles in visually guided movements (e.g., pinch and reach); interpositus, agonist-antagonist synergy and stretch reflexes at a single joint (Frysinger et al. 1984; Schieber & Thach 1985a; 1985b; Smith & Bourbonnais 1981; Wetts et al. 1985); fastigius, synergists in upright stance and locomotion (Antziferora et al. 1980; Thach et al. 1992). Thach et al. (1992) and Goodkin et al. (1993) reported that cerebellar lesions impair compound movements more than simple, and suggested that a cardinal role of the cerebellum is to combine

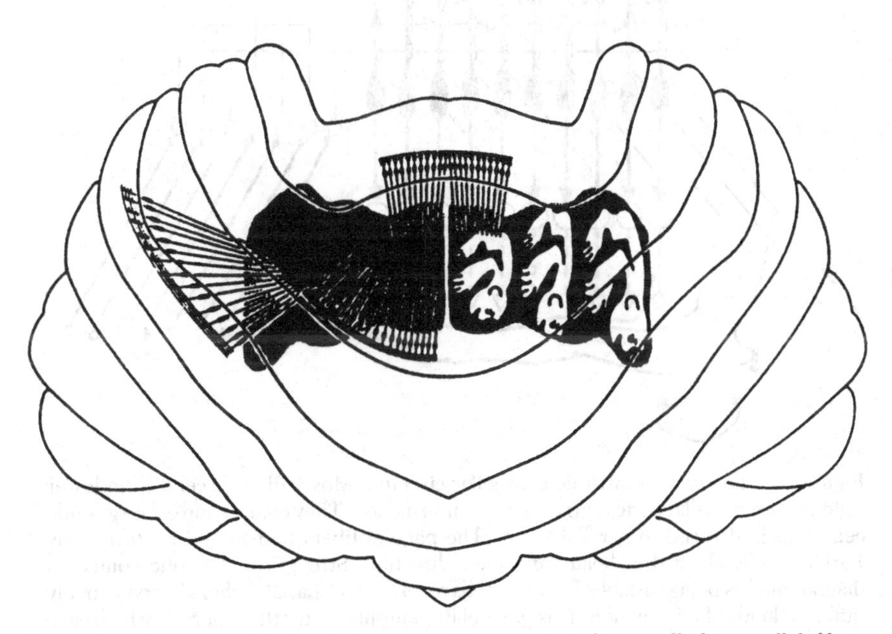

Figure 2. Diagram showing linkage into beams of Purkinje cells by parallel fibers. Beams project down onto the somatotopically organized nuclei. Purkinje cell beams thus link body parts together within each nucleus, and link adjacent nuclei together. Such linkage could be the mechanism of the cerebellar role in movement coordination (from Thach et al. 1992).

(through the learning mechanisms above) the elements of movement using the parallel fiber contacts on the long beam of Purkinje cells.

2.3. Proposed mechanism: Parallel fiber beams represent the context and combine the response

The proposal was that the conditions and "context" in which a movement is to be learned and performed is represented in the modulated discharge of the mossy fibers (Figs. 2, 3), which monitor not only sensory information, but also much of the ongoing activity of most of the nervous system. This afferent information is transmitted to the parallel fiber, which branches to contact thousands of Purkinje cells, which in turn project to the somatotopic motor representations within each of the deep nuclei. The parallel fiber is the critical middle layer between sensory and other input and motor output, representing both the context in which the movement is made, and also being a chief instrument for organizing the motor response.

2.4. The climbing fiber detects and corrects errors in performance and changes the strength of parallel fiber-Purkinje cell synapses, thereby creating novel context-response linkages and response combinations

When a new movement needs to be learned or an old one adapted, the climbing fiber (Fig. 3), which normally fires irregularly at a rate of around 1 Hz, and in no particular relation to movement, suddenly (driven by error between intended movement and actual movement – Albus) begins to fire (once) immediately after the error occurs, reliably time after time. The *effect* of this low-frequency but synaptically very powerful climbing fiber firing is to reduce (Albus) the strength of the synapse on the Purkinje cell of those parallel fibers that are active at the time (and helping to cause the inappropriate movement). What is left after practice, and repeated firings, are those parallel fibers whose action causes the correct movement, and gradually improving performance. Once the behavior has changed to correct behavior, the error that drove the change is eliminated, the climbing fiber returns to its random background firing, and the remaining potent parallel fibers are left to drive the system in the particular movement context.

2.5. Context-triggering, fan-in, and fan-out

Nearly a million parallel fibers contact the human Purkinje cell. Over mossy fibers, vestibular, somatosensory, visual and auditory sensory information arrives in medial and intermediate zones, and cerebral cortical (presumably "cognitive") information in the lateral zone cerebellar cortex. There, the information is conveyed to parallel fibers. But the parallel fibers are so long (Mugnaini 1983), that any one Purkinje cell could conceivably receive via parallel fibers the mossy fiber information from all three zones. The

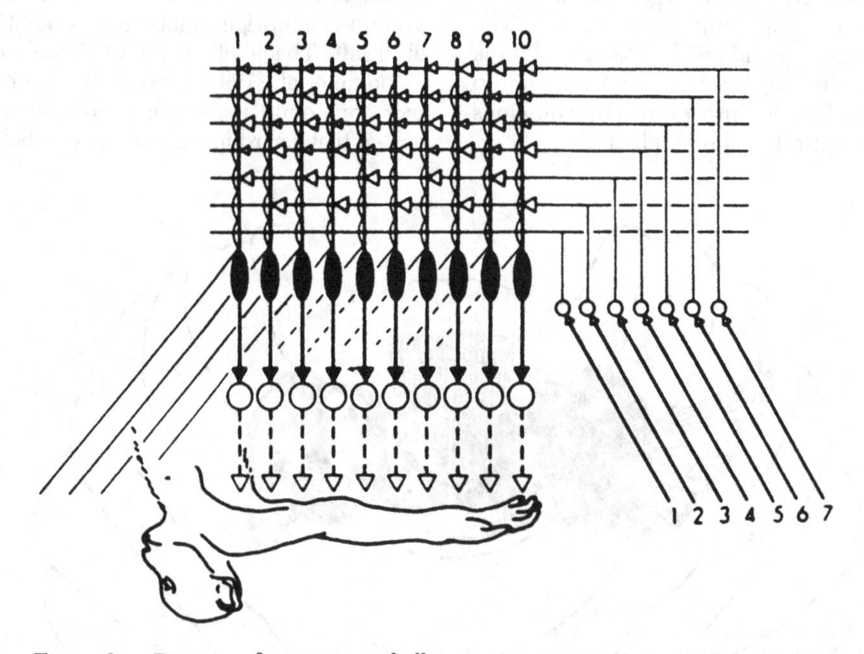

Figure 3. Diagram of intrinsic cerebellar circuitry. Mossy fibers ascend from lower right to the cerebellar cortex carrying input information. They contact and excite granule cells, which give rise to parallel fibers. The parallel fibers branch and contact many Purkinje cells along the "beam" in either direction. Strength of synaptic contact is diagrammed as being variable. Purkinje cells receive many parallel fibers, carrying many different kinds of information. This parallel fiber input creates the "context" which each Purkinje cell is capable of recognizing. Climbing fibers ascend from the lower left to excite Purkinje cells, one-on-one. Climbing fiber activity paired in time with mossy fiber-parallel fiber activity weakens that particular parallel fiber synapse. Climbing fiber activity may thus create and change the mossy fiber-parallel fiber activity context that drives the Purkinje cell (from Thach et al. 1992).

input – and the context that any one Purkinje cell "sees" – could include virtually all the sensory modalities, all the feedback from movement of each of the moving body parts, AND the feedforward of the plans for the control of each body part, as well as the environmental conditions as the subject perceives them and contemplates what to do about them.

2.6. Movement sequences

Learned context triggering also suggests the possibility of generation of sequences of behavior automatically across time. It has been said that "reflex chains" as Sherrington envisaged were improbable because of the time involved (see Keele 1981). Thus, in the sequence of notes A, B, C, D, E, F, and G, E cannot be triggered by feedback from the movement D, because it would take too long to run over real neural pathways. But if E could be triggered by the intent to play and contexts surrounding B or C, then each of the notes could be triggered by the pre-conditions for some of the earlier elements in the sequence, and the process might be made to play out with realistic speed. The phrase – or indeed an entire piece – might thus be generated automatically by preceding contexts. There is evidence that behavioral sequences do show such linkages, and are generated in "chunks" of linked elements, whose linkages are less dissociable than are the boundaries of the chunk (see Keele 1981). In this scheme, the triggering information would come from several different levels in the hierarchical scheme, and this is consistent with the observations from psychological experiments that different chunks are controlled from different levels.

2.7. Combining the actions of many muscles

Moving the skeleton is an engineer's nightmare. Over 100 bones are angulated at connecting joints which allow 2 or 3 degrees of freedom by over 600 muscles. In the simplest possible case, if one Purkinje cell controlled one muscle, and one muscle controlled only one joint, and each joint were controlled only by two muscles (agonist and antagonist), then moving three contiguous joints could be programmed by summing the weighted activities of the 6 Purkinje cells (and the nuclear cells to which they project). There is evidence that a single Purkinje cell may influence a single muscle (Simpson 1994). But the actual computations must be much more complex, because most muscles control two joints, and a single joint may be controlled by many more than two muscles. Yet, in principle, the actions of the muscles could be programmed by linking and weighting the activities of Purkinje cells (and their deep nuclear cells). This combining of muscles is but one level of somatomotor coordination (Thach et al. 1992a,b; 1993), and one that may be addressed at the level of the movement pattern generator (MPG).

2.8. Preventing errors due to interactive torques: Feedforward predictive control

There is yet another aspect of somatomotor coordination, beyond summation of muscle activities. For every action, there is a reaction. Nowhere are the consequences of this law more evident than in the efforts to coordinate the actions of a skeleton. Reaching out to touch an object throws the body backward at hip, knee, and ankle. Extending the elbow causes the shoulder to flex; flexing the shoulder causes the elbow to extend. Yet these reactions and inter-joint torques are effectively compensated by contractions of muscles of which we are usually completely unaware. The preventive actions are driven by afferent information from the moving limb; they are missing in patients with severe peripheral neuropathy (Sainburg et al. 1993) and cerebellar damage (Bastian et al. 1995). Yet these operations are not in the category of strictly linear feedback control, because the neural conduction is too slow for such a process to work. The preventive actions are driven by the movement and early enough in the movement for the corrective muscle torques to exactly match and nullify the passive interactive torques. Context-response linkage through learning is the way the cerebellum provides a mechanism for accomplishing this compensation. Each muscle's action is conditioned by other muscles' actions, and vice versa. In pinching, closure of the thumb on a grape is conditioned by closure of the index on the grape, and the sensory information monitoring the contact of each digit. The forces and positions of each digit are in part functions of each other: they are not controlled independently. While the exact "fanin" information on a Purkinje cell putatively controlling the thumb flexor is not known, the smear of sensory representation in the studies showing "fractured somatotopy" at the level of the mossy fiber terminals onto the granule cell is consistent with this idea (Nelson & Bower 1990). The amount and variety of input information and the number of granule cells and their synapses on Purkinje cells begin to make sense when one considers them as a candidates for a "lookup table" in this kind of learned automatic control.

3. Experimental support from animal studies

The different techniques mentioned in the introduction for studying systems neuroscience have been fruitfully employed to test the motor learning theories.

3.1. Ablation-behavior correlation

Cerebellar cortical ablation prevents or abolishes motor learning. Ablation of the cerebellar cortex has completely removed adaptation once it has been established, and has prevented any further adaptation (Ito et al. 1974; Robinson 1976; Yeo et al. 1984; cf. Ito 1984; Thach et al. 1992; Thompson 1990).

3.2. Neuron discharge-behavior correlation

Purkinje cell recording during motor learning gives results that are consistent with the Marr-Albus-Gilbert theories. During familiar movements in the awake alert animal, the mossy fiber-parallel fiber system drives the Purkinje cell to modulate its simple spike firing frequency over a range of 0–500/sec; the climbing fibers fire at around 1/sec, and in no relation to movement. During movement adaptation, the climbing fibers fire in relation to the movement, but then decrease as the adaptation is complete (Gilbert & Thach 1977; Ebner & Ojakangas 1992); simple spikes (caused by parallel fiber action) conjoined with natural climbing fiber activity change over performance and time as predicted by the Albus hypothesis (Gilbert & Thach 1977; but see Ojakangas & Ebner 1992, and reviews in Ito 1984; 1989; Thach 1980; Thach et al. 1992). In optokinetic reflex eye movements, climbing fiber discharge signals eye movement error (Simpson & Alley 1974).

3.3. Electrical stimulation conjointly of climbing fibers and mossy or parallel fibers reproduces the synaptic modification proposed to underlie motor learning

Paired stimulation of climbing fibers and mossy fibers gives results consistent with the Marr-Albus-Gilbert theories (Long Term Depression). Conjoint electrical stimulation of climbing fibers and parallel fibers repetitively over time decreases the strength of those parallel fiber-Purkinje cell synapses while sparing others (Ekerot & Kano 1985; Ito et al. 1982).

3.4. Classical conditioning is dependent on the cerebellum

One of the most influential contributions in recent years has been the work of Thompson and colleagues on a cerebellar role in classic conditioning of the nictitating membrane response in the rabbit (McCormick et al. 1981; Thompson 1986; 1990; Krupa et al. 1993; Topka et al. 1993). A puff of air to the cornea (the unconditioned stimulus) coupled with a tone (the conditioned stimulus) will lead after time and repetition to an eyeblink to the tone alone. This classic conditioning is prevented (Steinmetz et al. 1992) or at least greatly altered (Welsh & Harvey 1989) by cerebellar ablation. While there has been debate whether the critical lesion is in the interposed nucleus or in the cerebellar cortex (Yeo et al. 1984) this work was the first to claim a clearly cerebellar location for classical conditioning. It is not yet clear whether this represents the *de novo* creation of a new pathway from stimulus to response, or the adaptation of a latent one, such as the acoustic startle response (Ito 1984; Leaton & Supple 1986; cf. Mortimer 1973). More recently, it has been shown that the learning can take place and remain localized within the cerebellum even though it is temporarily kept from being expressed (Krupa et al. 1993). These observations critically uphold the Marr-Albus hypothesis.

4. Predictions about what to expect in cerebellar PET and lesion studies of motor learning

4.1. PET and functional MRI human brain activation

Radiologic brain imaging is the most exciting and promising approach to understanding human brain function of the past several decades. Not only can it safely confirm in humans the results of invasive studies of sensory and motor functions in animals, it can get at the emotional, perceptual, and cognitive functions in a way that animal studies cannot. Measurement of changes in blood flow or metabolism (the indirect indices of neural activity) can be correlated directly with known and controlled human mental activities. Human PET and animal invasive single unit recording studies are not competitive; neither is sufficient itself. Each supports and complements each other's critical strengths and weaknesses.

It is easy to point out the weaknesses of PET scanning. The subtleties behind the conventual displays of "hot spots of neural activity" surrounded by volumes of "quiescent brain" are more than meet the eye. The primary data are *not* measurements of neural activity, but rather of factors that are thought to correlate with neural activity. The primary data is positron emission or magnetic resonance that varies from background in proportion to local alterations in blood flow or sequestration/uptake of oxygen or glucose. The changes in blood flow or metabolism are usually small, 2% or so of baseline activity, and often require averaging across subjects and other statistical manipulations to be detected at all. Commonly the subject is scanned during a "rest" state and again during a behavioral "test" state, and the scans of the two states are subtracted to detect the differences, if any. Increased blood flow and glucose uptake are thought to be much greater in synapses than in somata or dendrites (Wooten & Collins 1981). Thus a "hot spot" on a PET scan reflects increased activity in the terminals of *inputs to* that area, more so than the activity of dendrites and somata within the area, and therefore the *output from* the area. For example, increases in climbing fiber firing should cause a "hot spot" on the PET scan in the cerebellar cortex. But this doesn't mean there is an increased output from the cerebellar cortex: from alert monkey and cat studies, the commonest immediate and long term effect of increased climbing fiber input is a *reduction* in Purkinje cell output from the cortex (Gilbert & Thach 1977; Ito & Karachot 1989; Ito et al. 1982). Furthermore, the increased blood flow or metabolic activity that is correlative with an increased nerve terminal firing is independent of whether those synapses are excitatory or inhibitory. If two inputs to a nucleus (e.g., dentate), the one excitatory (e.g., mossy fibers) and the other inhibitory (e.g., cortical Purkinje cells), are both active, the PET scan should show a "hot spot" in the nucleus. But the inputs might cancel, so as to lead to no change in the output from the nucleus. If only the Purkinje cell inhibitory input to the nuclei were active, the nuclei would still show as a "hot spot." But in this case, the normal tonic firing of the nuclear cell output would actually be reduced. The only way to infer increased or decreased activity in the nucleus *per se* is to look at its efferents and their synaptic termination on its downstream targets. For the dentate nucleus, these target sites would include the VL_c, VL_{ps}, VPL_o, X, MD, and CL subregions of the thalamus, the parvo-cellular red nucleus, the nucleus tegmenti reticularis pontis, the principle olive, and the superior colliculus.

But each of these areas receives inputs from areas *other* than the dentate. The influence of activity in dentate (or any other deep cerebellar nucleus) and thus cerebellar output depends on seeing increased blood flow or glucose or oxygen uptake in these structures that is *not* attributable to activation by some noncerebellar input.

Because of these many caveats, and because of the many differences in imaging experiments across labs, the main initial test of validity is one of *consensus*. What a number of PET scan studies have now shown is that there are changes in blood flow and metabolic activity in the human cerebellum during several examples of "motor learning." They have also shown that the cerebellum is active only in concert with a number of other parts of the cerebrum and brain stem, all together forming a network that appears to be active during motor learning. Perhaps most important, this network is more active during the learning than during the previously learned and now stereotyped "automatic" motor performance of the task. This final observation has been used to argue for a role that the cerebellum is specialized for motor learning *per se.*

4.2. Human brain ablation complements PET activation studies

As with strategies for imaging brain function, it is not the purpose here to review in any detail the pros and cons of ablation. It is the oldest technique for learning about brain function; it is still crucial in determining whether and how a brain part may implement a behavioral function. Ablation uniquely supports and checks the activation results of the PET or the animal unit study. Activation studies may show that an area is active during a behavior, but not that it is contributing to that behavior. Rather, it may instead contribute to some other coincidental behavior. Ablation of the part and a resulting change in behavior helps to support the causal connection. But like any technique, ablation has its limitations. Impairment of function following focal ablation unequivocally demonstrates involvement of a part in a function: it does *not* show that the function resides within the damaged part, and there alone, nor does it necessarily say *what* the part contributes to the function. Nonetheless, it may provide a clue. This is particularly so in humans, who through communication may permit a more detailed and introspective analysis of essentially what the normal behavior that ceases after the lesion (the negative deficit) and what abnormal behaviors may occur in its place (the positive deficit).

4.3. Specific predictions for PET studies, based on theory and the results of animal studies (Fig. 4)

1. At rest (without movement), motor, premotor, prefrontal cortex and SMA neurons are relatively inactive, firing at rates below 10 per second or so. These should register as "inactivity" on PET scanning.

The vestibular and spinocerebellar inputs to the medial and intermediate cerebellar cortex are tonically active at rates of 30–50/sec, and with the tonic activity of the intrinsic inhibitory neurons cause these portions of the cerebellar cortex to be intermediately active. The nuclear cells should also show moderate activity, since their Purkinje cell inhibitory inputs fire at their baseline frequencies of 50–70/sec, and their mossy fiber inputs are tonically active in a similar range of frequencies.

2. For learned, familiar movements, motor, premotor cortex, and SMA neurons should be fully active, firing at up to 200–300/sec.

In the cerebellar cortex, mossy fiber inputs to medial, intermediate and lateral cortex neurons will be active in relation to movement at 0–200–300/sec. This will probably show up in the PET scan as an increase in activity over the tonic baseline. The same will be true for the deep nuclei, receiving from both mossy fibers and Purkinje cells.

If the movement is indeed overlearned and performed "automatically," Brindley (1964) would have predicted that the prefrontal cortex would be inactive.

3. During learning, according to Brindley, prefrontal cortex for the first time will be fully active, as it thinks about, pieces together, and initiates the movement. In this phase,

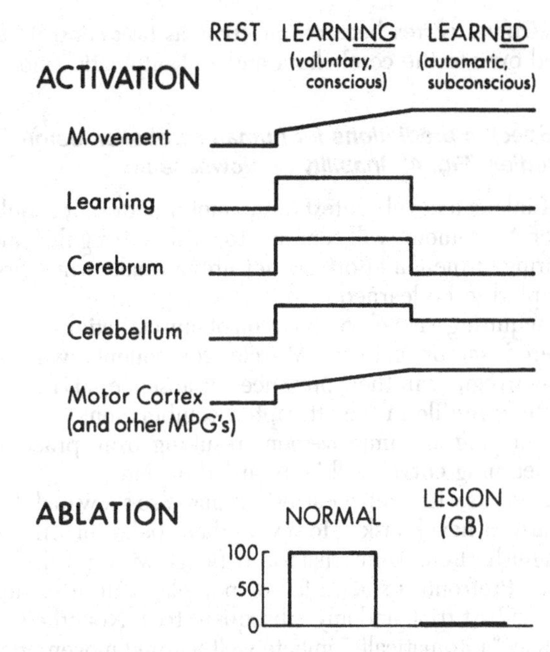

Figure 4. Predicted relations between movement, learning, and activity in thinking parts of cerebrum, cerebellum, and downstream movement generators according to Brindley-Marr-Albus-Gilbert Theories as they change over time and practice.

the movement is "consciously" made, with greater awareness of the details of the movement.

Cerebellar cortex – medial, intermediate, and/or lateral (depending on the nature of the task that is being learned) will show greatly increased activity, because of the increased discharge of the climbing fiber. Animal studies have confirmed the Marr-Albus prediction that under this condition the inferior olive and the climbing fiber will increase firing frequency from their low baseline, in order to "learn" the context in which prefrontal cortex is ordering up the performance, and seize control itself in an "automatic" mode. Since blood flow is proportionate to synaptic activity, and since the climbing fiber-to-Purkinje cell is one of the most massive synaptic structures in the nervous system, the cerebellar cortex should be highly active *in those areas controlling task performance.* This activity will continue only as long as the subject is learning – according to the theory and the evidence from animal experiments, it defines this phase.

Motor and premotor cortex and SMA activity will be the same whether commanded by prefrontal cortex in a "conscious" mode or by cerebellum in an "automatic" mode. These areas should be fully as active as during learned familiar performance.

4. During learned "automatic" performance, the activities become what they were for familiar performances. Intermediate and lateral cerebellar cortex will be moderately active (mossy fiber-parallel fiber-Purkinje cell synaptic activity controlling performance; climbing fiber activity at low-level baseline). Different sets of synapses will be active in controlling the newly acquired behavior, the overall level of activity will be the same as for the previous "familiar" performance.

Motor, premotor, and SMA cortex will be fully active, as the execution of the task requires their full participation.

Prefrontal cortex becomes inactive, as task control has passed over to the cerebellum and an "automatic" mode.

4.4. Specific predictions for human cerebellar lesion studies (Fig. 4): Inability or slowness in:

1. Linking a novel context to the motor pattern it should trigger. Movements will continue to be slow, irregular, and requiring of mental effort, as they are normally when first attempted to be learned.
2. Acquiring a novel combination of muscle actions in the triggered motor pattern. Muscle components will be grossly irregular in their presence (or absence) and time-amplitude profile in the attempted combination.
3. Showing any improvement resulting from practice. The "learning curve" will be retarded or flat.
4. Patients with prefrontal lesions may also show inability to learn a novel task. However, their behavior should distinguish them from that of patients with cerebellar lesions. Prefrontals should fail or be delayed in initiating even the first trial, and any subsequent trial. Nonetheless, they may "automatically" initiate well-learned movements, upon presentation of the appropriate context.

5. PET studies of motor learning in humans

5.1. Manipulation and "tactile learning"

The first PET studies of cerebellar O_2 consumption doing a learning paradigm were performed by Roland and colleagues (Roland 1987; Roland et al. 1988) (Fig. 5). The tasks consisted of "tactile learning" and "tactile recognition." In tactile learning, subjects used the right hand to manipulate small metal objects of similar size and weight, so as to learn to recognize them by touch. In tactile recognition, the learned objects were presented interspersed amongst similar "novel" objects. The goal was to distinguish the learned from the novel by manipulation and touch. Both task scans were compared to a "rest" scan where no movements or discriminations were made. Video analysis showed that the subjects manipulated objects on the discrimination task about two times faster than during the learning task.

Comparing the rest with both the learning and the recognition scans showed similar patterns of $rCMRO_2$ increase in the latter two conditions. *Bilaterally,* there were increases in "six" different areas in prefrontal cortex, supplementary motor, and premotor areas: anterior insula, lingual gyri, hippocampus, basal ganglia, and parasagittal portions of the anterior lobe of the cerebellum and lateral portions of the posterior lobe. *On the left,* there were increases in primary motor and sensory areas, the anterior superior parietal lobule, and the secondary somatosensory motor area. These areas are known from previous studies to participate in movement and sensation. In this study, there were "no differences in the anatomical structures participating in storage and retrieval."

But in comparing the learning and the recognition scans, "the $rCMRO_2$ was significantly higher in the neocerebellar cortex during tactile learning," while "the $CMRO_2$ increases in the left premotor cortex, supplementary motor area, and left somatosensory hand area were larger during tactile recognition." The authors attributed the higher $rCMRO_2$ during learning in the cerebellum to the increased climbing fiber activity that had been seen in animal

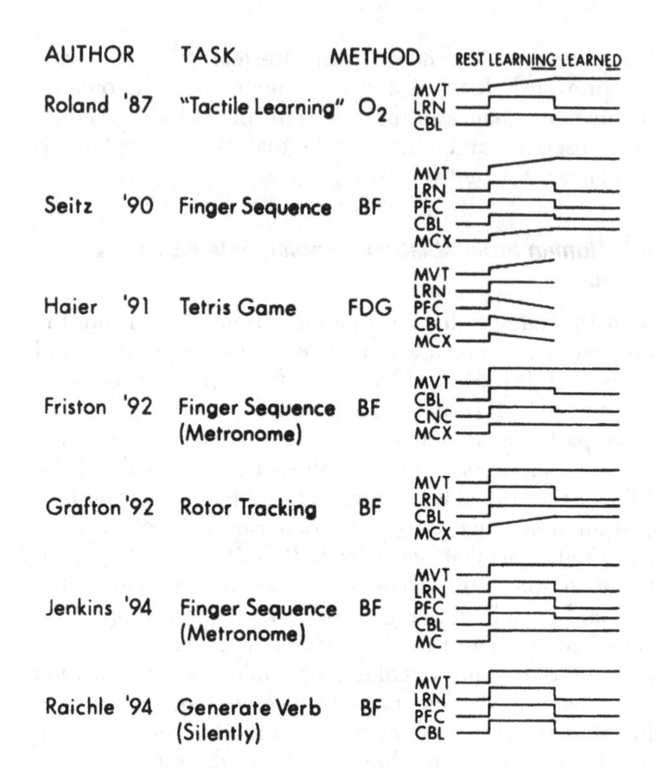

Figure 5. Schematic summary of results of PET studies during motor learning paradigms (see text). Authors, task, type of study, and changes in "activity" going from rest to learn*ing* to learn*ed* performance.

studies on motor learning (cf. Roland 1987, discussion; Gilbert & Thach 1977). They attributed higher $rCMRO_2$ during recognition in the various motor sensory areas to the higher rates of movement during that condition. Although movements increased in frequency through the "tactile learning" training procedure, they could identify no motor learning *per se.* They therefore attributed the learning to *tactile discrimination.* These results are diagrammed in Figure 5.

5.2. Learning sequences of single finger movements

Nevertheless, the first PET study specifically addressed to motor learning in man soon followed (Seitz et al. 1990) (Fig. 5). In the task to be learned, the right hand was used in touching the tip of the thumb (1) to the tips of the other four fingers in the sequences 3·2·2·4, 4·4·5·5, 5·5·4·4, 4·3·2·2, the sequence having been instructed verbally. One PET scan was made at rest, a second when subjects had had little practice ("initial learning"), a third after 50 min of further practice ("advanced learning"), and the last after another 50 min of training ("skilled performance"). Over the three movement scans movement speed nearly doubled. Errors dropped to "very few" in the third stage and to "none in less than 6 sec" in the fourth stage.

Areas of the brain to show changes in rCBF were similar to those in the previous study. During initial learning (compared to rest), increases were seen bilaterally in the inferior frontal gyri (= premotor of prior paper Roland et al. 1988?), in the right parasagittal anterior lobe of the cerebellum and the right lateral posterior lobe of the cerebellum, left greater than right premotor, supplementary motor and primary motor areas, left primary and supplementary somatosensory areas, anterior part of the superior parietal lobule, cortex in the intraparietal sulcus, and the left ventral posterolateral thalamus. Changes in subcortical structures included the basal ganglia, ventrolateral thalamus, red nucleus, region of the substantia nigra and of the pontocerebellar nuclei (oddly, mid-sections of putamen-globus pallidus, the red nucleus region, and the pontine nuclear region showed *decreases* in rCBF rather than *increases*).

As training proceeded through stages, there were progressive changes in the above pattern (Fig. 5). The bilateral activity in the inferior frontal gyri *dropped out* entirely. The authors noted: "At this time the subjects said they no longer needed to count internally the number of times the fingers should touch the thumb." This is a crucial insight, that could only have occurred in human studies: the significance will become clear. There was also diminution of the increases in somatosensory association cortices, more so on the right side (ipsilateral to the finger movements).

There was an increase in the rCBF in left (contralateral) primary motor hand area. The authors attributed this to increased speed and frequency of finger movements.

The increased rCBF of the right (ipsilateral) anterior lobe of the cerebellum remained at the same level from learning to learned performance. (There was no change in the activation in the left primary somatosensory hand area or in the ventrolateral and postero-lateral thalamus.) The authors interpreted the abrupt initial increase of the cerebellar activity at the *initial* stages of motor learning again to increased climbing fiber activity (seen in animal studies). The *decreased* rCBF seen initially in mid portions of the regions of putamen-globus pallidus, red nucleus, and pontine nuclei-diminished with practice: that is, rCBF *increased* in these areas from the drop during "initial learning" to the state during skilled learning (= learned performance).

This study again suggested that a battery of brain parts was involved *both* in motor learning and motor performance. Again it was not clear which part was responsible for what aspect of the learning, and what essentially was learned. Of note was the activity of some areas *at the start* of the learning process. For inferior frontal cortex, the increase was absolute: it dropped out entirely during practiced performance. For the cerebellum, it was relative: the increase in activity at the beginning was maintained in relation to a rising movement frequency and speed and a rising rCBF in motor cortex. The cerebellar increase was *inferred* to be relatively greater for the learning *per se* than for the performance *per se.* Inferior frontal cortex and the cerebellum then must have had preferential roles in the learning *per se.* The authors speculate that the inferior frontal inclusion and subsequent drop out may have been due to the trained use of remembered linguistic instructions (verbal rehearsal) before the performance became automatic. The maintained activity in the cerebellum was consistent with Brindley's idea of a cerebellar role in motor learning (see below). The initial state of volitional conscious control, where the cerebellum acquires control of performance through learning, gives way to a final state of automatic control, directed by the cerebellum.

In the subcortical areas, the drops in activity from rest to initial learning could be consistent with their known high firing frequency at rest. A basal ganglion role in the automatic control would be suggested by the putamen-pallidal resurgence, and a cerebellar role by the resurgence of red nuclear activity as the task became learned. I shall return to this point.

5.3. Learning to play the Tetris video game

The next work addressed learning of a complex visuo-motor task (Haier et al. 1992) (Fig. 5). This used fluorodeoxyglucose (FDG) uptake during learning and performance of the Tetris video game, compared and contrasted with the passive watching of numbers as they appeared on the video screen. The Tetris game required that subjects use the right hand to push keyboard buttons that would orient, move, and place groups of video screen blocks amongst larger groups of blocks so as to complete an overall block figure (a straight line of blocks across the bottom of the video screen). Subjects doing the Tetris task were first given three minutes of practice at the game to learn the rules and the goal. Then they had their ("naive") test trial in which they played for as high a score as possible for 35 min while FDG was injected. They then had the "naive" scan to determine the distribution and amount of sequestered FDG. After 4–8 weeks of 30–45 min per day, 5 days a week playing Tetris, they had their "practiced" test trial during which they played for 35 min while FDG was injected. Following this performance they had their "practiced" PET scan again to localize and quantitate sequestered FDG. With practice, the performance scores improved dramatically across all subjects. Subjects continued to improve with time to almost 50 days of practice.

Naive and practiced Tetris scans were then both compared to scans taken after 30 min of passively watching numbers appear on a video screen, and with each other.

Comparing Tetris (naive *and* practiced) with passively viewing numbers, the former showed increases in the right inferior frontal and right superior temporal gyrus, and the cerebellum. The latter showed greatest FDG uptake in left occipital, left supramarginal gyrus and right area 17, with lesser increases in left postcentral and bilateral precentral cortex.

Comparing scans of the naive (i.e., learning) with the practiced Tetris performance, the former showed more activity in the left superior frontal cortex, left anterior cingulate and right posterior cingulate gyrus, left anterior and middle cerebellar cortex, and right posterior cerebellar cortex. The latter scan showed more activity in the right area 18, right hippocampus, and left cingulate.

This study concluded that with practice there was a greatly diminished uptake of glucose across all the structures that had shown an increase during "naive" performance (Fig. 5). These structures included not only the cerebellum and the prefrontal cortical areas, but the primary motor cortex as well. This was despite the fact that the amount and frequency of movements had increased.

5.4. Learning sequences of single finger movements, controlled for frequency of movement

In a PET study of rCBF during motor practice and neurophysiologic adaptation in the cerebellum, Friston et al. (1992) compared three pairings of rest with

> right handed, brisk sequential finger to thumb opposition with each digit (2 to 5) in turn . . . To prevent gross performance changes over trials, the movements were entrained by a metronome at three per 2 sec (presented only in the task condition)

(Fig. 5). No measurements of task performance were made. The subjects were familiarized with the task 30 min before scanning, but were allowed to practice. Each motor activation started 30 sec before administration of radioactivity, and lasted 2 min. In all scans the eyes were closed.

There was no "global" (overall) change in brain blood flow across the six conditions. In the first motor practice scan, there were relative increases in regional blood flow in the following areas: left sensorimotor cortex and bilateral cerebellar cortex, right greater than left. Active to a lesser extent were premotor cortex (left greater than right), supplementary motor area (left greater than right), the left putamen, the left lateral thalamus, and the cerebellar nuclei. (There was also bilateral activation of the primary auditory areas and nearby insular cortex. This was attributed to the sound of the metronome.)

With practice over the sessions, there was an attenuation in the activity in the "right lateral cerebellum" and in the "medial cerebellar structures." They suggested that the lateral cerebellar adaptation was centered in the cerebellar cortex and that the medial location was deep to the cortex at the level of the deep cerebellar nuclei. Less extensive interactions (adaptations) were also seen in the right brainstem at the level of the inferior colliculus and in the left SMA. "There was no evidence for an interaction in the putamen or in the thalamus at this threshold" (Friston et al. 1992).

The authors concluded that during motor learning the cerebellum commences with higher activity, which then drops off with practice (Fig. 5). They, like Roland (1987) and Seitz et al. (1990), attributed this to the transient increase then decrease of climbing fiber activity reported in animal studies during motor learning (Gilbert & Thach 1977). Of note, the frequency of finger movements was controlled, and the sensorimotor cortex activity remained constant.

This study focused on activity in the cerebellum, and it is not clear whether they would have seen changes in prefrontal cortex.

5.5. Learning to track a rotating disk

A somewhat different result was seen in the following study of Grafton et al. (1992) (Fig. 5). They used PET to study rCBF during "procedural learning" in which subjects were asked to use the right arm and hand to move a stylus to track a 2 cm target on a 20 cm small disc rotating at 60 rpm. During the first and sixth (control) scans, the subjects tracked the target with eyes only (eye movement was not monitored). During the second through fifth scan, the subjects tried to keep the stylus on the moving target. In between test trials, there were practice sessions. Performance was monitored and dramatic improvement was documented with practice from 17% to 66% mean time across the four PET scans obtained during pursuit rotor performance.

"Motor execution was associated with activation of a distributed network involving cortical, striato-nigral and cerebellar sites" (Fig. 5). As performance improved with practice, *increases* in rCBF occurred in the left primary motor cortex, supplementary motor area, and pulvinar thalamus. By contrast, there was no change in an initial increase in rCBF in cerebellum. The authors contrasted this with the previous cited study, noting that the task was more difficult, and not enough practice had been allowed for performance to become "automatic." They concluded with an expression of their intent to study performance after prolonged practice to the point of automaticity to see if the cerebellar activity diminishes as previously.

5.6. Learning sequences of single finger movements, controlled for frequency of movement, and learning vs. performance

The most recent paper (Jenkins et al. 1993) used PET to measure rCBF during "motor sequence learning" (Fig. 5). The task was similar to that of Friston, 1992 and attempted to control and demonstrate motor learning and to distinguish it from performance. Thus subjects again had a period of rest, a period of performance of previously learned sequences of thumb-to-finger tapping at clocked intervals, and a period of trial and error "new" learning of a novel sequence.

During both new-learned (and learning) and pre-learned (and practised) performance: the contralateral sensorimotor cortex was activated and to the same extent for both conditions.

Prefrontal cortex (right greater than left) was active *only* during new learning.

"The cerebellum was activated by both conditions, but the activation was more extensive and greater in degree during new learning" (Jenkins et al. 1993). Similarly, lateral premotor cortex was more active during new learning.

By contrast, the supplementary motor area was more active during performance of the pre-learned sequence than of the learning sequence. Putamen-like primary motor cortex was equally active during the two conditions.

These authors again concluded that "the cerebellum is involved in the process by which motor tasks become automatic."

5.7. Summary

Summarizing these PET studies (Fig. 5), one shows higher cerebellar activity in "tactile" learning compared to practised performance, even though movement is greater in the latter (Roland 1987; Roland et al. 1988). In the second, cerebellar activity remained the same from motor learning to motor performance despite increased movement in the latter, from which the authors inferred a *relative* increase in relation to the learning phase (Seitz et al. 1990). In the three studies in which movement was controlled (Friston et al. 1992; Grafton et al. 1992; Jenkins 1994), cerebellar activity was greater in novel than in prelearned tasks (Friston et al. 1992; Jenkins et al. 1993) or remained elevated (Grafton et al. 1992).

Raichle et al. (1994) pointed out the difficulty in comparing these results, where the tasks and methods differed so widely. The crux is in keeping the movement performance constant and adequately dissociated from the motor learning. Since the cerebellum may be involved in both, dissociation is needed to demonstrate the extent to which it participates in each. In the above studies, this aspect is unclear. Of the three studies that controlled for the amount of motor activity, "one noted no change in *primary motor cortex* (Friston et al. 1992), one noted a decrease (Mazziotta et al. 1991) and the other noted an increase (Grafton et al. 1992). In two studies that did not control the amount of motor activity and, consequently, observed an increase in the number of acts performed during the same scanning session, one reported an increase in the activity in primary motor cortex (Seitz et al. 1990) and one reported a decrease (Haier et al. 1992). Two of the groups that controlled motor activity commented about changes in supplementary motor area (SMA) but reported opposite results; Grafton et al. (1992) reported an increase whereas Friston et al. (1992) tentatively reported a decrease. The varied results reported in these five imaging studies of motor learning in normal humans do not permit us to draw any conclusions about consistent changes in neuronal activity or organization in primary motor or supplementary motor cortex to be expected from practice and learning of various motor task or to anticipate our results (Raichle et al. 1994, p. 17).

Nonetheless, the studies do appear to show parallel changes in prefrontal cortex and cerebellum in going from a learning-ing to a learn-ed performance. This is exactly what Raichle et al. (1994) subsequently observed in their study of "non-motor" learning (Fig. 5). Cerebellar activity is increased *during* the learning, and is diminished or absent during the over-learned or automatic performance.

Yet Raichle's warning is appropriate: with this much variability across studies, what can one conclude? An obvious caveat is one given by this same group on the need to watch out for coincidental behavior and muscle activity that is not part of the main task performance. In a study of panic disorder, in which increased temporal blood flow was attributed to increased activity in limbic regions of the brain, it was subsequently found to localize instead to *temporalis muscles,* which were fortuitously active in the near vicinity (Drevets et al. 1992). The same could hold for regions of the brain that are active and covary with a particular task performance, but whose causal connection is to some coincidental behavior and muscle activity that is not at all necessary to task performance. In animal studies, one is always on the lookout for coincidental "suspicious behaviors" and consequent spurious correlations between unit activity and the primary task performance. And indeed, the cerebellum may be particular active in relation to synergistic associated movements (Schieber & Thach 1985a; 1985b; Thach et al. 1993). Perhaps too much trust is placed in human subjects following the investigator's instructions and wishes to the letter; certainly, spurious correlations seem to be a possibility (Drevets et al. 1992).

But here is where ablation studies can come to the rescue, and indeed, many have recently supported a cerebellar role in motor learning.

6. Cerebellar damage impairs motor learning

6.1. Adaptation of pointing while wearing magnifying lenses

Following the Brindley-Marr-Albus-Gilbert theoretical suggestions for a cerebellar role in motor learning and the early animal experiments showing impairment by lesion and unit discharge correlation with motor learning, the first study showing *human* relevance was that of Gauthier et al. in 1979 (Fig. 6).

In normal humans and in patients with "posterior fossa involvement," Gauthier and colleagues examined visuomotor performance in an eye-hand pointing task as it adapted to the wearing of magnifying lenses. When the target field was thus visually magnified, the normal response was to misreach by a distance proportionate to the magnitude of the magnification factor. With practice, and given knowledge of the error (open loop), normal subjects adapted sufficiently to compensate for about half the prism-induced error. Proof that a true adaptation (not just a change in strategy) had occurred came upon removal of the lenses: normal subjects pointed and overshot the target by a similar magnitude in the opposite direction. The patient had had prior superior vermal tumor removal, with accompanying loss of superior vermal and adjacent cortex, had had *transient* hydrocephalus, palato-pharyngo-paraspinal and diaphragmatic myoclonus. He had had a *persistent* cerebellar deficit that included saccadic hypermetria, eccentric gaze holding nystagmus, rebound nystagmus, scanning speech, symmetrical limb dysmetria, dysdiadochokinesia, Romberg sign, truncal titubation, and broadbased staggering gait. The patient was unable to recalibrate his gain during the visual-motor training, and he also had no post-lens exposure overshoot. The authors concluded:

> We believe that the persistent motor deficit of the cerebellar type displayed by patient 1 parallels the absence of visual-motor adaptation. Such adaptation clearly involves recalibration of visual-motor coordination. The same patient formed part of a study of saccadic overshoot dysmetria interpreted and modeled as a deficit in recalibration of oculomotor gain. Other cerebellar signs may have related explanations. (Gauthier et al. 1979, p. 160)

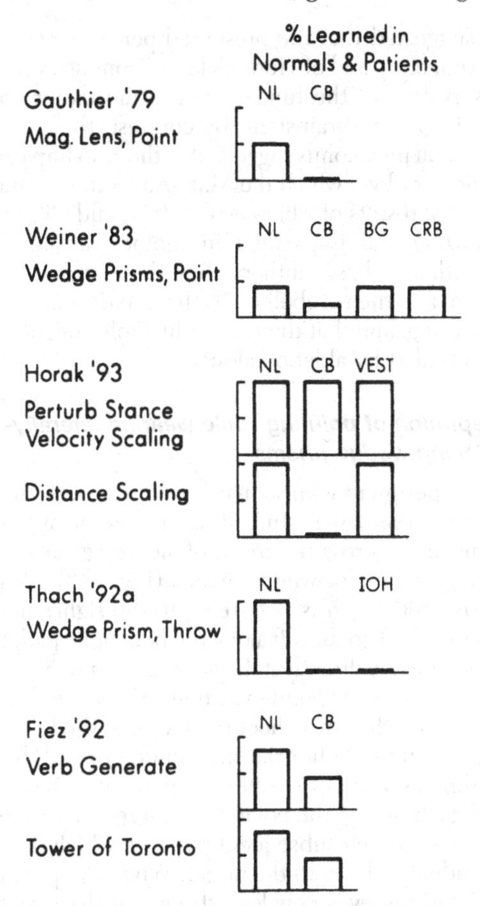

Figure 6. Schematic summary of studies of effect of cerebellar ablation on motor learning (see text). Authors, task, and comparisons of performance (0–100%) of normal and cerebellar damaged individuals.

They suggested that the preserved performance of the saccade trajectory and of Hering's law of conjugacy implied the preservation of the fundamental oculo-motor control mechanisms in the brainstem. By contrast, the prior episode of palatal myoclonus suggested to them an impairment of the inferior olive, which the Marr-Albus theory and the experimental results of Llinás et al. (1975) and Gilbert and Thach (1977) had implicated in motor learning. They agreed with all these authors that these inputs to the cerebellum and the cerebellum itself were thus involved in motor learning, and that their individual roles might not be further revealed by ablation alone.

6.2. Adaptation of pointing while wearing laterally-displacing wedge prisms

A similar experiment with similar results was performed by Weiner et al. (1983), who studied visuo-motor adaptation in a pointing task across a variety of neurological patients during the wearing of wedge prisms (Fig. 6). With these prisms, the optic path is bent (e.g., to the right), and the subject has to look to the left along the bent optic path to see the target which is directly in front of him. In pointing, the arm, hand, and finger point in virtually the same direction as the gaze, and thus overshoot the target to the left. But if the subject can see either the site pointed to and thus the error within a second or so after the point, or if he can see the hand itself during the point, then there is a progressive adaptation with each subsequent point, in which the point comes gradually closer to the target. When the prisms are removed and the eyes now look straight at the target, the adapted hand points to the *right* of the target in the opposite direction with a magnitude of error similar to the original error. This occurrence of overshoot and its persistence, requiring readaptation with practice, is what convinced the authors that a true adaptation had taken place. The authors pointed out that the phenomenon was described by Von Helmholtz in the last century, and its essential mechanism has been studied by a number of psychologists since. They summarized the then-existing beliefs as

> This adjustment is attributed to the combined effects of two processes: a *true visual* adaptation to the prisms and a *cognitive* correction in pointing when there is a perceived error, or the difference between pointing where the target is "seen" and where the target is thought to be (Weiner et al. 1983, p. 766). (emphasis mine)

The authors also reviewed various thoughts about the mechanism of the adaptation, which separated mainly into an altered sense of position or of posture (peripheral) and an altered perceptual or mental set (central). They also reviewed the various animal ablation experiments seeming to exonerate corpus callosum, hippocampus, anterior and posterior commissure, massa intermedia, optic chiasm and [they said] cerebellum (Bossom & Hamilton 1963). Bossom had reported reduced levels of adaptation after bilateral frontal lobectomy or bilateral caudate lesions (Bossom 1965). Baizer and Glickstein then briefly reported that cerebellar lesion in monkeys did indeed impair prism adaptation (1974).

Weiner et al.'s (1983) study involved normal humans and patients with cerebellar disease, Parkinson's disease, right and left cerebral hemisphere disease with affected corticospinal pathways, and Alzheimer's disease and Korsakoff's syndrome, the latter two with "declarative" memory defects. Only the cerebellar patients showed "significant reductions in the after effect" (Fig. 6), which the authors, like Gauthier et al., regarded as the best measure of true adaptation. The authors therefore likewise labelled the effect a visuo-motor adaptation.

> Poor adaptation in cerebellar patients may be due to impaired motor learning, defined as a change in motor program after environmental change. Adaptation to lateral displacement of vision requires new motor responses in response to induced alteration in visual input. (Weiner et al. 1983, p. 771)

6.3. Learning to trace complex figures and mirror tracing

This view was ostensibly supported by the experiment of Sanes et al. (1990) on normal controls and cerebellar patients with either focal or atrophic diffuse disease of cortex or with olivo-cerebello-ponto atrophy (OPCA). The tasks were two: (1) the tracing of a 5-sided complex figure repeated for 50 trials, and (2) the mirror-tracing of two different 4-sided figures, for 50 trials and 10 trials, respectively. The 5-sided figure was traced while being viewed directly; the two 4-sided figures only while being viewed as reflected in a mirror (and thus right-left reversed).

The details of the method and instruction are important.

> The patterns were sufficiently large that subjects had to move more than the fingers and wrist to accomplish the task. Typically, the movement involved shoulder and elbow joint rotations, as well as adjustments of the wrist, and occasionally of the fingers. Speed of execution was emphasized as *the most important movement parameter* (emphasis mine). Subjects were instructed to slide the stylus across the data tablet as rapidly as possible, but with the proviso that each movement segment begin and end in the small squares that enclosed each vertex of the pattern. Subjects were cautioned not to pause at the vertices, but rather to execute the movement in one *continuous sequential* [sic] movement [emphasis mine]. Subjects were informed that the lines connecting the vertices were intended as a general guide for the sequential movement. (Sanes et al. 1990, pp. 105–06)

A multivariate analysis was undertaken across patients and across conditions, in the hope that distinct abnormalities would stand out. Measured variables included: (1) movement time, measured for the entire four or five segment movement; and (2) tracking errors, calculated as the average, peak, and endpoint deviations of the tracing from the target lines, and acceleration changes that crossed zero. Reversals of acceleration were necessary *only* at endpoints; more than that reflected lack of "smoothness" of trajectory. The authors state that the patients made many *performance* errors, and that normalizing and averaging corrections were required to extract errors in learning from errors in performance.

6.4. Common features

Common features across groups and conditions included a tendency to speed up with practice, more so for the normals than the cerebellar groups. All groups continued to make errors of the various types across tasks from start to finish.

6.5. Differences

Differences across groups and conditions were the tendency on direct vision tracing for the normals and to a lesser extent the OPCAs (but *not* the cerebellar corticals) to *increase* the average and endpoint error (though not the peak and tangential acceleration zero-crossing errors). This was regarded as a natural phenomenon in accordance with Fitts's law of the speed/accuracy trade-off: as speed increases, accuracy *should* fall off. The failure of the cerebellar cortical patients to follow this law was regarded as some failure to adapt to the normal strategy.

Another difference occurred on the mirror tasks and consisted of the relative inability of the OPCAs to reduce their tangential acceleration zero crossings, as compared both to normals and to the cerebellar cortex patients. This was interpreted as a relative inability of these patients to learn a new skill. This purported deficit was likened operationally to associational conditioning, a failure to achieve a link between a new or different behavior to a given stimulus. The authors suggested an analogy to the work of Thompson and colleagues on conditioned eyeblink in the rabbit (McCormick & Thompson 1984; Thompson 1986; 1990).

The authors interpreted that both the normals and the cerebellar cortex patients adapted to the mirror task, because they increased their speed and reduced their tangential acceleration zero-crossing errors. The authors also interpreted that there was a carry-over from the 50 trials of mirror learning on the first figure to the ten trials of performance on the second figure. They gave as the most direct evidence for this the fact that no normals or patients (except one) made direction errors on the *first* movement on the figure.

Greatly to their credit is the authors' serious effort to define *what* indeed is learned in motor learning and *what* essentially is impaired by cerebellar disease. Their careful presentation of their methods, results, and interpretations allows one to interpret further. The first question one might raise is whether failure to obey Fitts's law is a specific neurologic deficit. The instructions given to the subjects to move as fast and as nearly continuously as possible (despite a segmented motor pattern task) clearly encouraged errors in approximating the target line. It is not at all clear why the OPCAs (who were slower) made fewer (when normalized) of these errors, but it does not seem quite appropriate to call the failure to make these errors a deficit.

The second question one might raise is to what extent any of the subjects really "learned" mirror drawing. Certainly, errors in performance persisted to the end. Further, in the claimed "carry-over" of mirror learning from the first mirror-traced figure to the second, one could question this interpretation. In mirror-tracing, movements parallel to the face of the mirror are not reversed, and no adaptation is needed to make them. Only movements toward or away from the mirror are reversed, and require adaptation. In the first mirror-traced figure, all lines had a component moving toward or away from the mirror. In the second figure, the first line to be drawn was *parallel* to the face of the mirror. This then was the *only* movement that a subject informed of the nature of the task but who was relatively poorly practiced could have made without errors. It is therefore not surprising that only one error (in a patient) was seen. The conclusion of "learning carry-over" seems questionable.

If the "carry-over" from the first mirror tracing task to the second is suspect, so then is the presence of an after effect or persistence of learning. An after-effect was clearly demonstrated in the prior papers on lens and prism adaptation, respectively. It is not clear what exactly was learned here, or if anything was truly learned. The alternative is that there was the conscious adoption of a strategy to reverse visuo-motor coupling proportionately as one moved from a plane parallel to the face of the mirror to one toward or away from it.

A further reason for questioning whether visuo-motor adaptation occurred is the relatively short time course claimed for it. An equivalent experience would appear to be that reported by Gonshor and Melvill-Jones (1976). In these experiments, dove prisms were placed over the eyes and worn continuously. These reversed the visual world in the left-right dimension without altering the up-down dimension, which would appear to be similar to but just opposite that of mirror drawing. What was measured was the phase and gain of the vestibulo-ocular reflex (VOR); other motor behavior was not studied systematically. Changes in the VOR occurred gradually over several weeks, and did not fully reverse until after five weeks of continuous wearing. At this time other motor behaviors appear also to have adjusted: one subject resumed driving his motorcycle! Whether he would have attempted this after 50 short trials is doubtful. In sum, the task of learning mirror tracing seems a good one, but one likely to take a much longer time (Fig. 6).

6.6. Learning to couple a novel postural response to a novel perturbation of stance

Another line of study attempting both to demonstrate a cerebellar role in motor learning and also the identity of the controlled variables is that of Horak (1990) and Horak and Diener (1993) on postural responses to the perturbation of stance (Fig. 6). In this study, subjects included normals, bilateral vestibular nerve lesions, anterior lobe cerebellar cortex degenerations, and OPCAs. Subjects were instructed to stand on a platform which was driven backward by a motor. Two perturbation parameters were varied independently, velocity, and distance of backward displacement: four velocities over a given distance and four distances at a given velocity. Any and all pairings of a velocity and a distance were repeated over a block of 10 critical trials, so that the subject had the opportunity to experience and adapt to those particular conditions. Only the last three trials were analyzed in a block of trials, after ample opportunity for adaptation. Postural (corrective) responses were measured as *torque* generated under the toes and balls of the feet, (to prevent falling forward), and the EMG of various muscles in the leg and trunk.

Normals adapted to each of four different velocities (fixed distance) and four different distances (fixed velocity) with proportionate differences in plantar flexor torque and EMG amplitude. Both the vestibular and the cerebellar patients also adapted to *velocity* of displacement, and with *gains* (velocity/torque) similar to normals. However, both patient groups showed excessive *bias*, generating hyperactive responses. That is to say, plots of velocity (abscissa) and torque (ordinate) were linear and of the same slope for all three groups, but had different intercepts on the torque ordinate, vestibular and cerebellar patients being progressively higher. Thus, translation of stimulus velocity to graded torque response clearly continued to take place in both patient groups and with identical and normal gains. What seemed missing was an inhibitory tonic bias on brainstem and spinal circuits. This could be understood as the loss of Purkinje cells in the case of cerebellar cortex disease (the most extremely abnormal). In the case of relatively milder excess bias in the vestibular patients, the mechanism is less obvious, but was attributed to somatosensory compensation for the vestibular loss (Bles et al. 1984).

As for variations in *distance of displacement,* the normals and vestibulars adapted, the vestibulars again with normal gain, but biased high (generating a fixed excessive amount of torque for each torque increment proportionate to greater displacement). But the cerebellar cortical atrophies were distinctly different, showing *no* gain (a torque/amplitude plot with zero slope) and a high bias (intercept displaced high on the torque ordinate). This led to hyperactive responses with overshoot of the proper endpoint position, without any stimulus-response scaling. Since the magnitude of the displacement (at fixed velocity) could *not* have been inferred from stimulus parameters at onset, the magnitude of the response had to be learned through trial and error. The adaptation is one of association and not of scale. The authors refer to it as the acquisition of *set,* which is specifically impaired in the cerebellar cases.

Oddly, the OPCA patients, though presumably similarly "ataxic" in walking and heel-knee-shin tests, showed *normal* scaling of distance vs. response.

Cerebellar patients showed excessive activity of antagonists, as well as of the agonists that led to the excessive responses and the hypermetria. This was interpreted as a compensatory cocontraction as if to restrain and reduce the response that would otherwise be even greater. Specifically, the authors denied that there was any disorder in the selection, sequencing, or latency of the muscles themselves. The interpretation was that this was exclusively a disorder of set (and bias), and not of coordination *per se.* Thus, the authors conclude

> the major effects of anterior lobe cerebellar damage on human postural responses involves impairment of response magnitude based on predictive control set and not on the velocity of feedback or the temporal synergic organization of multijoint coordination. Thus the anterior lobe of the cerebellum appears to play a critical role in modifying the magnitude of autonomic postural responses to anticipated displacement conditions based on prior experience.

As for bias and learning or "set" acquisition, this careful study shows that the cerebellum must be involved in these aspects of movement *at least.*

6.7. Adaptation of throwing while wearing laterally-displacing wedge prisms

Others have also shown that cocontraction is common in cerebellar disease (Hallett et al. 1975), which demonstrates an impaired selection of muscles appropriate to a task (cf. Thach et al. 1992; 1993 and below). Still other work has raised the old question of whether the role of the cerebellum is simply to provide control of bias (threshold) or gain (stimulus/response proportionality) of input-output relations in downstream movement generators, or whether it might in addition combine the elements within and across generators (Thach et al. 1992a). There can only be so much behavior that is hardwired within the nervous system. Any novel behavior, developed in response to novel environmental conditions, must obviously be "programmed" into the preexisting hardware. One kind of novel behavior is the novel synergic combination of muscle actions. The word "skill" is used for this novel acquired pattern of coordination. A paradigm that illustrates the learning of a synergy is the adaptation of eye-hand coordination in throwing a ball or a dart at a target while wearing wedge prism spectacles. In throwing at a target, the eyes fixate the target, and serve as the reference aim for the arm in throwing. The coordination between the held position of the eye and the synergy of the arm throw is a skill: it has to be developed and kept up with practice. If wedge prism spectacles are placed over the eyes with the base at the right, then the optic path will be bent to the right, and the eye will have to look to the left to see the target. The arm, calibrated to the line of sight, will throw to the left of target. With practice, the calibration changes, and the arm throws with each try closer to and finally on-target. Proof that gaze direction and eye position is in fact the reference aim for the arm throw trajectory comes when the prisms are suddenly removed and the arm throws. The eyes are now on-target, but the eye-arm calibration for the previously leftbent gaze persists; the arm throws to the right of target an amount almost equal to the original left error (Thach et al. 1992a). But with practice, the eye position and the arm throw trajectory are recalibrated back to the original setting: the throws move closer back to and finally on target. A good analogy is the relation between sighting and shooting a gun: the linkage between the sight and the bore trajectory is calibrated by adjustment, and kept true through practice.

But is this adaptation in the visual sensory domain, the perceptive, the cognitive, the motor, or somewhere in between? Our evidence that it is at least *largely* motor comes from its specificity for body part. When one arm is trained and the other arm is tested, the initial arm adaptation does not carry over to the opposite arm throws, but persists through to throws again by the initial arm, only to readapt with repeated throws (Thach et al. 1992c; cf. Prablanc et al. 1975). The lack of carry-over of the adaptation to the untrained arm, and the fact that throwing with the untrained arm does not degrade the adaptation in the trained arm, both speak for privacy of the storage to use of the trained body part.

But can one train the body parts to a specific task, without the adaptation spilling over to other tasks performed by the body parts? When one arm is trained on overhand throws and then tested on underhand throws, for most individuals the overhand training does not carry over to the subsequent underhand throws (Thach et al. 1992c). Yet the overhand training persists through to subsequent overhand throws, readapting with repeated throws. Prism adaptation is thus specific to the throwing arm and to the type of throw. This may be analogous to using a similar movement that is differently calibrated for two different contexts – such as hitting a baseball with a bat and a tennis ball with a racquet. One may modify the one activity, for example, using a heavier bat, and not have the training affect the other activity – one's tennis game. When the two movements become more nearly identical one is more likely to get carry-over. Some find it difficult to play tennis and squash alternately; more have trouble alternately playing squash and racquetball.

Can one learn to store more than the one gaze-throw calibration simultaneously? We asked subjects to throw 200 times while wearing the prisms and 250 without, daily 4 days per week for 7 weeks (Martin et al. 1993). We measured the progress on the 5th day of each week with 25 throws before, 100 throws during, and 75 throws after wearing the prisms. This made a total of 900 throws with prisms and 1100 throws without prisms each week for 7 weeks. Over time and practice, the first throw with the prisms landed closer to the target, and the first throw without the prisms (aftereffect) also landed closer to the target. By 7 weeks, throws are on-target for the first trial wearing and the first trial after removing the "known" prisms. This suggests that two adaptations (no-prisms and known-prisms) may be stored simultaneously and separately. This may be analogous to the fact that one can maintain eye-hand coordination while wearing one's spectacles and when not wearing them. It also accounts for the period of adjustment required to "get used to" a change in ones lens prescription.

What was the controlled variable in these dual calibrations? The gaze-throw angle for no-prisms was 0 degrees, that for the known-prisms of 30 diopters, about 15 degrees. When the two over-trained subjects donned the known-prisms, their gaze-throw angle immediately shifted to 15 degrees; the context of the prisms must have introduced a bias within this system. The bias was compartmentalized and specific: components of the gaze-throw shift consisted of changes of position of eye-in-head, head-on-trunk, and trunk-on-shoulder (Thach et al. 1995). In one subject, the relative proportions were 5, 5, and 5 degrees, respectively, for each component. In the second subject, the relative proportions were 6, 2, and 8 degrees, respectively, for each component. Thus, in each subject, varying amounts of tonic activity in specific muscles contributed to the compounded variable of static gaze. Each subject had learned two such patterns, which could be immediately changed from one to the other, depending on the behavioral context (prisms or no-prisms).

Baizer and Glickstein (1974) first showed in macaques trained to point to a visual targets while wearing wedge prisms that the adjustment mechanism was abolished by cerebellar lesion; Weiner et al. (1983) confirmed the result in patients with cerebellar disease, and showed further that *adaptation was not impaired in disease of corticospinal or basal ganglia systems.* We (Thach et al. 1991; Martin et al. 1995) have confirmed these results in the throwing task in patients with pure cortical cerebellar disease, and have also seen that patients with MRI-documented inferior olive hypertrophy and infarcts of olive output at the inferior cerebellar peduncle could not adapt, despite otherwise near-normal performance (Fig. 6). By contrast, patients with infarcts involving the dentate nucleus, despite severe ataxia, could adapt. This suggests that the adaptation mechanism could be dissociated at least in degree from those of coordination and performance.

6.8. What then is learned in "cerebellar motor learning"? What basically does the cerebellum control in the motor domain?

We have interpreted these studies as showing a marked capacity for storage of different types of adaptation for the same body part for different contexts and different movements (Thach et al. 1991; 1995). It is important that the different adaptations be kept private, without carry-over from one to another. We have argued elsewhere on theoretical grounds that the best place to store such context specific adaptations is somewhere far "upstream" (Fig. 4) from the controlling motor pattern generators and the motor neurons (Thach et al. 1992a). The reverse would be true for types of adaptation that should best apply to *all* uses of a body part. A weakening of a muscle would best be compensated by increased synaptic strengths of inputs to the alpha motor neuron. Damage of the labyrinth would best be compensated by increased synaptic strengths of vestibular nerve input to vestibular nuclei. In the various training paradigms that have been employed in animal studies, both or either kind (site) of adaptation may have been produced. Thus the controversy about the one exclusive of the other may result more from experimental method than from biologic restriction.

6.9. The experimental observations summarized here are compatible with the Brindley-Marr-Albus-Gilbert theory.

A number of PET studies show activity in prefrontal cortex ("voluntary, conscious activity") at the very beginning of the task learning, only to drop out as performance becomes practiced. Cerebellar activity is maximal throughout the learning phase, when learning theories predict and animal studies show increased discharge in the massive climbing fiber synapse onto the Purkinje cell. The activity drops sharply but *does not* stop as task performance becomes more automatic, theoretically under control of a "new" pattern of activity in parallel fiber synapses onto Purkinje cells. The "new" pattern of parallel fiber-Purkinje cell activity should not be greatly different from the "old" one in its metabolic needs, and blood flow should be the same or similar. These observations are compatible with two of the most critical predictions of Brindley, Marr, and Albus, and agree with the findings in animal studies.

The ablation studies of Horak are compatible with a bias adjustment mode of control, and also of context-response linkage. The observations of visual perturbations during pointing of Gauthier et al. (1979) and Weiner et al. (1983), of Sanes et al. (1990) in mirror drawing, and of our own in throwing argue for a compounding of controls across body parts – call it bias, gain, or what one will – that generate movement combinations of such variety and novelty as to easily qualify as "learned new movements." In the prism-altered eye-hand coordination studies, the essential variable that is changed is the angle between the line of gaze and the line of the point or throw. The gaze-throw angle is in turn made up of angles of eye-in-head, head-on-body, and body-on-arm. The angles in turn represent the different tonic activities in the agonist-antagonist muscle pairs controlling the body part in question. These activities are cerebellar-controlled so as to sum and allow precise "coordinated" motor acts.

7. Cognitive functions of the cerebellum

7.1. The cerebellum sits higher in the hierarchic organization of the central nervous system than it used to

From time to time, there have been suggestions that the cerebellum might play a role in one or another cognitive function (cf. reviews by Macklis & Macklis 1992; Rawson 1940; Schmahmann 1991). Ideas have ranged from memory (Poseidonius, 9th century A.D.; cf. Rawson 1940) to sexual potency (Gall in the 1800s; cf. Macklis & Macklis 1992). None of these suggestions have been incorporated in our knowledge of cerebellar functions, because they have not fit with other scientific facts. But there have been recent advances in our knowledge of connectivity, ablation syndromes, and natural activation correlations in animals and man which require us to reevaluate this question.

The so-called "motor association" areas are presumably concerned with motor planning – these include the supplementary motor area (medial area 6), the premotor area (lateral area 6), the frontal eye fields (area 8) and the accessory frontal eye fields of Schlag, and the motor speech areas in man (areas 44, 45). Anatomically, they receive from posterior areas associated with perception and awareness (see below), and project to the middle level motor pattern generators. They receive multi-modal sensory inputs, and send to different movement generators. These parts are active in animal recording and human PET studies during the movement, and their ablation impairs movement. For these reasons, they have been called the motor "association" areas of cerebral cortex.

It has recently become known that these areas may be active in anticipating or rehearsing a movement without actually performing it (the studies of anticipatory signals in monkey SMA and PMC by Evarts, Tanji, and Kurata (Tanji 1985; Tanji & Evarts 1976); of mimicry signals in monkey PMC by Rizzolatti (Di Pelligrino et al. 1992); and of mental motor rehearsal signals in human SMA by Roland (Roland 1987). The two roles in purely mental imaging of movement *and* in movement planning would appear to go together.

Recent work has added projections from the cerebellum to include virtually *all* levels of the motor system – spinal motor and inter neurons (cf. Asanuma et al. 1983a,b,c), the superior colliculi (May et al. 1993), and (via thalamus) the cerebral cortical "motor association" areas – premotor cortex, primary and secondary frontal eye fields, and now even areas 44, 45, and 46 (Lynch et al. 1992; Middleton et al. 1994; Schell & Strick 1983; Yamamoto et al. 1992). As such, the cerebellum is "upstream" from movement pattern generators at all levels. But since each of these movement generators has other prominent excitatory inputs, and since cerebellar ablation impairs but never abolishes movement, cerebellar functions have traditionally been characterized as regulatory and modulatory, rather than executive.

7.2. Does the cerebellum only modulate what others have begun, or can it run the whole show?

Nevertheless, these connections put the cerebellum in a position to excite any or all of the major motor generators, at every level from motor neuron to motor association cortex. Why has it thus been delegated to the role of regulator and modulator, rather than executor? The historic arguments include: (1) cerebellar lesions impair but never abolish movement, and often cause little or no observed motor defect; (2) until recently, cerebellar electrical stimulation has not been reliable in causing movement, and (3) cerebellar target motor structures each have an additional major excitatory input, which has been credited with the major driving effect. However, newer knowledge questions all three premises. First, focal cerebellar lesions do indeed abolish particular categories of movement without affecting others. The effect of the lesions depends upon its precise localization, and our knowledge of the particular region and its control functions. The effects may be disabling (cf. Thach et al. 1992 for review). Second, electrical stimulation of the output nuclei does reliably cause movement (cf. Thach et al. 1993 for review). Stimulation of the cortex may not cause movement, because of mixed excitatory and inhibitory effects. Third, cerebellar excitatory effects on a target structure are often stronger than those of the "second" excitatory input. This is known to be so for the red nucleus (cf. Toyama et al. 1970; Tsukahara & Fuller 1969). Also, the dentate nucleus fires before and apparently helps initiate output from motor cortex (Meyer-Lohmann et al. 1975). These facts shed a new light on the cerebellar control of movement and on the so-called cerebellar motor learning theories. They suggest that the cerebellum may operate at the highest level of direction and coordination.

7.3. Language and non-motor learning

In their PET study, Petersen and colleagues (1989) found activity in the right cerebellar hemisphere to increase during a language task. Subjects were given lists of words to read and say, hear, and repeat. Then they were given lists of nouns for each of which they had to "generate" and say an appropriate verb. Examples would be "nail" prompting a response of "hammer," or of "boat" a response of "row." Scans from the prior spoken tasks were then subtracted from scans during the "generate" task. This produce activation in the right cerebellar hemisphere, the left anterior cingulate gyrus, the left posterior temporal cortex, and in the left frontal lobe.

Raichle and colleagues (1994) have recently reported the effects on the PET scan of practice of the verb-generate task. In novel trials where word lists were presented for the first time, areas that specifically became active included left prefrontal cortex, left anterior cingulate gyrus, left posterior temporal cortex, and right lateral cerebellum. With continued practice, these areas dropped out, and other areas previously inactive became active. These areas included sylvian-insular cortex bilaterally and the left medial peristriate cortex. They commented:

> These results indicate that two distinct circuits can be used for verbal response selection and normal subjects can change circuits of the brain used during tasks performance following less than 15 minutes of practice. One critical factor in determining the circuitry used appears to be the degree to which a task is learned or automatic. (Raichle et al. 1994, p. 8)

7.4. Mental movement

There is growing evidence that premotor cerebral cortex and the cerebellum participate in imagined movement. I have already referred to the word-generate tasks studied by Petersen et al. 1989. These first showed activity in the right cerebellum hemisphere during silent speech – but only on some tasks. Decety et al. 1990 then showed that the cerebellum is activated during imagined movement – a game of tennis. Surprisingly, silent counting also produced activation of the cerebellum. It is not clear whether the regions were the same. This was confirmed by Ryding et al. 1993.

Again, with PET as for animal recording studies, one cannot be certain the brain activity was not causally related to some unobserved activity such as movement of body parts, which we have found to be particularly the case with the cerebellum (Thach et al. 1993). Therefore ablation studies form an important control and confirmation that the observed phenomenon is indeed correlated with and causally connected to the putative behavior.

7.5. Cerebellar deficits: impaired language generation, planning in games; impaired learning and error detection in both

Fiez et al. (1992) supplied this confirmation in their study of a patient who had had an infarction of the right cerebellar hemisphere (posterior inferior cerebellar artery, PICA, distribution) (Fig. 6).

His language was not noticed to be abnormal during his hospital admission, and after discharge he returned to his law practice. Thereafter he noted only an

> increase of "slips of the tongue" [which, based upon his self-explanation, appeared to be semantic paraphasia]; a decline in his "instant recall abilities," e.g., the ability to instantly associate the clients' names with details of the cases; and a shortened attention span, which he attributed to the cessation of smoking. When questioned, RC1's [the subject] wife also mentioned his "slips of the tongue," but did not report any other problems or changes in his personality.

Specific tests included memory quotient, digit span, tapping span, delayed match to sample, Boston Diagnostic Aphasia Exam, Boston Confrontation Naming, Token Task, Wisconsin Card Sorting Test, Picture Arrangement, Word Fluency (CFL test). His scores were at or above normal on all of these.

Yet on the verb generation task, his performance was abnormal in two ways. First, he did not improve his reaction time as a function of practice. Whereas normals reduced theirs by approximately 30%, he showed little or no reduction. Second, he made errors by choosing inappropriate words. Examples included: for the test word "money," the incorrect response "market" (a control gave "spend," the most common response across controls); for the test word "razor," the incorrect response "sharp" (the control gave "shave," the most common response across controls). Whereas controls made between 10 to 20% such errors, the patient made between 40 to 75% on subsequent repetitions. He did not correct and seemed unaware of these errors.

On the Tower of Toronto Task, he required more moves than normals to complete the task, and failed to improve to the same extent with practice. On the first block of trials, normals averaged about 30 moves, while the patient required over 60 (the minimum necessary was 15). After 5 blocks of trials, normals had reduced to 20 moves, and the subject to 45. Normals continued to improve over blocks 6–10, while the patient plateaued.

The authors summarized:

> performance on standard tests of memory, intelligence, "frontal function," and language skills was excellent, [but] he had profound deficits in two areas: (1) practice related learning; (2) detection of errors. Considered in relation to cerebellar contributions to motor tasks, the results suggest some functions performed by the cerebellum may be generalized beyond a purely motor domain. (Fiez et al. 1992, pp. 155)

How "profound" the deficit was is a matter of scale and of focus. The failure to improve in reaction time and in the Tower game would pass as normal unless specifically looked for and quantitated against controls. The word generation errors appeared to stand out to interested observers, including his wife. Even so, many might consider the deficit rather subtle. Certainly, he could perform the tasks, and did show improvement with practice, albeit at a measured and grossly reduced magnitude.

In sum, one PET and one ablation study of "non-motor learning" gave results surprisingly similar to those of "motor learning." In PET, similar parts "lit up" during the learning, which "dropped out" during learned performance (prefrontal cortex and lateral cerebellum). Some parts involved in learned performance were not active during learning (sylvian-insular cortex bilaterally and the left medial peristriate cortex). In ablation, there was impairment (but not elimination) of task learning.

7.6. Sequences

Inhoff et al. 1989 tested the ability of cerebellar patients to generate sequences of one, two, and three movements. In normals, the reaction time increased sharply with the number of movements in the sequence. This indicates that some preprogramming has occurred (somewhere in the nervous system), whose duration is proportionate to the length of the sequence. Cerebellar patients failed to show this result, in proportion to the degree of their disease. Those with mild disease showed less delay in reaction time with each added segment in the sequence. Those with severe disease had almost the same reaction time independent of the length of the sequence. The authors commented that it was as though they were "decomposing" the movements in time, as Babinski, 1899 and Holmes, 1939 had demonstrated for compound movements. That is to say, they generated each of the series as if dealing with it singly, rather than in a unit.

Fiez et al. (1992), in the single patient, and subsequently Grafman et al. (1992), in a series of patients with cerebellar disease, showed difficulties in performing the Tower games of Toronto and Hanoi, respectively. Both studies showed that patients with cerebellar disease achieved lower scores (took more moves) and made more errors (more wrong moves) than did normals. Fiez et al. give quantitative information showing that the impaired subject made about twice as many moves and errors as compared to normals. Grafman et al. only say that the impairment was statistically significant. It is important to point out that after cerebellar disease the subjects could still plan ahead well enough to play and finish the game. Whatever the cerebellum contributed, it was apparently useful, but not necessary – *in these tasks.*

7.7. Timing

A very robust finding has been that of Ivry and Keele and colleagues (Ivry et al. 1988; Keele & Ivry 1990) on mental timing. In subjects asked to compare control and test tone bursts of different durations (of the order of half a second), patients with lateral cerebellar infarcts scored poorly, as if random. The deficit was felt not to be due to making timed occult movements: the laterals did not show deficits in motor timing, and those with intermediate zone disease showed deficits in motor timing but not in perceptual estimates. The result was interpreted to imply that the lateral cerebellum was the clock of the nervous system, independent of motor activities.

7.8. Autism and attention

There is a considerable literature that connects the cerebellum with schizophrenia, autism, and attention. Berman et al. (1974), Weinberger et al. (1980), and Berntson & Torello (1982) have shown that brains of schizophrenic patients often show cerebellar atrophy in the posterior vermis. Floeter & Greenough (1979), showed that "deprived rearing" of infant monkeys produced both an autistic condition and a lack of development in Purkinje cell dendrites. Yet these studies have not dissociated the mental from the known motor abnormalities that human schizophrenics and deprived-reared monkeys both show. Thus it remains to be seen what the correlation means.

7.9. Other

Appollonio et al. (1993) have recently performed a number of tests on their series of patients with cerebellar atrophy. Tests included those of general intellectual ability, different aspects of memory (effortful, automatic, and implicit), speed of information processing, and verbal fluency (both category and letter fluency tests). Cerebellar patients were significantly impaired only in "tasks requiring the use of *executive* [emphasis mine] functions, such as the initiation/perseveration subtest of the Mattis Dementia Rating Scale or the fluency tests, and on memory measures requiring greater processing effort." The authors concluded that "the impairment is secondary to a deficit in executive functions." I shall comment on this shortly.

Bracke-Tolkmitt et al. (1989) previously had conducted a broad neuropsychological study of cognitive functions in their patients with cerebellar disease. IQs were found to be slightly lower in the cerebellar group, but most tests were in the normal range. They were significantly abnormal on 3 tests. These included the immediate re-drawing of the Benton figures, the learning of verbal paired associates, and in learning arbitrary associations between colors and abstract words. Again, the impairment was that of reduced – not abolished – performance. At their worst, in the word-color learning, and the Benton figure recall, subject scores were about half those of normals.

These cognitive contributions of the cerebellum could seem to be very like the popular conception of its motor contributions – "fine control" and "coordination." In the mental as in the motor performance, the ability to act is never entirely lost, but only degraded. This in turn could conceivably result from a general "tuning up" function across many brain parts, as originally proposed by Holmes (1939) for movement. Somehow, this is not intuitively very satisfying.

8. What does the cerebellum specifically contribute to cognition?

8.1. What does the cerebellum contribute to mental movement?

The Leiners have argued for a cerebellar influence on frontal lobe mentation. Much of their careful reasoning depends on there being anatomical projections from the cerebellum via thalamus to cerebral frontal association cortex. Whereas evidence for these connections was at the time rather scant, evidence for the connections has steadily increased. Anatomical connections from cerebellum to far frontal association areas were proposed in monkey by Sasaki et al. (1976) using electroanatomical methods. In man, a phylogenetically new and unique posterolateral part of dentate was proposed by Hassler (1950) and Leiner et al. (1991) to project via thalamus to far frontal cortex. The Leiners inferred that the cerebellum could contribute to whatever was processed in these areas. Though they did not specify the type of mental operation performed in these areas, they did specify regions that are now identified as being higher order motor. And while they did not say exactly what the cerebellum might provide to these areas, because of the uniformity of the cerebellar architecture and the likelihood that the algorithms were similar or identical to those used for motor control, they used the terms "coordination" and "skill."

Let us return to the fact that the cerebellum is active during mental movement (Decety et al. 1990 and Ryding et al. 1993), which in itself could have been a spurious correlation; the cerebellum could have been active in relation to some unobserved synergic muscle activity that was associated with – but was not necessary for – the performance of the task. But the fact is now clear that ablation of the cerebellum *does* impair mental task performance. So somehow the cerebellum is involved, and the question is "how"?

8.2. Context triggering and combination

For motor learning, I have pursued the idea that the cerebellum may link a behavioral context to a motor response. The response may be a combination of many downstream neural elements firing together. Both the context-response linkage and the response composition would be achieved through trial and error learning. After practice, the occurrence of the context would trigger the occurrence of the response. This would explain how combinations of muscles may become active all at once, especially in skilled movements outside the capabilities of the motor pattern generators, so as to produce coordinated behavior appropriate to a specific context.

I have extended the idea to include the premotor parts of the brain, to which the cerebellum is now known to project, and which are known to play a role in mental movement and cognition. These areas are active in the planning of movements that are then executed, and they plan movements that are not to be executed. They "think" movements. From recent evidence, the cerebellar output extends even to what has been characterized as the ultimate frontal planning area, the "prefrontal" cortex, area 46. The cerebellum may be involved in combining these cellular elements, so that, through practice, an experiential context can automatically evoke an action plan. The plan would be in the realm of thought. The plan either could – or need not – lead to execution. Again, the specific cerebellar contribution would be one of linkage of the context to a specific response, the combination of the response from simpler elements, and these accomplished through repeated practice. The prefrontal and premotor areas could still plan without the help of the cerebellum, but not so rapidly, automatically, or so precisely linked to context. At some level of task complexity, cerebellar damage would reveal itself in behavioral errors. This would have nothing to do with other cognitive activities – visual, auditory, attention, and so on, and areas of the brain to which the cerebellum does not project.

8.3. Mental rehearsal of motor performance

It has been well documented that mental rehearsal improves motor performance (see Jeannerod 1995). This is common knowledge amongst musicians, athletes, chess players, actors, and lecturers. Repetitively playing through the performance in one's mind can remarkably improve the next actual performance. Is there any way in which the cerebellum might play a role in motor learning in which only mental movement is practiced?

A hitherto curious and unexplained oddity in the structure of the primate motor system is the evolutionary change in the connectivity of the red nucleus. As mentioned previously, the magnocellular (phylogenetically older) red nucleus gives rise to a prominent rubrospinal tract in carnivores, and receives from motor cortex and the cerebellar interposed nucleus. It is supposed to play a role similar to that of the corticospinal pathway in providing voluntary control of small distal muscle groups. By contrast, the parvocellular (phylogenetically newer) red nucleus receives from premotor cortex (area 6) and the cerebellar dentate nucleus. Its output is thought to go not to the spinal cord but exclusively to the principal portion of the inferior olive, which in turn projects back to the lateral hemisphere of the cerebellum. It seems odd that a system capable of firing at high sustained frequencies should funnel into and dead-end in the conspicuously low frequency inferior olive. The suggestion had been made that this might fit into the general cerebellar role of motor learning (Kennedy 1990), but a tighter rationale had not been provided. Mental motor rehearsal could well be the specific framework for its role in learning motor performances that are only practiced internally and not overtly expressed. The contexts would be brain states corresponding to prior elements of the performance, the combined responses would be of premotor neuron assemblies in area 6 (and 8, 9, 44, 45, 46?), the linkage would be established through repetitive practice. It is not clear what mental performance errors might consist of, or how they might be detected.

The cerebellar contribution to mental (and motor) performance would thus become critical when the task (1) involves imagined movement that is rapidly and automatically triggered by context, (2) contains a number of linked neural (or body part) components, and (3) when these properties are being adapted or newly acquired. Paradoxically, the context-response coupling and the response combinations would be "unconscious" aspects of thought. They would have been at a level of awareness only during the learning phases. Ironically, this would be something that we have learned and no longer "know" anything about because it has been given over to the cerebellum for the implementation of automatic motor control, actual or imagined.

8.4. "Generated" language

It has been known that cerebellum lesions produce dysarthria – Holmes localized them to the midline vermis, Lechtenberg and Gilman (1978) to the midline and just to the left of it. The deficit is one of articulation – speech is variable in pronunciation (as if all the component muscles of speech can not be got together at once in the proper combination), and slow (as if all the components can not be got together linked in a sequence, and are instead emitted one at a time). It is not known to be context or task specific.

There is the fairly common condition of "cerebellar mutism" seen uniquely in children after cerebellar surgery (e.g., Rekate et al. 1985). In earlier descriptions, it was unclear whether this condition was one of abulia, autistic withdrawal, aphasia, aphonia, or anarthria. As more cases have been reported, the features have become clearer. Except for one case reported in a young adult, the syndrome has been confined to children. Characteristically its onset is delayed (in over half the reported cases) as much as 4 days after the injury. Often it does not occur with the initial surgery, but after some complication such as infection, hemorrhage, or reoperation. There may (or may not) be frank signs of brainstem involvement. Opinion is divided as to whether this condition results specifically from cerebellar damage, or due to some remote delayed effect – for example, anarthria of brainstem origin. The problem in definitely implicating the cerebellum arises from the facts that: (1) most cerebellar surgery – in all respects equivalent – is unaccompanied by the mutism, (2) the delay after the cerebellar damage, and (3) the restriction to childhood. By contrast, the practicing neurologist and neurosurgeon, who see many acute and subacute cerebellar injuries in adults, never see them produce mutism, autism, or schizophrenia.

Quite a different situation is found with the verb generating tasks in the studies of Petersen et al. (1989), Raichle et al. (1994), and Fiez et al. (1992). There, localization is distinctly to the right cerebellar hemisphere, the phenomenon occurs with some speech and language tasks and not others, and the effect wears off with familiarity.

8.5. Is variegated vocal sound production a basis of language?

Lieberman (1969) has suggested that language began phylogenetically as an increased ability to make different vocal sounds. Proposed to originate with phylogenetic developments in the primate larynx, it paralleled the ability to fractionate movements and the dexterous use of the right hand, which is dependent on the development of the primate neo-motor cortex. The argument runs that a capacity for making a great number of sounds (movements) gives one a vocabulary and dictionary of as yet meaningless items to which one can then begin to assign symbolic identities. With the passage of time, the brain evolved to use the motor substrates for communication, symbolic representation, and some forms of thought.

8.6. "Automatic" speech

We begin to speak before we understand language. Beginning with babbling in infancy, we proceed through "rote learning" of nursery rhymes, nonsense poems, and jingles without necessarily understanding them. We learn and can recite "Jabberwocky" as movement and not at all as language.

A somewhat similar case is mnemonic sayings. We rote-memorize something that has so little linguistic or logical connection among the elements that it is learned as a movement. We can listen to what we say in order to get at what we otherwise can't remember. For example, "Thirty days hath September . . ." allows us to remember how many days there are in each month. But it is not something we know. It is buried in a rote-learning movement sequence. Physicians have and use a similar formula to recall in order the 12 cranial nerves; those few American students who can recall all the Presidents in the correct order will have committed the first syllable of each name to a catechism to verbal (motor) memory. One can suggest that recitation of the alphabet, multiplication tables, are similar in nature.

8.7. Speech and gesture

Another common observation is the coupling of gesture, especially of the face and hands, with speech. At what level in behavior and in the brain is this coupling? Do we gesticulate when we make mental speeches? Do we think of a word more quickly by thinking of it as a whole with its associated gesture?

8.8. Is such rote-learned, automatic language dependent on any one brain-part to the exclusion of others?

People with lesions of the left frontal lobe operculum may suffer loss of "propositional" speech, yet retain relative amounts of other utterances – nursery rhymes, songs, the alphabet, multiplication tables. This is not to say that all of these sayings are "stored" in the cerebellum. No study yet has shown loss of nursery rhyme, childhood jingle, or verbal mnemonic from cerebellar ablation. Yet Bellugi's studies (Bellugi et al. 1990) have suggested a role for the cerebellum in automatic speech. Children with Down's syndrome and a diseased cerebellum are limited in their development of speech, with relative sparing of other intellectual activities. By contrast, children with Williams' syndrome have an atrophic cerebrum with a relatively preserved cerebellum, and are conspicuous for their precocious "cocktail conversation." Socially responsive, fluent, correct in language content and appropriate to context, their language is nevertheless peculiarly devoid of propositional content.

In this light, one is tempted to interpret the PET localization of "language" in the right cerebellar hemisphere (Petersen et al. 1989) as indicative of a fundamental process in automatic speech. The process of attaching a verb (a thought movement) or rhyme (a rehearsed movement) to a noun-context. And in Fiez's study, damage of the area results loss of automatic utterances of "the right word," and inability to make new associations.

8.9. Sequences

If it is true that the cerebellum participates in context-response coupling through learning, and response formation, then the same mechanism could explain sequential behavior. Because of the richness of the input to the cerebellum, it is conceivable that any internal state at almost any level of the central nervous system hierarchy could serve as a context. If the cerebellar targets are themselves high within the behavioral "planning" parts of the brain, plans for movement component "A" could trigger movement component "B" long before "A" was ever actually executed.

Such a mechanism could also account for the improvement in performance through mental rehearsal, and "non-motor" learning. The primary sequence becomes automatic, and we can proceed to build higher order structures on top of it. Through mental and actual practice, the moves of chess men may become so well learned that one may think at higher levels of strategy in planning the next moves, even through to thinking through an entire game. Being able thus to anticipate errors would help to prevent them.

8.10. Timing

Melvill-Jones and Watt (1971) showed that certain motor acts have precise and stereotyped timing. In studies of hopping and stepping, they found that a group of students hopped at a frequency of 2 Hz, with little variation. They observed that this was the same frequency as the then popular dance music. They measured the latency of the segmental and also of the "functional" or long loop stretch reflex in the gastrocnemius at 120 msec. They surmised that the reflex, triggered by the setting down of the foot, was optimally timed to assist in the lift-off of the next foot rise. The reflex and body mechanics were therefore suggested essentially as being a "clock" for the body in this activity.

Decety and Michel (1989) reported that imagined movements take proportionately the same time as when actually performed. This is somewhat surprising, since part of the mechanism proposed by Melvill-Jones and Watt must have depended on the mass-mechanics of the system, including inertia, elasticity and viscosity of tissues, and muscle contraction times, which would influence the resonance properties at least as much as the neural conduction times. This in turn suggests that the movements themselves are modeled within the nervous system, and that there is, as Jeannerod has recently argued, a complete neural motor "representation" (Jeannerod 1994). Whatever the centers and connections are that complete this representation, its capability for precise timing could well serve as a clock for "non-motor" activities.

What Melvill-Jones and Watt showed for hopping is equally true for many other activities. We say "one thousand one, one thousand two . . ." to measure exposures in the darkroom, which have been observed to come very close to the times given by a clock. We sing to pace and synchronize bodily activities of dancing, marching, rowing, hauling, and many work activities. We can remember the meter of the song, because it is timed to the movement activity. Oddly, despite the physical differences between mental movements and actual movements, the timing is surprisingly similar, as Decety and Michel have shown. The question then becomes "Is there any activity that is mentally timed that does *not* have a movement referent?" One could imagine that in an attempt to time a process, one set up some internal standard that consisted of a mental motion of some part of the body. Since movements take time, and since the time taken by mental movements is equivalent to that taken by actual movements, the mental movement becomes a "clock" against which any external event may be compared. This could be a phrase, a tapping of the foot, a rocking of the body, or the blink of an eye – consciously, or not quite consciously employed. Implicit is that the movement have some period appropriate to the duration of what is to be measured. The "beat" need not involve the inferior olive! (See Keating & Thach 1995)

8.11. Does the cerebellum control everything mental?

Some researchers have suggested that the cerebellum may have a very general influence on cognition, as if to participate in anything or everything mental. See the reviews of Botez et al. 1985; 1988; 1989; Ito 1990; Leiner & Leiner 1989; Leiner et al. 1986; 1987; 1991; Macklis & Macklis 1992; Schmahmann 1991.

Akshoomoff & Courchesne 1992, Akshoomoff et al. 1992, and Murakami et al. 1989 have raised questions about a cerebellar contribution to attention. These observations have been based on a few cases who had disease of other parts of the nervous system as well.

Meyer believed that schizophrenia was essentially a learning disability, a maladapted response to social contexts. Bleuler believed that in this disease, the personality, rather than having the elements properly combined, was dissociated into its constitutive elements. There are the unquestioned correlations listed above between cerebellar atrophy, schizophrenia, autism, "cerebellar mutism," and experimentally socially deprived rearing. Nevertheless, at the present time, none of these correlations satisfy the criteria set out at the beginning for demonstrating cause and effect. (1) The focal anatomy of these conditions is not well enough understood to know whether the cerebellum projects to it or not. (2) Cerebellar ablation in the normal individual has not reproduced these conditions. (3) No cerebellar PET or single unit activity pattern has been particularly correlated with these aspects of behavior. This is not to say that these objections will not be overcome in the future.

8.12. How might one go about testing such a theory?

One might begin with PET and ablation studies where mental task performance requires mental movements. It would be of interest to break these down into categories of context-triggered responses, of simple versus compound responses, and of the learning of each. Many of the tasks described herein that appear to engage the cerebellum fall into one or another of these categories. Examples include games where the player has to "play ahead to see the errors," and develop correct playing strategies. A good checkers or chess player can play an entire game in his head; a really good player, several games at once. The goal would be to identify and dissociate the context-response coupling from the response formation from the learning. The control for this could include mental tasks involving imagination in a spatial domain but not movement – such as solving abstract math problems, hearing music, seeing maps and faces, and learning them. But even these control tasks might be deceptive, as it might prove to be very difficult to "think" without employing mental movement.

These tests could be done with human PET and fMRI scans and in human ablation studies. However, tests of mental timing would require greater sensitivity to dynamics. If the cerebellum and premotor cortex were involved in periodic rhythmic activity, simulating movement of body parts, such as might be used as the mechanism of a biological clock in the physiological motor domain, electrical analysis of the circuit (EEG, ERP, macro- or microelectrode) would be required to look for signs of such periodic activity.

9. Summary

What PET and ablation studies tell us about the role of the cerebellum in motor learning in humans is that the cerebellum is but one part of a larger system that includes primary, supplementary, and premotor cortex, basal ganglia, red nuclei, and especially the prefrontal cortex. PET studies show that the cerebellum and prefrontal cortex participate in "motor learning" and "motor performance" in different ways. Ablation studies show that the cerebellum may play a small part or a large part in "motor learning," depending on what is being learned. Ablation may slightly impair or abolish motor learning, depending on the task. Both types of study raise but have not themselves resolved the question of what essential factor the cerebellum contributes to movement that may be learned.

The greater contribution of human functional imaging and ablation studies has been to show that the cognitive functions may require the involvement of the cerebellum in imagined movement. Our thesis is that imagined movement is similar or identical to the early initiatory phases of actual movement. The cerebellum plays the same specific role in "coordinating" these imagined movements that it does in actual movements. I have supported the Brindley-Marr-Albus-Gilbert theories and suggest that these cerebellar pathways are used to build through trial and error learning behavioral context-response linkages, and to build up appropriate responses from simpler constitutive elements.

As cerebellar ablation does not abolish actual movements, but only their fine control and coordination, neither does it abolish imagined movements, but only their automaticity, speed of response, stereotypy, and the ability to improve them with practice. It is presumably due to the same inexactness in the use of the central representation of movement that, after cerebellar damage, errors are made in the movement-associated mental activities that are required for some "cognitive" task performances.

ACKNOWLEDGMENTS

Marcus Raichle and Steven Petersen revealed subtleties of PET methodology; Marc Jeannerod, Michael Arbib, Giacomo Rizzolatti, and Hideo Sakata led me through mental motor imaging; Julie Fiez, Peter Strick, Peter Gilbert, Jeffery Keating, Howard Goodkin, Amy Bastian, and Tod Martin helped with these attempts at explanation. Nine unnamed referees corrected many errors. Robert S. Dow, Henrietta Leiner, Alan Leiner, David Marr, James Albus, and Masao Ito did most of the original thinking.

Open Peer Commentary and Authors' Responses

Table 1. *Commentators for special cerebellum issue*

Commentators	Target article authors							
	Linden	Crépel et al.	Kano	Vincent	Houk et al.	Simpson et al.	Smith	Thach
Arbib, M.A.	[DJL]	[FC]	[MK]	[SRV]	[JCH]	[JIS]	[AMS]	[WTT]
Baudry, M.	[DJL]	[FC]		[SRV]				
Bekkering, H., Heck, D., & Sultan, F.					[JCH]			
Bindman, L.J.	[DJL]	[FC]		[SRV]				
Bower, J.M.		[FC]			[JCH]		[AMS]	[WTT]
Calabresi, P. Pisani, A., & Bernardi, G.	[DJL]	[FC]						
De Schutter, E.	[DJL]	[FC]	[MK]	[SRV]	[JCH]	[JIS]	[AMS]	[WTT]
Dean, P.					[JCH]			
Dufossé, M.						[JIS]	[AMS]	
Feldman, A.G., & Levin, M.F.					[JCH]	[JIS]	[AMS]	[WTT]
Fiala, J.C., & Bullock, D.	[DJL]	[FC]	[MK]	[SRV]	[JCH]	[JIS]	[AMS]	[WTT]
Flament, D., & Ebner, T.J.								[WTT]
Gielen, C.							[AMS]	[WTT]
Gilbert, P.F.C.					[JCH]			[WTT]
Gomi, H.							[AMS]	
Haggard, P.								[WTT]
Hallett, M.								[WTT]
Hartell, N.A.	[DJL]	[FC]		[SRV]				
Hepp, K.					[JCH]	[JIS]		
Hesslow, G.					[JCH]			
Hirano, T.	[DJL]	[FC]		[SRV]				
Hore, J.							[AMS]	
Houk, J.C., & Alford, S.	[DJL]	[FC]	[MK]	[SRV]	[JCH]			

(*continued*)

Table 1. *(Continued)*

Commentators	Target article authors							
	Linden	Crépel et al.	Kano	Vincent	Houk et al.	Simpson et al.	Smith	Thach
Jaeger, D.			[MK]		[JCH]	[JIS]		[WTT]
Kano, M.		[FC]						
Kawato, M.					[JCH]	[JIS]	[AMS]	[WTT]
Kiedrowski, L.				[SRV]				
Latash, L.P., & Latash, M.L.					[JCH]		[AMS]	[WTT]
Miall, R.C., Malkmus, M., & Robertson, E.M.					[JCH]	[JIS]		[WTT]
Mori-Okamoto, J., & Okamoto, K.	[DJL]	[FC]		[SRV]				
Okada, D.	[DJL]	[FC]		[SRV]				
O'Mara, S.M.								[WTT]
Paulin, M.G.					[JCH]		[AMS]	[WTT]
Roberts, P.D., McCollum, G., & Holly, J.E.					[JCH]	[JIS]		
Schmahmann, J.D.					[JCH]			[WTT]
Sultan, F., Heck, D., & Bekkering, H.					[JCH]			[WTT]
Swinnen, S.P., Walter, C.B., & Dounskaia, N.					[JCH]		[AMS]	[WTT]
Thompson, R.F.	[DJL]	[FC]	[MK]	[SRV]	[JCH]	[JIS]	[AMS]	
Timmann, D., & Diener, R.F.								[WTT]
van Donkelaar, P.					[JCH]		[AMS]	[WTT]
Van Galen, G.P., Hendriks, A.W., & DeJong, W.P.							[AMS]	
Weiss, C., & Disterhoft, J.F.					[JCH]	[JIS]		[WTT]
Wessel, K.								[WTT]

Open Peer Commentary

Commentary submitted by the qualified professional readership of this journal will be considered for publication in a later issue as Continuing Commentary on this article. Integrative overviews and syntheses are especially encouraged.

Spanning the levels in cerebellar function

Michael A. Arbib
Center for Neural Engineering, University of Southern California, Los Angeles, CA 90089-2520. **arbib@pollux.usc.edu**

Abstract: We ask what cerebellum and basal ganglia "do," arguing that cerebellum tunes motor schemas and their coordination. We argue for a synthesis of models addressing the real-time role and error signaling roles of climbing fibers. "Synthetic PET" bridges between regional and neurophysiological studies, while "synaptic eligibility" relates the neurochemistry of learning to neural and behavioral levels. [CRÉPEL et al.; HOUK et al.; KANO; LINDEN; SIMPSON et al.; SMITH; THACH; VINCENT]

1. Does the cerebellum control muscles or tune motor schemas? SMITH (sect. 2.2, para. 5) tells us that the mutant mouse Lurcher, in which no Purkinje cells survive beyond early adulthood, "show deficits in both the ability to simultaneously (e.g., asynergia) and sequentially (e.g., dysdiadocokinesia) command the desired muscle synergies." However, the spinal cat can walk on a treadmill if properly supported and stimulated, and so I would argue that cerebellum serves to adjust the spinal motor schema for walking rather than "commanding" the muscle synergies (For clarity, I will reserve "synergy" for this sense of "muscle synergy," and "motor schema" for a task-specific "program" of coordinated motor control.) Earlier, SMITH notes "The locomotion was very ataxic and the frequent interruptions from a loss of equilibrium accounted for the absence of modulation in the contralateral limb"; moreover (Smith, personal communication), Lurcher mice can coordinate their limbs for swimming. This all suggests that the problem for Lurcher mice is not generating the locomotor synergies per se (these combine to yield the motor schema for walking), but in coordinating them with anti-gravity synergies. Smith offers the hypothesis that "two populations of Purkinje cells [might act] . . . together . . . to execute the reciprocal switching of excitation from one agonist group to another, or alternatively, to allow agonists and antagonists to co-contract (sect. 2.7, last para.)." However, noting, for example, Orlovsky's (1972) demonstration that, during walking, stimulation of Deiters's nucleus yields increased activation of an extensor muscle only when the extensor is being actively employed, Boylls and I (Arbib et al. 1974; Boylls 1975) argued that the cerebellum modulates "tuning parameters" of synergies (motor schemas) located elsewhere in the brain, rather than acting directly on motoneuron pools.

THACH argues for a three-fold role for cerebellum: (1) Brindley's hypothesis that, after learning, the occurrence of a context (represented by a certain input to the cerebellum) will trigger (through the cerebellum) the appropriate motor response, with the "learned" movement being automatic, rapid, and stereotyped; (2) the ability of cerebellum to form novel combinations of downstream elements; and (3) the extension of the second idea to the premotor parts of the brain and their role in cognition. The previous paragraph argues that the "elements" in (2) may be parameters of complex motor schemas rather than single muscle contractions; this makes (3) more plausible since motor schemas are but one example of schemas considered as units of cognition as well as perception and action (Arbib & Caplan 1979).

However, I doubt that the cerebellum "takes over" from other regions. Recently, Nicholas Schweighofer and I have developed a series of models which provide useful insights into adaptation of saccades (Schweighofer et al. 1996) and dart throwing (Arbib et al. 1995). In each case, the cerebellum acts by modulating a motor schema (in brainstem and premotor cortex, respectively) to yield a compound system with increased control accuracy. This system maintains errors within the limits that can be kept "below the conscious level" even for rather fast movements. Our work in adaptive control of reaching (Hoff & Arbib 1992) suggests that motor control does not involve replacement of a feedback system by a feedforward system; rather, once a feedback system is well-tuned it can expect smaller errors and can thus proceed effectively at velocities high enough to appear "ballistic" unless perturbations yield an "unexpectedly" large error. The Schweighofer and Arbib models seem at variance with the adjustable pattern generator (APG) model which is at the heart of the article by **HOUK et al.** in which motor control is mediated through a set of APGs, each of which generates an "elemental burst command" (sect. 3), but more in line with the saccade models (sect. 3.7) such as the model of Dean et al. (1994) in which the cerebellum adjusts the gain factors of the brainstem saccade generator.

In sect. 7.2, THACH argues that the cerebellum can "run the whole show" since, for example, electrical stimulation of output nuclei can reliably cause movement. However, the cerebellar cortex inhibits the cerebellar nuclei, and so if the cerebellum "runs the show," then the nuclei must be in charge. THACH gives no account of studies of nuclear activity which would support this view. Given a "cooperative computation" view of the brain in which many regions are constantly interacting to shape overall behavior, and given much classic data (e.g., Holmes 1939) that the overall plan of a movement survives cerebellar damage, it seems more consistent to hold that, for example, cerebral cortex and basal ganglia cooperate to explicitly combine the "pieces" that make up a skilled behavior, while the cerebellum adjusts parameters to adapt and coordinate components of the movement to yield a seamless whole. Thus, in some cases, these "parameter adjustments" may yield an overt action, while in other cases they can only be observed behaviorally if the appropriate phase of a motor schema is active. In support of the competing claims of the basal ganglia to "executive status," consider the akinesia of Parkinson's patients and the functions of basal ganglia in suppressing inappropriate movements, suppressing forthcoming movement during preparation, and in switching between overt and covert behaviors (Hikosaka et al. 1993).

2. Climbing fibers: Real-time role and error signaling. In responding to **SIMPSON et al.,** I must confess a dichotomy in my approach to modeling the role of climbing fibers. In the Boylls's model, there is no learning. Rather a post-complex-spike lowering of a Purkinje cell's activity yields disinhibition of nuclear cells (Murphy & Sabah 1970); Boylls showed how patterned activity of climbing fibers could then assist the creation of a "working memory" of motor parameters encoded in the activity of subcerebellar loops (Tsukahara 1972). The recent Schweighofer models posit no "real-time" role for the climbing fibers, but instead use them only as "error detectors" (more on this below). But I still believe that climbing fibers have real-time as well as learning roles; the next generation of models must reflect this.

The rest of this commentary addresses issues involved in bringing studies of the role of cerebellum at the gross level (linking behavior to ablation and imaging studies) to ever finer analyses at the neurophysiological and neurochemical levels.

3. Linking animal models to human brain imaging: Synthetic PET. While we have not applied this to models of cerebellum, we (Arbib et al. 1995) have introduced a new computational technique, Synthetic PET, to use simulations with neural models based on single-cell neurophysiology for the prediction and analysis of results from PET studies of a variety of human behavior. In each case, the problem is to find an integrated measure of activity in each simulated neural group which provides a predictor for the 3D volume of the image to which the neurons in this group correspond. The key hypothesis (shared by THACH) is that PET is correlated with the integrated synaptic activity in a region, and thus reflects in part neural activity in regions afferent to the region studied, rather than intrinsic neural activity of the region itself. However, the method is general, and can potentially accommodate other hypotheses on single cell correlates of imaged activity, and can thus be applied to other imaging techniques, such as functional MRI, as they emerge. Of particular interest is whether there is an added "metabolic cost" for synaptic plasticity as distinct from activation of a stable synapse. In other words, does the greater cerebellar activity seen during certain forms of learning signal only the increased synaptic activity of climbing fibers, or does it contain a signal for synaptic plasticity per se?

4. Linking behavior to neurochemistry: Synaptic eligibility. It is stressed by **HOUK et al.** that models of motor learning must confront the credit assignment problem, which is the difficulty of directing training signals to the appropriate sites in the network, and at the appropriate moments in the training process, in order for learning to be adaptive. They note that network theorists "typically address temporal credit assignment by assuming a trace mechanism that provides a short-term memory of preceding synaptic events until the arrival of the corresponding training information" (sect. 4.1). The Schweighofer models have offered a variation on this idea, describing the trace by a second-order differential equation designed to yield peak "eligibility" for change of parallel fiber → Purkinje cell synapse if a climbing fiber signal occurs "at the right time" after parallel fiber activity at the synapse. Here, "right time" refers to the time between cerebellar involvement in generation of a motor command, and receipt of an "error signal" (perhaps 200 msec later) on a climbing fiber. Moreover, we have now offered a detailed neurochemical model of such eligibility. This poses the following questions to **CRÉPEL et al., LINDEN,** and **VINCENT:** What are the prospects for developing precise descriptions of the kinetics of neurochemical mechanisms underlying the plasticity of these synapses so that models such as ours (Schweighofer & Arbib, to appear) can be tested quantitatively? The question must also be put to **KANO,** whose study of "long-lasting potentiation" addresses the point made by **HOUK et al.** and in our own work, that LTD alone cannot be the only learning mechanism for these synapses, since otherwise all synapses would eventually be driven to a state of minimal responsiveness.

5. Who teaches the teacher? Models of motor learning, **HOUK et al.** stress, must confront the problem of defining the information that is likely to be available for guiding the learning process in the organism. Looking at the adaptation of dart throwing to prisms (as analyzed by **THACH** and modeled by Arbib et al., 1995) throws the question of the climbing fiber error signal into focus. For saccade adaptation, it is plausible to postulate that evolution provided inferior olivary cells responsive to retinal error and projecting to the part of cerebellum involved in saccades. However, it is implausible that evolution should provide innate olivary circuitry to signal the position of a dart relative to the bullseye. Thus, even when all the questions of "Motor learning and synaptic plasticity in the cerebellum" are resolved, the question of "Who teaches the teacher?" will remain.

Similarities and contrasts between cerebellar LTD and hippocampal LTP

Michel Baudry

Neuroscience Program, University of Southern California, Los Angeles, CA 90089-2520. **baudry@neuro.usc.edu**

Abstract: In this commentary, I review the articles by Crépel et al., Linden, and Vincent to compare and contrast the similarities and differences between hippocampal LTP and cerebellar LTD. In particular, the role of glutamate receptors in the induction of plasticity, the participation of different second messenger pathways, and the mechanisms involved in expression/maintenance of these forms of synaptic plasticity are discussed. [**CRÉPEL et al.; LINDEN; VINCENT**]

The target articles by **CRÉPEL et al., LINDEN,** and **VINCENT** discuss the cellular mechanisms underlying LTD in cerebellum and illustrate the difficulties inherent in this type of review that raise more questions than they provide answers for. As the work on cerebellar LTD is still in its infancy as compared to the maturity of the work on hippocampal LTP, it might be useful to point out obvious similarities and contrasts between these two examples of synaptic plasticity. In doing so, it is possible that new ideas will emerge and provide testable hypotheses. In my commentary, I will use the three reviews mentioned above to compare and contrast hippocampal LTP with cerebellar LTD and to discuss three issues: the question of the induction mechanisms and the role of different types of glutamate receptors; the question of the cascades of biochemical reactions triggered by the induction mechanisms; and the question of the expression/maintenance mechanisms.

(1) Induction mechanisms. In hippocampal LTP, it is clear that there are multiple forms of LTP, some of which are NMDA receptor-dependent and others which are NMDA receptor-independent (Grover & Teyler 1992; Nicol & Malenka 1995). It is important to note that several forms of LTP appear to coexist at the same synapses. The role of the glutamate metabotropic receptor in LTP induction is still under debate (Bortolotto et al. 1994; Manzoni et al. 1994; Selig et al. 1995), but the most likely interpretation of the data is that activation of the metabotropic receptor results in an up-regulation of NMDA receptor function through a PKC-dependent phosphorylation process (Ben-Ari & Aniksztejn 1995). The suggestion by Crépel et al. of the existence of different forms of LTD (which are defined as LTD_{AMPA} and LTD_{mGlu}) is well in line with the LTP story. As in hippocampal LTP, increased intracellular calcium appears to be the critical step in the induction process. It is interesting to note that, in the current view of plasticity in hippocampal and cortical neurons, a relatively small rise in intracellular calcium leads to long-term depression of synaptic transmission, whereas a higher calcium influx results in LTP (Bear & Malenka 1994). The lack of NMDA receptors on Purkinje cells might possibly account for a relatively small rise in intracellular calcium and thus in LTD. In this regard, it is surprising that the ability of the mossy fiber-granule cell synapse (which has the same complement of glutamate receptors as a hippocampal CA3-CA1 synapse) to exhibit LTP is rarely mentioned.

(2) Biochemical cascades. One of the most debated issues in hippocampal LTP was centered around the pre-versus-post-synaptic localization of the modification underlying the increase in synaptic efficacy (Baundry & Davis 1994). As the induction mechanism is located postsynaptically, advocates of the presynaptic site had to assume the existence of a retrograde messenger released postsynaptically and acting presynaptically to modify the properties of neurotransmitter release. Although arachidonic acid was first proposed (Williams et al. 1989) it is NO which has in recent years generated the most controversy (Barnes et al. 1994; O'Dell et al. 1991; Schuman & Madison 1991). As NO was first identified as a chemical messenger in the cerebellum, it is somewhat ironic that its role in cerebellar LTD remains controversial. What seems clear is that, in the cerebellum, NO is not a retrograde messenger (in the sense that it is not released from Purkinje cells) but rather an anterograde messenger. In contrast to hippocampal LTP, there is little disagreement that cerebellar LTD is due to postsynaptic modifications, thereby eliminating the need for a retrograde messenger.

Multiple calcium-dependent cascades have been implicated in hippocampal LTP, including the activation of calcium-dependent proteases (calpains) (Baudry & Lynch 1993); calcium-dependent phospholipases (PLA2) (Massicotte & Baudry 1991); Protein Kinase C (PKC) (Linden & Routtenberg 1989); cGMP-dependent protein kinase (Zhuo et al. 1994); and calcium/calmodulin kinase II (CamKII) (Malenka et al. 1989). Moreover, phosphatases have been shown to play a critical role in depotentiation as well as in long-term depression in hippocampus (Mulkey et al. 1993; O'Dell & Kandel 1994). In addition, a role for cAMP and protein kinase A (PKA) has also been discussed in relation to the magnitude of the potentiation phenomenon (Arai & Lynch 1992; Greengard et al. 1991; Weisskopf et al. 1994). The fact that so far only PKC and cGMP-dependent protein kinase have been implicated in cerebellar LTD is somewhat surprising, as it would seem more likely that LTD requires activation of multiple biochemical cascades. It is indeed intriguing that phorbol esters generally produce increased responses in hippocampal networks (presumably mediated by an increase in transmitter release) but produce a depression in cerebellar networks.

(3) Expression/maintenance mechanisms. Although the question is debated, hippocampal LTP expression involves at least in part a modification of AMPA receptors (Maren & Baudry 1995). Recent studies have also indicated the existence of silent synapses in hippocampus, which exhibit functional NMDA receptors but non-functional AMPA receptors; LTP at these synapses would consist in the transformation of nonfunctional into functional AMPA receptors (Isaac et al. 1995; Liao et al. 1995). Several reports also implicate adhesion molecules and integrins in the maintenance of hippocampal LTP (Luthi et al. 1994; Xiao et al. 1991). In cerebellar LTD, there seems to be a consensus that expression is also due to a modification of AMPA receptors; most models point to a phosphorylation of the receptors. This is somewhat surprising, as current understanding of AMPA receptors indicate that phosphorylation of the receptors by PKC, PKA, or CamKII results in an increased rather than a decreased responsiveness of the receptors (Greengard et al. 1991; McGlade-McCulloh et al. 1993). Clearly then, more work is needed to resolve this apparent contradiction. The role of adhesion molecules and integrins has not been addressed in studies of cerebellar LTD. Finally, structural modifications have repeatedly been reported in hippocampal LTP, and there has been hints that structural modifications are also involved in cerebellar LTD.

In conclusion, it is clear that cerebellar LTD shares many features of hippocampal LTP and possibly of hippocampal LTD. It will be interesting to compare and further contrast the distinctive features of these forms of synaptic plasticity in order to understand the mechanisms involved in the storage of different forms of information in the brain.

What has to be learned in motor learning?

Harold Bekkering,[a] Detlef Heck,[b] and Fahad Sultan[c]
[a]*Max-Planck-Institute for Psychological Research, Department of Cognition and Action, D-80802 Munich, Germany;* [b]*Washington University Medical School, Department of Anatomy and Neurobiology, St. Louis, MO 63110-1031;* [c]*California Institute of Technology, Division of Biology, Pasadena, CA 91101.* **bekkering@mpipf-muenchen.mpg.de; heck@thalamus.wustl.edu; fsultan@bbb.caltech.edu.**

Abstract: The present commentary considers the question of *what* must be learned in different types of motor skills, thereby limiting the question of what should be adjusted in the APG model in order to explain successful learning. It is concluded that an open loop model like the APG might well be able to describe the learning pattern of motor skills in a stable, predictable environment. Recent research on saccadic plasticity, however, illustrates that motor skills performed in an unpredictable environment depend heavily on sensory (mostly visual) feedback. [**HOUK et al.**]

The target article by **HOUK** and colleagues provides an anatomically based adjustable pattern generator (APG) model to explain the adjustment of motor actions in primates. Instead of evaluating the model mainly on its neuro-anatomical and physiological merits, we would first like to consider the question of *what* must be learned in different types of motor skills; thereby limiting the question of what should be adjusted in the APG model in order to explain successful learning.

One way to classify (the learning component of) movement skills, concerns the extent to which the environment is predictable throughout performance (e.g., Schmidt 1991). Skills performed in stable (hence predictable) environments, have been called closed skills. A good example of a closed skill performance is professional typists typing a text on a computer keyboard. They know their environment, that is, the keyboard, and can perform the required motor acts without the need of any visual feedback. When typists want to write a word, in **HOUK et al.**'s view, they have a composite movement command – presumably, processed in a set of parallel fiber weights, which causes Purkinje cell (PC) firing – in order to terminate the movement command at the desired endpoint for the finger. Because the world is stable and limited, the typist could find out about the causes of an inferior olive (IO) signalled typing-error by comparing the sensory feedback with the perceptual trace of the actually performed movement (Adams 1971).

On the other hand, however, many so-called open skills, are performed in a variable environment; as a consequence, the forthcoming motor action must be adjusted in relation to (changes in) the environment. For instance, when playing beach volleyball, players must make adjustments in their motor program for the amount of force in the arm muscles needed to serve the ball into the game, while taking into account environmental factors such as the variable wind. Knowledge of results, that is, visual feedback about whether the ball was served in or out of the playing field, is used to evaluate the action performed and is taken into account before the initiation of the next movement sequence. Earlier research showed that without knowledge of results there was no motor learning at all in, for instance, a simple limb pointing task (e.g., Trowbridge & Cason 1932).

Presumably, the best practiced voluntary human motor action is a saccade – the fast movements of the eye that are used to bring a new part of the visual surrounding on to the foveal region of the eye. The discharge from the motoneurons that drive the eye muscles to generate a saccade is characterized by a high frequency burst, or pulse, which creates a large force for a short period of time in order to overcome the forces of resistance that act on the eyeball, and, in addition, to attain the typical, large acceleration of a saccade (Robinson 1964). The eye is held in its new position by a tonic, lower frequency discharge from the motoneurons, called the step; this serves to counter the elastic forces of the tissues in which the eyeball is suspended, forces that would otherwise drive the eye back to a position in which all the elastic forces were in equilibrium. Because saccadic eye movements are so fast that there is normally no time for visual feedback to guide the eye to its final position (see also Carpenter 1988), saccades are often thought to be pre-programmed and thus ballistic in nature. Such an open-loop action should fit into **HOUK et al.**'s APG pattern-recognition model perfectly, and, indeed, the target article provides a model to do so. Although there is accumulating evidence that the cerebellum is crucial for gain settings within the saccadic eye system (e.g., Keller 1989; Optican & Robinson 1980; Straube et al. 1995), several reasons make it unlikely that saccadic adjustment will occur in a fixed-gain saccadic circuit as suggested in the **HOUK** et al.'s target article.

First, the eye muscles that move the eye, as well as the tissues in which the eye is suspended, are subject to change due to growth, aging, disease, or fatigue (e.g., Lemij 1990). Nevertheless, saccades remain accurate, which implies that the eye muscles need to be recalibrated frequently in order to make accurate saccades possible throughout a person's life. Second, a closer look at the saccadic output, reveals that although the eye can sometimes reach the goal accurately within one saccade (Kapoula & Robinson 1986), it usually undershoots the goal slightly and sometimes overshoots it (Becker 1972). The corrections made after an undershoot or overshoot range from being completed fairly quickly (dynamic overshoot and undershoot), to several hundred milliseconds (glissadic overshoot and undershoot), to even the generation of a secondary saccade (e.g., Carpenter 1988) – a variability in output, which is very hard to explain in terms of a pattern-recognition model. Third, the finding that displacing the target for a saccade during the eye movement will eventually change the saccadic eye movement (e.g., Abrams et al. 1992; Deubel et al. 1986; McLaughlin 1967) is likewise hard to explain in terms of pattern recognition. That is, in such a double-step paradigm, subjects will initially miss the displaced target and then make a secondary, corrective eye movement to fixate it. However, the eye soon adapts and goes directly to the final displaced target position. It is fascinating to see that when the intrasaccadic shifts are not too large, the subjects completely fail to perceive them – probably, because of the phenomenon of saccadic suppression (e.g., Bridgeman et al. 1975).

A solution for overcoming the learning-capability problems for accurate saccades has been suggested by Kawato (1990). This simple feedback controller has recently been implemented in a neural net model for adaptive control of saccadic accuracy by Dean et al. (1994). A likely candidate for calibrating the saccadic system is the visual information that enters the brain at the end of a saccade. In this view, the oculomotor system attempts to minimize retinal error in the location of the image of the target on the retina relative to the fovea (e.g., Mays & Sparks 1980; Robinson 1975). Subsequently, this error signal is supposed to teach the cerebellum to adjust appropriately the gain of the brain stem burst generator's internal feedback loop, which can in turn alter the size of the burst sent to the motorneurons. Bekkering et al. (1995) found that a gain adjustment within the oculomotor system, was transferred significantly to the hand motor system. That is, they found that accompanying hand movements were also shorter in test trials when the saccadic system was adapted to target displacements, compared to identical test trials in a condition where the saccadic system was not adapted. This finding suggests that the feedback controller might evoke a common gain parameter adjustment for all related motor systems. Supporting evidence for this notion has recently been found by de Graaf et al. (1995) in the domain of eye-head adaptation.

In conclusion, the assumption in the APG model that the sensory nature of IO signals serve as the basis for the generation of training information might well be valid for skills performed in a stable environment. However, many primates' motor skills – for instance, saccadic eye movements – are performed in an unstable environment; hence training information depends heavily on sensory (and in this case mostly visual) feedback.

How and where does nitric oxide affect cerebellar synaptic plasticity? New methods for investigating its action

Lynn J. Bindman

Department of Physiology, University College London, Gower St., London, WC1E 6BT, United Kingdom. **lkiedr@psych.uic.edu**

Abstract: The role of nitric oxide in cerebellar synaptic plasticity is controversial. Two recent papers on nitric oxide (NO) and synaptic plasticity are discussed which use new methods for investigating the actions of NO and suggest that under certain conditions it is an essential anterograde messenger. In the light of current thinking about the diversity of expression of hippocampal LTP, should a presynaptic contribution to cerebellar LTD be re-examined? If NO acts orthogradely, might its product cGMP have a retrograde action? [**CRÉPEL et al.; LINDEN; VINCENT**]

1. New methods for examining the action of NO in synaptic plasticity. Contradictory results from different laboratories are a common feature of research on aspects of long-term depression (LTD) and long-term potentiation (LTP). The study of the relation of nitric oxide (NO) to synaptic plasticity in cerebellum is no exception. I should like to consider the evidence concerning NO presented in this issue of *BBS*, and in two recently published papers with new methods of investigating the NO by cGMP pathway.

VINCENT points out that NO is formed in basket and granule cells in the cerebellum. LTD of the response to applied AMPA can be induced in isolated cerebellar Purkinje cells in *culture*, which suggests that NO cannot play an essential role in this form of plasticity. However, these observations should not be extrapolated to mean that NO cannot play an essential role in cerebellar LTD in intact cerebellum. Here there is contradictory evidence for the involvement of NO in LTD even in the *slice* preparation (Daniel et al. 1993; Glaum et al. 1992). Many positive results have been obtained by **CRÉPEL**'s group. However, a control is needed to show that the decline of the cerebellar PF-evoked EPSPs following the dialysis of NO donors into Purkinje cells is not caused by toxic side effects. Thus it would be reassuring to see that after LTD was produced by bath application of 8-bromo-cGMP plus pairing, the application of NO donors such as 3 mM SIN-1 intracellularly or 8 mM SNP extracellularly, could no longer produce a slow decline of EPSPs in the cell (cf. Fig. 5, **CRÉPEL** et al. (this issue) and Daniel et al. 1993).

A new experimental approach to the study of the actions of NO has appeared recently and hence is not cited. It involves the use of caged NO in a form that is unlikely to diffuse out of the Purkinje cells (Lev-Ram et al. 1995). The requirement for postsynaptic depolarization of the Purkinje cell together with NO liberation shows that Ca^{2+} dependent processes additional to the activation of NOS are crucial. Several lines of evidence show NO, liberated presynaptically but acting within the Purkinje cell, plays a convincing role in cerebellar LTD.

A selective inhibitor of soluble guanylyl cyclase, the oxadiazoloquinoxaline derivative (ODQ) has now been developed (Boulton et al. 1995). In the hippocampal CA1 area, in these authors hands, there is an equal effect of NOS inhibitor and of ODQ, in depressing the size and time course of LTP. Injected intracellularly it could be useful in determining whether it is cGMP within the Purkinje cell that is important for LTD, an area of controversy as noted by **VINCENT** in his introduction.

2. Some sources of variable results on the induction and expression of LTP and LTD. In the hippocampus, investigations into a role for NO have been motivated by the search for a retrograde messenger, since LTP is induced postsynaptically but evidence shows it can be expressed in part presynaptically. In recent years some of the reasons for obtaining controversial results concerning many aspects of LTP have been elucidated in hippocampus. First, different induction procedures may elicit different types of expression of LTP: (e.g., Kullmann 1994; Liao et al. 1995 produced hippocampal LTP of AMPA response only, while Aniksztjen & Ben Ari 1991 produced LTP of only the NMDA receptor component). This is not an obvious source of discrepant results in cerebellum.

Second, it is recognized that the type of experimental preparation may be crucial for certain aspects of expression of synaptic plasticity: **LINDEN** points out that cultures of cerebellum do not express β-adrenergic receptors (assessed by electrophysiological methods) which are found in normal cerebellum. By analogy, the absence of normal NO release close to isolated Purkinje cells in cerebellar cultures might also lead to changes in the ability of these neurones to respond to transcellular NO (Linden 1994). Such differences in preparation might account for different findings in culture and slice, but not between Daniel et al. (1993) and Glaum et al. (1992).

Third, the initial state of the preparation affects the way in which synapses respond to what is apparently the same LTP induction procedure: for example, the probability of transmitter release before LTP induction determines whether a primarily pre- or postsynaptic change ensues (Larkman et al. 1992). Presynaptic tetanic stimulation can elicit LTP of small EPSPs that is expressed largely as an increase in presynaptic release, as assessed by quantal analysis and a reduction in paired pulse facilitation (Kuhnt & Voronin 1994).

There is clearly good evidence in the cerebellum that postsynaptic changes in AMPA receptor mediated currents occur in cells, in culture and in slices, when LTD is induced postsynaptically. This is not sufficient evidence to exclude a presynaptic contribution to LTD when pairing of synaptic inputs from PF-CF occurs, and indeed **LINDEN** is careful not to do so.

How can presynaptic involvement be assessed? Paired pulse facilitation (PPF) is generally agreed to be a presynaptic phenomenon, but many laboratories claimed that in the hippocampus, PPF does not change in LTP. However, Schulz et al. (1994) and Kuhnt and Voronin (1994) show, using careful choice of experimental conditions, that a decrease in quantal content in PPF is correlated with the magnitude of LTP, and hence presynaptic processes are implicated in the expression of LTP. I suggest that PF test shocks at a strength that permits clearcut but submaximal PPF should be used to investigate whether the expression of PF-Purkinje cell LTD of EPSPs is coincident with some enhancement of the PPF, and hence decreases in presynaptic release.

It is not only the analogy with diverse sites of expression of LTP in hippocampus that causes me to make this suggestion, but the description by **VINCENT** of the marked changes in the extracellular soup that follow many stimuli that are used for LTD induction. There is an increase in extracellular cGMP, and other endogenous chemicals. Is it time to investigate their possible retrograde action on transmitter release?

Perhaps it's time to completely rethink cerebellar function

James M. Bower
Division of Biology, California Institute of Technology, Pasadena, CA 91125.
jbower@smaug.bbb.caltech.edu

Abstract: The primary assumption made in this series of target articles is that the cerebellum is directly involved in motor control. However, in my opinion, there is ample and growing experimental evidence to question this classical view, whether or not learning is involved. I propose, instead, that the cerebellum is involved in the control of data acquisition for many different sensory systems. [**CRÉPEL et al., HOUK et al., SMITH, THACH**]

Out with the old. The predominance of the view that the cerebellum is directly involved in motor control is clearly reflected in these target articles, as it is in the field in general. However, I believe there is increasing evidence that this view may not be correct. First, as is well known (**THACH,** this issue; Bower 1992), considerable motor coordination and learning are possible following lesions of large sections, or all of the cerebellum. Given the unique structure of cerebellar circuitry, and the unique position of the cerebellum within the nervous system, it seems unlikely that other brain regions are "taking over." Instead, quoting from the commentary by **THACH,** the cerebellum appears to be "useful, but not necessary" (sect. 7.6, para. 2) in a wide range of brain functions.

Second, recent imaging experiments in humans must seriously call into question classical theories of direct and continuous cerebellar involvement in movement control (Bower 1995). For example, as reviewed by **THACH,** a number of these studies report no cerebellar activity during the performance of learned motor skills as well as strong cerebellar activity when no overt movement is involved. Recently, we have found no activity in the dentate nucleus during fine finger movements of the sort assumed to be under cerebellar control in these articles (Gao et al. 1996).

Finally, as asserted by **HOUK et al.,** any theory of cerebellar function must eventually be fundamentally related to the actual physiological and anatomical structure of the cerebellum itself. In this regard, all cerebellar motor control theories assume that parallel fiber inputs directly drive Purkinje cell output. However, both physiological (Bower & Woolston 1983) and modeling (De Schutter & Bower 1994b) data suggests that the parallel fiber system may actually be "modulatory" in nature, with the ascending branch of the granule cell axon providing the primary stimulus-related drive on cerebellar Purkinje cells (Jaeger & Bower 1994). With respect to the Marr/Albus learning theory in particular, this means the climbing fiber does not "select" which mossy fiber inputs influence the Purkinje cell. Instead, Purkinje cells appear to respond to the input arriving in the granule cell layer immediately below them, modulated by ongoing parallel fiber and stellate cell inputs. I would point out that this change in thinking about the role of the parallel fiber system also requires a reinterpretation of the significance of LTD as described in several of the target articles.

The view from the whiskers of a rat and the fingers of a primate: A new theory of cerebellar function. Based on our own experimental and theoretical efforts in somatosensory regions of rat cerebellar cortex (Bower & Kassel 1990), I have proposed that the cerebellum may actually be involved in coordinating the acquisition of sensory data and not in motor control per se (Bower 1996). In this view, the influence of lateral regions of primate cerebellar cortex on fine finger movements is seen as a means to coordinate the acquisition of somatosensory information from the fingers, just as the influence of the flocculus on ocular-motor nuclei assures proper retinal movement to reduce retinal slip (Paulin et al. 1989). Similarly, I suggest that, through gamma motor neurons, the posterior vermal cerebellum coordinates the acquisition of proprioceptive information received from muscle stretch receptors that is then used by motor control centers to coordinate movement. Just as in the case of the somatosensory and visual systems, control of stretch receptors should significantly enhance motor performance which is nevertheless under the direct control of other brain structures. In each of these systems, cerebellar control is useful but not necessary for any particular behavior.

Lessons, imaging studies, cognition, and the cerebellum. Space allows only a brief discussion of the evidence I believe supports a general role for the cerebellum in sensory data acquisition. First, we have recently confirmed in humans our prediction that the dentate nucleus should not be activated by finger movements unless the fingers are being used for somatosensory data acquisition (Gao et al. 1996). The proposed role for the cerebellum in proprioceptive data acquisition is also consistent with the interesting recent experiments by Horak and Diener (1993) described but not really taken to heart by **THACH** and **SMITH.** In these experiments, cerebellar patients did not, in fact, lack the synergistic coordination of muscle activations assumed by **THACH** and **SMITH** to require the cerebellum. Instead, they were unable to establish an "anticipatory set" for impending movements. I suspect that this "set" is related to premovement cerebellar control of the state of the muscle spindle afferents.

Human cerebellar imaging studies represent perhaps the most exciting new data related to cerebellar function. It is my view that these experiments are the most likely to finally force an expansion of our thinking beyond the constraints imposed by the 100 year old view of a direct cerebellar involvement in motor control (Bower 1995). Unfortunately, **HOUK et al.** and **SMITH** basically ignore this work while **THACH** seems bent on maintaining some version of classical cerebellar motor control theories. In my view, this bias somewhat obfuscates several common and important features of these new results including: (1) the surprisingly wide range of behavioral tasks that activate the cerebellum; (2) the consistent relationship seen between cerebellar activity and apparent task difficulty; and (3) the actual reduction in cerebellar activity after new behaviors (including motor) are learned.

I believe that each of these results are difficult to understand in the context of classic motor control theories despite **THACH**'s efforts. This is especially true of the consistent reduction in cerebellar activity during the time when the cerebellum is supposed to have "taken over" control from the cerebral cortex. However, these results are quite consistent with a cerebellar role in data acquisition. First, involvement of the cerebellum with almost all sensory systems could explain the wide range of human behaviors that elicit cerebellar activity. Second, if we assume that the cerebellum is more or less involved depending on the need for fine data control, then the cerebellum should be more active with more difficult tasks. Third, because network learning is often associated with the ability to generalize over sensory data, once other structures have learned a new task the need for sensory data control and thus cerebellar activity should be reduced.

The sensory data acquisition hypothesis also provides a different interpretation for several common deficits found in cerebellar patients across tasks as described by **THACH**, including longer performance times and poor speed improvement with practice. In both cases, these deficits could be produced by an increase in the length of time it takes to compute with less well controlled sensory data. A similar explanation may apply to the special case in which a discrimination explicitly involves timing. Any difficulty in using acquired sensory data should especially disrupt timing dependent discriminations. In keeping with these interpretations, Tachibana et al. (1995) have recently proposed that changes in cerebral cortical evoked potential recordings in cerebellar patients during cognitive tasks may reflect "a difficulty in pattern recognition of the stimuli presented" (abstract).

A critical role for modeling. Finally, I completely agree with **HOUK et al.** that any cerebellar theory must be intimately related to the actual anatomical and biophysical properties of its neurons and networks and that actually simulating models is essential to this process. However, a critical distinction must be made between models that explore the computational properties emerging from the biophysical details of neural structure and models whose sole purpose is to demonstrate the biological plausibility of a particular preexisting idea. **HOUK** et al.'s stated desire to demonstrate "models of the cerebellum and motor learning" (abstract), places their models firmly in the latter category. Unfortunately, the primary objective of "existence proof" simulations is to assign physiological significance to known cellular and network components, not to test underlying biophysical assumptions. In contrast, biophysical models that are first tuned to replicate functionally neutral biophysical phenomena like Purkinje cell responses to current injection (De Schutter & Bower 1994a) can then be used to highlight previously unknown structure/function relationships (cf. De Schutter & Bower 1994c). In our own case, the modulatory function of the parallel fibers that has emerged from our models has substantially influenced the formation of our theory of cerebellar control of sensory data acquisition (Bower 1996). Collaborative efforts currently underway to construct a biophysically realistic model of cerebellar cortex appear likely to provide further tools with which to loosen the hold of classical theories on cerebellar function.

Long-term changes of synaptic transmission: A topic of long-term interest

Paolo Calabresi, Antonio Pisani, and Giorgio Bernardi

Clinica Neurologica, Dipartimento Sanità, Università di Roma "Tor Vergata," 00173, Rome, Italy. **pisani@tovvx1.ccd.utovrm.it**

Abstract: Long-term modifications of synaptic efficacy at excitatory synapses are considered as a putative cellular substrate for learning and memory processes. Cerebellar long-term depression (LTD) occurring at parallel fibers (PF)-Purkinje neurons (PN) synapses is thought to represent a cellular model of motor learning. Crépel et al. and Linden describe this phenomenon by utilizing different tissue preparations. [**CRÉPEL et al., LINDEN**]

A sustained depression of synaptic transmission at PF-PN (parallel fiber, Purkinje neuron) synapses characterizes the so-called cerebellar long-term depression (LTD). Over the last decade, a great effort by researchers has resulted in a detailed (and sometimes controversial) characterization of the mechanisms underlying the generation of this plastic event. There are some puzzling points which remain a matter of debate. We will still consider the methodological differences that distinguish the work of **CRÉPEL et al.** and **LINDEN.**

Tissue preparation. In attempts to approach synaptic plasticity, it is desirable to maintain the circuitry of the structure as much as possible. This is achieved more easily using slice preparations. **LINDEN** actually provides a precise analysis of the limitations of the cell culture preparation, specifically in the study of LTD and of multisynaptic connections involved in various aspects of information processing and storage. Moreover he considers the absence of the extrinsic modulatory networks, such as the noradrenergic and the serotoninergic systems to be a limitation of culture preparations. Nevertheless, the use of cell cultures represents a powerful approach for the study of postsynaptic mechanisms of LTD without presynaptic contamination.

Sodium ions. In an attempt to study the initial trigger for LTD induction, the role of sodium (Na^+) influx was evaluated by both groups. The results are somehow conflicting. **CRÉPEL**'s group (Crépel & Jaillard 1991) reported that the degree of membrane depolarization influenced the direction of synaptic strength from LTD to LTP. LTP was observed when PF-mediated EPSPs were coupled with a moderate depolarization of the recorded cell, inducing a sustained firing of Na^+ spikes. In contrast, pairing of PF-mediated EPSPs with calcium (Ca^{2+}) spikes evoked by strong depolarization caused LTD (Crépel & Jaillard 1991). These findings are in apparent contrast with the hypothesis proposed by Artola and Singer (1993), which considers a rise in intracellular Ca^{2+} important for LTD and LTP; however, the findings do suggest that LTP requires higher levels of intracellular Ca^{2+} concentration than LTD. We have approached the study of the role of Na^+ in striatal LTD by utilizing QX-314-filled electrodes (Calabresi et al. 1994). QX-314 is a lidocaine derivative that blocks voltage-dependent Na^+ channels without affecting EPSPs amplitude. In the presence of QX-314, tetanic stimulation failed to induce LTD in striatal neurons. Some cells filled with QX-314 before tetanic stimulation were depolarized to a membrane potential of $-30/-20$ mV. In this condition the tetanus induced LTD. We concluded that firing of Na^+ spikes during the tetanus is important mainly because it determines the activation of voltage-dependent Ca^{2+} channels.

LINDEN's group, on the other hand, reported that Na^+ influx is necessary for LTD induction (Linden et al. 1993). It was shown that Na^+ influx occurs mainly through AMPA receptor-gated channels, although a contribution of voltage-dependent Na^+ channels is not excluded.

Ca^{2+} ions. Experimental evidences suggest that Ca^{2+} plays a prominent role in the induction of cerebellar LTD, as for most of the forms of synaptic plasticity described so far. However, two major issues are still unclear: (1) the source of Ca^{2+} and (2) the site of Ca^{2+} influx.

(1) Ruling out the contribution by NMDA receptors, which are down-regulated with age in PNs, both **LINDEN** and **CRÉPEL** et al. agree that the Ca^{2+} rise occurs mainly through voltage-dependent channels. In this respect, a clear-cut experiment would be the application of organic CA^{2+} blockers selectively affecting different subtypes of Ca^{2+} channels. In striatal LTD, bath-application of nifedipine, a blocker of L-type Ca^{2+} channels, prevents the induction of LTD (Calabresi et al. 1994). A recent report provides evidence that in cultured rat PNs, inositol trisphosphate-induced Ca^{2+} mobilization is involved in LTD induction. Ca^{2+} release may result from the activation of mGluR1, or it may follow the influx of Ca^{2+} through voltage-dependent channels (termed Ca^{2+}-induced Ca^{2+} release). The latter phenomenon seems to be important in Ca^{2+} signalling of PNs, where the peak Ca^{2+} rise increases more than linearly with voltage step duration (Llano et al. 1994).

(2) The second issue, the site of Ca^{2+} entry, mainly concerns the cellular localization of mGluR1 subtypes (α, b, and c). Conflicting results arise from studies of Ca^{2+} imaging in PNs. Some groups reported a somatic signal, others observed that Ca^{2+} mobilization occurs in dendrites. **LINDEN** notices how the developmental stage of culture preparations modifies the characteristics of Ca^{2+} responses. It is likely, as the authors suggest, that differences are linked to the growth of the dendritic tree. This aspect would deeply influence the study of synaptic plasticity in culture preparations. It has recently been proposed that in cultured rat PNs, glutamate application together with intracellular Ca^{2+} increase, induced by utilizing a photolabile Ca^{2+} chelator, but without depolarization, is sufficient to induce cerebellar LTD (Kasono & Hirano 1994).

Glutamate receptors. It is well established that adult PNs bear only non-NMDA receptors. **LINDEN** proposes that coativation of AMPA and mGluR1 is necessary for LTD induction, since application of antagonists for either AMPA or mGlu receptor subtype blocked LTD induction. Until recently the role of mGluR1 was still uncertain because of the lack of selective antagonists; however, recent reports have provided evidence that mGluR1-lacking transgenic mice do not express LTD (Shigemoto et al. 1994). Moreover, the developmental profile of mGluRs should be taken into account, since Ca^{2+} responses, as mentioned by **LINDEN**, varies noticeably with age. Likewise, AMPA receptor subunits, their different pattern of assembly, also in view of the developmental stage, renders the story for LTD even more complex. **CRÉPEL et al.** suggest the existence of two separate forms of LTD: a LTD_{AMPA}, obtained by pairing PF stimulation, Ca^{2+} spikes firing, and bath application of trans-ACPD, an agonist of mGluRs. Another form called LTD_{mGlu}, was obtained either by pairing Ca^{2+} spike firing with the application of trans-ACPD without PF stimulation, or when CNQX, antagonist of AMPA receptors was bath-applied at the time of the pairing protocol, suggesting that in the latter form, AMPA receptors are not necessary. The authors conclude that for both types a rise in Ca^{2+} through voltage-dependent channels is required, but for LTD_{mGlu} the Ca^{2+} increase would be mediated by mGluR activation. It might be argued, however, that in both cases the final outcome is an increase in intracellular Ca^{2+}. We feel that the use of thapsigargin (which depletes aspecifically most of Ca^{2+} internal stores) is not sufficient to discriminate two forms of LTD.

Second messengers. There is general agreement about the idea that in cerebellar LTD the activation of a Ca^{2+}-dependent cascade of biochemical events would lead in turn to a long-lasting desensitization of AMPA receptors. The involvement of Ca^{2+}-dependent protein kinase C has been demonstrated by both **CRÉPEL**'s group in slices and by **LINDEN**'s group in cell cultures. In contrast, there is substantial disagreement concerning the role of the nitric oxide (NO)/cyclicGMP route in cerebellar LTD. Experiments performed either in the current-clamp mode from slices (Crépel & Jaillard 1990) or in the whole-cell patch-clamp mode from thin slices (Daniel et al. 1993) support the involvement of NO/cGMP in the induction of this phenomenon. **LINDEN**, however, reports that in cell cultures he could not reproduce these effects on his model of cerebellar LTD. Both the authors suggest a number of reasons why these discrepancies may occur. Since PNs do not possess NO synthase, it is believed that neighboring cellular types are responsible for the production of this gas. This, in our view, might represent a limitation for the study of LTD in a cell-culture preparation. In addition, PF's synaptic inputs onto PNs might not be particularly represented. **LINDEN** suggests that the type of stimulus utilized may be the cause for this discrepancy: in fact, PF stimulation is utilized in experiments performed in slices, while in cell cultures a glutamate pulse substitutes for PF stimulation. If the hypothesis that PF terminals contain NO donors is true, then it would be reasonable to see this difference. The poor representation of terminals might also be against a presynaptic site of action for LTD.

One cannot build theories of cerebellar function on shaky foundations: Induction properties of long-term depression have to be taken into account

Erik De Schutter

Born-Bunge Foundation, University of Antwerp—UIA, B2610 Antwerp, Belgium. **erik@kuifje.bbf.uia.ac.be**

Abstract: The theories of cerebellar function presented in this *BBS* special issue cannot be reconciled with the established induction properties of cerebellar LTD. At the same time, the authors presenting their research on cerebellar LTD do not appear very interested in function. System physiologists and cell biologists need to collaborate more if we are to make progress in understanding cerebellar function. [**CRÉPEL et al.; HOUK et al.; KANO; LINDEN; SIMPSON et al.; SMITH; THACH; VINCENT**]

THACH introduces his paper by describing the steps necessary to understand brain structure and function (sect. 1, para. 1). The only non-behavioral evidence he mentions is connectivity. This seems to align him with most of the neural networks field (Hertz et al. 1991), which does not consider cellular properties to be relevant to the functioning of neural circuits either. Probably **THACH** would agree that properties of specific neurons, including, for example ion channel kinetics and types of synaptic receptors expressed, are relevant to brain function. But his target article is typical of cerebellar system physiologists, who often neglect single neuron physiology in their reasoning.

This is symptomatic of the almost complete separation between the fields of research on synaptic plasticity and of system analysis of cerebellar function, which is quite noticeable in this special issue of *BBS*. Although the issue is called "Controversies in Neurosciences," most contributors neatly keep to their side of the fence. The cell biologists (**CRÉPEL et al., LINDEN,** and **VINCENT**) describe their particular views on how long-term depression (LTD) is induced, often covering the possible metabolic pathways involved in great detail. But they do not address the issue of the functional significance of synaptic plasticity in the cerebellum. In fact, **LINDEN** (abstract) explicitly says this is not the purpose of his article, while **CRÉPEL** et al. (sect. 1, para. 4) limit their discussion to conventional references to Marr (1969) and Albus (1971). **KANO** (sect. 8, para. 3) briefly describes a possible function of the potentiation of inhibitory inputs. But one cannot call the suggested role of LTD-enhancement very imaginative.

It cannot be said that cerebellar system physiologists lack imagination, as is apparent from the widely different theories on cerebellar function presented by **HOUK et al., SMITH,** and **THACH.** Their theories seem to have only three ideas in common: each claims to have extensive experimental support for their particular proposal, the cerebellum is involved in motor learning, and LTD is the neuronal mechanism of this learning. How is it possible that starting with the same body of experimental knowledge, scientists can arrive at such divergent ends? The simplest explanation is that the problem of cerebellar function is underconstrained, or, in other words, that the available data allows for many different interpretations. This is certainly true in some aspects; for example the contribution of other brain regions to motor control has not been completely defined either. Unfortunately this is not the only cause; instead it seems that several system physiologists are not aware of experimental evidence reported in the last five years. In fact, **SMITH** and **THACH** both implicitly assume that cerebellar LTD will provide the substrate for the kind of learning that has been proposed by Marr (1969), Albus (1971), and Ito (1984). However, as reviewed recently (De Schutter 1995), several aspects of cerebellar LTD induction do not fit well with the Marr-Albus-Ito theories. I will refer here only to the most serious problem, referred to as the "temporal credit assignment problem" by **HOUK** et al. (sect. 4, para. 7) and also described by **LINDEN** (sect. 3, last para.). It has been shown both in decerebrated animals and in slice that climbing fiber stimulation must precede parallel fiber stimulation to obtain LTD induction (Ekerot & Kano 1989; Schreurs & Alkon 1993; Karachot et al. 1994). This is impossible to reconcile with the proposed role of the climbing fiber as an error signal (Albus 1971). Karachot et al. (1994) have actually demonstrated that no LTD is induced in slice when inputs are timed like they are presumed to occur in the classical conditioning of the nictating membrane reflex (Thompson 1988; see also **HOUK** et al., sect. 3.6, para. 1 and **THACH**, sect. 3.4).

The contribution of **SIMPSON et al.** in this volume is refreshing. They conclude that the function of the climbing fiber input in the rabbit's vestibulocerebellum remains a mystery (sect. 6). I think this is a fair description of cerebellar function in general. Moreover, I'm afraid that the field will not make much progress unless we bring the cell and system physiologists closer together. I already mentioned the problems associated with fitting cerebellar LTD into a theory that assumes that the climbing fiber input is an error signal. But the opposite also happens too frequently. Many theories of cerebellar function not presented in this volume simply neglect LTD altogether (e.g., Bloedel 1992; Welsh et al. 1995). **HOUK et al.** (sect. 4.1, para. 8) at least take the trouble to analyze the problems associated with attributing a function to cerebellar LTD. But their suggested solution, an additional, hereto experimentally unrecognized way to induce LTD, is unlikely to be true. Considering the complexity of the network in which the cerebellum is embedded (Ito 1984) and our lack of knowledge about many properties of this network, including even the neural code (Ferster & Spruston 1995), it is too early to generate specific theories on how the cerebellum participates in motor control or cognition. This does not exclude theories on a more basic level. For example, detailed modeling of single cells can provide new insights. **HOUK** et al. (sect. 3.3, para. 4) have used this approach to analyze bistability of the Purkinje cell dendrite (Yuen et al. 1995), but the validity of their extremely simplified model is questionable. Detailed modeling of the Purkinje cell (De Schutter & Bower 1994a) has revealed important properties of this neuron's active dendrite. For example, parallel fiber inputs get amplified so that even inputs contacting the most distal parts of the dendrite have the same access to the soma as more proximally located inputs (De Schutter & Bower 1994b). The modeling work has also suggested a new hypothesis on the function of cerebellar LTD (De Schutter 1995). Because these hypotheses are defined at the neuronal level, they immediately suggest experiments which can falsify the theory. This is something that I found lacking in the target articles by **HOUK et al., SMITH,** and **THACH.**

In conclusion, system physiologists need to take cellular properties more into account. To achieve this goal both sides will need to contribute. For example, at the system level the firing properties of the neurons need to be studied in more detail. This can lead to some surprising results, as shown by **SIMPSON et al.** in this volume. At the cell biology level too much emphasis has been placed on identifying receptors and metabolic pathways. This has spawned a literature full of details which are often specific to the experimental conditions used, for example, cell culture (**LINDEN**) or the effect of trans-ACPD in slice (**CRÉPEL et al.**, sect. 1.3). Unfortunately this does not tell us much about the induction requirements of LTD in the context of behavioral learning. For example, it seems that about 100 climbing fiber-parallel fiber pairings at the natural firing frequency of climbing fibers (1 to 4 Hz) are sufficient to induce LTD (Ekerot & Kano 1989; Ito 1989; Schreurs & Alkon 1993). But, while 100 pairings seems a reasonable number to learn something, it is unlikely that in real life such pairings would occur at 1 Hz. For example, if LTD is involved in motor learning one might expect one pairing during each failed motor execution; which, depending on the task, might take tens of seconds to several minutes. Nobody has ever reported if LTD could be induced in slice with a low frequency of pairing (0.01 to 0.10 Hz).

Saccades and the adjustable pattern generator

Paul Dean

Department of Psychology, University of Sheffield, Sheffield S10 2TP, England. **p.dean@sheffield.ac.uk**

Abstract: The adjustable pattern generator (APG) model addresses physiological detail in a manner that renders it eminently testable. However, the problem for which the APG was developed, namely, limb control, may be computationally too complex for this purpose. Instead, it is proposed that recent empirical and theoretical advances in understanding the role of the cerebellum in low-level saccadic control could be used to refine and extend the APG. [**HOUK et al.**]

One goal of the adjustable pattern generator (APG) model of cerebellar learning is to explain how the firing patterns of cerebellar output cells are produced by "appropriately timed bursts and pauses of PC discharge" (**HOUK et al.**, sect. 3, para. 2). Temporal details of cerebellar output were not a feature of the original Marr-Albus formulations (sect. 2, para. 1). The adherence of the APG to electrophysiological observations offers interesting possibilities for its testing and refinement.

A problem arises, however, in choosing the data best suited for this purpose. The original APG, with its emphasis on corticorubro-cerebellar loops (sect. 3.3, para. 1), was designed to explain the role of the cerebellum in limb control (Houk et al. 1990). Unfortunately, limb control is a task of notorious computational complexity: limbs have multiple joints and correspondingly complicated kinematics and dynamics, and are subject to varying loads. The brain appears to solve problems of this kind by distributed control (cf. Houk & Barto 1992; Houk et al. 1993), in which a given motor task is effectively decomposed into simpler subtasks, each handled by a specific neural subsystem. Which parts of the cerebellum handle which subtasks in limb control? Difficulties in answering this question contribute to the problems of applying an APG array to the control of a realistic limb (Berthier et al. 1993).

But if we accept the cerebellar microcomplex argument (e.g., Ito 1984), data about any movement in which the cerebellum is involved can be used to test the APG model.

> The microscopic similarity of local circuitry coupled with macroscopic variety in input-output connectivity has suggested to a number of authors that the cerebellum is organized in a modular fashion . . . Each module is presumed to operate in the same manner on whatever inputs come to it. Modules in different regions of the cerebellum are able to perform different functions in a behavioural sense by virtue of their specific input-output connections (Houk et al. 1990, sect. 2.1).

Such freedom of choice allows exploitation of the favorable features of eye movements. As explained elsewhere (e.g., Robinson 1991), the eye has a single "joint," constant load, and muscles whose motoneurons lie inside the cranial vault. These features have facilitated experimental and theoretical investigations of eye-movement control subsystems, and the results constitute a fruitful source of constraints on models of the cerebellar microcomplex. Any eye-movement offers these advantages (as do related movements such as eye-blink): the focus here is on saccades.

The cerebellum intervenes in saccadic control at a number of levels, of which most is known about the level nearest to the oculomotoneurons (referred to in sect. 3.7 para. 1 as "the level of a brainstem burst generating network"). The relevant cerebellar region has been identified (Noda 1991) as the oculomotor vermis (primarily lobule VIIa) and its output via the fastigial oculomotor region FOR (caudomedial fastigial nucleus). Although many details remain to be clarified, there is a general consensus about the function of this region. It appears that the oculomotor vermis adjusts saccadic commands from higher levels (superior colliculus, frontal eye fields) so that they produce fast, accurate, conjugate movements for any starting position of the eye in the orbit (e.g., Goldberg et al. 1993; Keller 1989; Optican 1985). Learning is involved in this process, as indicated by the changes in saccade characteristics that follow muscle weakening or target jumping (Optican & Robinson 1980; Robinson et al. 1995).

The extensive anatomical and physiological data available for the oculomotor vermis and FOR provide some idea about how the adjustments might be carried out. For example, the timing of FOR bursts has been related both to saccade parameters (Fuchs et al. 1993; Noda 1991), and to the firing patterns of Purkinje cells and mossy fibres in the oculomotor vermis (Ohtsuka & Noda 1995). It is possible to incorporate timed FOR bursts into a brainstem model of horizontal saccade generation so that saccadic dynamics are accurately reproduced (Dean 1995). It would seem to be an interesting exercise for the APG to account quantitatively for these electrophysiological data, modeling for example the difference in firing patterns between ipsi- and contralateral saccades.

Quantitative application of the APG model to the saccadic system may also contribute to the debate on training signals (sect. 4.3).

In general terms, the use of corrective-saccade parameters to drive adaptive change (cf. Berthier et al. 1993) has computational advantages: (a) it absolves the low-level vermal system from needing access to the complex calculations required for initiating a corrective saccade (Harris 1995); (b) it solves the problem of translating a sensory training signal into a motor one (Dean et al. 1994), a general point emphasized by Kawato's feedback-error learning scheme (sect. 4.3, para. 3).

However, as far as the details of this training signal are concerned, "the climbing fiber inputs to saccade-related regions of the cerebellar cortex remain somewhat of a mystery" (Houk et al. 1992, p. 464). The statement in section 4.3 para. 4 that "the CFs that innervate the saccadic region of the cerebellum do not encode motor signals, although they do respond to proprioceptive inputs" may be a little strong. The relevant electrophysiological data were obtained in cats, and whether the precise region corresponding to the primate oculomotor vermis receives a proprioceptive climbing fiber input seems not to have been established (cf. Schwarz & Tomlinson 1977). The primate anatomical data strongly suggest that the oculomotor vermis receives input from a region of the inferior olive that in turn receives from the superior colliculus (references in Dean et al. 1994). The information carried by this projection is unknown. Although "(u)nderstanding the training information carried by climbing fibres is a crucial issue that will require careful experimental attention in the future" (Houk et al. 1992, p. 464), the APG model might nonetheless be used to investigate which parameters of a corrective saccade would be necessary to achieve realistic levels of saccadic accuracy (cf. Dean et al. 1994).

How can the cerebellum match "error signal" and "error correction"?

Michel Dufossé

Laboratoire CREARE, UPMC, case 23, 75252 Paris Cedex 05, France.
michel.dufosse@snv.jussieu.fr

Abstract: This study examines how a Purkinje cell receives its appropriate olivary error signal during the learning of compound movements. We suggest that the Purkinje cell only reinforces those target pyramidal cells which already participate in the movement, subsequently reducing any repeated error signal, such as its own climbing fiber input. [SIMPSON et al.; SMITH]

The Marr-Albus-Ito model of cerebellar learning is based on the Rosenblatt perceptron (1958). This consists of a single layer of plastic synapses located between the second layer of granular cells and the third layer of Purkinje cells, where the first layer is the precerebellar nuclei cells whose axons are the mossy fibers.

Unlike the more recent multi-layer perceptron, which includes several layers of plastic synapses and a "back-propagation" learning rule, the original perceptron has an external teacher. The inferior olive is the cerebellar teacher in this neurobiologically plausible model, and the cerebellar cortex can overcome the original perceptron's limitations (Minsky & Papert 1969) because it contains an exceptionally large number of neurons that form the intermediate granular layer and provide many different combinations of mossy fiber input signals at the glomerular stage. Linear separation, which would be impossible in the low-dimensional input layer space, is then possible in this high-dimensional granular space, which is one hundred times larger (Marr 1969).

Recordings from single units in alert animals (Dufossé et al. 1978; Gilbert & Thach 1977) and acute preparations (Ito et al. 1982) have shown that each Purkinje cell receives a climbing fiber error signal from a teacher-cell located in the inferior olive. But the crucial question of how the olivary error signal needed for the LTD process of a given Purkinje cell can modify the downsteam motoneuronal activities to produce an error correction, which can then decrease the error signal of this specific Purkinje cell, remains to be answered.

We developed a hypothesis based on the connectionist idea of "parsimony," which always tries to minimize the need for any genetic hardware. There are three successive, but largely overlapping, phases in the learning of compound movements; these result from the differences in the rates at which the different brain structures adapt. They are assumed to correspond to the cerebrospinal, the cerebellar-cerebral, and the cerebellar-rubral pathways. The cerebellum fine-tunes the cerebral processes learned in the initial phase during the second learning phase. Even though each cerebellar longitudinal microzone projects over a wide cerebral zone, a given Purkinje cell will only reinforce the strongly active pyramidal cells related to the actual or intended movement by heterosynaptic LTP-LTD plasticity at the cerebral level. As a result, this cerebellar effect increases the actual cerebral output, unloading the cerebrum from its internal recurrent processing task. This putative mechanism of cerebral-cerebellar dialogue involves two independent learning processes that take place in the cerebral and the cerebellar cortices (Burnod & Dufossé 1991).

Let us examine three examples. First, the increase in climbing fiber activity that occurs at the onset of a fast ballistic movement (Mano et al. 1986) may indicate that "movement is not fast enough!" By cerebellar disinhibition, it will increase the activities of pyramidal cells involved in the motor command. This climbing activity will never vanish, even after long-term training. Second, any unexpected event (see the target article by **SIMPSON et al.**), such as "running into an obstacle," is also an olivary-nucleus-mediated error. This repeated signal may gradually increase the cerebral control of the muscular braking process (see the target article by **SMITH**). A fairly small number of cerebellar modules is sufficient to modify the concomitant cerebral motor activities. Third, the cerebellum may contribute to a straight manual trajectory. Dornay et al. (1996) showed that a single global constraint of minimum muscle tension change is sufficient to solve the four so-called "ill posed" problems of arm movements: hand path and trajectory formation, coordinate transformation, and the calculation of muscle tensions. Olivary signals may simply originate from a global summation of fusorial information given by arm muscles, where each piece of fusorial information indicates a mismatch between the actual (muscular fiber) and desired (gamma fusimotor) lengths (Houk & Rymer 1981), and each Golgi tendon afferent indicates a force change (Jami 1992).

This conceptual framework does away with the "mystery" of error signal generation. The putative cerebral-cerebellar mechanism first selects the target pyramidal cells, then produces the motor error correction, and later reduces the error signal. No hardware is needed and all olivary afferents are potential cerebellar error signals. Central or peripheral signals project topographically to the inferior olive, then to related cerebellar cortical beams, and further to their learned cerebral targets. The origin of error signals in the most lateral cerebellar modules may be cerebral overloading during movement or succession of movements (Frolov et al. 1993).

Grasping cerebellar function depends on our understanding the principles of sensorimotor integration: The frame of reference hypothesis

Anatol G. Feldman and Mindy F. Levin

Centre de recherche, Institut de réadaptation de Montréal, Montréal, Canada H3S 2J4; Institut de génie biomédical, Université de Montréal, Montréal, Canda H3C 3J7; Ecole de réadaptation, Université de Montréal, Montréal, Canada H3C 3J7. **feldman&ere.umontreal.ca; levin&ere.umontreal.ca**

Abstract: The cerebellum probably obeys the rules of sensorimotor integration common in the nervous system. One such a rule is formulated: the nervous system organizes spatial frames of reference for the sensorimotor apparatus and produces voluntary movements by shifting their origin points. We give examples of spatial frames of reference for different single- and multi-joint movements including locomotion and also illustrate that the process of motor development and learning may depend critically on the formation of appropriate frames of reference and the organism's ability to manage them. We suggest that a solution to the problem of sensorimotor integration may not be trivial and may actually change the mental and experimental paradigms used in the understanding of the cerebellum and other brain structures. [**HOUK et al.; SIMPSON et al.; SMITH; THACH**]

The cerebellum plays an essential role in movement production (**HOUK et al.; SIMPSON et al.; SMITH; THACH**) but how this is accomplished remains unclear. The variety of models of cerebellar function (**HOUK et al.**) shows a growing interest of neuroscientists in this brain structure but simultaneously demonstrates the absence of consensus on its specific functions. Although we will not attempt to solve the cerebellum puzzle, we suggest how a solution may be approached.

The cerebellum is the recipient of vestibular, visual, proprioceptive, and other sensory information. It also receives independent influences at the level of Purkinje cells and output nuclei. In other words, the cerebellum combines signals describing physical characteristics of the organism and its interactive environment with relatively independent signals from other parts of the brain. The cerebellum, however, is not the only structure that combines sensory-dependent and independent inputs. A similar combination of inputs (called here sensorimotor integration) is characteristic of almost every neuron and system involved in movement regulation (muscle spindles, motoneurons, spinal interneurons, motor and premotor cortical neurons, etc., etc.). This implies that sensorimotor integration is based on general, albeit hidden, rules which the cerebellum may equally obey. The discussion of cerebellar function in the more global context of sensorimotor integration is actually absent in the target articles. We will suggest a basic rule of sensorimotor integration using the λ model for motor control. In canonized representations, the model suggests that shifts in the equilibrium or steady state of the system underlie voluntary movements (see Feldman & Levin, 1993; 1995). The λ model also suggests a specific hypothesis on sensorimotor integration in the production of muscle activation, force, and movement. We will mainly focus on this aspect of the model recently extended to multi-muscle and multi-joint motor tasks (Feldman & Levin, 1993; 1995), called the "frame of reference hypothesis."

Generally, frames of reference or, synonymously, systems of coordinates, are used to describe, for example, the position and movement of a body in space. These coordinates are defined relative to another body or a point called the origin of the system of coordinates. Mathematical descriptions of a system's behavior on paper use frames which may be called symbolic or non-physical in the sense that changes in the position or geometric characteristics of the frames do not affect the real behavior of the system. In contrast, natural or physical frames constrain the behavior of the system. For example, a hermetically sealed vessel may be considered as a physical frame of reference for the gas molecules contained within it since the movement of molecules is constrained by the vessel.

The central idea of the λ model is that using afferent systems, the nervous system organizes spatial frames of reference for the sensorimotor apparatus. These frames are physical so that the generation of neuronal and muscle activity and forces is frame-dependent. Active movements may be produced by shifting the origin points of the frames. For our sealed vessel, a shift in the frame of reference consists of moving the vessel from one point to another in which case all molecules follow the shift.

We propose that proprioceptive inputs to a motoneuron define its spatial frame of reference and its origin point (threshold muscle length λ) is specified by, among others, independent control inputs (Fig. 1A). This frame is physical because it constrains motoneuronal behavior: the motoneuron is recruited or derecruited at the threshold length and changes its activity depending on the threshold and current muscle length (x). Thus, motoneuronal behavior is frame-dependent. Moreover, by changing the threshold muscle length, that is, the origin point, control inputs may prescribe the range of muscle lengths in which the motoneuron may participate in the generation of forces required for movement and posture.

The motoneuronal frame of reference hypothesis has several attributes which may be common for different frames used for movement production. First, a frame is created by the convergence of sensory information (visual, vestibular, auditory, proprioceptive, etc.) and independent control inputs on common threshold elements. Since sensory systems are sensitive to variables associated with the environment, they relate the frame of reference to the physical world. Due to sensory influences, threshold membrane potentials are associated with positional-dimension variables which correspond to the origins of the spatial frames. Thus, the generation of activity and forces is frame-dependent. Shifting the frame of reference by control inputs compels the system to modify its activity. Consistent with the Bernsteinian view (Bernstein 1967), the hypothesis suggests that kinematic, electromyographic, and force patterns are not pre-determined by control signals but emerge from the dynamic interaction of the system's components with external forces within the designated frames of reference.

For a single joint, there are two positional frames of reference for the activation of flexor and extensor muscles respectively (Fig. 1B; Feldman & Levin 1995). In one, the origin is associated with the threshold angle (R) at which the transition of flexor to extensor activity or vice versa is observed. The other frame surrounds the transition angle with an angular zone (C) in which flexor and extensor muscles are co-active (if $C > 0$) or silent (if $C < 0$). Control inputs to motoneurons may shift the origin point (R) of the first frame and/or change the width (C) of the second. These control influences are called the R ("reciprocal") and C ("coactivation") commands, respectively. These commands are independent of each other and may be used in different combinations by the nervous system to produce different single-joint movements (Levin et al. 1992). Recent experimental tests of the frame of reference hypothesis have dealt with the timing of the commands, strategies utilized by the nervous system to correct movement errors, as well as the effects of perturbations on the EMG and kinematic patterns in the fastest elbow and wrist movements (Feldman et al. 1995; Weeks et al. 1996).

We would like to briefly compare the effects of the R and C commands. Of the two commands only the R command is able to produce shifts in the equilibrium state of the system by translating the respective frame of reference in spatial coordinates (Feldman & Levin 1995). A positive C command may contribute to the net joint stiffness defined as the slope of the static torque/angle characteristic specified by the R and C commands (cf. **SMITH**). Like EMG activity, stiffness is an emergent property resulting not only from the R and C commands but also from the dynamic interaction between the system's components themselves and external forces. Applied in combination with an R command, the C command may contribute to movement speed. While changes in stiffness and speed are also dependent on the R command, the C command may affect them without influencing the timing of equilibrium shifts.

SMITH discusses movement production using a traditional approach in which control processes are considered in terms of variables such as stiffness, reciprocal and coactive EMG patterns for agonist and antagonist muscles in different motor tasks. According to the frame of reference hypothesis, the main limitation of such approaches is that the spatial aspects of control processes cannot be described in terms of these variables. Since R and C commands are position-dimensional determinants of their respective frames of reference, they are independent of EMG patterns (but not vice versa). Thus, control processes, occurring at a higher (hidden) level, cannot be directly related to EMG output.

The names reciprocal and coactivation, in our approach, refer to their typical EMG effects; these command effects, however, are

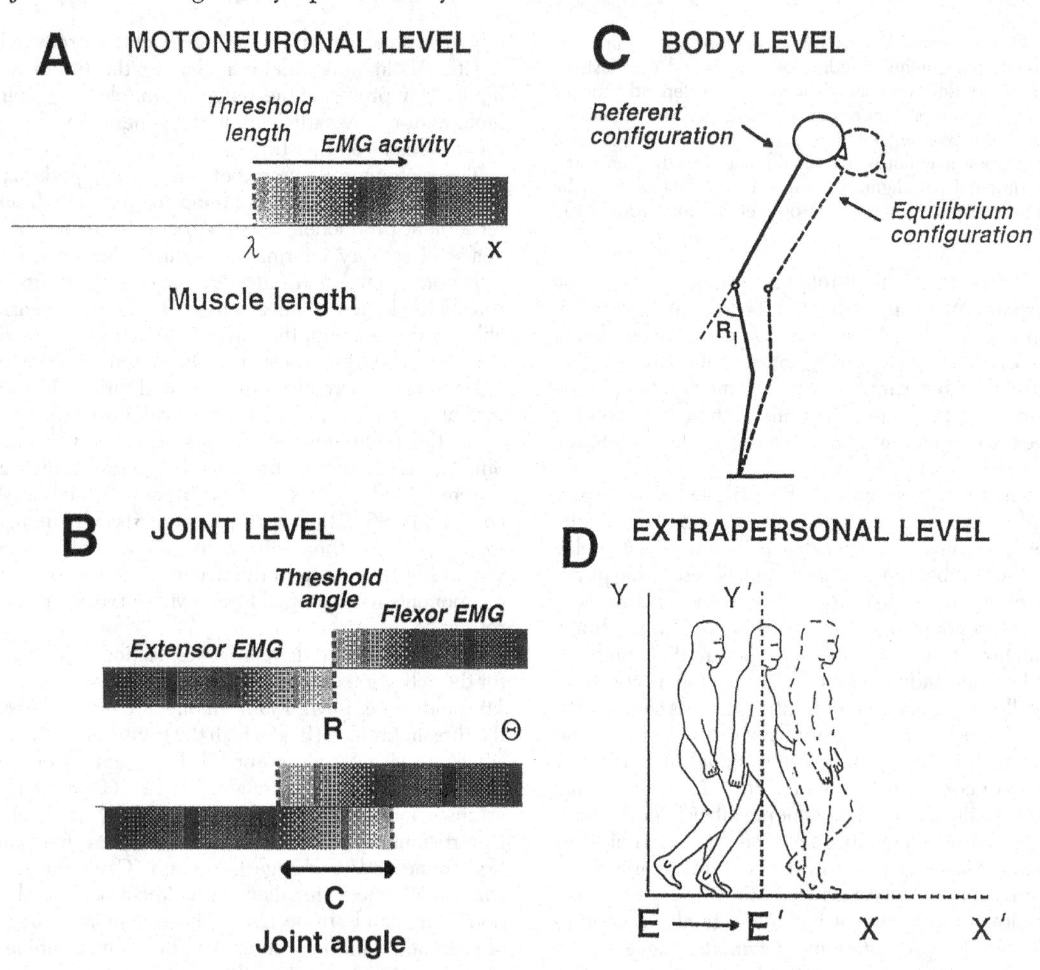

Figure 1 (Feldman & Levin). Spatial frames of reference at different level of motor control. A: Motoneuronal recruitment starts at a threshold muscle length (λ) and increases (as shown by shaded area) as the actual length (x) increases. B: Two positional frames of reference for activation of flexor and extensor muscles of a joint defined by a transitional angle (R) and the coactivation zone C. C: Referent body configuration (solid lines) defined by the threshold angles (R_i) for each degree of freedom of the body. Actual, and in particular, equilibrium configuration (dashed lines) result from neuronal and mechanical interactions of the systems with external forces in the frame of reference designated by control signals. D: Extrapersonal frame of reference associated with external 3-dimensional space. Shifts in the origin point of the frame (E → E′) may be used, in particular, to elicit a step or gait.

not unique. For example, although a coactivation zone may be present ($C > 0$), the load may be balanced outside this zone so that only agonist muscle activation occurs. Muscle activity also depends on movement speed and the timing of the R command. For example, even if the current joint position is inside the coactivation zone, whether or not agonist and antagonist coactivation occurs also depends on the movement speed and timing of the R command. The distribution of activity among different agonist muscles (synergists) may also be affected by the C command (Feldman & Levin 1995).

It may be even more complicated to make inferences about control processes based on stiffness studies since the ambiguity of the relationship between the C command and stiffness is complicated by the participation of the R command. More importantly, stiffness is specified in spatial frames of reference defined, according to the λ model, by position-dimensional R and C commands. The spatial aspects of stiffness regulation were not discussed by SMITH although these aspects rather than stiffness magnitude may be more closely related to what it is the cerebellum actually controls. In addition, the regulation of stiffness is an extremely controversial issue because of the absence of consensus on the definition and measurement of the phenomenon. SMITH's target article does not rectify the problem although a critical analysis would be helpful in consideration of the hypothesis that the cerebellum participates in the regulation of joint stiffness.

The frame of reference hypothesis is easily generalized to all muscles and degrees of freedom of the body. The R command becomes a vector composed of individual threshold angles, R_i, for each degree of freedom:

$$R = (R_1, R_2, \ldots, R_n).$$

It represents a referent body configuration (Fig. 1C) at which active muscle forces are zero if similarly defined vector C command is zero. Any deflection of the body from this configuration by external forces elicits an active counteraction. The referent configuration also represents the origin of a multi-dimensional frame of reference in which actual or equilibrium configuration

$$\theta = (\theta_1, \theta_2, \ldots, \theta_n)$$

is established. The actual configuration depends on the current dynamic interactions between the system's components and external forces. Generally, $R \neq \theta$ as, for example, in leaning the body (Fig. 1C) but this discrepancy is not considered as a positional error. The actual configuration is in error if it does not meet some requirement, for example, is different from a desired boy configuration. Error corrections are produced by a change in the referent configuration with subsequent comparison of the results with the desired configuration.

Experimental data on the neurons involved in the production of the united body frame, as well as on their sensory and control input and output projections to motoneurons have not been sought. Possible candidates may be propriospinal neurons and neurons of descending tracts and their pluri-segmental projections to motoneurons of multiple muscles of the body. To unify the body frame, control systems should widely project to these neurons, although each neuron may receive relatively local sensory inputs (e.g., from proprioceptive afferents of a small number of muscles). Reflex intermuscular interaction may also play a significant role in the organization of the united body frame (Nichols 1994).

Atop the hierarchy of spatial frames of reference may be an extrapersonal frame utilized for actions of the body in external 3-dimensional space (Fig. 1D). Such a physical frame may be based on sensory information from vestibular, visual, and proprioceptive receptors. The physical character of the frame implies, again, that the nervous system has the capacity to shift the frame and thus provoke an active movement of the body to another position in external space. A discrete shift of the frame will elicit neuronal and muscle activity resulting in the transition of the body from the initial to a final equilibrium configuration reached in another part of space. In other words, depending on the direction of the shift, a step forward or backward will be produced. Indeed, a continuous shift of the frame will elicit gait. One may also assume that, with increasing speed of the shift, gait will be transformed into running (cf. Kelso 1995). A jump up may be produced by a rapid vertical shift of the frame. For a forward jump, a vertical shift should be combined with a horizontal shift.

Shifts of an extrapersonal frame of reference may be used for a part of the body, for example, for production of pointing movements. The origin point of such a frame may be associated with a point in actual space (e.g., the corner of the table) above which the subject moves his arm. Pointing movement is produced by shifting the origin point of the frame in the appropriate direction. As in the case of the referent body configuration, the possible discrepancy between the referent and actual directions of movement is not considered as an error. An error exists when the actual movement direction of the endpoint differs from that which was intended. The correction of the error is accomplished by a modification of the direction of the shift in the frame. This control strategy was tested in a computer model (Feldman & Levin 1995).

Explanations of different motor acts may be based on the frame of reference hypothesis. The hypothesis may be used to suggest that motor learning and development may be associated with the formation and management of appropriate frames of reference (cf. **HOUK et al.; THACH**). For example, one may suggest that the ability to stand in a toddler appears only after the formation of the united body frame is complete (Feldman & Levin 1995; cf. Thelen & Smith 1994). The ability to produce a step or string of steps (gait) arises only after the formation of a global extrapersonal frame. Similarly, a two-year old is able to walk and jump upwards but not to jump while propelling himself forward. This implies that the child is at the stage of development when he is able to produce separate but not simultaneous horizontal and vertical shifts in his frame of reference.

Whether or not the frame of reference hypothesis is correct, solving the problem of sensorimotor integration may not be trivial. Ignoring this problem in discussions of the motor function of the cerebellum may lead us to non-realistic models even if such models use neural network simulations or servo-theories. It was surprising, in particular, that **HOUK,** who has previously stressed the importance of considering sensorimotor integration for the understanding of the function of any brain structure chooses here, in his target article, to discuss cerebellar models outside this general framework.

The frame of reference hypothesis provokes questions concerning the cerebellum that have never been raised before. Which frames does the cerebellum support? How are they developed in the process of learning? How is the temporal shift of the frame produced? What is the specific role of climbing fibers in this process (cf. **SIMPSON et al.**)? And so on. Since these questions have not previously been asked, there are no experimental data to provide the answers. There is, however, the hope that the understanding of the general rules of sensorimotor integration may actually change the mental and experimental paradigms in our approaches to the cerebellum and other brain structures.

Timing implications of metabotropic mechanisms for cerebellar learning

John C. Fiala and Daniel Bullock[1]

Department of Cognitive and Neural Systems, Boston University, Boston, MA 02215-2411. **danb@cns.bu.edu**

Abstract: A major theme of the systems physiologists is the critical timing function of the cerebellum. However, the biophysicists do not appear to directly address the biophysical basis of the adaptive timing competence implicated in the physiological and behavioral data. Thus, the bridge between the macroscopic and microscopic data bases seems to be incomplete in a critical area. We report successful results from an attempt to add the missing part of the bridge. It comes in the form of a model of how the second messenger system activated by parallel fiber inputs – the mGluR channel – can literally bridge the temporal gap between CS and CR, both in standard conditioning tasks and more generally. [**CRÉPEL et al.; HOUK et al.; KANO; LINDEN; SIMPSON et al.; SMITH; THACH; VINCENT**]

The wealth of cerebellar data and theory marshalled by the authors of the eight target articles in this *BBS* special issue is far too extensive to address in a comprehensive way within a brief commentary. Still, it is important to ask whether the macroscopic view of the systems neurophysiologists can be seen to cohere with the microscopic view of the biophysicists. One way to answer this question is to identify a single but ubiquitous behavioral task that by itself can illustrate the need both for the basic design of the cerebellar network and for the complex biophysics of plasticity in Purkinje cells. Our proposal for such a single adaptive function, treated in varying degrees by **HOUK et al., THACH, SMITH,** and **LINDEN** (among many other cerebellar scientists), is to solve the problem of adaptively picking out motorically relevant parallel fiber (PF) signals from among vast numbers of irrelevant signals, and of using the selected signals at appropriately delayed times to improve response generation. By simultaneously considering temporal association paradigms, such as the classical eyeblink conditioning task utilized by Thompson, Steinmetz, and others, as well as the specialized biophysics summarized in four of the target articles, we believe that the complexity exemplified by **CRÉPEL et al.**'s "dual paths" to LTD becomes more comprehensible.

How does an unconditioned stimulus (US), occurring seconds after the conditioned stimulus (CS), induce a learned cerebellar response to the CS that only slightly anticipates the US? As shown in Steinmetz (1990b), direct stimulation of mossy fibers can serve as a CS and direct stimulation of climbing fibers as a US, such that a conditioned response (CR) emerges following repeated pairings, and the CR topography is mirrored by a timed, anticipatory increase in interpositus activity. Moreover, the interstimulus interval (ISI) between direct brain stimulation CS and US may be as long as 2000 msec but no shorter than about 80 msec for conditioning to occur (Steinmetz 1990a). The most parsimonious explanation is that Purkinje cells' sensitivity to correlation between temporally distant inputs induces a learned response which manifests after an appropriate delay from CS onset. If so, then the mechanisms of synaptic response and plasticity must contain a temporal component capable of spanning hundreds, even thousands of milliseconds.

Does our current understanding of synaptic plasticity as described in the articles by **CRÉPEL et al., VINCENT, LINDEN,** and **KANO** adequately account for this temporal component? We believe that the data, set in a suitable theoretical framework, do support this temporal component. Below, we briefly describe one possible framework and its consequences for the temporal dimension of cerebellar learning.

Parallel fibers stimulate the intracellular release of calcium in cerebellar Purkinje cell dendritic spines by activating metabotropic glutamate receptors (mGluRs). The calcium responses to mGluR activation in cerebellar Purkinje cells are governed by essentially the same mechanisms as light-activated calcium release in invertebrate photoreceptors (Richard et al. 1995), albeit on a slower time scale. A characteristic of this type of metabotropic response is that response latency is dependent on the amount of receptor activation. Variations in transmitter concentration or number of receptors lead to variations in the latency of the intracellular calcium responses. We have modeled the metabotropic response pathway, and have shown that delayed calcium responses spanning the range of conditionable ISIs can be realized by this mechanism (Fiala et al. 1995). The pool of variously delayed calcium responses to PF inputs provides one half of a Darwinian mechanism for adaptation. The mechanism is completed by a selective learning process that normally allows only correctly delayed variants to be expressed in behavior.

Figure 1 shows one way that cooperation between long latency metabotropic responses and a variant selecting process might produce learned timing at the parallel fiber-Purkinje cell synapse. The figure shows how climbing fiber (CF) and PF signals (top) can affect two ion channels (bottom) that control postsynaptic transmembrane current. Parallel fiber stimulation of mGluR1 receptors activates PKC by the production of DAG and the release of Ca^{2+} intracellularly. The PKC-driven phosphorylation of AMPA receptors appears to be responsible for the induction of LTD, as discussed by LINDEN. Therefore, mGluR responses normally induced by PF stimulation are important for LTD induction.

Climbing fiber stimulation is the second pathway involved in LTD. Climbing fiber activation produces a rise in [cGMP] in Purkinje cell dendrites, perhaps by indirect activation of NO synthase in the juxtaposed terminals of basket and stellate interneurons. Elevated cGMP levels activate PKG which phosphorylates G-substrate. Phosphorylated G-substrate inhibits protein phosphatase-1 (PP-1). PP-1 dephosphorylates the AMPA receptor. Dephosphorylation of G-substrate may be produced by calcineurin, a Ca^{2+}-activated protein phosphatase.

A robust and maintained level of phosphorylation of specific channel proteins is obtained when the increase in the mGluR1-mediated Ca^{2+} and DAG signals coincides with the increase in cGMP induced by CF activation. As noted above, different patches of membrane will vary in the delay that characterizes their mGluR pathways. LTD in the model can be produced by a CF-induced [cGMP] increase occurring hundreds of msec after the onset of the CS and the PF signal it causes (Fig. 2A), but not by CF

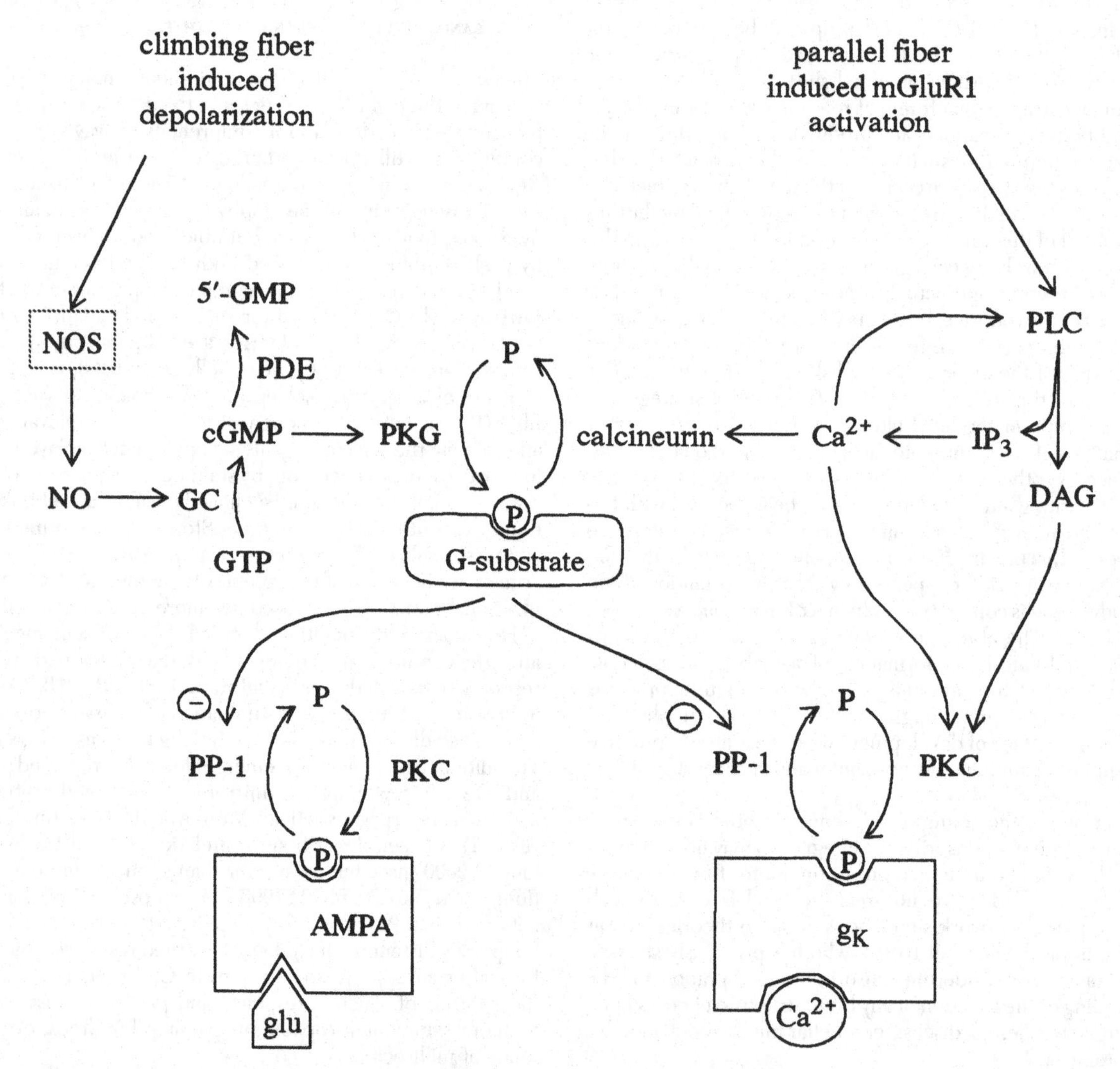

Figure 1 (Fiala & Bullock). Processes mediating learning in a model of a timed response in cerebellar Purkinje cells. In the two lower phosphorylation cycles, the two PKC symbols are meant to refer to a common PKC signal/pool generated by the mGluR cascade.

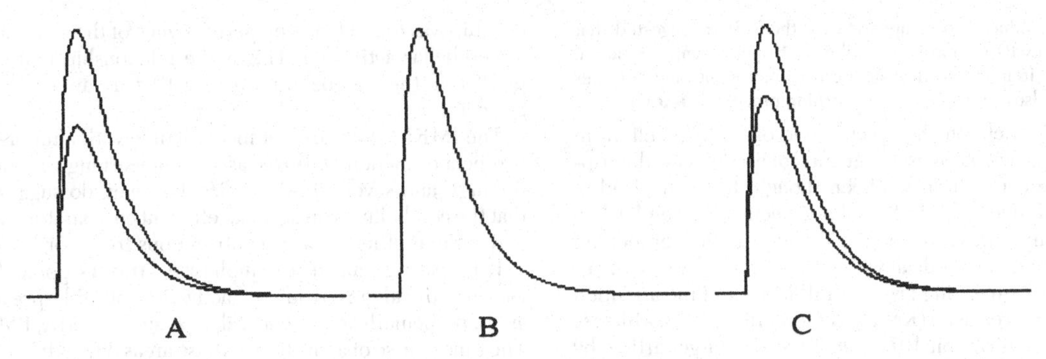

Figure 2 (Fiala & Bullock). Simulated LTD based on model of Figure 1. A: LTD (lower trace) following climbing fiber input 400msec after the onset of parallel fiber activity. B: No LTD observed with climbing fiber input at onset of parallel fiber activity. C: LTD (lower trace) following elevation of $[Ca^{2+}]$cyt to 6mM for 0.9sec starting at the onset of parallel fiber activity.

activity simultaneous with CS/PF signal onset (Fig. 2B) because the mGluR path delay makes non-coincidental the times of arrival of the CF- and CS/PF-induced signals at the critical biophysical convergence point. Extinction of the learned response results from dephosphorylation produced by activation of the mGluR pathway alone, without a coincidence CF signal.

As described by LINDEN, CF inputs are not required for LTD. It suffices to significantly depolarize Purkinje cells (3 seconds at +10mV) in conjunction with application of glutamate. According to the Figure 1 model, an abnormally large rise in $[Ca^{2+}]$cyt can evoke LTD in the absence of a CF-stimulated rise in [cGMP], particularly if the calcineurin signal saturates at a total level of activity lower than that of PKC. This is due to the fact that calcium activates PLC which produces DAG. The combination of high levels of calcium and DAG drive PKC phosphorylation beyond that recoverable by the release of protein phosphatase activity by calcineurin (Fig. 2C).

LTD in some preparations occurred only when CF stimulation preceded PF stimulation or followed it by less than 20–30 msec (Ekerot & Kano 1989). Use of classical conditioning-like timing did not produce LTD in these cases. This suggested to some that AMPA receptor LTD was not related to behavioral learning in classical conditioning. The present model shows why these data may not warrant such a conclusion. In particular, the model assumes that CF activation under normal conditions does not induce substantial increases in cytoplasmic free Ca^{2+} in spines. If allowed, such increases would activate PKC by the stimulation of DAG through a Ca^{2+}-dependent PLC pathway. This would constitute interference by the selection process with the variant generation process. Instead, current data indicate that spine heads are insulated from these Ca^{2+} increases by the activity of inhibitory interneurons *in vivo* (Callaway et al. 1995).

Consistent with this hypothesis, LTD induced by CF stimulation preceding PF stimulation is dependent on the blockade of cortical inhibition by picrotoxin or bicuculline. When inhibition is not blocked, little or no LTD can be observed with this paradigm (Schreurs & Alkon 1993). Blockage of inhibition increases the activation of Purkinje cell dendrites by parallel fibers and allows substantial calcium increases to occur. Thus, the type of LTD due to excessive intracellular calcium increases is likely to occur in these cases. This type of LTD does not exhibit the temporal characteristics of the natural case. This is because the latency of the PF-induced metabotropic calcium response is dependent on low initial levels of calcium in the dendritic spines. When the intracellular calcium level is artificially elevated, PKC is activated in all spines with no latency. So this type of LTD can be expected to occur with CF activity simultaneous with or preceding PF activity.

These data on AMPA receptor LTD return us to the question of the need for dual messengers and the relative role of the two phosphorylation targets in behavioral learning. Note that of the two channels affected by phosphorylation-mediated learning, only the conductance of the K-channel is calcium dependent. Therefore, the delayed calcium response in the mGluR cascade is expressed in performance by its phasic effect on the learning-modified calcium-dependent K-conductance. In particular, timed phasic enhancement of this conductance can hyperpolarize the Purkinje cell and transiently release its nuclear target cells from inhibition. Simulations in Fiala et al. (1995) have shown that a population of model Purkinje cells containing the Figure 1 biophysics and conditioned in the manner of Steinmetz (1990a) can give rise to the kind of anticipatory CR topography that emerges, during learning, in the interpositus nucleus. Applications of this learned timing competence to context-dependent, anticipatory, multi-channel (synergic) motor error cancellation (along lines similar to those imagined by THACH, Gilbert, Ito, Kawato, SMITH, and many others cited in the target articles) is straightforward, on the assumption that CF discharges wax and wane with, and are time-locked to, detections of movement errors, an assumption supported by SIMPSON et al.

In summary, by focusing on the partly neglected need to bridge the time between CS and US and on the possibility that time-critical pausing may occur via a metabotropic pathway's effect on a calcium dependent K-channel, we have attempted to build a stronger bridge between the systems and cellular data bases covered in the target articles. A final point concerns cognition. Because after learning the AMPA channel will act with reduced "gain" from the time of CS onset, it could serve a response priming function by reducing cortical inhibition on target nuclear zones. The later, time-delayed, action of the secondary messenger could serve a response release function. This two-phase property is in agreement with part of the HOUK et al. model, and suggests that the cognitive tack recommended by THACH may prove quite fruitful.

NOTE

1. Address correspondence to Daniel Bullock.

The cerebellum as comparator: Increases in cerebellar activity during motor learning may reflect its role as part of an error detection/correction mechanism

D. Flament[a] and T. J. Ebner[b]

[a]*Rush University, Department of Physical Medicine and Rehabilitation and Molecular Biophysics and Physiology, Chicago, IL 60612;* [b]*University of Minnesota, Departments of Neurosurgery and Physiology, Minneapolis, MN 55455.* **dflament@rpslmc.edu; ebner001@maroon.tc.umn.edu**

Abstract: The role of the cerebellum as a comparator of desired motor output and actual performance may be most important during learning of a novel motor task, when movement errors are common and corrective movements are produced to compensate for them. It is suggested that PET and recent fMRI data are compatible with such an interpretation. Increased activity in motor cortical areas during motor learning indicates that these areas also contribute to the learning process. [THACH]

THACH's target article on the specific role of the cerebellum in motor learning provides an excellent and comprehensive description of how the cerebellum has been thought to be involved in motor control. Indeed, this historical perspective is taken back to the ninth century A.D. However, the theory that the cerebellum may act as a comparator is dealt with only superficially (sect. 2.4). This is somewhat surprising given its influence and the attention that this model has received (Kawato & Gomi 1992; 1993; Miller & Oscarsson 1970; Oscarsson 1973; 1980; see also target article by **HOUK et al.**). In addition, much of the PET work described in this article can be interpreted to support the hypothesis that the cerebellum is involved in error control.

The general consensus from PET studies of motor learning is that cerebellar activity is greater during the performance of movements that are actively being learned than when previously-learned movements are made (sect. 4.1, para. 5). As THACH points out, however, the results of these PET studies are not as homogeneous as such a statement might imply, and there are many contradictory findings. These are described in the target article and have also been addressed in a monograph by Ebner et al. (1996). One point of note, not mentioned by THACH, is that some of the contradictory results may be due not only to the different motor tasks studied, but to the fact that some motor learning tasks appear to have been dominated by declarative learning. For example, tactile recognition and learning of a finger movement sequence are undoubtedly highly dependent on declarative processes, particularly in the early stages of learning.

With these caveats in mind, one can interpret the general PET finding of increased cerebellar activation during motor learning. The early stages of motor learning are typically associated with performance errors. These errors must be detected and recognized as deviating from the desired goal. A corrective signal must then be generated to compensate for, or correct, the movement errors. The high level of cerebellar activity during motor learning is compatible with Oscarsson's view that the cerebellum may be detecting and correcting these errors. When performance improves and learning has been accomplished the cerebellum is less activated as fewer and fewer errors are detected and the need for corrective movements is diminished.

The cerebellum's role in motor learning was recently studied using functional MRI and a complex motor task that required a visuo-motor transformation (Flament et al. 1994). Subjects were trained to perform a step-tracking task, using a joystick to superimpose a cross-hair cursor onto visual targets displayed on a screen. The relationship between movement of the joystick and cursor was either (1) normal, (2) reversed in both the x and y direction, or (3) varied randomly between 4 conditions. The normal relationship was well-learned by all subjects; the reversed relationship was new to them and had to be learned; the random relationship could not be learned and for each movement the correct joystick movement direction had to be determined by trial and error. An index of performance based on the movement trajectory and the number of successful movements was calculated for each condition.

The results showed that cerebellar activity was highest during the random task and the early stages of the learning task. In both of these conditions the performance index was very low (i.e., poor performance). These conditions were characterized by many movement errors and frequent corrective movements. During the random task the subjects had a 3 in 4 chance of initiating their movement in the wrong direction, thus requiring at least one corrective response, and frequently several more. As performance improved over time during the learning task, cerebellar activity decreased. After prolonged practice at the learning task, cerebellar activity declined to very low levels, apparently becoming disengaged from the processing loop. A significant, inverse relationship was found between performance of the task and the level of cerebellar activation. This is the relationship that one would predict if the cerebellum were acting as an error detector/corrector.

The fMRI results are not incompatible with alternate views of cerebellar function or its role as a motor learning structure. These findings, and several PET studies, however, do suggest strongly that the cerebellum can act as an efficient comparator and that this may be one of its primary contributions to motor learning.

It is also relevant to re-emphasize THACH's point that motor learning paradigms result in activation of widespread cortical areas, particularly motor cortical areas such as SMA, PM, and M1. The time course of activation in these areas differs from that of the cerebellum, as shown by both PET studies and our own fMRI data (Flament et al. 1995). We have found that activity in motor cortical areas generally increase at the onset of learning and remain elevated for longer periods than the cerebellum. Furthermore, the inverse relationship between performance and activation that characterizes cerebellar learning was not found in motor cortical areas. It is tempting to speculate that the persistent activation of cortex represents some form of modification of an internal model responsible for the correct performance of the newly learned task. This interpretation might relegate to the cortex the role THACH attributes to the cerebellum. Additional studies will be required before such an interpretation can be accepted.

Cerebellum does more than recalibration of movements after perturbations

C. Gielen

Department of Medical Physics and Biophysics, University of Nijmegen, 6525 EZ Nijmegen, The Netherlands. **stan@mbfys.kun.nl**

Abstract: We argue that the function of the cerebellum is more than just an error-detecting mechanism. Rather, the cerebellum plays an important role in *all* movements. The bias in (re)calibration is an unfortunate restrictive result of a very successful and important experiment. [SMITH, THACH]

The target articles in this issue provide a superb up-to-date overview of the neuroanatomy and neurophysiology of the cerebellum. However, relatively little attention has been paid to studies on the role of the cerebellum in various motor tasks and on the implications of these studies.

One of the working hypotheses about cerebellar function was based on the idea of an error-detecting mechanism, which could somehow recalibrate the system in order to eliminate any motor errors. This provided the basis for models on the cerebellum by Marr (1969) and Albus (1971). This hypothesis received further support from the observation that (re)calibration after perturbation of the motor system (usually induced by lesions) is not possible when the cerebellum is not active (Optican & Robinson 1980).

As already pointed out in the target article by SMITH there is good evidence that the cerebellum does more than simply (re)calibrate the motor system. For example, cerebellar lesions also give rise to abnormal arm movements when no learning or (re)calibration is required (Becker et al. 1990; Diener et al. 1993; Hallett et al. 1975; Terzuolo et al. 1973). This suggests in our view that the cerebellum is essential not only for adaptive recalibration, but also for normal motor functions.

The statement that cerebellum plays a role both in (re)calibration of the motor system and in normal movements can be most convincingly illustrated for the oculomotor system. Recalibration of the saccadic system can easily be imposed in adaptation experiments in which a visual target appears which serves as the goal of a saccadic eye movement. There is a well known paradigm in which the visual target is displaced further from the start position during the saccadic eye movement. As a consequence, the amplitude of the saccadic eye-movement is too small to reach the *new* target. The subjects are never consciously aware of the displacement during the saccade because of the well known phenomenon of saccadic suppression of visual information during saccades. After a few of these saccade-triggered displacements, the amplitude of the saccade gradually increases in order to adapt to the intra-saccadic displacement. This adaptation is confined to a limited area around the adapted saccade vector. Frens and van Opstal (1994) demonstrated that after visually induced adaptation, auditory-evoked saccades are also adapted to the visually induced displacement.

Since the superior colliculus (SC) plays an important role in the generation of saccadic eye movements, Goldberg et al. (1993) investigated the role of the SC in the adaptation process. Neurons in the deeper layers of the SC reveal a burst of activity for saccadic eye movements with a particular direction and amplitude. The amplitude of this burst of activity decreases when the direction or amplitude of the saccade deviate more from the optimal saccade amplitude and direction. The range of saccadic eye movements, for which a cell is active is called the cell's movement field. Goldberg et al. found that SC saccade-related cells revealed the same burst-like activity after adaptation if the adapted saccade was no longer in its original movement field. This implies that the range of visual targets, which elicit saccades, differs from the movement field for these cells. They suggested that some system, parallel or downstream of the SC, modifies the neuronal signal from the SC in order to make it suitable for the required adapted saccade. More specifically, they suggested that the cerebellum might be involved. This hypothesis was supported by the finding that lesioning of the cerebellum prevented saccadic adaptation (Goldberg et al. 1993). Surprisingly, electrical stimulation in the SC *after* adaptation elicited saccades with the same amplitude as was observed *before* the adaptation. These results were interpreted as suggesting (1) that the superior colliculus programs saccades in terms of visual location of the target, not the physical amplitude; and (2) that the cerebellum, which does not receive sensory input about visual target position during electrical stimulation, is responsible for the calibration of the properties of the oculomotor plant. This implies that the cerebellum plays an important role for all saccadic eye movements, not only after adaptation. This is well in agreement with the observation that lesioning of the cerebellum also affects normal limb and eye movements.

Recently, Melis and van Gisbergen (1995) found that target displacements during electrically-induced saccades in the SC give rise to adaptation also. At least as important was their finding that after adaptation for saccades induced by electrical stimulation, adaptation also occurs in visually elicited saccades. Their results agree with the hypothesis by Goldberg et al. (1993) that the SC activity codes visual target displacement and that the adaptation for the plant occurs later, downstream of the SC. The main conclusion is that the adaptation center apparently receives both sensory input as well as input of motor commands.

These experiments raise the question of whether a similar model can be proposed for limb movements. For example, it has been shown (Ashe et al. 1993; Georgopoulos et al. 1984) that neurons in primary motor cortex (M1) may be active after presentation of a visual target for arm movements, even when the arm movement does not yet occur. More particularly, if the monkey is trained to move in a direction 90 degrees rotated from the visual target, the population vector in M1 rotates to the final target position before a movement is initiated (Georgopoulos & Massey 1987). These observations suggest that motor cortex codes a "desired" movement direction, rather than actual movement direction and that the properties of the effector system are accounted for by a system parallel or downstream of motor cortex. Although this may seem to be a speculative hypothesis, it is an obvious hypothesis based on the experimental evidence, which is open to falsification.

How an important observation can mislead us. Until the early eighties the main view on the function of reflexes (both the well known tendon reflex or short latency reflex and the long-latency reflex) was that their main function was to correct for external perturbations of posture or movement. The important observation that large reflex activity was induced by fast lengthening of a muscle gave rise to the notion that the main function of reflexes was to correct for external perturbations (for a review see Houk & Rymer 1981). Later studies have shown that although reflexes play an important role in compensating for external perturbations (the present view is that in general they cannot provide complete compensation), they also figure in normal movements such as walking (Stein 1991). In summary, reflexes are part of the normal activity in the motor system. What happened in the study of reflexes is that a very important and successful experimental observation unfortunately placed too much emphasis on a single aspect of reflexes.

In our view a similar development may be found with regard to the function of the cerebellum. One of the first studies on cerebellum demonstrated its important role in recalibration of movement commands. This was a significant finding, but does not eliminate other possibly important functions of the cerebellum. There are enough observations to indicate that the cerebellum plays a role in natural movements and that cerebellum does more than just recalibrate movement commands. This is also implied in the model by Kawato and Gomi (1992), who consider the cerebellum as an adaptive parallel fine-tuning device to adjust for the complex properties of the effector system.

A comparison between cerebellar activity and neuronal activity in primary, supplementary, and premotor cortex did not demonstrate clear differences in neuronal activity during learning and learned or automatic movements (Raichle et al. 1994). Based on this finding THACH concludes that many of the current hypotheses about cerebellar function are due to erroneous extrapolation of specific experimental results to human motor performance following artificial investigators' instructions and spurious correlations. We fully agree with this conclusion and we share with him the opinion that for most movements cerebellum is necessary for adequate motor performance, not only to correct for perturbations.

How and what does the cerebellum learn?

Peter F. C. Gilbert

Department of Anatomy and Developmental Biology, University College London, London WC1E 6BT, United Kingdom. **p.gilbert@ucl.ac.uk**

Abstract: Information for motor learning is likely to be stored by groups of Purkinje cells in the cerebellum as suggested by **HOUK et al.**, though their proposal that Purkinje cells act as bistable elements is open to question. **THACH's** excellent article breaks new ground on what type of information the cerebellum may store. I have queries on whether the cerebellum can generate "thought" or is likely to engage in the purely mental learning of movements. **[HOUK et al.; THACH]**

HOUK et al. give an interesting review of a number of theories of how the cerebellum could be involved in motor learning. With widespread acceptance of the concept of motor learning by the cerebellum the details of the various theories assume more importance in light of possible experimental verification. The first cerebellar-learning theories (Albus 1971; Marr 1969) focused on information storage by single Purkinje cells, whereas it is clear from the circuitry of the cerebellar cortex that parasagittally oriented groups of Purkinje cells must be the memorizing units (Gilbert 1974).

The single-cell learning theories cannot readily be extended to encompass learning by groups of cells (Gilbert 1974). One proposal for storage by a group of cells is the adjustable pattern generator (APG) theory developed by Houk and colleagues and described in detail in **HOUK et al.**'s target article. The APG theory is similar to an earlier one by myself (Gilbert 1974; 1975). I showed how a parasagittally oriented array of Purkinje cells projecting to a nuclear cell (equivalent to an APG in Fig. 4 of Houk et al.) could be taught by the climbing fiber input to produce a required frequency of firing of the nuclear cell when a particular parallel fiber input was presented to the Purkinje cells. Frequencies of firing of the individual Purkinje cells in response to a particular parallel fiber input would be successively adjusted by the climbing fiber signals until the correct nuclear cell output frequency was achieved. Under this theory Purkinje cells involved in a learned movement would vary continuously in firing frequency in a reproducible manner during performance of the movement in accordance with what is found for awake performing animals (Thach 1970). In contrast the APG theory proposes that Purkinje cells are bistable elements with either on- or off-states (sect. 3.3) but this is not borne out by the experimental results of recording from Purkinje cells (Thach 1970).

HOUK et al. deal with *how* the cerebellum learns and **THACH** considers the question of *what* is learned. Whilst in agreement with the main thrust of THACH's target article, I have comments on two issues:

(1) **THACH** refers to the automatic generation of an action plan (which would not necessarily be executed) by the cerebellum as being in the "realm of thought" (sect. 8.2). I would reserve the terms "thinking" or "thought" for the operations of the "conscious" system (sect. 4.3) and not apply them to any of the operations of the cerebellum which in response to particular contexts generates its learned action plans or responses automatically and *subconsciously*. These same action plans or responses may also be generated consciously whilst being learned by the cerebellum and could then appropriately be described as being in the "realm of thought." This semantic issue is similar to that for the debate on whether the cerebellum has a "cognitive" role (Glickstein 1993; Leiner et al. 1993). Again I would maintain that the term "cognitive" should be reserved for operations of the "conscious" system (sect. 4.3) and not be applied to operations carried out subconsciously by the cerebellum (even though these operations may be identical in all other respects).

(2) The proposed role for the parvocellular red nucleus input to the inferior olive in "learning motor performances that are only practiced internally and not overtly expressed" (sect. 8.3) seems unduly narrow. Surely very few of us engage in purely mental rehearsal to improve motor performance. Instead, rehearsal usually consists of a *combination* of mental and motor activity, such as trying to improve a golf swing or practicing the piano. The cortical input channeled to the inferior olive via the parvocellular red nucleus could ensure more rapid learning by enhancing the olivary firing frequency during the early stages of learning (Gilbert & Thach 1977). It should be noted that purely mental rehearsal of movements is also inefficient on theoretical grounds as the contexts signalled by parallel fiber inputs to Purkinje cells storing the movement would not be as well determined as in the case where the movement was also made (Gilbert 1974).

In conclusion I believe **THACH**'s target article makes an important contribution in extending our ideas about the potential scope of cerebellar learning in light of recent PET scanning and other results. What does seem remarkable is the accuracy of the predictions of the early theoreticians Brindley and Marr (Brindley 1969; Marr 1969) who speculated so presciently about this role for the cerebellum.

Is stiffness a byproduct or a target?

Hiroaki Gomi

NTT Basic Research Laboratories, Information Science Research Laboratory, Morinosato, Atsugi, Kanagawa, Japan. **gomi@idea.brl.ntt.jp**

Abstract: To examine the model of the biological motor control proposed by A. Smith, this commentary briefly introduces some result of stiffness measurements during multi-joint arm movements, and, in the light of these results, discusses how likely it is that stiffness is a control parameter and that the cerebellum solely codes the stiffness. [**SMITH**]

Stiffness change during multi-joint movement. By developing a new manipulandum and a new estimation algorithm, we have recently succeeded in measuring the dynamic stiffness in multi-joint arm movements where human subjects performed planar point-to-point movements of one-second duration from the left to the right side of their workspace (40 cm) with their right arms (Gomi & Kawato 1995). The stiffness during multi-joint movement varied greatly in time, like during the single-joint movement of Bennett (Bennett et al. 1992). The shoulder single-joint stiffness greatly increased at the start (22–30 Nm/rad) and at the end (30–36 Nm/rad) of the movement. This increase probably reflects the muscle tension required to accelerate and decelerate for the movement. The decrease in shoulder stiffness (20–24 Nm/rad) in the middle of the movement might correspond to the switch from extension activation to flexion activation as observed in single-joint stiffness (Bennett et al. 1992). The range of elbow joint stiffness during multi-joint movement (5–21 Nm/rad) was close to but slightly higher than the range of elbow stiffness during single-joint stiffness (3–14 Nm/rad) (Bennett et al. 1992; Bennett 1993).

The equilibrium trajectories predicted, based on the measured stiffness and the equilibrium-point control hypothesis, were clearly different from the actual ones, especially in those velocity profiles. This result indicates that the equilibrium position strongly depends on the nonlinear arm-dynamics because stiffness was not high enough to keep the equilibrium trajectory close to the actual trajectory, which is contrary to the equilibrium-point control hypothesis in which the equilibrium trajectory is assumed to be planned independently of the arm dynamics.

Also, the ratios between joint stiffness components (i.e., shoulder, elbow, and double joint stiffness) were altered during movement, and the directions of hand stiffness (characterized by the stiffness ellipse) changed in different environments (direction-constrained movement, and viscosity-constrained movement). Moreover, while a multi-joint posture was held (i.e., the static condition), the ratio between joint stiffness components could also be altered when subjects were asked to voluntarily change the muscle activation level about each joint without external force.

Is the stiffness a control parameter? **SMITH** pointed out the importance of stiffness and proposed the hypothesis that the cerebellum codes the joint stiffness rather than other motion-related signals such as EMG, torque, position, or its derivatives. Before discussing whether the cerebellum code stiffness or not, I would like to address the factors realizing stiffness. The dominant factors are not only, as mentioned in **SMITH**'s target article, (1) the purpose of movement, (2) a cost function of optimization and a proficiency level, and (3) external environment, but also (4) the muscle inherent property: the fact that stiffness increases as muscle tension increases. Stiffness changes during movement do not directly mean that stiffness is a control variable. During free movement, stiffness should be altered because muscle tensions are varying and their imbalance generates the driving torque. In this case, the stiffness would not be a control variable, but rather a by-product derived from the inherent muscle property. Conversely, stiffness could be a control variable in a hand-stand task as noted in **SMITH**'s article. A hand-stand requires a quick correction of posture because of its inherent instability, indicating that stiffness of the arm must be increased. In this case, stiffness might be a control variable rather than a byproduct. It is also often required that a certain stiffness is set in interactions with external environments.

Does the cerebellum code the stiffness parameter? As explained in these examples, stiffness could be a byproduct rather than a target. Is there any advantage to always coding the stiffness explicitly in the cerebellum? From the computational point of view, it might be less likely to assume that the cerebellum codes only stiffness, which is a crucial parameter in the equilibrium-point control hypotheses. The reason for this is that the equilibrium trajectories predicted from our experimental data were different from the actual trajectories because of low stiffness. This result indicates that not only stiffness needs to be taken into account in movement planning but also the dynamics of the controlled object, even if the planning output is an equilibrium point trajectory. I would like to emphasize that a similar planning complexity exists when obtaining a movement plan by means of computing the inverse dynamics of the controlled object. Another consideration about what is planned in the cerebellum comes from the clinical studies. The fact that cerebellar patients cannot execute movements naturally implies that they lost an important ingredient of motor control. If it were only stiffness control, an equilibrium point trajectory plan which, for this argument, is hypothesized to take place in another brain structure, would actually be able to compensate for the loss of stiffness control by changing the trajectory plans to take the cerebellar deficit into account. The lack of this effect seems to suggest that the cerebellum codes an important element of the final motor command.

Neural activities in the cerebellum further support this idea. The observation of the P-cell activities independent of movement direction (Fortier et al. 1993) could be interpreted as P-cell inhibition of excessive or inappropriate inputs in all irrelevant movement directions to the cerebellar nucleus. The preferred direction of P-cells contributes to produce the muscle tension imbalance which generates torque patterns for smooth coordinated movements. Actually, many P-cells were directionally activated during single-joint movements (Frysinger et al. 1984), and P-cell firing frequencies were predicted from smooth eye movements elicited by large visual scene (Shidara et al. 1993).

From those discussions, my interpretation of a motion control strategy is: (1) Learners initially try to imitate the kinematics of a demonstrated skillful movement. Their stiffness might be high enough to reduce the unknown effects of dynamics of limbs and the external environment, as shown in the simulation of multi-joint movements (Flash 1987). Unskilled movements would still be clumsy and would easily be exhausting. (2) As the brain acquires internal models of the controlled object, stiffness decreases in an optimization process to avoid fatigue or to achieve some targets. The stiffness would be controlled according to the constraints of the external environment and the requested tasks.

From this discussion, the role of the cerebellum in movement control is that it learns muscle activation patterns for smooth, accurate, and effortless movements, and that it controls not only joint stiffness but also torques, as generated by muscle tension imbalance. In any case, stiffness is important information in biological motor control, as described in SMITH's target article. To understand the control mechanism, it might be necessary to build a model that can predict the stiffness from important factors such as motion purpose, strategies, and external environmental conditions.

What can and what cannot be adjusted in the movement patterns of cerebellar patients?

Patrick Haggard

Department of Psychology, University of College London, London WC1E 6BT, England. **p.haggard@ucl.ac.uk**

Abstract: This commentary reviews the case of a patient who could alter the coordination of her prehensile movements when removal of visual feedback reduced her kinetic tremor, but could not coordinate her hand aperture with her hand transport within a single movement. This suggests a dissociation between different subtypes of cerebellar context-response linkage, rather than a single, general association function. [THACH]

How general is the cerebellar context-response linkage mechanism proposed in THACH's target article? In a recent study of coordinated reaching and grasping movements in a unilateral cerebellar patient (Haggard et al. 1994), we found a dissociation between two different kinds of motor adjustment, one which could be made following cerebellar lesion, and one which could not. This reply uses this dissociation to clarify what kinds of motor learning the cerebellum may provide.

Our patient, KA, was left with profound intention tremor and hypermetria, largely confined to the right arm, following surgical removal of an ependymoma of the IVth ventricle. She also had hypotonia of the right limbs, and a profoundly ataxic gait. We tested the patient's ability to make coordinated reaching and grasping movements. In normal prehension (Jeannerod 1981), the hand preshapes to grasp the object in a way which is tightly spatially coupled with hand's approach to the target (Haggard & Wing 1995). This normal pattern was seen in our patient's unaffected left hand. The right hand opened much wider than the left, to compensate for the profound intention tremor of the right arm: a greater hand aperture gave the patient a greater chance of grasping the target object.

We then compared maximum hand aperture during reaches made in normal lighting conditions, and in complete darkness. The normal left hand opened wider in the no vision condition, replicating the finding of Wing et al. (1986). The affected right hand, in contrast, showed an interesting pattern of movement in the no vision condition: there was a significant decrease in the severity of the intention tremor in the no vision condition (Beppu et al. 1987). Therefore, removing visual feedback increased the level of certainty that the patient could have about movements of her right arm, even though it decreased the level of certainty she could have about movements of her normal left arm. The maximum right hand aperture was accordingly reduced in the no vision condition. The patient thus has a preserved ability to monitor the effects of the movement context on reaching accuracy, and adjust hand aperture accordingly. Therefore, the cerebellum does not seem to be required in order to make functional strategic adjustments to motor patterns.

In contrast to the successful adjustments, KA could not coordinate opening and closing of the hand with the forward movement of the arm in any given trial. In normals, the motor system controls hand aperture so as to keep its spatial relation to hand transport constant (Haggard & Wing 1995). Therefore, plots of hand aperture against hand transport on repeated trials normally show a decrease in variability as the hand approaches the target, as hand aperture is adjusted so that it bears the appropriate spatial relation to hand transport. In this sense, the hand transport can be treated as a context, and coordination as a linkage which selects an appropriate instantaneous hand aperture for that context.

We observed a significant decrease in variability of the hand aperture against hand transport plot as movements progressed from start to target for KA's normal left hand. However, a significant increase in variability was found for the affected right hand. This pattern was found for movements both with and without vision (Fig. 1). Therefore, cerebellar damage does affect the context-response linkage, or coordination, between the hand transport and hand aperture components of prehension.

Our patient, then, could adjust her movement patterns to compensate for a change environmental context, but could not adjust them within a single trial to preserve an appropriate coordination of an action as a whole. This result suggests that the context-response mechanism in the cerebellum is not entirely general: some forms of linkage are lost following cerebellar damage while others are preserved. What informational features of the second, within-trial, form of response adjustment make it specifically vulnerable to cerebellar damage? In prehension, an important distinction between these two kinds of adjustment is that the first (modulating maximum aperture according to environmental conditions) relies on a strategy arising from knowledge of results of previous movements, and anticipation of the likely success of forthcoming movements. The spatial coordina-

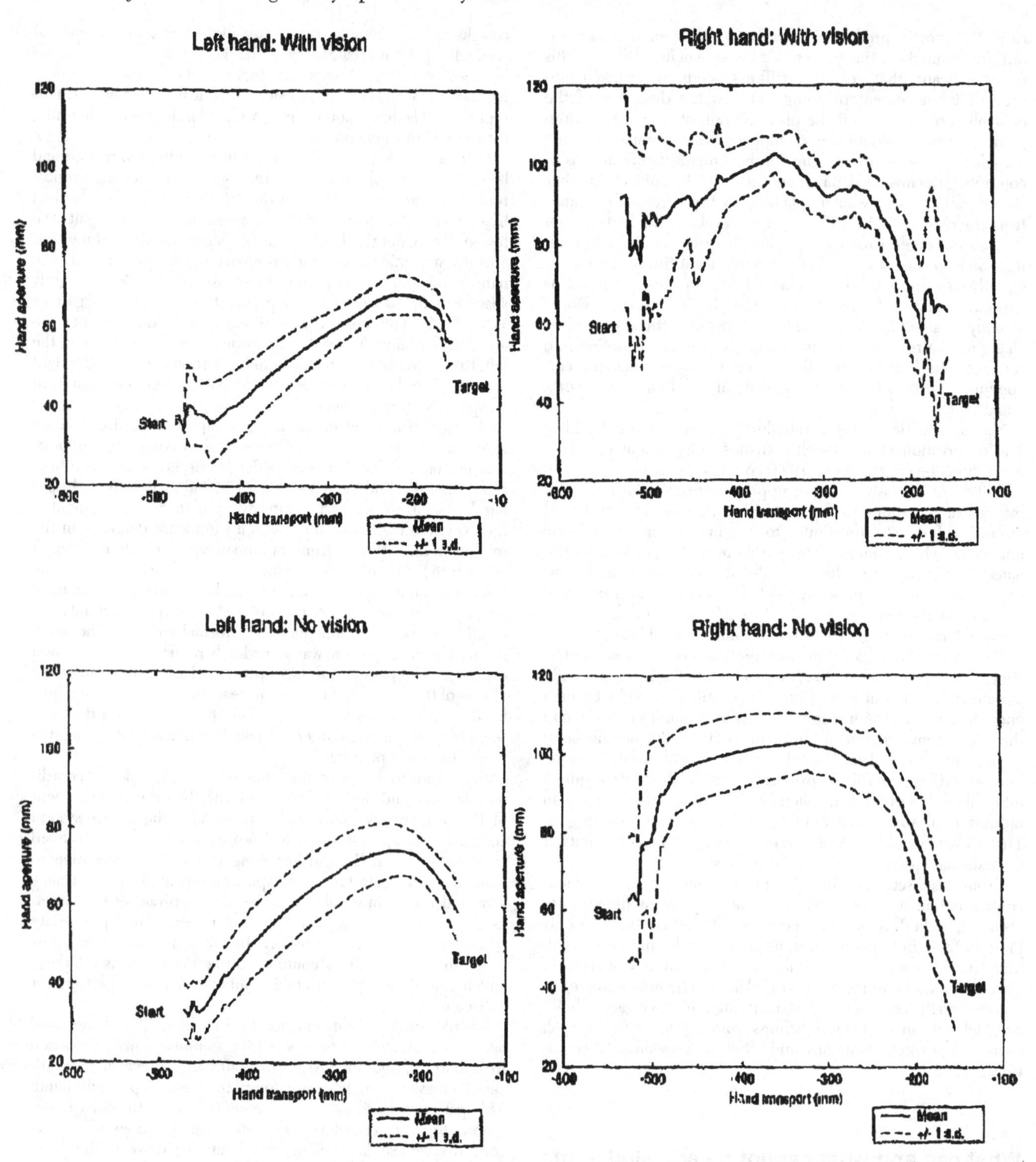

Figure 1 (Haggard). Mean (solid line) +/−1 standard deviation (dashed lines) of the spatial relation between hand aperture (finger-thumb distance) and hand transport (thumb position along start-target axis) for left (unaffected) and right (affected) hands, with and without visual feedback, in patient KA. The target is located at −150 mm on the abscissa. The vertical distance between the solid and dashed lines gives an index of the regularity of repeated movements at each stage of hand transport. Reprinted from *Neuropsychologia* (1994), with kind permission from Elsevier.

tion of aperture and transport in a single trial in contrast, is not based on strategies and knowledge of results, but instead resembles a servo system.

In this view, the cerebellum could usefully contribute to any behavior or neural operation which required rapid adjustment of control parameters to maintain an optimal motor output. But the cerebellum is not a necessary element in behaviors which involve prior heuristic selection of one response from an extensive range of options. If the motor function of the cerebellum is modeled as "context-response linkage," then the cerebellar linkage seems designed for optimizing ongoing responses, rather than planning or selecting "smart" movement patterns in advance.

The role of the cerebellum in motor learning is limited

Mark Hallett
Human Motor Control Section, NINDS, NIH, Bethesda, MD 20892-1428.
hallett@codon.nih.gov

Abstract: Motor learning can be divided into adaptation and skill learning. The cerebellum plays the major role in adaptation learning, which can be described as context-response linkage. Motor skill learning is development of a new ability such as a sequence of motor actions. Skill learning can occur in patients with cerebellar lesions. Cortical mechanisms play a major role in development of skills. [THACH]

THACH has provided a masterly synthesis of the data about the role of the cerebellum in motor learning. Moreover, he has suggested an overarching hypothesis to explain the data and define the fundamental processing in the cerebellum, "context-response linkage." The cerebellum is clearly important for motor learning and plays a critical role, but it does not do it all. The brain is a parallel-processing machine with a great deal of redundancy. It is often difficult to define specific roles for parts of the brain and certainly difficult to specify hierarchical levels, but to the extent possible, the cerebellum's position seems to be at a relatively low level.

My colleagues and I have argued elsewhere that it is reasonable to think that there are at least two types of motor learning, which might be called adaptation learning and skill learning (Hallett et al. 1996; Sanes et al. 1990). Skill learning is the development of a new ability, a completely new sequence of muscular activation, such as riding a bicycle or playing a new piece on the piano. A new skill is a new motor program. Adaptation learning is fine tuning of the motor program, getting all the forces applied at the right time, and maintaining constant readjustments in the face of the changing external world. It is possible to isolate adaptation learning experimentally, but skill learning requires associated adaptation learning. The evidence suggests that the cerebellum plays a critical role in adaptation learning, but is not the principal site of skill learning.

When the cerebellum is damaged, patients complain of incoordination. All of the components of the motor program are present, but they are not put together smoothly. They lack the ability to adapt their program to the current environmental demands. This fits THACH's hypothesis; they cannot link their response to the context. They complain of loss of automaticity, but this does not mean that movements become automatic because of cerebellar processing alone. Patients with basal ganglia lesions and cortical lesions also complain of loss of automaticity. Loss of any component of the brain requires increased attention.

Studies of relatively pure adaptation learning appear to depend heavily on the cerebellum. Examples reviewed by Thach include a number of experiments with prism adaptation, but also experiments with changing visuo-motor gain in a tracking task can be quoted (Deuschl et al. 1996). We have done experiments with learning of a complex, two-dimensional trajectory using the upper extremity (Hallett et al. 1996; Topka et al. 1991). Both normal subjects and patients with cerebellar disorders have trouble at first, but then learn to reduce error at about the same rate. Looking at the data on an absolute scale, the patients made many more errors all along, but this might be ascribed to a difficulty with adaptation that they could not circumvent.

It is likely that several regions of the cortex play important roles in skill acquisition. Motor skills can be completely lost with cortical lesions. For example, lesions of the dominant parietal lobe may give rise to ideomotor apraxia, which is characterized by the inability to put the components of a motor program together.

Involvement of the motor cortex itself with the learning process is suggested by some of the PET scan studies reviewed by Thach. Using mapping of the motor cortex with transcranial magnetic stimulation (TMS), we have shown that the map of muscles involved in the heavily practiced, skilled motor task of Braille reading is enlarged (Pascual-Leone et al. 1993a). Moreover, the map of muscles involved in learning to play the piano increases with practice (Pascual-Leone et al. 1993b).

Many of the physiological studies of motor learning in humans have been confounded by too much complexity. The serial reaction time task, SRTT, has proven to be a useful tool for the study of skill motor learning by isolating a single factor, that of sequences of actions. In studies with TMS mapping during the course of SRTT learning, we demonstrated that the map of muscles involved in the task first increased in area and then diminished after explicit knowledge was obtained (Pascual-Leone et al. 1994). We have duplicated this result with the pattern of alpha desynchronization on the EEG that peaks when learning becomes fully explicit (Zhuang et al. 1995). In a PET study of the SRTT, Grafton et al. (1995) found participation of the motor cortex that gradually increased during implicit learning, but then declined during explicit learning. This pattern, confirmed with three different modalities, corresponds with THACH's model for involvement in learning. In our studies of performance of sequences of different lengths, other areas of the cortex, the parietal and prefrontal regions are active, and thus also likely play a role in sequence learning. The recent elegant studies of Tanji and Shima (1994) in primates during tasks with motor sequences show that cells in the SMA have highly specific patterns in different phases of the task. In other experiments, we have found that the cerebellum supports SRTT implicit learning (Pascual-Leone et al. 1993c), but clearly it is only one component of a complex network, and the role of each component is yet to be defined.

There is, therefore, good reason to think that cortical mechanisms are very important for the context-response linkage of sequences. In some circumstances, the cerebellum may well help in the learning of sequences, but its major role in running a learned sequence is likely coordination and not the sequence itself. The cerebellum supports skill learning, but the major action is elsewhere.

Two separate pathways for cerebellar LTD: NO-dependent and NO-independent

Nick A. Hartell
Laboratory for Synaptic Function, Frontier Research Program, Riken, Wako-shi, Saitama, 351-01, Japan. **n.a.hartell@aston.ac.uk**

Abstract: Strong stimulation of parallel fibers (PFs) leads to localized calcium influx and a long-term depression (LTD) of PF responses at the site of stimulation. These same stimuli also induce depression at spatially distant synapses to the same cell that is mediated by the diffusible messenger nitric oxide (NO). The demonstration of two distinct forms of LTD provide an explanation for the absence of the NO/cGMP pathway in cultured cells. [CRÉPEL et al.; LINDEN; VINCENT]

NO contributes to LTD in slices but not in culture. Despite an increasing number of reports to the contrary (Crépel & Jaillard 1990; Daniel et al. 1993; Hartell 1994a; 1994b; Ito & Karachot 1992; Lev-Ram et al. 1995; but see Glaum et al. 1992; Shibuki & Okada 1991; 1992), VINCENT suggests that the nitric oxide/cyclic guanosine monophosphate (NO/cGMP) pathway is not important for the induction of cerebellar long-term depression (LTD), primarily on the basis that a NO/cGMP independent form of LTD can be observed in isolated, cultured Purkinje cells (Linden 1994; Linden et al. 1992). Since cultured Purkinje cells receive only sparse innervation compared to slice or *in vivo* preparations, and given that the source and target of NO are likely to be different cell types, a more reasonable explanation for the absence of NO-mediated LTD in culture is that two different forms of LTD exist, as suggested by CRÉPEL. One form requires NO and is absent in culture and the other, common to both systems, requires activation of metabotropic glutamate receptors/protein kinase C and AMPA receptors, and calcium elevation.

Two forms of LTD exist. In cerebellar slices, moderately strong, repetitive stimulation of parallel fibers (PFs) at 1 Hz leads to an

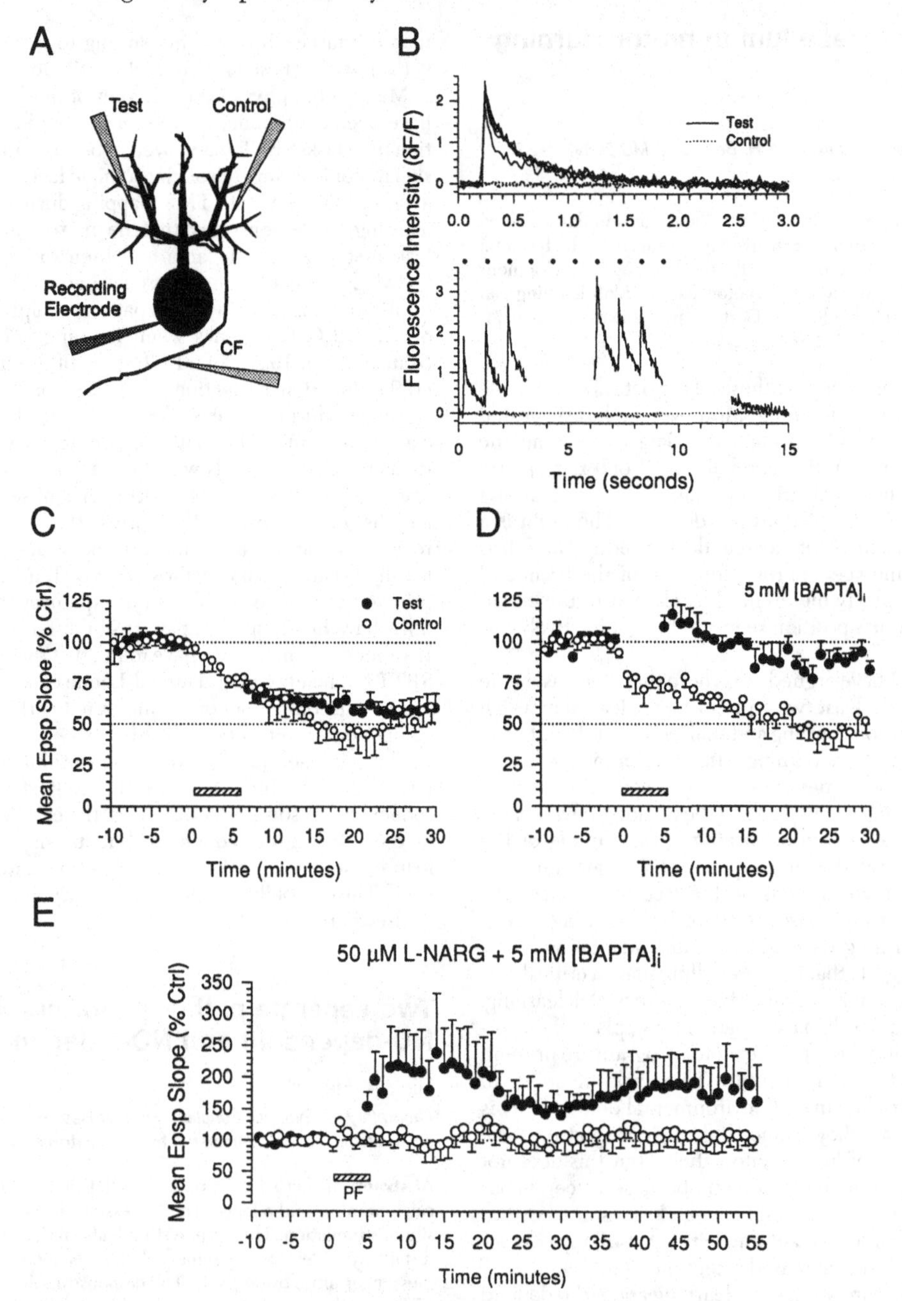

Figure 1 (Hartell). Mechanism of parallel fiber induced LTD. A: Schematic illustration of the recording conditions. B: Strong stimulation of PFs at the test site induced increases in calcium restricted to only the test site (solid lines). 1 Hz stimulation led to an accumulation of calcium. C: The effects of 5 minutes of 1 Hz, high-intensity stimulation to one of two separate PF pathways (Test, filled circles, n = 5). After 5 minutes, stimuli were resumed at pre-test, control levels. For clarity, the responses to this pathway evoked during the test stimuli are omitted. Responses to the control input (open circles) were elicited at constant stimulus intensity and at 0.2 Hz throughout. Means and standard errors of EPSP slopes are plotted against time (n = 5). The hatched bars indicate the periods during which the intensity and frequency of stimulation to the test input were raised. Experiments outlined in C were repeated with 5 mM BAPTA (D, n = 5) in the recording pipette and additionally in the presence of 50 μM LNARG in the perfusate (E). Adapted from Hartell, in press.

elevation of calcium in discrete regions of the Purkinje cell dendritic tree (Eilers et al. 1995 and Fig. 1B). These parameters of stimulation are sufficient to induce a LTD of PF EPSPS (Fig. 1C; Hartell, in press) that is similar to that observed in culture (Linden et al. 1992; 1995) since it is largely prevented by inclusion of calcium chelators to the patch pipette (Fig. 1D) but not blocked by inhibition of either NOS or PKG. In addition to this NO/cGMP independent form of LTD, the same stimuli also induce a robust depression of responses in spatially distant PF inputs to the same Purkinje cells (Fig. 1C), whose level of stimulation was not altered. This depression is independent of postsynaptic calcium (Fig. 1D) but is effectively blocked by inhibitors of NOS and PKG. NO/cGMP-mediated depression was not observed at the site of strong stimulation under postsynaptic chelation of calcium because it is masked by an underlying potentiation of responses (Fig. 1E), a finding consistent with previous studies indicating that under conditions of low postsynaptic calcium, LTD is transformed into LTP (Hartell 1994a; Shibuki & Okada 1992). Thus strong stimulation of PFs leads to the generation of NO and this acts both locally and at distant sites to reinforce LTD.

Under synaptic stimulation, these two pathways may act in concert with the relative proportions of each depending on the degree of PF stimulation and level of calcium influx. In culture, where the PF input is bypassed completely, NO-mediated LTD would not be expected to play an important role. LTD induced by pairing PF stimulation with either stimulation of CFs or intradendritic injections of cGMP was similarly blocked by inhibitors of PKC, PKG, or mGluRs (Hartell 1994a; 1994b), suggesting both pathways were activated by these parameters of stimulation and both were necessary to fully induce LTD.

These data provide an explanation for the clear differences between LTD in slices and in culture and provide strong evidence that the PF is a likely source for NO production. The high level of PF stimulation necessary for both calcium elevation and LTD induction and the lack of input specificity of the depression suggest that PF induced LTD, and the recruitment of the NO-pathway may be important as a neuroprotective mechanism rather than underlying long term storage of memory.

Programming the cerebellum

K. Hepp
Institute for Theoretical Physics, Eidgenössische Technische Hochschule, CH 8093 Zürich, Switzerland. **khepp@itp.phys.ethz.ch**

Abstract: It is argued that large-scale neural network simulations of cerebellar cortex and nuclei, based on realistic compartmental models of the major cell populations, are necessary before the problem of motor learning in the cerebellum can be solved. [**HOUK et al.; SIMPSON et al.**]

By relating structure to function in an important, evolutionarily stable part of the vertebrate brain the Marr-Albus-Ito hypothesis has had a major impact on neurobiology. A large number of excellent investigations have been carried by either the desire to refute or to refine the original hypothesis. Unfortunately, as **SIMPSON et al.** point it out in the conclusion of their admirable critical survey, the function of the climbing fibers remains largely an enigma.

However, our ignorance about the functioning of the cerebellum is much more general, since we have no quantitative understanding of the biophysics of computation of the cortical network. What is the use of ten thousands of Purkinje cells in the flocculus of the rabbit, where clever engineers would regulate the rather simple visuo-vestibulo-oculomotor transformations by just a few adjustable pattern generators (APG) or gain elements? What is the information flow on single parallel fibers which span several microzones and excite almost every Purkinje cell they cross? On one hand "gaze velocity" Purkinje cells in the monkey show a linear superposition of a recurrent eye velocity and a vestibular head velocity signal (Lisberger & Fuchs 1978), in an analog mode. On the other hand the finding of Babour (1993), that on the order of 50 simultaneously active granule cells are sufficient to generate an action potential in a Purkinje cell, points to the necessity of sparse coding or strong counteraction of excitation by inhibitory interneurons.

For a better understanding of cerebellar physiology a major computational effort is necessary to explore biophysically realistic compartmental models of the major cerebellar cell types and their interaction. Prototypes of Purkinje cells (de Schutter & Bower 1994a; 1994b; 1994c), granular cells (Gabbiani et al. 1994) and vestibular (cerebellar) nuclear cells (Quadroni & Knöpfel 1994) have recently been published, and data for modeling Golgi and stellate cell are available (Midtgaard 1992). In such simulations the computational load is enormous, and the analysis requires strong cooperation, since nobody can alone explore the physiologically interesting part of the solution space. However, interesting subproblems can be worked out at preliminary stages of this program:

(1) Biophysical properties of realistic Purkinje cells, like bistability (Yuen et al. 1995), which in the simple version of the APG model (Houk et al. sect. 3.3 and Fig. 5) has to my knowledge never been observed in the firing patterns of eye-movement related Purkinje cells in the monkey.

(2) The transfer characteristics of the input loop between mossy fibers, granular, and Golgi cells, predicting the impulse traffic along parallel fibers, for example, in visuo-vestibulo-oculomotor integration, where much is known about the incoming mossy fiber firing patterns.

(3) The dynamics of the output loop between Purkinje, nuclear, and olivary neurons, which might shed some light on the function of the climbing fibers.

Experience from these subproblems can immediately be confronted with experiment and constrain more abstract models. Simplified, but essentially realistic cell prototypes and connectivity can be developed at this stage. This should allow larger circuits to be tested using massively parallel supercomputers without getting lost in the data forest. Before solving the problem of learning in and around the cerebellum (Pastor et al. 1994) we must learn how to program the cerebellum.

Positive cerebellar feedback loops

Germund Hesslow
Department of Physiology and Neuroscience, Lund University, S-223 62 Lund, Sweden. **germund.hesslow@mphy.lu.se**

Abstract: Houk et al. suggest that excitatory output from the cerebellar nuclei is fed back into the cerebellum, thus generating a prolonged activity in the cerebellar nuclei which is then molded by inhibitory Purkinje cells. Recent evidence, however, shows that a temporary break in this loop causes a pause in a conditioned eyeblink response, but does not terminate it. [**HOUK et al.**]

It has been known for many years that cerebellar output pathways terminate on structures which provide input to the cerebellum, but the role of this positive feedback has remained controversial. One possibility is that such positive feedback loops might determine the temporal characteristics of responses which are generated by the cerebellum. A certain cerebellar output, which initiates a movement, might be channeled back into the cerebellum and constitute the stimulus for a new output, which generates the next component of the movement, and so on until the whole movement is completed. A good, although relatively simple, example of a temporally organized movement which might be controlled in this way is the classically conditioned response. A conditioned eyeblink for instance, so far the only kind of conditioned response known to involve the cerebellum, may have a duration of a couple of hundred milliseconds; one way of understanding its temporal topography is to assume that it consists of several components, each of which acts as a stimulus for the next. This idea is fittingly reminiscent of the early behaviorist concept of response chaining.

In their target article **HOUK et al.**, propose essentially this idea. A positive feedback loop is set up which provides a prolonged activation of the cerebellar nuclei. Activity in the loop is then controlled by inhibitory Purkinje cells, which also terminate the reverberating activity at an appropriate time.

There is a simple way of testing the hypothesis that the time course of a conditioned response is continually determined by cerebellar output, and that is to introduce a brief interruption of this output. If the suggestion by HOUK et al. is correct, a temporary break in the loop should lead to a termination of the response. If, on the other hand, the conditioned stimulus in combination with internal cerebellar mechanisms is sufficient to generate a temporally patterned response, then a brief interruption of cerebellar output would affect the response only while this interruption lasts.

This test has actually been performed in decerebrate cats which were trained to give eyeblink responses to a conditioned stimulus consisting of a train of stimuli to the forelimb (Hesslow 1994b). A brief electrical stimulus applied to the area of the cerebellar cortex which controls the orbicularis oculi muscle will activate the inhibitory Purkinje cells and briefly suppress activity in the interpositus nucleus, thus interrupting any ongoing positive feedback loop.

When such a stimulus was applied in the middle of a conditioned eyeblink response, there was a brief pause in the EMG activity recorded from the eyelid, but this activity soon resumed so that the late part of the conditioned response was relatively unaffected. Although repetitive cortical stimuli could sometimes terminate the conditioned eyeblink, the typical effect of a single stimulus was a pause in the response which lasted for 25–75 msec.

This would seem to constitute pretty strong evidence against the positive loop hypothesis. It could of course be argued that the excitatory loop may not have been completely suppressed by the cortical stimulation, but that this only weakened the activity in the nuclei so much that the conditioned response was suppressed, but not so much that the loop activity was terminated. This is highly unlikely in view of the fact that other experiments strongly suggest that cortical stimulation can silence the underlying nuclear cells completely (Hesslow 1994a).

Even if positive feedback does not seem to play an essential role in the generation of conditioned eyeblink responses, it is clearly still possible that such mechanisms are essential in more complex movements.

ACKNOWLEDGMENTS

This commentary was supported by grants from the Swedish Medical Research Council (project no. 09899) and the Knut and Alice Wallenberg Foundation.

Molecules involved in cerebellar long-term depression (LTD) and mutant mice defective in it

Tomoo Hirano

Department of Physiology, Faculty of Medicine, Kyoto University Sakyo-ku, Kyoto 606-01, Japan. **hirano@med.kyoto-u.ac.jp**

Abstract: Recent studies have demonstrated that a Purkinje cell specific ionotropic glutamate receptor δ2 subunit is involved in long-term depression (LTD). Identification of molecules involved in LTD has led to the production of mutant mice defective in LTD, motor control, and motor learning. There are disagreements about whether or not and how nitric oxide (NO) is involved in LTD. [CRÉPEL et al.; LINDEN; VINCENT]

Further identification of molecules involved in long-term depression. The cellular and molecular mechanism of cerebellar long-term depression (LTD) is addressed in the target articles of LINDEN and CRÉPEL **et al.** We have recently made some contributions in this area. The most important finding is that a Purkinje cell specific ionotropic glutamate receptor subunit δ2 is involved in the LTD. Throughout the central nervous system only in the Purkinje cell, is the δ2 subunit mRNA expressed, and the δ2 protein is highly expressed in the distal dendritic spines where parallel fibers form synapses. Treating cerebellar culture with the oligonucleotide antisense to the δ2 mRNA suppressed the induction of glutamate responsiveness LTD that is inducible by coupling glutamate application with depolarization (Hirano et al. 1994). This result was confirmed using mutant mice deficient in the δ2 subunit raised by the gene targeting technique. Purkinje cells cultured from mutant mice did not express glutamate responsiveness LTD (Hirano et al. 1995) and Purkinje cells in cerebellar slices prepared from the mutant mice failed to show the LTD for parallel fiber-Purkinje cell synaptic transmission that is inducible by coupling parallel fiber stimulation with Purkinje cell depolarization (Kashiwabuchi et al. 1995). We concluded accordingly that the δ2 subunit is involved in LTD. We also showed that function inactivating antibodies against mGluR1 metabotropic glutamate receptor suppressed glutamate responsiveness LTD (Shigemoto et al. 1994). This result was confirmed and extended by other investigators using mGluR1 deficient mutant mice (Aiba et al. 1994; Conquet et al. 1994). The function inactivating antibodies can be used as specific inhibitors. These studies showed that culture preparations are useful in identifying molecules involved in LTD, because applying molecular biological techniques such as antisense treatment or applying small amounts of high molecular weight antibodies are feasible. Culture preparations will also be suitable for DNA transfection experiments.

The involvement of Ca^{2+} influx through voltage-gated Ca channels has been demonstrated by blocking LTD with intracellularly loaded Ca^{2+} chelators (see LINDEN, sect. 3 and CRÉPEL **et al.,** sect. 1.2). We addressed whether or not the intracellular Ca^{2+} increase without depolarization is sufficient to induce LTD and showed that intracellular Ca^{2+} release from caged Ca^{2+} (nitr5) coupled with glutamate application is sufficient to do so (Kasono & Hirano 1994). We also suggested that IP3 presumably produced by mGluR1 activation is involved in LTD. Heparin, which inhibits binding of IP3 to its receptor, suppressed LTD induction; IP3 released from caged IP3 facilitated LTD induction (Kasono & Hirano 1995). As substantial time has passed since the target articles were written; both Linden and Crépel et al. may wish to add explanations about their more recent findings concerning molecular mechanisms of the LTD.

NO (nitric oxide) and LTD. The involvement of nitric oxide (NO) in LTD is a matter of dispute. There are data and opinions both for and against the idea, and even among researchers who support the NO involvement, there are disagreements about how and where NO is produced (see LINDEN, sect. 5, CRÉPEL **et al.,** sect. 1.4, and VINCENT, sect. 2). Some consider that NO is produced by climbing fiber activation, others that it is produced by parallel fiber activation (Lev-Ram et al. 1995). One problem with the idea that NO is produced by climbing fiber activation is that NO synthase (NOS) has not been detected in either climbing fiber terminals or Purkinje cells (VINCENT, Introduction). Another problem is the fact that climbing fiber activation can be substituted for direct depolarization of a Purkinje cell. How depolarization of a Purkinje cell leads to NO production is unclear. A recent paper claiming that NO released from parallel fibers plays a role in LTD induction (Lev-Ram et al. 1995) seems to be more accurate regarding the NOS location, but there are problems. The results contradict the previous finding that NO can substitute for climbing fiber activation and that coupling glutatmate application to a Purkinje cell with depolarization induces the LTD. In the latter condition NO does not seem to be involved. Schilling et al. (1994) recently reported that NOS is expressed only in subsets of adult granule cells. The patchy distribution of NOS positive granule cells suggests that NO may not play an essential general role in LTD induction. One would like to know the target article authors' recent data and their current opinion on this issue. Experiments on slice preparations from nNOS deficient mutant mice (Huang et al. 1993) may provide crucial data about this issue.

Mutant mice defective in LTD. As mentioned earlier, mutant mice deficient in δ2 subunit and mGluR1 have been raised. Cerebellar LTD is retarded in both mutant mice, and they both showed clear ataxia and deficit in motor learning (Aiba et al. 1994; Conquet et al. 1994; Funabiki et al. 1995; Kashiwabuchi et al. 1995). Thus, the identification of molecules involved in LTD has led to the production of mutant mice without LTD, but with deficits in motor learning and motor control. Although a causal relationship between LTD and motor learning cannot be inferred because of some abnormalities over and above the LTD, these animals will be useful not only in the further analysis of molecular and cellular mechanism of LTD, but also in the attempt to correlate LTD with motor learning and control.

Cerebellar arm ataxia: Theories still have a lot to explain

J. Hore
Department of Physiology, Medical Science, University of Western Ontario, London, Ontario, Canada N6A 5C1. **jhore@physiology.uwo.ca**

Abstract: Theories of cerebellar arm function should be able to explain the now well-characterized disorders that occur during cerebellar dysfunction in fast arm movements made at single joints. Contrary to claims in Smith's target article, cocontraction is not one of these disorders. Although loss of the joint stiffness function proposed by Smith is consistent with some findings, it does not explain all of these disorders, and therefore gives only a limited view of cerebellar function. [SMITH]

A characteristic of those who speculate on the function of the cerebellum is that they inevitably justify their theory by relating it to "ataxia" or to particular elements of ataxia that occur following a lesion of the cerebellum. This is not difficult since the literature, which dates back to the last century, describes a vast array of disorders in arm movements, leg movements, eye movements, muscle tone, gait, posture, and so forth. A further characteristic of these speculators is that they are usually selective, choosing only those disorders that fit their theory and ignoring others that do not.

In his target article SMITH shows himself to be a worthy follower of this time honored tradition. Thus he concludes that failure to achieve optimal stiffness in a time-varying manner will result in – you guessed it – ataxia. Further, to justify his idea that the cerebellum modulates the cocontraction of agonist-antagonist muscles, he selectively dips into the literature to find examples of disordered cocontraction in the EMG of patients with cerebellar damage. Unfortunately, the examples he chooses are not representative of most of the literature on this topic and consequently give a misleading picture. While it is true that "cerebellar lesions are associated with disturbances of agonist-antagonist muscle relations" (sect. 2.1), this disturbance is not one of cocontraction as he suggests.

Cocontraction is not a major feature of cerebellar ataxia at single joints. Most workers have found that cocontraction, that is, the simultaneous occurrence of EMG activity in agonist and antagonist muscles, is not a major feature of cerebellar arm ataxia. This is true of the study of Hallett et al. (1975) which is misquoted by SMITH. This study reported that, when performing a fast elbow flexion, cerebellar patients showed a qualitatively similar triphasic pattern to that of normals (i.e., reciprocal agonist-antagonist activity). While it is true that Rondot et al. (1979) show a figure of biceps and triceps bursts firing together during elbow flexion in a patient with a "cerebellar lesion," no details were given about the lesion or in how many patients this pattern was observed.

In more recent work, fast arm movements were studied at single joints of 9 patients who were selected on the basis of having lesions restricted to the cerebellum (Hore et al. 1991) and in 13 patients who had cerebellar deficits in the arm with no other neurological signs (Hallett et al. 1991). The results of the two studies were strikingly similar: both found prolonged agonist EMG activity and accompanying accelerations that were prolonged and decreased in magnitude. Neither reported that cocontraction was a major feature.

A finding that may fit the theory. One finding that may fit the theory comes from reversible cooling experiments in monkeys. Hore and Flament (1988) found that some muscle-like neurons in motor cortex showed a decrease in (reciprocal) inhibition and a more gradual decline of premovement tonic discharge during cerebellar nuclear dysfunction. This may be related to the more gradual decline of antagonist EMG activity prior to movement onset that was found in 12 of 16 cerebellar patients (Hallett et al. 1975).

Such results can be explained by the SMITH theory as loss of the mechanism whereby Purkinje cells inhibit deep nuclear cells thereby evoking "a cascade of effects culminating in the disfacilitation of motoneuron pools resulting in antagonist muscle relaxation" (sect. 2.7). The theory assumes that the cerebellum instructs motor cortex, though whether this is by means of a programming role or a tuning-up role (Hore 1993) is still not clear.

In spite of this decrease in reciprocal activity in motor cortex there was still reciprocal activity in the EMG of these monkeys (and in cerebellar patients). Presumably this results from the ongoing activity of other (relatively) normal mechanisms that generate reciprocal activity, for example, the Ia inhibitory interneuron.

What theories of cerebellar control of arm movements have to explain. Anyone with an interest in the cerebellum cannot help but be impressed with the myriad of functions that are currently ascribed to this structure. From adjusting reflexes to a role in cognition – the cerebellum is claimed to do it all. Clearly, no current theory is able to explain all these functions satisfactorily. Nevertheless, those theories that apply to arm movements should be able to explain the characteristics of the arm disorders that occur during cerebellar dysfunction rather than predicting that "ataxia" will occur with loss of the theoretical function. Much work in monkey and man that has been performed in different laboratories over the last 20 years has clearly defined these disorders for the case of fast movements at single joints (for references see Diener & Dichgans 1992; Hore et al. 1991). These disorders include: (1) an increase in reaction time accompanied by a delay in discharge of motor cortex neurons; (2) a slowed initial movement onset accompanied by less abrupt onset of agonist EMG activity and loss of phasic activity of some motor cortex neurons; (3) prolonged agonist EMG activity and prolonged acceleration; (4) delayed onset of antagonist EMG activity accompanied by changes in phase and gain of antagonist-related motor cortex neurons; (5) a terminal tremor that is sensitive to the mechanical state of the limb.

It is not clear how loss of synergies for the optimal time-varying control of joint stiffness would cause all of these disorders. Instead it is likely that the disorders result from loss of a variety of functions performed by the cerebellum. Until a theory is proposed that gives insight into all of these functions, and thereby all of these disorders, I for one will remain unconvinced. Theories still have a lot to explain.

ACKNOWLEDGMENT
The author's research is supported by the Canadian MRC.

Computational significance of the cellular mechanisms for synaptic plasticity in Purkinje cells

James C. Houk and Simon Alford
Department of Physiology, Northwestern University Medical School, Chicago, IL 60611-3008. **physiology@nwu.edu**

Abstract: The data on the cellular mechanism of LTD that is presented in four target articles is synthesized into a new model of Purkinje cell plasticity. This model attempts to address credit assignment problems that are crucial in learning systems. Intracellular signal transduction mechanisms may provide the mechanism for a 3-factor learning rule and a trace mechanism. The latter may permit delayed information about motor error to modify the prior synaptic events that caused the error. This model may help to focus future cellular studies on issues that are particularly critical for a computationally viable concept of cerebellar plasticity. [CRÉPEL et al.; HOUK et al.; KANO; LINDEN; VINCENT]

The latest data on the cellular mechanisms for synaptic plasticity in the cerebellum are nicely summarized by four of the target articles in this *BBS* special issue (CRÉPEL et al., KANO; LINDEN, VIN-

CENT). It is possible and desirable to formulate computational models of these data in a form that constitutes learning rules if one is to gain an insight into the means by which the cerebellum contributes to motor learning. Figure 9 of another target article (**HOUK et al.** sect. 3.1) attempts to do this, based on evidence that was previously available, but here we have attempted to improve upon this by incorporating the latest cellular data, as summarized in the above mentioned articles.

From a computational perspective, one of the more important issues that the cerebellar mechanism for synaptic plasticity must confront is the problem of proper credit assignment (**HOUK et al.**). This refers to the difficulty of directing training signals to appropriate sites in the network and at appropriate moments in the training process, in order for learning to be truly adaptive (Houk & Barto 1992; Minsky 1963). Efficient credit assignment generally begins with a three-factor learning rule in which weight change depends on (1) a presynaptic factor that marks active (and therefore participating) synapses as being eligible for modification, (2) a postsynaptic factor that identifies neurons that have actively participated in controlling a behavior, and (3) a training factor that rewards or punishes the network on the basis of performance. Furthermore, efficient credit assignment requires more specialized training information than merely global reward or punishment signals. Assuming that individual climbing fibers (CFs) transmit specialized training information, the different signals must then be directed to appropriate sites in the network. The zonal organization of CF input directs training information to specific functional modules in the cerebellar cortex, which should facilitate this structural part of the credit assignment problem. Finally, and of particular consequence to this commentary, one must address the temporal part of the credit assignment problem, which concerns the delivery of training information at appropriate times. Because of transmission lags and sluggish dynamics, information about erroneous performance is always delayed (typically by 100 msec or more) so that it follows the synaptic events that actually control the erroneous performance. CF signals detecting errors in performance need to modify those earlier synaptic events if learning is to be adaptive.

The intracellular signal transduction mechanisms that implement synaptic plasticity have properties that may be appropriate for mitigating some of these credit assignment problems (Houk & Barto 1992; Houk et al. 1995; Sutton & Barto 1981). The right side of Figure 1 summarizes the latest information about cerebellar intracellular signaling in a manner that emphasizes potential relationships to a three-factor learning rule and credit assignment. Those processes that are localized to individual spines are demarcated by a shaded box to emphasize the critical role of individual synapses in determining the computational competence of a learning rule. The AMPA receptor that mediates parallel fiber (PF) excitatory postsynaptic potentials (EPSPs) is the centerpoint of this model spine. Following the summary provided by **CRÉPEL et al.** (Fig. 6), the sensitivity of the AMPA receptor to glutamate neurotransmitter is regulated in a push-pull manner by the phosphorylation state of receptor protein. Phosphorylation is catalyzed by protein kinase C (PKC) and dephosphorylation is catalyzed by a local phosphatase. When the AMPA receptor is phosphorylated, it becomes less sensitive to glutamate, which is the cause of long-term depression (LTD) at PF synapses.

PKC needs to be activated before it can promote receptor phosphorylation, and this critical step appears to require both intracellular Ca^{++} and the activation of mGluR1 receptors by glutamate (**CRÉPEL et al.; LINDEN**). Of these two cofactors, mGluR1 receptor activation is likely to be the more specific and is thought to endow synapse specificity in the learning rule (**HOUK et al.; LINDEN**). Although mGluR1 receptors are located in the postsynaptic membrane, they function as transducers of presynaptic activity in individual parallel fibers (PF* in Fig. 1). We therefore assume that mGluR1-mediated effects specify the presynaptic factor in the learning rule.

The second factor in the learning rule is postsynaptic activity. The glutamate released by the individual PF* will promote spine depolarization through its action on AMPA receptors; however, additional depolarization of the adjacent dendrite appears to be required to produce enough of the postsynaptic factor (**HOUK et al.**). Figure 1 suggests that the combined depolarization caused by PF* plus cooperative inputs from many other PFs on the same dendrite produces a plateau potential which then facilitates spine depolarization. The latter would activate Ca^{++} channels in the spine, thus leading to the increase in spine Ca^{++} that is required as a cofactor in PKC activation (Fig. 1). From a computational standpoint, this contributes a dendrite-specific postsynaptic factor to the learning rule.

The third factor in the learning rule, the training information conveyed by CFs, needs to be directed to spines throughout the entire Purkinje cell and also to spines in adjacent cells that participate in the same cerebellar module. Furthermore, there must be some mechanism for overcoming the delayed nature of this training information. Figure 2 illustrates the hypothetical time course of local events in a spine in comparison with some of the global events that are important in the cerebellar control of movement.

The example in Figure 2 assumes that the subject makes a primary movement that undershoots the target position, followed by a secondary corrective movement. The upper trace shows the discharge frequency of a parallel fiber, PF*, that is responsive to the position of the limb. (The normal graded response of the PF is illustrated, as opposed to the synchronous volley that is used in most studies of LTD.) Assume further that the increase in PF* firing contributes to the abrupt onset of a dendritic plateau potential in the Purkinje cell. Firing of this Purkinje cell then contributes to the inhibition that causes cerebellar nuclear discharge to abruptly fall off, as shown in the CBN trace. The CBN cells, and the motor cortical and rubral cells that they innervate, are the source of motor commands for voluntary movements of the limb (Houk et al. 1993). The abrupt termination of these CBN commands decelerates and, after some delay, terminates the primary movement. Since there is an error, a subsequent corrective movement, presumed to be mediated by an extracerebellar mechanism (cf. Berthier et al. 1993), moves the limb to the target.

Climbing fibers that are responsive to the proprioception of the movement are inhibited during the primary movement, thus preventing CF discharge (**HOUK et al.**). However, these fibers appear to be responsive to corrective movements (Gilbert & Thach 1977), which is the postulated mechanism whereby a CF is able to signal the occurrences of errors in performance (Berthier et al. 1993). Note, however, the appreciable time delay T between the onset of the dendritic plateau potential that contributes to the premature termination of the primary movement and the CF signal that detects this error in performance. The delay between these events causes the temporal credit assignment problem discussed earlier.

Studies of cerebellar plasticity generally have not attempted to manipulate the relative timing of the experimental manipulations that are used to elicit LTD, and the poverty of such data has made it difficult to model temporal credit assignment. Recently, however, Chen and Thompson (1995) demonstrated that delaying CF activation by 250 msec after a PF volley facilitates the appearance of LTD, suggesting that there may be a cellular mechanism that compensates for T.

Another new finding that should be incorporated in the model learning rule is the involvement of nitric oxide (NO) in LTD, although the significance of this substance is still being debated (**CRÉPEL et al.; VINCENT**). The model outlined in Figure 1 accepts the recent demonstration that the delivery of NO to the cytoplasm of Purkinje cells is a critical step in LTD (Lev-Ram et al. 1995). Furthermore, it couples this finding to the schema proposed in CRÉPEL et al. (Fig. 6) whereby NO activates cGMP and G-substrate which then inhibits the phosphatase that acts on AMPA receptors. The steps leading up to the formation of NO are less clear, but seem to be linked primarily to CF discharge with a lesser

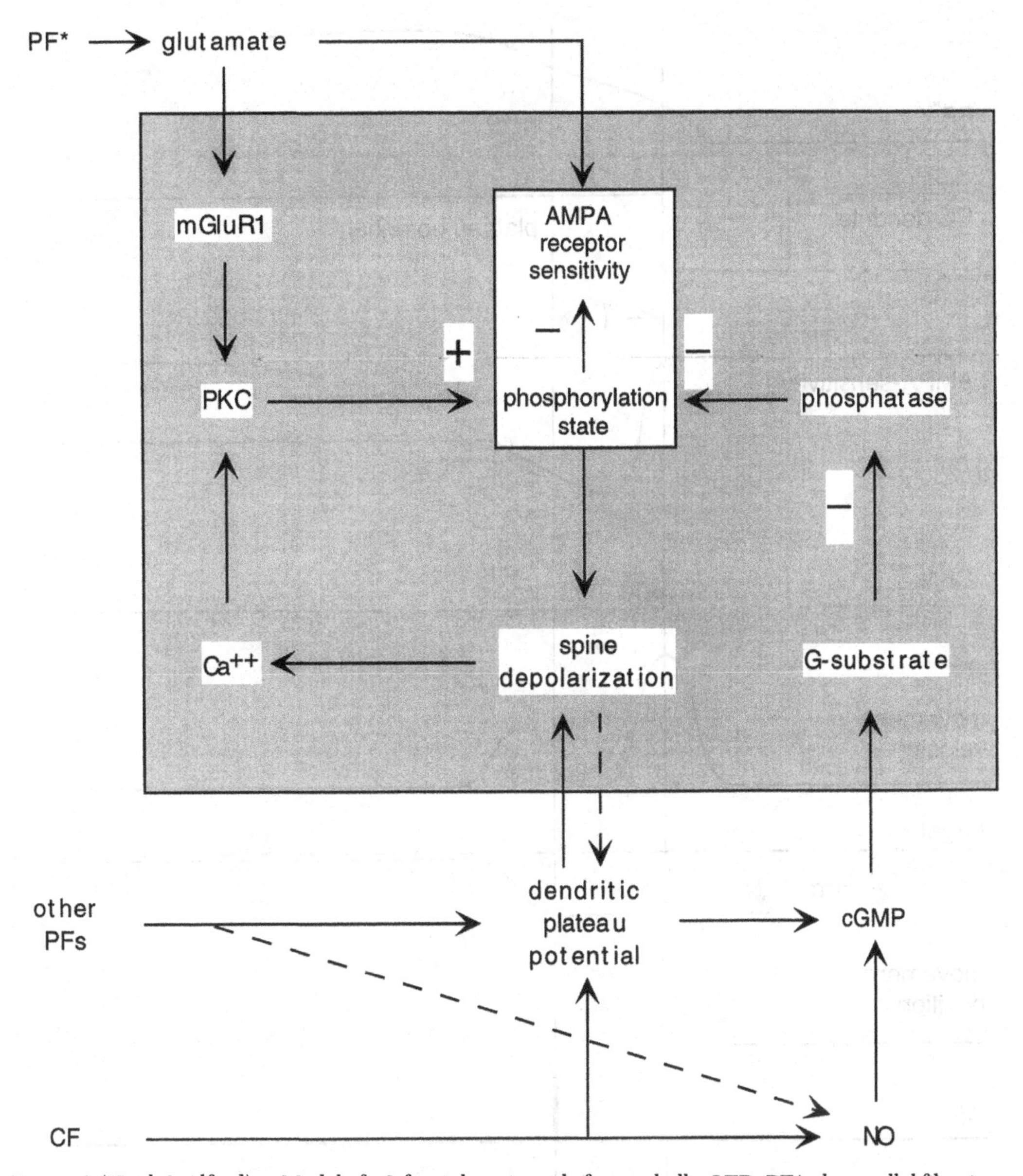

Figure 1 (Houk & Alford). Model of a 3-factor learning rule for cerebellar LTD. PF*, the parallel fiber input to an individual spine, is the presynaptic factor. PF* combines with cooperative inputs from many other PFs to produce a dendritic plateau potential, the postsynaptic factor in the learning rule. Training information conveyed by the CF input contributes the third factor in the proposed learning rule. The shaded box demarcates the borders of a synaptic spine. Abbreviations are defined in the text.

dependence (dashed arrow in Fig. 1) on PF activity (Shibuki & Okada 1991).

The features of the NO mechanism that seem most important from a computational standpoint are the following. (1) NO is highly defusible, which would allow a CF to exert its training influence on spines throughout the entire Purkinje cell it innervates as well as the spines of other Purkinje cells within the same module. This would provide appropriate structural credit assignment for a unified training action on an entire functional module. (2) NO has a very brief (~50 msec) period of action in Purkinje cells (Lev-Ram et al. 1995), which would limit the duration of its training influence to not more than the most recent corrective movement. (3) NO should act to stabilize prior phosphorylation of the AMPA receptor, which would allow the assignment of credit to those prior synaptic events that caused the original phosphorylation. We postulate that this stabilization mechanism could be a crucial factor in temporal credit assignment.

The proposed mechanism for temporal credit assignment is further analyzed in the time plots of Figure 2. The production of the plateau potential in conjunction with PF input causes a reduction in the sensitivity of the AMPA receptors that is short-lived when ample phosphatase is available to dephosphorylate the receptor (dashed trace). However, if a movement error results in a CF spike, this leads to a brief expression of NO, followed by perhaps a longer expression of G-substrate. Since the latter inhibits the phosphatase, this interrupts the dephosphorylation of AMPA receptors, thus stabilizing their depression. In this manner, NO might function to convert a short-term depression into LTD. A key feature of this hypothesis is the relatively slow onset and long duration postulated for AMPA receptor phosphorylation. This time course spans the interval T required for proper temporal credit assignment.

In contrast with the input specificity of the above model, **KANO** reviews a mechanism for potentiation of inhibitory synapses that

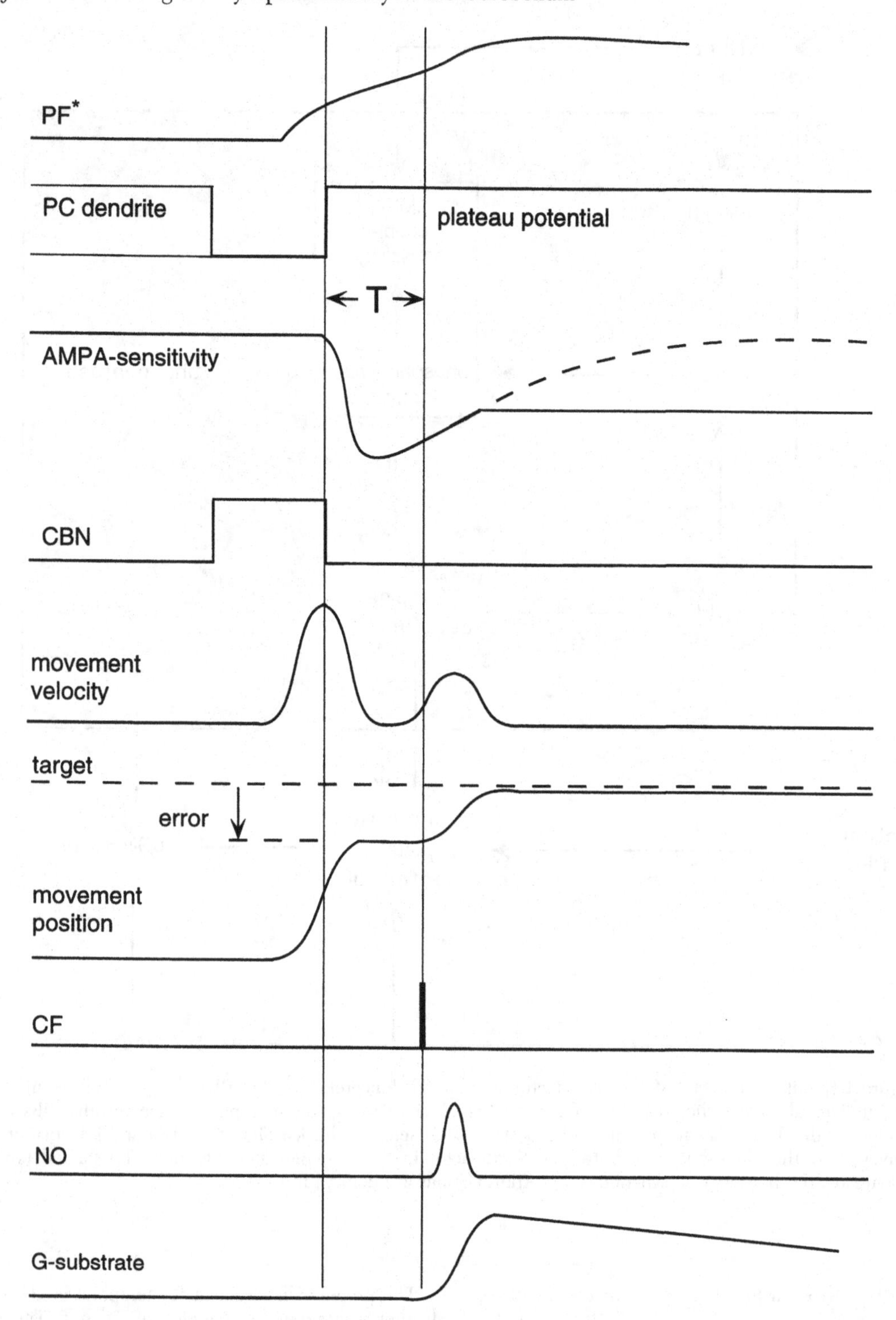

Figure 2. (Houk & Alford). Hypothetical time course of local and global events that are important in temporal credit assignment. T is the delay that must be overcome to achieve temporal credit assignment. The combination of PF* and the onset of the dendritic plateau potential produce the initial depression in AMPA-receptor sensitivity. The plateau potential also contributes to the abrupt cessation of CBN discharge, which terminates the motor command (prematurely in this case) that is sent to the neuromuscular system. The resultant undershooting error is detected due to the CF response to the corrective movement. NO release by CF input activates G-substrate to stabilize the depression in AMPA-receptor sensitivity, thus converting it into LTD. In this manner, these hypothetical events might resolve the temporal credit assignment problem.

appears to lack input specificity. According to these findings, inhibitory GABAergic synapses made by basket cells onto Purkinje cells are potentiated if the basket cell fires while the Purkinje cell it innervates is depolarized. Any of the many PFs that excite this basket cell would subsequently have an enhanced inhibitory action. Although this mechanism lacks input specificity, it has output specificity, since inactive Purkinje cells, ones already inhibited by the firing basket cell, would not receive enhanced inhibition. It seems ideally suited for regulating the level of excitability in Purkinje cells. This could insure that properly assigned credit in the training of PF synapses would have as sensitive as possible an effect on Purkinje cell firing.

While there are many convergent findings regarding the mechanism of cerebellar LTD, some controversy clearly remains. It is likely that this represents, in part, redundancy within the protocols by which this process is elicited. In turn, the experimental demonstration of this redundancy may follow from the experimental paradigms used to induce LTD. Intense activation of a particular pathway may result in LTD when in the animal the pathway receives a less intense or an asynchronous input. It will be exciting to see, in the near future, the impact of timing of inputs to Purkinje neurons, and particularly whether this timing would be critical to the cascade of events that occur to initiate LTD when the animal is learning a motor task. Under these circumstances redundancy in the system may collapse to a necessary sequence of events, perhaps not unlike the model presented in this commentary.

Constructing a theory of cerebellar function in limb movement control is premature

D. Jaeger

Division of Biology 216-76, California Institute of Technology, Pasadena, CA 91125. **dieter@bbb.caltech.edu**

Abstract: Cerebellar theory has been unable to converge on a single concept of cerebellar function. Problems arise because we have little knowledge about how motor control works in the structures with which cerebellum interacts and because our knowledge of cerebellar activity patterns is very rudimentary. Rather than putting forth more underconstrained high-level theories, it may be time to start an interactive process between modern physiological experimentation on multiple levels and realistic computer modeling of physiological variables to reproduce the actual system. [KANO; HOUK et al.; SIMPSON et al.; THACH]

Opinions on cerebellar function differ. The target articles on cerebellar function in limb-movement control in this issue (**HOUK et al., SMITH, THACH**) proclaim different ideas about the computational task of cerebellum. HOUK et al. place their adjustable pattern generator model in the framework of selecting/deselecting motor programs that control elemental movements, **SMITH** proposes that cerebellum regulates joint stiffness by controlling co-contraction of agonist-antagonist muscles, and **THACH** describes cerebellar activity in the motor domain as coordinating the activity of single muscles to produce smooth movements (sects. 2.8, 6.9). Thus, even though all of these proposals revolve around motor control, there is no agreement as to the actual control mechanisms by which cerebellum exerts its influence.

Cerebellar output is received by structures we do not understand. The confusion as to the level at which cerebellum controls movement is not surprising, given that our understanding of motor control is lacking even at the level of the spinal cord. Maybe the best knowledge on this topic available at this time comes from the work of Bizzi and colleagues (Bizzi et al. 1995), which suggests that muscle synergies are already organized in the spinal cord to produce smooth movements to a stable endpoint. The code by which high-level motor structures "communicate" with such spinal cord networks is unknown, however, and remains a great challenge. None of the target articles in this issue address this problem. Thus, asserting definite statements as to how cerebellum controls movement appears premature, which is reflected in the fact that we get three different views in three articles, and would get yet different views if other authors were added.

The physiology of cerebellar activity patterns is poorly understood, too. A second set of problems with cerebellar theories arises from the fact that 27 years after Marr's proposal of parallel fiber beams controlling elemental movements (Marr 1969), the actual dynamics of cerebellar network activity are still poorly understood. The target article by **SIMPSON et al.** is illuminating with respect to climbing fiber activity, which is by far the most accurately measured input pattern influencing cerebellar activity, and yet its influence on cerebellar output cannot be determined. The proposed beams or patterns of parallel fiber activation that form pillars in the edifices of most cerebellar theories (including the ones by **HOUK et al.** and **THACH**) remain unmeasurable, and crucial parameters like numbers of fibers involved, spatial and temporal spread of activation, frequency of beam activation, and so on, are entirely unknown. In fact, when beams of Purkinje cell activation were tried to be measured, they were not found (Bower & Woolston 1983). The lack of knowledge about natural patterns of parallel fiber activation is paralleled by a lack of knowledge about the activation pattern of intrinsic inhibitory circuits in cerebellar cortex. **HOUK et al.** are to be commended for the inclusion of breaks in Purkinje cell activity in his model, as inhibition of Purkinje cell activity is largely neglected by other theories. Nevertheless, the computational properties of such breaks depend on parameters like the statistics of inhibitory connectivity, the strength of the parallel fiber – stellate/basket cell – Purkinje cell link, and the response dynamics of Purkinje cells to inhibition. Again, these parameters have not been measured to date. The target article by **KANO** indicates that such inhibitory pathways are plastic as well, further supporting their potential importance in cerebellar function.

Our lack of understanding cerebellar dynamics extends to almost any aspect of cerebellar activity. Important features include the spatial and temporal parameters of Purkinje cell inhibition of cerebellar nuclear cells, and the dynamics of interaction between extrinsic input and Purkinje cell inhibition at the level of the cerebellar nuclei. Further complexity in activation patterns is introduced by the nonlinearities of single cell properties. For example, a prolonged response pattern in a Purkinje cell can be evoked by a single pulse of inputs (Jaeger & Bower 1994).

When the parameters are not constrained, many theories can be proposed, and many models can be constructed. As described above, there is a fundamental lack of physiological data on cerebellar activity patterns *in vivo*. It seems to me – being slightly facetious – that cerebellar theorists in general have picked a number of known coarse-grain anatomical connections, picked a functional hypothesis that sounded plausible and interesting, added a number of ad hoc assumptions and voilà – a theory. The problem with this process appears to be twofold. First, the choice of functional hypothesis is underconstrained, due to our lack of understanding what cerebellar output does to motor control. And second, due to the lack of data on actual cerebellar activity patterns, models based on each theory can be constructed without being constrained by the actual system. For instance, **HOUK et al.** postulate that Purkinje cells become refractory to further input after switching to a new firing state in response to an instruction stimulus (sect. 3.3, para. 2). There is no evidence that a Purkinje cell becomes refractory to input over an extended period of spiking. Thach does not present a control model of the functionality he proposes, and therefore does not reach the level of ad hoc assumptions with respect to cerebellar activity patterns. Essentially his contribution is a functional hypothesis rather than a control theory. I do not doubt, however, that a matching control model could be constructed. In fact, I believe that due to the lack of constraints a working model reflecting the gross anatomy of cerebellum can be built in support of any hypothesis of cerebellar function.

Back to the experimenters workbench: Feedback between physiology and realistic modeling. To me the above arguments clearly indicate that at the present time a correct theory of cerebellum cannot be derived from known facts, and if by chance someone made a correct theory, there would be no way to distinguish it from incorrect ones. Rather than producing more different theories by the process summarized in the preceding paragraph, I think our experimental and modeling techniques have advanced to a point where we can do better. The new process involves an interactive approach between physiology and constructing realistic models, which map measurable physiological variables 1:1 with model parameters. The necessary physiology will require years to come, and involve modern imaging and recording techniques at the single cell and network level. On the modeling side, working compartmental models of single cells need to be put into networks that perform actual processing of input information. Such models need to be updated and made more elaborate as the physiological data become available. In turn, the model can be used to generate predictions about important physiological interactions, and guide the experimenter to the experiments that most crucially address our understanding of the system. Ultimately, of course, a theory and model of cerebellum cannot be achieved without a parallel education of the structures to which cerebellum is connected.

New players for cerebellar long-term depression

Masanobu Kano

Laboratory for Neuronal Signal Transduction, Frontier Research Program, Riken, Wako-shi, Saitama 351-01, Japan. **mkano@postman.riken.go.jp**

Abstract: In 1994 and 1995, several new reports appeared regarding cellular mechanisms of LTD. These concern (1) involvement of mGluR1, (2) involvement of glutamate receptor δ2 subunit (GluRδ2), (3) involvement of Ca^{2+} release from intracellular stores, and (4) origin of nitric oxide (NO). I would like to add these new results to the review article by Crépel et al. [CRÉPEL et al.]

The target article by CRÉPEL **et al.** provides a good comprehensive review for cerebellar long-depression (LTD), a form of synaptic plasticity widely believed to underlie motor learning. Their scheme (Fig. 6) incorporates experimental results of several different groups and describes hypothetical mechanisms of LTD. In 1994 and 1995, however, several new results were reported, including studies on gene knockout mice, which I would like to introduce in this commentary. I would also like to address the issue of the origin of nitric oxide (NO) in the induction of LTD.

(1) Further support for the involvement of mGluR1 in LTD induction. Shigemoto et al. (1994) raised specific antibodies against mGluR1 that effectively blocked glutamate-induced IP3 formation *in vitro*. They showed that these antibodies completely abolished LTD of glutamate responsiveness in cultured Purkinje cells. Aiba et al. (1994) and Conquet et al. (1994) produced mutant mice deficient in mGluR1 by embryonic stem cell gene targeting technique. The mGluR1 mutant mice are clearly ataxic, and are impaired in both motor coordination and motor learning. The morphology of the cerebellum is not apparently disturbed. Electrophysiological analysis using acute cerebellar slices from mGluR1 mutant mice revealed that voltage-gated Ca^{2+} currents of Purkinje cells are normal and both parallel fiber-mediated and climbing fiber-mediated EPSCs are functional. By marked contrast, LTD of parallel fiber responses is clearly deficient in the mGluR1 mutant mice. These results, together with the finding that an antagonist of mGluRs (MCPG) blocks LTD in acute slice preparation (Hartell 1994), corroborate that mGluR1 is necessary for LTD induction in both culture and acute slice preparations.

(2) The glutamate receptor δ2 subunit (GluRδ2) is necessary for LTD induction. The ionotropic glutamate receptors consist of six families among which the functions of the δ subfamily are poorly understood. The GluRδ2 is selectively localized in cerebellar Purkinje cells, suggesting its possible involvement in cerebellar LTD. Hirano et al. (1994) applied antisense oligonucleotides against the GluRδ2 mRNA to rat cerebellar culture and reported that this treatment suppressed the induction of LTD of glutamate responsiveness. Kashiwabuchi et al. (1995) produced GluRδ2 deficient mice by a gene targeting technique. The GluRδ2 mutant mice have impairment in motor coordination, defects in synapse formation along both parallel fibers and climbing fibers to Purkinje cells, and deficient LTD in acute cerebellar slice preparation. Hirano et al. (1995) further confirmed that LTD of glutamate responsiveness was deficient in cultured Purkinje cells from GluRδ2 mutant mice. These results suggest that the presence of GluRδ2 on Purkinje cell membrane is necessary for LTD induction in both culture and acute slice preparations.

(3) CA^{2+} release from intracellular stores appears to be involved in LTD. Cerebellar Purkinje cells are rich in both IP3 and ryanodine receptors, which form Ca^{2+} release channels in intracellular organelles. Several lines of evidence indicate that both IP3 and ryanodine receptors contribute to Ca^{2+} signaling in Purkinje cells. Two recent reports suggest that these intracellular Ca^{2+} release channels are involved in LTD induction in cultured Purkinje cells.

Kazono and Hirano (1995) reported that LTD of glutamate responsiveness is suppressed by heparin that blocks IP3 binding to its receptor. They also showed that AMPA application and depolarization of Purkinje cells induced LTD only when they were paired with photolysis of caged IP3. These results suggest that IP3-induced Ca^{2+} mobilization is involved in LTD induction in cultured Purkinje cells.

Kohda et al. (1995) recently reported that LTD of glutamate responsiveness is suppressed by thapsigargin (a specific inhibitor of Ca^{2+}-ATPase on the endoplasmic reticulum), ryanodine, and ruthenium red (both are inhibitors of ryanodine receptors). These drugs had no suppressive effects on voltage-gated Ca^{2+} channel currents in Purkinje cells. Their results suggest that ryanodine receptor-mediated Ca^{2+} release is necessary for LTD induction in cultured Purkinje cells.

In acute slice preparation, however, Hemart et al. (1995) reported that thapsigargin did not affect LTD of parallel fiber mediated EPSCs. Further tests are necessary whether IP3 or ryanodine receptor-mediated Ca^{2+} release is necessary for LTD in acute slice preparation.

(4) Origin of NO in the induction of LTD. Nitric oxide (NO) production is involved in LTD induction in acute slice preparations. However, as pointed out by CRÉPEL **et al.,** it is not known whether NO is produced by Purkinje cells or by parallel fibers and/or basket cells. Lev-Ram et al. (1995) recently reported that photolysis of caged-NO inside the Purkinje cells mimicked parallel fiber stimulation in synergizing with Purkinje cell depolarization to induce LTD. They claim that NO is an anterograde messenger produced in parallel fiber terminals. In their scheme of LTD, however, it is unclear where mGluR1 and PKC are incorporated. Further studies are necessary regarding the mechanisms of NO action during LTD induction.

The common inverse-dynamics motor-command coordinates for complex and simple spikes

M. Kawato

ATR Human Information Processing Research Laboratories, Kyoto 619-02, Japan. **kawato@hip.atr.co.jp**

Abstract: Recent advanced statistical analysis of complex spikes has revealed that their instantaneous firing rate within a time bin of a few milliseconds carries information if many trials are averaged, as happens in motor learning. The firing rate encodes sensory error signals in the inverse-dynamics motor-command coordinates, and these are exactly the same coordinates as for simple spikes. This strongly supports the most critical assumption of the feedback-error-learning model and argues against several hypotheses about the functions of the complex spikes. [**HOUK et al.; SIMPSON et al.; SMITH; THACH**]

***1. Instantaneous firing rate of complex spikes encodes motor-command error:* SIMPSON et al. *and* HOUK et al.** What is the information carrier of complex spikes (CS)? And what information is represented by it? These questions are central to the target articles by **SIMPSON et al.** and **HOUK et al.** The very low firing rate of the CS makes it unlikely that information is transmitted by their instantaneous firing rate. A common view is that either an individual spike or the accumulated spike count during a relatively long period (e.g., 100 msec) carries information. Despite this widely accepted view, Kobayashi et al. (1995; 1996) demonstrated that the temporal patterns of the CS firing rate within a 2-msec time bin during ocular following responses (OFR) in the monkey ventral paraflocculus (VPFL) were well reconstructed by the inverse-dynamics representation of eye movements. A large number of traces (200–1500) were averaged and a "generalized linear model" (Kawato 1995) was utilized for this analysis.

The high-frequency temporal pattern of CS was a mirror image of the simple spike (SS). For each cell, a negative correlation existed between the coefficients in the inverse dynamic representation of the CS and SS. An independent analysis using cross-correlation between the SS and CS showed that only 1% of the SS modulation can be explained by the direct short-term modulation of the SS by the CS. Thus, the observed mirror-image relationship is most probably the result of long-term effects such as LTD or LTP. We found that the optimum directions of the visual stimulus for the SS and CS were always opposite and were in the same horizontal-vertical motor-command coordinates. Furthermore, the temporal patterns of the SS and CS were similarly well reconstructed from a generalized linear model of the retinal slip. On these grounds, Kobayashi concluded that the instantaneous firing rate of the CS carries a sensory error (retinal slip) signal in the same spatial coordinates as the SS. The SS are known to encode dynamic components of the necessary motor commands (Shidara et al. 1993).

HOUK et al. state in section 4.3 that the basic problem with the feedback-error-learning model may be its assumption that IO activity reflects motor error. I define motor error as an error signal induced by movements, represented in motor-command coordinates, and originating at least partly from sensors. If the SS should be found to be a component of motor commands, the above is the inevitable condition for LTD or LTP to be computationally effective in motor learning. This most essential assumption is gaining experimental support. Kobayashi's data argues against hypotheses that the CS (1) represent the occurrence of unexpected events by individual impulses, (2) short-term modulate the simple-spike firing, (3) operate as motor commands independent of the simple spike firing, or (4) synchronize the activities of many Purkinje cells (**SIMPSON et al**). The reasons against (1) and (2) were stated above. The reason against (3) and (4) is that the complex spikes are just the reverse of the dynamic motor command. This means they have the wrong polarity for on-line control, and for many trials do not even appear, making them unable to contribute to control. The CS's information content with brisk temporal resolution in the required motor-command coordinates becomes apparent only if averaged for many trials, which is possible only in the context of motor learning.

***2. Desired trajectory as input to the cerebellum:* HOUK et al.** One of the most important predictions of the cerebellar feedback-error-learning model, that is, that the inverse dynamics model of a controlled object resides in the cerebellar cortex, requires that some mossy fiber inputs represent the desired trajectory information (sect. 3.4 of **HOUK et al.**). This is evident in VOR, where the semicircular canal output encodes the head velocity, which is exactly the sign-reversed desired trajectory information for the compensatory eye movement. Thus, the question concerns phylogenetically newer parts of the cerebellum. Kawano (Kawano et al. 1994; Takemura et al. 1996) compared firing patterns of the MST (medial superior temporal) area of the cerebral cortex and the DLPN (dorsolateral pontine nucleus) of the brain stem with those of the VPFL during OFR. MST-DLPN-VPFL is known as the main pathway for control of OFR as well as for smooth pursuit eye movement (SP). Only for a small portion of neurons in the MST and DLPN, could firing patterns be well reconstructed from the eye movement using the inverse dynamics model. These firing patterns did not encode the necessary dynamic motor commands even for neurons with successful reconstruction. Thus, the mossy-fiber inputs to the VPFL from MST-DLPN are certainly not the motor commands, but rather encode visual information about the desired eye movement. We can conclude that sensory-motor transformation occurs at the parallel-fiber-Purkinje cell synapse, or more computationally speaking, that the cerebellar cortex is the major site of inverse-dynamics transformation.

For reflexes such as VOR or OFR, the desired trajectory information is primarily given by sensors. For voluntary movements, the desired trajectory must be calculated in the brain from movement goals such as the target position. In SP, one of the best studied voluntary movements, desired trajectory generation is essential when the retinal slip is zero (i.e., perfect tracking). Although several models have suggested that this is achieved below the cerebellar cortex by the so-called gaze-velocity positive feedback loop (sect. 3.8 of **HOUK et al.**), recent unit recordings from MST, DLPN, and VPFL both during OFR and SP seem to reject this mechanism, and indicate that the desired trajectory is already generated at the MST. This is because the MST firing, DLPN firing, VPFL firing, and the eye movement always covary even under the electrically achieved 0-retinal-slip condition and under the target-blink condition both for OFR and SP (Newsome et al. 1988; Kawano et al. 1992; 1994). Although **HOUK** stated that there is no evidence of signals that specify desired trajectories (sect. 3.4) even for limb movements, Kalaska et al. (1990, Figs. 6B and C) showed that some neurons in the parietal association cortex encode not only the movement direction but also the velocity profile, which is the trajectory information.

I do not argue against the possibility that the cerebellum is involved in trajectory formation. Rather, if we combine our previous proposal that both the forward and inverse models reside in the cerebellum (Kawato et al. 1987) with the more recent FIRM trajectory generation neural network model which uses the two types of models (Wada & Kawato 1993), it is natural to expect that the cerebellum plays an essential role in the desired trajectory calculation. The APG model does not provide any concrete computational mechanism for learning invariant features of multijoint-arm movements, such as roughly straight hand paths in Cartesian space and bell-shaped speed profiles, in a recurrent network when only the target error information is provided. In this case, trajectory learning is confronted with computationally very difficult temporal and structural credit assignment problems because the above trajectory features must be embedded into synaptic weights of the recurrent network while satisfying the end-point target condition. It is extremely difficult to imagine how the necessary error signal during movement execution could be computed from the final error information below IO. Our FIRM approach resolves all of these difficulties in a computationally transparent and biologically plausible way. It must be noted that several artificial neural-network learning algorithms, such as the recurrent back propagation or forward-inverse modeling approach previously used for trajectory learning, cannot easily be mapped onto the cerebellar circuitry in a biologically plausible way.

***3. Coordinate frame of cerebellar motor command:* SMITH.** The fact that only vertical and horizontal Purkinje cells were found in the monkey VPFL during OFR and SP supports the hypothesis that the cerebellum encodes its motor command outputs in slightly more abstract coordinates than those of muscles, such as joint torques, equilibrium-point trajectories or stiffness, because the vertical direction is different from the extraocular muscle action lines' direction whereas the horizontal direction is the same. Gomi and Kawato (1995) found that subjects do have the ability to significantly adapt the shapes and orientations of stiffness ellipses for different task constraints (A, free movement; B, straight-path constraint realized by servo-control of manipulandum, C. B plus, imposed viscous force field), even while executing the same multi-joint arm movements. That is, A, B, and C were with approximately the same joint angle trajectories, and A and B were with approximately the same joint torque. These data seem to support **SMITH**'s proposal (sect. 1).

Caution needs to be used when interpreting observed firing patterns of Purkinje cells. Schweighofer (1995; Schweighofer et al. 1996) constructed a cerebellar microcircuitry model based on feedback-error-learning while considering time delays (contrary to criticisms in section 3.4 of **HOUK et al.**) for the motor learning of visually guided reaching with a two-link six-muscle arm model. Although the desired trajectory and sensory feedback were all encoded in either joint or muscle coordinates, some simulated Purkinje cells showed single peak activities during movements in all 8 directions. I hesitate to say that these neurons encoded something like stiffness, although it is the only variable which I can imagine to behave in that manner.

4. How could the computational principle for sensorimotor transformation be extended to cognitive functions?: **THACH.** As Ito (1993) suggested, the most straightforward way to extend computational theories, such as the feedback-error-learning or internal model hypothesis which were developed for the sensorimotor-transformation functions of the cerebellum to cognitive functions, is to replace the controlled object in the external world, such as the eye, arm, or speech articulator, by a module in the cerebral cortical circuit. Let us suppose that module A acts on module B to obtain B's output which is similar to A's input. We can say that A is functioning as B's inverse. While this original process is under conscious control, it could be automatically achieved if one microzone of the cerebellum becomes A's substitute and an inverse model of B. Similarly, we can imagine B's forward model (simulator, predictor). The bidirectional architecture using both the forward and inverse models has been found attractive for a number of computations, ranging from trajectory formation, and integration of vision modules, to even higher cognitive functions. These computational models seem to have a relation with the "learning context-response linkage" of **THACH.**

Which cerebellar cells contribute to extracellular cGMP?

Lech Kiedrowski

The Psychiatric Institute, the University of Illinois at Chicago, Chicago, IL 60612. **kied@iris.rfmh.org**

Abstract: Vincent proposes that the extracellular cGMP found in cerebellum after glutamate receptor activation is released mainly from Purkinje cells because in these neurons the presence of guanylate cyclase has been shown using monoclonal antibodies. It is uncertain, however, whether Purkinje cells are the only source of extracellular cGMP in the cerebellum. This commentary examines the possibility that glial and cerebellar granule cells may also participate in cGMP synthesis and release. Moreover, the hypothesis of transcellular metabolism of citrulline and arginine is discussed. [**VINCENT**]

In his target article, **VINCENT** presents the very interesting idea that the increase of extracellular cGMP concentration in cerebellum (Luo et al. 1994) is mostly due to the cGMP release from Purkinje cells in which the presence of guanylate cyclase has been well determined in immunocytochemical studies using monoclonal antibodies (Ariano et al. 1982; Nakane et al. 1983). The release of cGMP indicates new possibilities of yet unknown extracellular functions of this nucleotide. However, it still remains uncertain whether cGMP is released from Purkinje cells, exclusively.

Some studies on *Nervous,* mutant mice, which lack most Purkinje cells, show a decrease of guanylate cyclase activity (Mao et al. 1975; Schmidt & Nadi 1977), while others do not (Wood et al. 1994). Therefore, it should be considered that other cerebellar cells may also contribute to cGMP synthesis and release. The glutamate receptor agonist-induced increase of cGMP levels in cerebellar glial cells is well documented (Garthwaite & Garthwaite 1987; de Vente et al. 1990) and might be a result of either cGMP synthesis, or the probenecid-sensitive cGMP uptake, or both. In primary cultures of cerebellar astrocytes, we have observed that the NO donor, sodium nitroprusside, induces a robust cGMP accumulation (Kiedrowski et al. 1992), indicating expression of guanylate cyclase. It is likely that guanylate cyclase is expressed in glial cells, since the cultures were prepared from 8-day old rats; at this age Purkinje cells are already differentiated (Woodward et al. 1971), and do not survive the procedure of tissue culture preparation. It can be argued, that astrocytes express guanylate cyclase only *in vitro* but not *in vivo.* However, in cerebellar slices from adult rats, the polyclonal antibodies against guanylate cyclase from rat brain yield positive immunostaining of glial cells and all neurons (Zwiller et al. 1981). Therefore, one cannot exclude that some isoforms of soluble guanylate cyclase, different from the one expressed in Purkinje cells, may be expressed in glial and other cerebral cells.

cGMP may be also synthesized in cerebellar granule cells because hemoglobin, which does not penetrate plasma membrane, inhibits cGMP accumulation much more effectively in cerebellar slices exposed to NO-donors than to glutamate receptor agonists (Southam & Garthwaite 1991). This result indicates that cGMP is synthesized also in the cells that produce NO, and very likely cerebellar granule cells are included. The cGMP increase in cerebellar granule cells may be very short lasting due to the fact that elevation in $[Ca^{2+}]_i$, necessary to activate NO synthase (NOS) (Bredt & Snyder 1990) will also activate Ca^{2+} and calmodulin-dependent phosphodiesterase (Mayer et al. 1992). On the other hand, in cells not expressing NMDA receptors such as astrocytes, this phosphodiesterase will not be activated by $[Ca^{2+}]_i$, and the increase in cGMP concentrations will last much longer. To inhibit cGMP metabolism and simplify data analysis, inhibitors of phosphodiesterase such as IBMX are frequently included in the experimental protocol; however, this may not always be sufficient since some isoforms of neuronal Ca^{2+} and calmodulin-dependent phosphodiesterases are resistant to this inhibition (Mayer et al. 1992). It appears that at present, the available data do not allow one to exclude that cerebellar cells other than Purkinje cells might release cGMP.

Based on the fact that the antibodies raised against argininosuccinate synthetase (ASS) from liver (Arnt-Ramos et al. 1992; Nakamura et al. 1991) and argininosuccinate lyase (ASL) from liver (Nakamura et al. 1990) do not stain cerebellar granule cells, the author also suggests the very interesting possibility that transcellular metabolism of citrulline may occur in cerebellum. In the primary cultures of cerebellar granule cells incubated with $[^3H]$arginine, we observed (Kiedrowski et al. 1992) that shortly after activation of the NMDA receptor a plateau of $[^3H]$cytrulline concentration is reached. We determined that NOS was active during the plateau and interpreted the data to indicate that this plateau in the $[^3H]$cytrulline level was due to the equilibrium in the rates of reactions $[^3H]$arginine → $[^3H]$cytrulline → $[^3H]$argininosuccinate → $[^3H]$arginine. Our results suggest that both ASS and ASL are present in these cultures. It remains uncertain however, whether these enzymes are expressed in cerebellar granule cells, since these cultures are contaminated (3–5%) with glial and other cells (Nicoletti et al. 1986). If citrulline is not converted to arginine in cerebellar granule cells, our results indicate that it might be released to extracellular medium and taken up by the cells that express ASS and ASL. The mechanism of such specific citrulline release remains to be established. However, it is very puzzling that cerebellar granule cells might not be able to convert citrulline back to arginine, while other NOS expressing cells can do so (Hecker et al. 1990; Wu & Bronsan 1992). It should be verified whether some isoforms of ASS and ASL, not recognized by antibodies raised against enzymes purified from liver, may be expressed in these neurons. It is a fairly common occurrence that enzymes of low homology may catalyze the same reaction in different cells. NOS itself is a good example of such functional enzymatic polymorphism; in fact, antibodies against neuronal NOS do not cross-react with inducible NOS (Bredt et al. 1990). Therefore, negative immunohistochemical data obtained with antibodies directed against enzymes purified from liver cannot be used to negate the existence of catalytic activity in neurons. Clearly, the hypothesis of the transcellular metabolism of citrulline is very interesting, but requires additional experimental support.

The notions of joint stiffness and synaptic plasticity in motor memory

Lev P. Latash and Mark L. Latash

Biomechanics Lab, The Pennsylvania State University, University Park, PA 16802. **mll11@psu.edu**

Abstract: We criticize the synaptic theory of long-term memory and the inappropriate usage of physical notions such as "joint stiffness" in motor control theories. Motor control and motor memory hypotheses should be based on explicitly specified hypothetical control variables that are sound from both physiological and physical perspectives. [HOUK et al.; SMITH; THACH]

On the notion of joint stiffness in motor control. The fields of motor control and neurophysiology of movement frequently use terms and notions from other fields of science endowing these terms with new meanings which sometimes contradict their original definitions and introduce confusion. This commonly happens to the notion of stiffness. The inaccuracies in using this term lead to such undefined or even impossible expressions as "limb stiffness," "endpoint stiffness," "dynamic stiffness," or "negative stiffness." Let us briefly explain this point (for more details and references see Latash & Zatsiorsky 1993):

Stiffness in physics means the property of an object *to deform* under the influence of an external force, *to generate force* against the deformation, and *to store potential energy* during the deformation with the possibility of its release. Note that this definition is applicable only to deformable objects and that deformation is not the same as displacement. Joints and limbs do not deform during movements (they rotate and move) while muscles and tendons do. Muscles and tendons may be viewed as nonlinear springs whose properties are defined by both peripheral elasticity and reflexes. However, a system containing several nonlinear springs, some of which have reflex time delays, cannot be substituted with one spring and assigned a value of "stiffness."

Thus, "joint stiffness" is jargon. It is not a physical notion and, as such, it cannot be unambiguously measured with physical methods. Many studies introduced a measure for the property of joints to resist externally imposed *displacements,* commonly based on simplified second-order models. This is done ignoring the physical nature of forces within the system and assigning to it a measured or calculated property ("stiffness") which is in principle inapplicable to the system. in such cases, the measured quantities are likely to reflect not only the properties of the system but also the nature and parameters of the method of measurement. It is not surprising that different studies arrived at considerably different values of "joint stiffness." Thus, the "technical problem of stiffness measurement" (SMITH, sect. 2.8, para. 3) is not technical but conceptual and cannot be overcome.

Consider the endpoint of a limb in an equilibrium (Fig. 1). The elastic properties of muscles and tendons create a potential well in which the endpoint resides. A characteristic of the slope of the well walls is frequently measured in movement studies and called "stiffness." However, *it is not stiffness.* The central nervous system (CNS) can change the properties of the potential field through changes in muscle reflexes leading to changes in the muscle elastic properties. For example, it can move the potential well from the original position (A in Fig. 1) to another position (B in Fig. 1) leading to movement of the endpoint. It can also change the shape of the walls of the original well without displacing its bottom. This may lead to changes in the stability properties of the endpoint. This simple model allows us to describe the process of control with a couple of variables related to the equilibrium position of the endpoint and to the configuration of the potential field. These processes, however, cannot be described as "optimal time-varying control of joint stiffness" (SMITH). Expressions "joint stiffness" and "limb stiffness" sound very attractive and intuitive. However, to our knowledge, these notions have not been and, moreover, cannot be defined.

On the synaptic mechanisms of memory. We do not know how the hypothetical control variables are represented and memorized in the CNS. It is commonly and rather arbitrarily assumed that motor memory resides in synapses. In particular, long-term memory has been associated with presynaptic changes in synapses used during the event. These assumptions have been based on comparisons of neurophysiological changes and behavior. However, behavior may reflect many factors that are not memory, such as, arousal, attention, state of the effectors.

Using synapses for storing long-term memory is very noneconomical. Using a synapse in remembering an event makes it occupied and useless for future memories. Numerous "disposable synapses" are required to memorize a single event. The amount of memory of a grown-up person suggests that even the astronomical number of synapses within the CNS may be insufficient if such a crude and straightforward mechanism is used. Apparently, memorizing an event requires *a pattern of activity in complex neuronal formations whose organization and neuronal composition may vary.* The notorious variability of patterns of skilled movements (Bernstein 1967) is the strongest argument that such a pattern cannot be represented as a combination of activity of a number of individual neurons induced by stable changes in individual synapses.

One of us performed a study of spinal memory using, as a model, the phenomenon of stable asymmetry of monosynaptic reflexes which can be induced by ablation of one cerebellar hemisphere in the rat (Latash 1979; cf. Chamberlain et al. 1963). The asym-

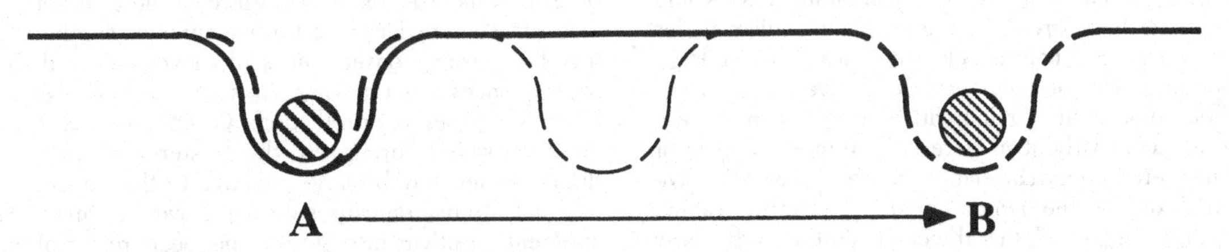

Figure 1 (Latash & Latash). The endpoint of a limb is in an equilibrium in a potential well (A). Stability of the equilibrium relates to the slope of the walls. It cannot be described as "stiffness," however. Movement (from A to B) may result from a shift of the potential well.

metry persisted after total spinal transection rostrally to the level of recording if the time interval between the two surgeries was sufficient. After the second surgery, the reflex asymmetry could theoretically originate from traces of the first surgery (memory) stored either in motoneurons, or in presynaptic afferent terminals, or in interneurons controlling the monosynaptic reflex arc. Experiments with warming up after deep local cooling demonstrated an initial restoration of reflexes without asymmetry followed by restoration of the asymmetry. These observations, together with experiments with pharmacological agents pointed at interneurons as the site of long-term changes underlying the asymmetry (probably inhibition of inhibitory interneurons). Note that, after the spinal transection, presynaptic influence to interneurons were removed and therefore, the site of memory traces must be postsynaptic, probably using macromolecules within the cell bodies. There are major differences between the phenomena of memory in the spinal cord and in the brain mainly due to different degrees of participation of whole-brain mechanisms. However, we believe that the synaptic theory of memory should be critically reconsidered.

Sensory prediction as a role for the cerebellum

R. C. Miall, M. Malkmus, and E. M. Robertson
University Laboratory of Physiology, Parks Road, Oxford OX1 3PT, United Kingdom. **chris.miall@physiol.ox.ac.uk**

Abstract: We suggest that the cerebellum generates sensory or "state" estimates based on outgoing motor commands and sensory feedback. Thus, it is not a motor pattern generator (HOUK et al.) but a predictive system which is intimately involved in motor behavior. This theory may explain the sensitivity of the climbing fibers to both unexpected external events and motor errors (SIMPSON et al.), and we speculate that unusual biophysical properties of the inferior olive might allow the cerebellum to develop multiple asynchronous sensory estimates. [**HOUK et al.**; **SIMPSON et al.**; **THACH**]

HOUK et al. have developed an extensive set of ideas of the cerebellum as an adjustable pattern generator, in which the key role for the cerebellar cortex is to modulate and terminate motor commands being driven by positive feedback within brainstem circuits. They dismiss the suggestion of the cerebellum as a sensory predictor or state estimator (Miall et al. 1993; Paulin 1993) with little comment on the data that supports such theories. We think the weight of these data point towards sensory predictions and are difficult to reconcile with a role of adjustable pattern generator. However, it is worth pointing out that a sensory predictor is not a sensory analyzer (Bower & Kassel 1990). We do not claim the cerebellum is concerned with processing sensory information per se, but is vital for the processing of sensory inputs by other brain structures in the context of movement.

Perhaps the first point to make is one stressed by Paulin (1993) – that the comparative anatomy of the cerebellum is difficult to explain if its role is that of generating motor commands. It seems that those species with advanced sensory-motor abilities, in which reafference resultant on motor behavior must be analyzed, have larger cerebellar cortices than their body or brain mass would predict. Cetaceans have greatly extended paraflocullar nodes (Riley 1928); echolocating bats and electric fishes also have large cerebellar volumes. The primates have a massive expansion of lateral cerebellar cortex, in parallel with the expansion of neocortex, but this occurs without an increase in number of joints or muscles. Thus there is poor correlation between cerebellar size and the complexity of the motor apparatus. **THACH** follows Flourens (1824) in suggesting that the cerebellum is particularly involved in coordinated movement of many different joints, and there is data implicating its role in coordination between different motor structures (hand-eye coordination, for example). The APG model (**HOUK et al.**) does not seem to have a role here – its output cannot easily be used for coordination, unless it holds separate APGs for every coordinative structure or synergy.

Our hypothesis is that the cerebellum provides an estimate of the current state of the motor system, which the proprioceptive and teleceptive systems cannot do because of their intrinsic processing and conduction delays. State estimation has many uses (Miall & Wolpert 1996); thus all neural functions which depend on state estimation (motor planning, mental imagery, internal feedback control, cancellation of reafference, coordination) could be linked to the cerebellum, and these multiple uses explain the expanded area of the primate cerebellum.

Of course it is only in man that one can properly address questions of cognitive function, and of the mental planning or imaging of movements without execution. Again, suggestions that the cerebellum may have a key role in these processes (**THACH**) are difficult to account for on the basis of motor command generation. And it is striking that the output of the cerebrocerebellum is not restricted to "downstream" motor executive areas, as one would predict from Houk's theory. Instead there are extensive connections to frontal areas and to parietal areas.

Let us turn to some specific features of **HOUK et al.**'s theory. The APG model is based on positive feedback, which can be difficult to control. Positive feedback loops can be easily pushed into excessive activation. For a motor command, a graded response is sought, and yet small changes in the responses of the neurons in the loop would be magnified into larger and larger variations in the loop output. Thus brainstem positive feedback loops would be rather unreliable, and the task of inhibitory modulation by the cerebellar cortex difficult. However, if it is accepted, the APG hypothesis makes a strong prediction: mass stimulation of the cerebellar cortex (e.g., with transcranial magnetic stimulation) should halt all movement in its tracks. We do not know if there is data to support this. Finally, the APG model suggests a reciprocal relationship between Purkinje cell responses and motor commands. This can indeed be seen, but in many, many papers there are reports of quite powerful activation of P-cells throughout movement, rather than just at its termination.

We believe, like **SIMPSON et al.**, that a key to understanding the function of the cerebellum is given by the climbing fibers. If the climbing fibers provide a sensory rather than motor signal, as **HOUK et al.** accept, and if this signal is used to train the cerebellum, then it seems much more likely that the cerebellum learns within a sensory rather than motor framework. Kawato and colleagues (1992a; 1992b) have accepted that one should use a motor error signal to generate a motor output; yet one sees pronounced somatosensory sensitivity of olivary cells in the absence of movement.

So let us now address the response properties of the IO, which as **SIMPSON et al.** describe, remain a puzzle. They cannot easily be characterized as simple error signals, nor do they have a straightforward relationship to movement. The theory that the cerebellum makes estimates of sensory states requires a teacher that can signal mismatches between present estimates and reality. The climbing fibers may do that. They signal something like an error during motor tasks (**SIMPSON** et al.), but this may reflect the mismatch in expected and actual sensations when the motor task changes. They also signal purely sensory events, for example the passive stimulation of the skin (Gellman et al. 1985) which cannot be a movement error. It can, however, be a sensory prediction error. The cerebellar predictor may function continually, predicting the sensory consequences of movement, and the sensory consequences of not moving. Thus any unexpected external event is a failure of sensory predictions. Given the low levels of sensory inputs expected during rest, these externally generated sensory discrepancies may be large, relative to the sensory errors that external stimulation causes during active movement. Hence the apparent sensitivity during rest. The recent report of time-locked climbing fiber activity during rhythmic movement (Welsh et al. 1995) might reflect the sensory differences from moment-to-moment when licking a feeding tube that delivers a water drop.

In our original model of the cerebellum (Miall et al. 1993) we suggested one or more predictors, operating in exteroceptive (visual) and proprioceptive coordinates. Malkmus (in preparation) has suggested that the cerebellum could generate multiple asynchronous predictions, families of predictions at many different temporal offsets from the current state of the motor system. Some of these might be long-lead predictions used in motor planning, others shorter lead for internal error correction, and some even synchronous with reafference, to allow its cancellation.

Of course, we have implicitly accepted that the climbing fibers carry a training signal, and this remains an assumption. However, one can ask what training signal would be best to allow the development of a sensory prediction? Malkmus suggests that the inhibitory output from the cerebellum might be propagated and delayed through serial connections within the inferior olive, so that multiple asynchronous comparisons between cerebellar output and reafference take place. The appropriate cerebellar outputs would be trained by the combination of Hebbian learning at the parallel fiber/Purkinje cell synapse with LTD in response to CF activation. Simulations have shown that the combination of these two learning rules provides a powerful learning paradigm. These simulations also suggest that even after successful training, the maintenance of that prediction requires a constant level of IO activation (constant over a long time scale), and that individual climbing fiber responses can be expected to follow episodes of higher than average P-cell activity; this again differs from Welsh's interpretation of CFs (Welsh et al. 1995). It does not seem to be necessary to actually store the complete pattern belonging to a given prediction. Rather, cerebellar cortex could store the difference between the current state and the predicted state, thus considerably reducing the amount of information stored. The deep nuclei would then use this difference information to create the prediction. Finally, sagittal microzones within the cerebellar cortex would hold similar but temporally shifted patterns of information. If these ideas hold up, they provide an exciting explanation for the well known but purely understood relationship between the inferior olive and the cerebellum.

ACKNOWLEDGMENT

This work was supported by the Wellcome Trust and MRC.

Further evidence for the involvement of nitric oxide in trans-ACPD-induced suppression of AMPA responses in cultured chick Purkinje neurons

Junko Mori-Okamoto[1] and Koichi Okamoto[2]

Departments of [1]Physiology and [2]Pharmacology, National Defense Medical College, Tokorozawa, Saitama 359, Japan.

Abstract: In addition to SNP and SIN-1, SNAP suppresses AMPA responses. This suppression is antagonized by carboxy-PTIO in cultured chick cerebellar Purkinje neurons. Intracellular application of cGMP shows a long-lasting suppression of AMPA responses mimicking the cerebellar LTD. These recent results demonstrate that NO can induce LTD-like suppression of AMPA responses and intracellular cGMP and cGMP-dependent protein kinase participate in this suppression. [CRÉPEL et al.; LINDEN; VINCENT]

In our recent experiments with cultured cerebellar Purkinje neurons, we have further confirmed that NO is involved in the trans-ACPD-induced suppression of AMPA responses and further found that this relatively short-lasting suppression becomes a long-lasting one like cerebellar LTD as long as the intracellular cGMP level is kept high. Results of our following two experiments strongly indicate that LTD is induced specifically by NO itself.

Experiment 1. SNAP-induced suppression of AMPA responses. In our previous study (Mori-Okamoto et al. 1993), we used SNP and SIN-1 as NO donors. As pointed out by VINCENT, SNP releases not only NO but also ferricyanide ions having a nonspecific neuronal effect. Since SIN-1 liberates NO and superoxide (O^{2-}), it is possible that the peroxynitrite ($ONOO^-$) formed from NO and O^{2-} may have affected AMPA responses. We accordingly used another NO donor, SNAP (S-nitroso-N-acetylpenicillamine), to confirm the involvement of NO itself in the suppression of AMPA responses. As shown in Figure 1A, AMPA responses were suppressed by SNAP and this suppression was reversibly antagonized by the specific NO-scavenger, carboxy-PTIO. Moreover, carboxy-PTIO also antagonized the suppressive effect of SIN-1, SNP, and trans-ACPD on AMPA responses. The SNAP-induced suppression of AMPA response was blocked by LY83583, a selective inhibitor of soluble guanylyl cyclase.

These results with SNAP and carboxy-PTIO strongly indicate that NO from an NO donor activates soluble guanylyl cyclase in Purkinje neurons and leads to the formation of cGMP and then to the activation of cGMP-dependent protein kinase.

Experiment 2. Effect of CO. CO, like NO, is known to activate soluble guanylyl cyclase, and the CO-producing enzyme, heme oxygenase, is localized in cerebellar granule cells and Purkinje neurons (Verma et al. 1993). When Tyrode's solution, saturated with 100% CO gas bubbling, was superfused onto cultured chick cerebellar neurons, AMPA responses were slightly and only transiently suppressed. In view of this very weak effect of CO as compared with NO, it is unlikely that CO is the major activator of soluble guanylyl cyclase in Purkinje neurons. The results of these two experiments provide further evidence for the involvement of NO in trans-ACPD-induced suppression of AMPA responses. Thus, at the present stage, we do not have any positive evidence against the involvement of NO.

Experiment 3. LTD by intracellular cGMP. This experiment demonstrates that the suppression of AMPA responses in cultured chick Purkinje neurons is converted to LTD, when the intracellular level of cGMP is kept high.

The Mori-Okamoto et al. (1993) study had shown that extracellularly applied 8-bromo-cGMP suppresses AMPA responses in a manner similar to trans-ACPD and the NO donors, SNP and SIN-1, and also that this 8-bromo-cGMP-induced suppression is antagonized by KT5823, an inhibitor of cGMP-dependent protein kinase, suggesting the involvement of soluble guanylyl cyclase and cGMP-dependent protein kinase. However, the suppression of AMPA responses induced by externally applied 8-bromo-cGMP lasted for 30–60 min at most. Thus, we have examined the effect of intracellularly applied cGMP on AMPA responses.

cGMP was applied directly into a Purkinje neuron through the whole-cell recording pipette containing 0.2–0.5 mM cGMP and 0.1 mM IBMX (isobutyl-1-methylxanthine). As shown in Fig. 1B, AMPA responses were gradually depressed in about 5 min and this suppression lasted for 80 min or longer without showing a tendency to recover. In contrast to the finding of LINDEN et al., extracellularly applied cGMP never suppressed AMPA responses in our cultured chick Purkinje neurons.

These recent results have solidly demonstrated that intracellular cGMP and the subsequent activation of cGMP-dependent protein kinase clearly contribute to the trans-ACPD-induced suppression of AMPA responses in cultured chick cerebellar Purkinje neurons. Since NOS has not been demonstrated to be present inside the Purkinje cell, the origin of NO is still to be identified. The target of activated cGMP-dependent protein kinase might be G-substrates, but this has not been directly demonstrated. Moreover, there is the possibility that depolarization induced LTD and chemically induced LTD are mediated by different mechanisms, VINCENT and CRÉPEL et al. Until these questions are answered, all we can say with confidence at present is that trans-ACPD-induced suppression of AMPA responses is mediated by elevated intracellular cGMP and activated cGMP-dependent protein kinase.

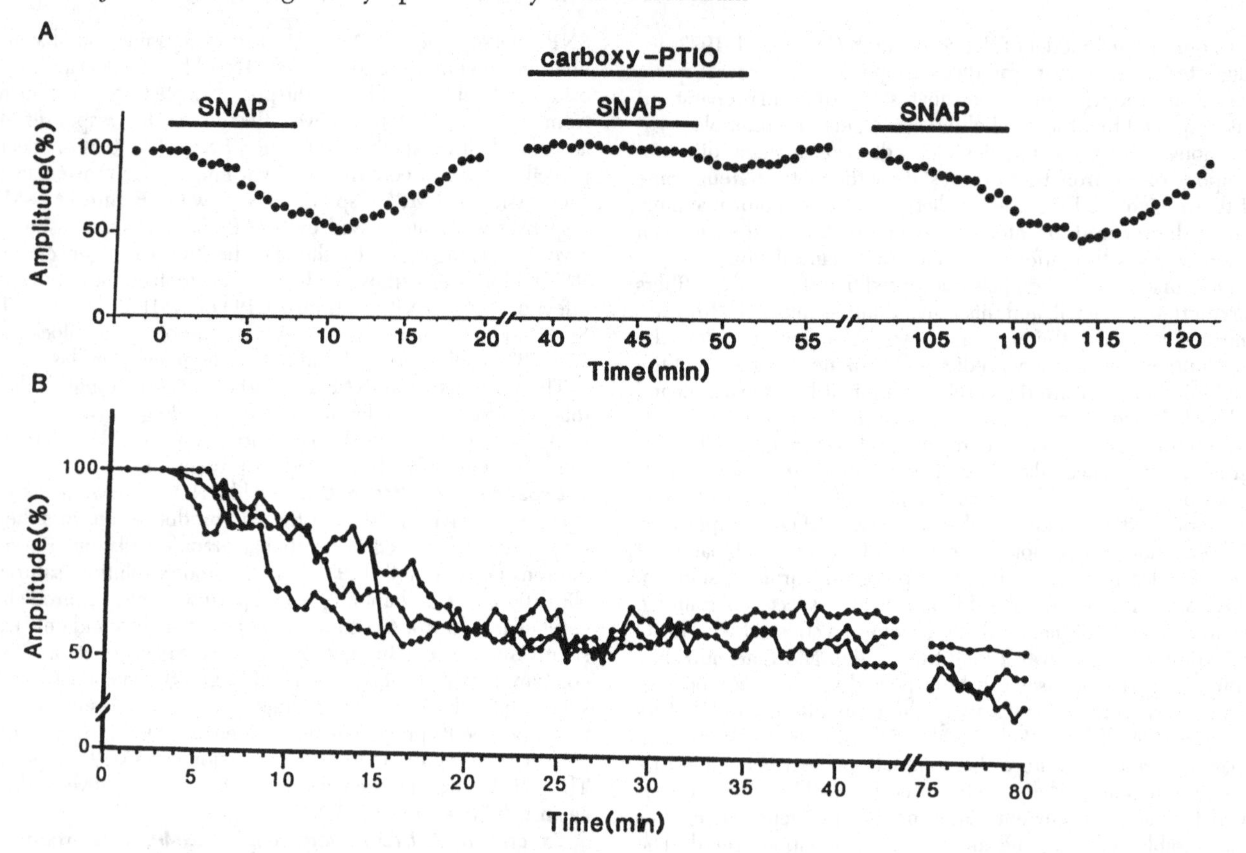

Figure 1 (Mori-Okamoto & Okamoto). (A) The suppressive effect of SNAP on AMPA responses, and the antagonistic effect of carboxy-PTIO thereon. (B) Long-lasting suppression of AMPA responses by intracellular application of cGMP. In both graphs showing the time course of the suppression of AMPA responses, each dot indicates the relative peak amplitude of the whole-cell current induced by iontophoretically applied AMPA. In (A) carboxy-PTIO (10 μM, applied for 13 min) antagonized the suppression of AMPA responses by SNAP (25 μM, for 8 min). In (B) cGMP was applied into the recording neuron by diffusion from the patch pipette containing cGMP (0.2–0.5 mM) and IBMX (0.1 mM). AMPA responses were gradually depressed and this suppression persisted for at least 80 min. Records from 3 neurons are shown.

Nitric oxide is involved in cerebellar long-term depression

Daisuke Okada

Laboratory for Synaptic Functions, Frontier Research Program, The Institute of Physical and Chemical Research, Wako, Saitama 351-01, Japan.
okadadai@postman.riken.go.jp

Abstract: The involvement of nitric oxide in cerebellar long-term depression is supported by the observation that nitric oxide is released by climbing fiber stimulation and by pharmacological tool usage. Two forms of long-term depression – one depending on large calcium increase and another on nitric oxide – should be distinguished by their physiological relevance. [CRÉPEL et al.; LINDEN; VINCENT]

Introduction. One of the controversies reviewed by CRÉPEL et al., LINDEN, and VINCENT is the involvement of nitric oxide (NO) in cerebellar long-term depression (LTD). Several laboratories have reported NO involvement (Crépel & Jaillard 1990; Ito & Karachot 1990; Lev-Ram et al. 1995; Shibuki & Okada 1991), whereas others found NO-independent depression (Hemart et al. 1995; Linden & Conner 1992). Before drawing any conclusions, we must look more precisely into where NO is generated and why certain forms of LTD require NO whereas others do not.

Source of NO. Applying an electrochemical NO probe (Shibuki 1990) to rat cerebellar slices, Shibuki & Okada (1991) observed that white matter stimulation evoked NO release; this was abolished by hemoglobin and N$^{\omega}$-monomethyl-L-arginine as well as by TTX, lowering extracellular calcium, and glutamate receptor antagonists, showing a need for glutamatergic transmission. The NO release was abolished in slices prepared from rats whose inferior olive had been lesioned with 3-acetylpyridine (Shibuki & Okada 1991), indicating the necessity of intact climbing fiber (CF) for this NO release. Involvement of CF activity in cerebellar cyclic GMP (cGMP) elevation has been reported (Biggio & Guidotti 1976; Southam & Garthwaite 1991). CF activity may explain the extraordinarily high basal cGMP levels in the cerebellum (de Vente et al. 1989). These data strongly support the idea that CF activity releases NO, though the responsible cell has not yet been identified. Neuronal NO synthase (nNOS) was found in the granule and basket cells but not in Purkinje cells and CF. Since NO release is observed in slices, a circuit passing through deep cerebellar nuclei, as proposed by VINCENT is unlikely. The following possibilities could explain CF-dependent NO release. First, an NO-producing enzymic activity other than nNOS such as cytochrome P450 (Boucher et al. 1992) might exist in Purkinje cells. Second, CF activity could increase extracellular K^+ which then depolarizes basket cells or Bergman glia (Murphy et al. 1993) containing nNOS. Finally, CF collaterals connected to basket cells have been confirmed electrophysiologically (Bloedel & Roberts 1970) and could activate NOS in basket cells.

Since LTD is induced by costimulation of CF and parallel fibers (PF), we have to know which fiber generates NO in LTD induction. One might argue for PF (Lev-Ram et al. 1995), because PF is closer to the postsynaptic spine, the putative LTD locus. However, in general, the diffusion distance of NO should depend on the duration and density of nNOS activity and concentrations of scavengers. Strong inputs may generate great amounts of NO which may affect farther locus. Indeed, Wood & Garthwaite (1994) estimated that NO can diffuse as far as 100 μm, hence CF activity cannot be excluded as an NO source in LTD.

Involvement of NO in LTD induction. Glaum et al. (1992) reported that sodium nitroprusside (SNP) did not affect PF EPSP amplitude at 100 μM and suggested that it may be toxic at 1–3 mM. However, 3 mM SNP did not cause significant depression when applied alone, but when paired with PF stimulation it evoked LTD (Shibuki & Okada 1991). Instability of NO under an oxygenated environment (Stamler et al. 1992) might account for the ineffectiveness of 100 μM SNP in Glaum's experiment. In my hands, when SNP was dissolved in deoxygenized buffer at high concentration and a small aliquot added to oxygenated buffer containing slices, concentrations above 1 μM significantly and dose-dependently increased cGMP concentration in cerebellar slices, in agreement with Luo & Vincent (1994). The effect of 30 μM SNP was comparable with that of 1 μM AMPA (100 pmol cGMP/mg protein; Okada, unpublished observations). Thus, our observation that SNP paired with PF stimulation evoked LTD (Shibuki & Okada 1991) strongly supports the involvement of NO in LTD. Other pharmacological tools should also be used with care. L^{ω}-nitro-L-arginine specifically inhibits nNOS in cerebellar slice, but only if preincubated for at least 15–20 min (Okada 1992). Evidence to support the involvement of endogenous carbon monoxide (CO) in cGMP formation (Nathanson et al. 1995) is so far circumstantial at best, because zinc protoporphyrin-9 is on no account a specific inhibitor of CO generation (Ignarro et al. 1984; Meffert et al. 1994).

PF activation paired with calcium spikes evoked NO-dependent LTD (LTD_{PF}; Crépel & Jaillard 1990). On the other hand, Daniel et al. (1992) showed that a metabotropic glutamate receptor (mGluR) agonist paired with calcium spikes evoked LTD-like depression (LTD_{mGLU}) that required calcium from intracellular stores but not NO (Hemart et al. 1995). Linden & Conner (1992) also showed that mGluR activation paired with a 4-min depolarization evoked LTD-like depression independent of NO. K^+-induced depolarization evoked LTD-like depression that depended on two separate pathways involving protein kinase C and NO, respectively (Crépel et al. 1994). Thus, it is likely that two independent mechanisms can induce LTD-like depression. LTD_{mGLU} involved intracellular calcium release mechanisms which are calcium-dependent, therefore calcium spikes will cause a relatively greater calcium level in the LTD_{mGLU} paradigm. One might also expect a large calcium increase by Linden's protocol. On the other hand, although strong PF stimulation (over 20 mV of PF-EPSP; Eilers et al. 1995) can result in calcium increase in the Purkinje cell dendrite, the lower intensities of PF input usually used in LTD experiments may increase calcium far less effectively. LTD_{mGlu} and LTD_{PF} may well occur at different calcium levels within spines. Intracellular calcium regulates various calcium-dependent enzymes whose threshold calcium concentrations differ. Even a subtle difference in calcium level may promote different signalling cascades. Thus, if a large calcium rise occurs, LTD may take place without NO or bypass NO, whereas at physiologically relevant stimulus conditions accompanied by a smaller calcium change, LTD requires NO. It is therefore crucial to estimate the calcium rise in the spine by coincident activation of a single granule cell and an inferior olivary neuron.

ACKNOWLEDGMENT
I am grateful to Dr. Nick A. Hartell for discussion.

The cerebellum and cerebral cortex: Contrasting and converging contributions to spatial navigation and memory

Shane M. O'Mara
Department of Psychology, University of Dublin, Trinity College, Dublin 2, Ireland. **smomara@mail.tcd.ie**

Abstract: Thach's target article presents a remarkable overview and integration of animal and human studies on the functions of the cerebellum and makes clear theoretical predictions for both the normal operation of the cerebellum and for the effects of cerebellar lesions in the mature human. Commentary is provided on three areas, namely, spatial navigation, implicit learning, and cerebellar agenesis to elicit further development of the themes already present in Thach's paper. [THACH]

THACH integrates a very diverse literature on the cognitive functions of the cerebellum into a remarkable theoretical synthesis. I would like to extend the scope of the review to elicit comment on three further areas: first, to spatial navigation and navigational strategies and their interaction with brain areas which control and execute motor movements; second, to the area of implicit learning in patients with anterograde amnesia; third, to the question of motor control in cerebellar agenesis patients.

(1) Spatial navigation. Several areas of the brain are involved in spatial navigation (the coherent and directed motion of the body through space): these include the hippocampus (O'Mara 1995; O'Mara et al. 1994), the vestibular system (Glasauer et al. 1994; Wiener & Berthoz 1993) and certain cortical areas such as the parieto-vestibular insular cortex in monkeys (Grusser et al. 1990), presumed vestibular cortex in humans (Grusser & Landis 1991), and parietal cortex in rats (McNaughton et al. 1994). It is interesting that much of the connectivity of the cerebellum with higher cortical areas cited by THACH seems to be mirrored by connections from the hippocampal formation; these areas include prefrontal cortex, suggesting that there should be some information integration in structures that are connected in common to the cerebellum and hippocampus. These possible interactions are suggestive of possible mechanisms for spatial planning and spatial navigation on the one hand, and actual locomotion on the other.

Navigation requires at least two elements: first, spatial planning, which involves deciding where and when to go to particular places and second, the actual execution of intended trajectories. The former involves a number of the areas cited above and the latter requires the interactions of these areas with motor control areas which must include the cerebellum. How does this interaction occur? No current theory of spatial navigation deals with this interaction in any meaningful way. At least one theory (local view theory, McNaughton & Nadel 1990) in order to account for the place-specific firing of hippocampal neurons in freely moving rats (see O'Mara, 1995, for review), presumes that the hippocampus is involved in the recall of motor sequences, a view which seems to supplant the functions of the cerebellum because it requires the hippocampus to encode motor sequences. Specifically, McNaughton & Nadel (1990) suggest that "a location is nothing more than a set or constellation of sensory/perceptual experiences, joined to others by specific movements" (p. 366). This view predicts that hippocampal place cells should reflect motor sequence information, which is probably not correct, as the overwhelming evidence indicates that hippocampal place cells are behavior-independent (O'Mara 1995). A tidier scheme might have the hippocampus interacting with higher-order motor areas and prefontal planning areas which in turn interact with the cerebellum.

The view that the head approximates an inertial guidance platform for locomotion and navigation is becoming increasingly accepted (Wiener & Berthoz 1993). The head maintains a stable posture during a wide range of movements such as running, walking, jumping, and so on (Berthoz & Pozzo 1988). In a sense, the body is almost hung from the head in order to make contact with a substrate and this contact in turn allows locomotion to occur. The rhythmic motor patterns of walking must then be engaged and coordinated with the navigational goal. How might this occur?

(2) Implicit learning in patients with anterograde amnesia. A distinction is commonly drawn in memory research between implicit (procedural) and explicit (declarative) memory (Gaffan 1992; Rolls & O'Mara 1993; Squire 1992; Squire & Zola-Morgan 1991). The former is nonconscious, there is a lengthy time before asymptotic performance, and it is not explicitly verbalizable; the latter is conscious, and verbalizable information is readily accessed and learned in at most a few trials. Explicit learning depends on an intact hippocampus, whereas implicit learning does not. The prototypical implicit learning tasks all seem to be motor tasks of one sort or another (Squire 1992; riding a bicycle; pursuit rotor learning; mirror writing; simple classical defense conditioning). A large literature demonstrates the importance of the integrity of cerebellar circuits in simple classical conditioning; presumably lesions of the cerebellum result in large deficits in the acquisition of these other motor tasks as well. Is the deficit, however, simply a performance impairment that is evident under these conditions, with deficits in implicit learning also likely to result from damage to the motor circuits of the basal ganglia? Amnesic patients who have acquired specific motor skills are usually not aware that they have acquired such motor skills but they clearly demonstrate on testing that they have them. **THACH** provides the complementary example (sects. 6.3 and 6.6) of patients with cerebellar damage who have difficulties in motor skill acquisition; these patients should differ from amnesic patients in being aware of having encountered such tasks previously and remembering their poor performance on these tasks.

Crick and Koch (1995) have predicted that humans are aware of activity in any brain area if, and only if, there is a projection from that area to the frontal lobes (including prefrontal cortex). Their hypothesis refers primarily to visual processing but it presumably generalizes to other types of neuronal information processing; the cerebellar projections to prefrontal cortex cited by **THACH** appear to limit the general applicability of their theory, for otherwise amnesics would be aware of their procedural motor skill learning. Are humans aware of cerebellar activity?

(3) Motor control in cerebellar agenesis patients. Comparatively few cerebellar agenesis patients have come to light and it appears that those few may have or do have a rudimentary or vestigial cerebellum (Glickstein 1994). The critical importance of the cerebellum for a variety of motor and verbal tasks is obvious from the few such cases that are available. Contrary to popular thinking, "people born without a cerebellum are slow to walk, to talk and probably remain very clumsy throughout their lives" (Glickstein 1994, p. 1211). Only those individuals presenting a deficit are likely to come to neurological attention, however, so there is an element of selection bias in the cases that have been presented to date. **THACH** presents a series of predictions for human cerebellar lesion studies (sects. 4.2 and 4.4) and contrasts these with the effects of prefrontal lesions (cf. Gaffan & Harrison 1989). How are cerebellar agenesis patients likely to differ from patients who sustain lesions either during the course of postnatal development or as adults? Presumably these patients will have developed at least some compensatory strategies to cope with their deficits; perhaps they also find cerebellar agenesis easier to cope with than do patients who sustain lesions later in life.

Cerebellar theory out of control

Michael G. Paulin
Department of Zoology and Centre for Neuroscience, University of Otago, New Zealand. **mpaulin@otago.ac.nz**

Abstract: The views of Houk et al., Smith, and Thach on the role of cerebellum in movement control differ substantially, but all three are flawed by the false reasoning that because information passes from the cerebellum to movements the cerebellum must be a movement controller, or a part of one. The divergent and less than compelling ideas expressed by these leading cerebellar theorists epitomize the fruitlessness of this paradigm, and signal the need for a change. [**HOUK et al.; SMITH; THACH**]

1. Introduction. The function of cerebellum has not been established. It is not necessary for movements but appears useful for moving quickly and accurately. This does not imply that cerebellum regulates motor output. By analogy, information passing through the windscreen is not necessary for moving a vehicle but is useful for moving it quickly and accurately. It could be said that windscreens are for motor control because they facilitate motor control. This would be trivial, but not merely so. Characterizing a windscreen's function like this would raise a barrier to understanding its broader role.

2. APG? After reminding us that it is not clear what the cerebellum does, **HOUK et al.** state that it regulates premotor networks. They postulate that elemental movement commands are generated by positive feedback in distributed neural loops regulated by Purkinje cells. Their APG (adjustable pattern generator) model appears consistent with overall anatomy and physiology, but their argument seems to be that it is possible to imagine that it might be true, and they believe that it is. It is instructive to contrast this with an engineering approach. Kalman has explained how to solve the simplest problem addressed by control theory, designing a feedback regulator. He comments: "Even in this simplest case, the 'canonical wiring diagram' of the regulator is much too complicated to be intuitively understandable" (Kalman 1969, p. 66). Analyzing the simplest artificial control systems is beyond the imagination of the individual who developed the theory used to design them. **HOUK et al.** claim to have analyzed neural control circuitry by inspired guesswork. They might be right, but I am sceptical. Their first piece of reasoning about cerebellar function – why the cerebellum must be a regulator rather than a generator of neural commands (sect. 1) – is patently wrong, even ignoring that there is no reason for considering only these possibilities. Their claim is that the cerebellum must be a regulator because Purkinje cells are inhibitory. But cerebellar output is mediated through excitatory nuclei with reciprocal connections to other structures. The effect of inhibition in these loops could be very different from regulating signals within them, as demonstrated in references which Houk et al. cite in section 2.1, particularly Boyll's (1975) thesis.

Having outlined their model, **HOUK et al.** (sect. 1.1) ask, "Why are the movement-related discharges recorded from [different parts of the loop] so similar to each other?" The answer they suggest is that "elemental commands are generated as a collective computation in this . . . network." Engineering design is instructive again. Usually all parts of an analog control circuit are active simultaneously. Calling this "collective computation" only adds alliteration. Signals in different parts of artificial control systems are similar. This does not mean that the parts have similar functions, for example, the signal transmitted to a feedback regulator specifying the intended trajectory resembles the signal from the plant indicating the actual trajectory. Specifying a servo command and observing plant behavior are different tasks, performed by subsystems with different architectures. Various parts of a regulator contain qualitatively similar signals. Suggesting that similar behavior implies similar function merely illustrates the pitfalls of attempting dynamical systems analysis without adequate theory.

The cerebellum is involved in motor learning. I agree that in order to understand this involvement it is essential to specify precisely what is being learned (sect. 4), but the APG model is not even in the right ballpark. It leads **HOUK et al.** to ask questions which, as I have indicated, can be answered more simply, more clearly, and more generally using simpler, clearer, and more general reasoning. Computer modelling is useful for understanding brain function, however it is important to use it as a tool not only for organizing facts but also for identifying and challenging assumptions. Without a critical component, computer modelling is mere storytelling. While fundamental general questions remain unanswered it might be wise to not even attempt this exercise.

3. Joint stiffness? It is also the belief of SMITH that cerebellar function is understood in broad outline and that it remains only to work out the details, but this seems to entail something different from the task envisaged by HOUK et al. SMITH documents the importance of joint stiffness as a parameter of limb mechanics, then hypothesizes that the cerebellum might control joint stiffness. However, as SMITH notes, there is no evidence to implicate the cerebellum directly in this role, and there is evidence that joint stiffness *per se* is not controlled by the nervous system at all. SMITH claims that if technical difficulties can be overcome then it should be possible to establish whether cerebellum plays "any role" in controlling joint stiffness. However, because joint stiffness is coupled to other mechanical parameters and state variables of limbs it must vary during movements. SMITH does not explain how to disentangle the predictions of his hypothesis from the consequences of any cerebellar involvement in movement control.

4. Movement initiator? For HOUK et al., evidence that the cerebellum is a regulator rather than an initiator of movements is so compelling that it requires no argument, but **THACH** argues from the same evidence to the opposite conclusion. The key evidence for THACH is that cerebellum appears to lie upstream rather than downstream from other structures involved in movement control. Having committed himself to a hierarchical view (sect. 1.1), in which if A sends a message to B and then B does something, then A is controlling B, he concludes that cerebellum initiates movements (sect. 7.1).

If B normally acts after receiving a message from A, then it would seem that A provides information which B normally uses. When A does not transmit this information (correctly) then B may still act but not as quickly or as accurately. For example, cerebellum could participate in observing dynamical states of targets and body parts. Dynamical state information permits rapid, accurate movements, but is not needed for movements *per se.* It is useful for other tasks, therefore it would be a mistake to characterize state estimation as a motor control task. This alternative hypothesis is not merely logically possible, but consistent with evidence indicating cerebellar involvement in other tasks (Paulin 1993). **HOUK et al.**, and **SMITH**, ignore this evidence. Thach massages a little of it painfully into the existing paradigm, but does not consider the possibility that the paradigm is wrong. This possibility does not depend on whether the alternative hypothesis which I have suggested is correct.

5. Conclusion: None of the above. The target articles by **HOUK et al., SMITH,** and **THACH** epitomize the state of the art in cerebellar theory and modelling more than a quarter of a century after *The Cerebellum as a Neuronal Machine* (Eccles et al. 1967). It is a state of logical and conceptual confusion. Here are three different stories about what the cerebellum does and how it does it. The history of science advises us that when scientists fail to reach consensus we should look for errors in a broader consensus already reached. Thus, individually these articles convey little of importance, but collectively they signal the future of cerebellar theory. It is out of control.

Cerebellar rhythms: Exploring another metaphor

Patrick D. Roberts, Gin McCollum, and Jan E. Holly
R. S. Dow Neurological Science Institute, Portland, OR 97209.
proberts@reed.edu

Abstract: The behavior of the climbing fiber system in the cerebellum is viewed in terms of resonances and rhythms. Building upon the anatomical modules in the target article by Simpson et al., the rhythmic behavior of the system is analyzed using a discrete approach. Rhythmic behavior requires oscillations of the olivary cells, but does not necessitate synchrony of complex spike activity. [**HOUK et al.; SIMPSON et al.**]

Scientific understanding must often navigate through tangles of seeming contradictions. A standard example is that of the nature of light. Is light a particle or a wave? In certain situations, light acts clearly as a particle, while in other situations it acts clearly as a wave. By now, physicists have dispensed with being disturbed by the situation and have proceeded to alternate between the particle and wave descriptions when appropriate with ease. It is a well-established fact that light is, in some sense, both a particle and a wave and that this is not a paradox.

As arguments over the role of the cerebellum continue, the resolution will probably not lie in the victory of one group over another, but in the development of novel approaches. The key is that "approaches" is plural. Complex systems have been characterized by the need to use several different approaches to arrive at a complete understanding (Segel 1995). The central nervous system is certainly complex, and only plural viewpoints will guide our path to understanding the functional role of its many components.

To eventually fit the pieces of the cerebellar picture together, the distinction must be made between facts and metaphors. Scientists are proficient in metaphor, without necessarily realizing it. However, metaphors are most useful when the scientist does realize that is what it is. Literal words may include "change," "pause," "correlate" (used with caution), "spike," etc., whereas words used metaphorically include "teach," "learn," "unexpected," "pattern generator," "calculate," and "control." Does an underwater balloon "calculate" the water depth with its volume? Does an individual walking down the street "control" the air next to the body? For the nervous system, many words are used as metaphors to aid the human scientist in forming a conceptual picture.

Care must be taken when using computational metaphors as in the article by **HOUK et al.** The use of artificial neural networks continues to provide a great deal of insight because of the malleability of learning algorithms, but it is this advantage that can be misleading in applications to biological systems. Since artificial networks are able to simulate most nonlinear dynamical systems, simulations provide an irrefutable hypothesis. The behavior of the network may not depend on its architecture; it may be simulating a system that was not imagined by its designers.

Another metaphor suggested by the cerebellar data is resonance. Taking seriously the subthreshold oscillation that is well supported by the olivary cells' internal physiology leads to a metaphor of many oscillators, oscillating at physiological resonance, with a phase reset by stimulation. This cellular resonance approach yields, on an ensemble scale, both rhythmic and "unexpected event" behaviors (McCollum 1995).

On an ensemble level, we have also developed a method to classify the rhythmic behavior of dynamic biological networks (Roberts 1995) that can be used to address questions about the climbing fiber system of the cerebellum. Given the known synaptic connections and cellular properties of the anatomical modules described by **SIMPSON et al.** (see this commentary's Fig. 1A), one may delineate all possible rhythmic patterns for a single module or several modules coupled together. Although individual modules can readily oscillate independently, this behavior is suppressed when modules are linked by gap junction in the olive, or convergence in the cerebellar nuclei.

The results of this investigation are displayed in the Figure 1 (B and C). The rhythms generated by the cellular and the synaptic properties of the circuit form clusters in the space of all possible rhythms. The members of each cluster are superimposed where the nuclear cell sets the beginning of each phase (details of this method are given in Roberts 1995). Suitable time courses are assigned to each mechanism and the overall time of the cycle is normalized. The depth of each cell's activity is represented by how dark that sector of the phase is colored as the scale to the right of each phase diagram in the figure shows. Darker shading during any part of the cycle indicates that more reinforcing mechanisms are available to drive a particular cell's activity. Thus, one may interpret the shading as an indicator of the probability of cellular activity during each point of the cycle.

In Figure 1B, there is only one cluster of 32 rhythms generated

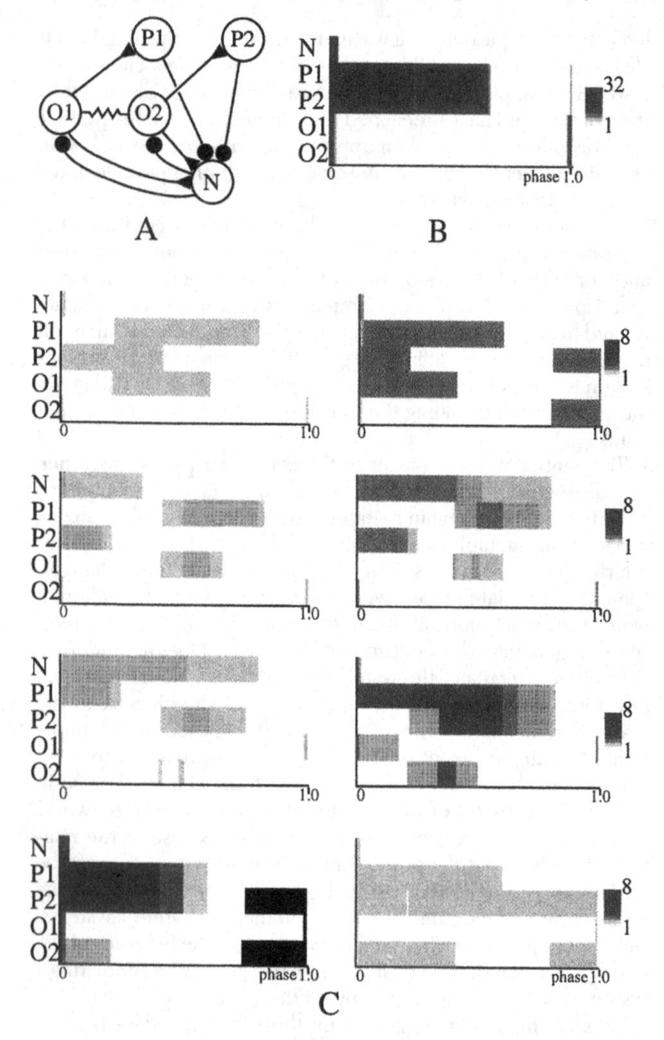

Figure 1 (Roberts et al.). Phases of neuronal activity in cerebellar anatomical modules. (A) Circuit diagram of two anatomical modules. Abbreviations: O1 and O2 are two olivary cells, P1 and P2 are two Purkinje cells, and N is a cell in the cerebellar nuclei. Synapses: Filled circles are inhibitory synapses, triangles are excitatory synapses, and the zig-zag (resistor) is a gap junction. (B) Phase diagram for superposition of rhythms in the cluster generated by circuit in A where the olivary cells are oscillating. (C) Phase diagram for superposition of rhythms in the 8 clusters generated by the circuit without the gap junction between the two olivary cells. See text for details.

by the circuit, with the added assumption that the olivary cells are oscillating. Without such oscillations, no rhythms are generated. These results suggest that the circuitry of the climbing fiber system can maintain rhythms if the neurons of the inferior olive are oscillating. The source of olivary oscillations is not addressed, but the rhythms favor synchrony of complex spike activity along parasagittal strips.

SIMPSON et al. report that there is no correlation between rhythmicity of complex spike activity and synchronous firing, so we repeated the analysis without the gap junction between the olivary cells and show the results in Figure 1C. In this case there are 24 rhythms that fall into 8 clusters. Clearly, strict adherence to synchrony is broken as only one cluster of 8 rhythms shows the two Purkinje cell firing complex spikes together. It should be emphasized that since a network of five cells can have 362,880 possible rhythms, rhythmic behavior of the cerebellar modules is quite weak. This may explain why the phenomenon has been so elusive to experimental probes.

Since rhythmic behavior and synchrony appear to be inherent in the anatomical modules, it is difficult to draw conclusions about their functional significance. The behavior may simply reflect a property of the local circuitry rather than serving a global organizational function. Until a greater consensus about the function of the climbing fiber system is achieved, there is little that can be said with assurance about this aspect of the system. Such a consensus can come about by recognizing the complementary aspects of different approaches and metaphors, rather than focusing on their conflicts.

Dysmetria of thought: Correlations and conundrums in the relationship between the cerebellum, learning, and cognitive processing

Jeremy D. Schmahmann
Department of Neurology, Massachusetts General Hospital and Harvard Medical School, Boston, MA 02114.
schmahmann@helix.mgh.harvard.edu

Abstract: This commentary concerns the cerebellar contribution to cognition. It addresses the relevant historical background, elaborates upon the associative and paralimbic incorporation into both feedforward and feedback limbs of the cerebrocerebellar system, and relates this to the newly described cerebellar cognitive-affective syndrome. It then addresses the degree to which available anatomic data support current theories regarding nonmotor learning in the cerebellum. [**HOUK et al.; THACH**]

Historical background. Contemporary methods in cognitive neuroscience, and particularly functional imaging, have helped validate and provide direction to the new hypotheses concerning the putative cerebellar contribution to cognitive processing (Botez et al. 1989; Leiner et al. 1986; 1993; Schmahmann 1991; Schmahmann & Pandya 1987; 1989; Thompson 1988). It is valuable to recognize, however, that these concepts did not arise *de novo* in the last decade. Clinicians in the last century were intrigued by the association of cerebellar diseases and changes in mental activity (for references and review see Dow & Moruzzi 1958 and Schmahmann 1991), and investigators throughout this century have explicitly postulated a role for the cerebellum in sensory, autonomic, and behavioral phenomena (for example, Abbie 1934; Berman et al. 1978; Frick 1982; Heath 1977; Reis et al. 1973; Snider 1950; Watson 1978; Zanchetti & Zoccolini 1954). Interested researchers may find useful insights in these writings.

The feedforward system: The corticopontine pathway. Thach draws on cerebrocerebellar connectivity studies to support the postulated relationship between the cerebellum and higher function. The feedforward limb of this circuit, and particularly the corticopontine pathway, deserves further emphasis and is worth reviewing briefly. Recent investigations have demonstrated that the anatomic and functional heterogeneity of the behaviorally relevant associative and paralimbic cortices are reflected in topographically organized projections to the basilar pons. Pontine connections emanate from posterior parietal areas critical for directed attention, visual-spatial analysis, and vigilance in the contralateral hemispace (Glickstein et al. 1985; May & Anderson 1986; Schmahmann & Pandya 1989); from temporal and occipital association areas in the dorsal visual stream concerned with the periphery of the visual field and object motion (Tusa & Ungerleider 1988; Fries 1990; Schmahmann & Pandya 1991; 1993); from cortex in the multimodal upper bank of the superior temporal sulcus known to be important for tasks such as face recognition (Schmahmann & Pandya 1991); and from auditory association areas in the superior temporal gyrus and supratemporal plane (Glickstein et al. 1985; Schmahmann & Pandya 1991). Prefrontal cortex, critical for a number of complex cognitive operations, provides substantial efferents to the pons arising in area 8 in the arcuate concavity, area 46 around the principal sulcus, and area 10 at the frontal pole, and from the medial prefrontal cortex (areas 9 and 32), with a small contribution from area 45 considered to be homologous to the language area in humans (Glickstein et al. 1985; Künzle & Akert 1977; Schmahmann & Pandya 1995). Paralimbic structures projecting to the pons include the cingulate gyrus concerned with motivation and affect (Vilensky & Van Hoesen 1981), and the posterior parahippocampal gyrus implicated in spatial memory (Schmahmann & Pandya 1993). The pons receives subcortical input from deep layers of the superior colliculus (Harting 1977) which are implicated in attention, and from the medial mammillary bodies (Aas & Brodal 1988) which play a role in new learning, and there are direct connections between the cerebellum and the hypothalamus, locus coeruleus, raphe nuclei, and brainstem dopaminergic systems (Dietrichs 1984; Marcinkiewicz et al. 1989; Snider 1975; Snider et al. 1976). These associative and paralimbic corticopontine projections are distributed amongst unique constellations of neuronal groups dispersed throughout characteristic sets of pontine nuclei, and they are conveyed to the cerebellum via the pontocerebellar pathway. This detailed anatomy has yet to be matched by correspondingly precise physiology, but these studies reveal that the cerebellum receives information that is critical for cognitive processing and is thereby afforded the opportunity to modulate that information.

The feedback system. Thach's discussion of the cerebello-cerebral feedback focuses on the dentate nucleus communication with the frontal lobe. Thalamic projections from cerebellum are derived from the interpositus and fastigial nuclei as well as from the dentate. Additionally the "motor" thalamic nuclei project not only to motor related areas and the prefrontal periarcuate cortex, but also to the superior temporal polymodal region and to the posterior parietal cortex. Furthermore, traditionally "nonmotor" nuclei (intralaminar, paralaminar medial dorsal), which also receive cerebral input, project to the parietal, temporal, and prefrontal association areas as well as to the cingulate and parahippocampal gyri (Schmahmann 1994). Direct anatomic confirmation of these previously identified pathways is now available with respect to area 46 in the prefrontal cortex (Middleton & Strick 1994), but further direct anatomic and physiologic confirmation is awaited for other higher order areas. It does appear, though, that the associative and paralimbic cerebral areas that project to the cerebellum via the pons may in turn receive projections back from the cerebellum in a rather tightly linked feedforward/feedback system.

Theoretical relevance of these pathways. The interpretation we have given to these connections is different from that discussed by THACH in that it focuses not only on the feedback limb of the cerebrocerebellar circuit, but on the feedforward limb as well. Our conceptual approach holds that the cerebellum modifies behaviorally relevant information that it has received from the cerebral cortex via the corticopontine pathway, and it then redistributes this now "cerebellar-processed" information back to the cerebral hemispheres. In this manner the cerebellum does more than influence prefrontal cortical areas, but rather it is an integral component of the distributed neural circuitry subserving multiple domains of cognitive processing.

Clinical relevance. Based on these anatomic observations in the nonhuman primate, and drawing on the analogy of overt motor deficits observed following certain types of cerebellar injury, we have proposed the concept of "dysmetria of thought" arising from lesions of the cerebellum (Schmahmann 1991). By this we mean that mental processes generated in the absence of intact cerebellar modulation are imperfectly conceived, erratically monitored, and poorly formed. These hypotheses have found corroboration in functional neuroimaging studies (e.g., Jenkins et al. 1994; Parsons et al. 1995; Peterson et al. 1988; and many of the studies referenced by THACH), and also in the clinical context as described in our recent report of a newly defined "cerebellar cognitive-affective syndrome" (Sherman & Schmahmann 1995). This syndrome, documented in twenty two patients to date, is characterized by deficits in affect including either blunting or inappropriateness, varying levels of attention, difficulty with planning, strategy formation, and alternating sequences, perseveration, diminished verbal fluency, visuospatial disorganization, and impairments of visuospatial memory. We hypothesized further that there is a topography of cognition within the cerebellum (Schmahmann 1991) by virtue of the organization of the cerebrocerebellar circuitry and the results of behavioral studies (e.g., Berman et al. 1978; Heath 1977; Reis et al. 1973; Zanchetti & Zoccolini 1954). In this schema the flocculonodular lobe, vermis, and fastigial and globose nuclei may be considered the equivalent of the limbic cerebellum, regulating autonomic responses, emotion and affectively important memory; whereas the lateral cerebellar hemispheres and the dentate and emboliform nuclei may be concerned with the modulation of thinking, planning, strategy formation, spatial and temporal parameters, learning, and linguistic processing. This proposed topographic organization of behavior within the cerebellum has yet to be demonstrated at a clinical level, and a larger series of patients will be required for this purpose.

Does the Marr-Albus model hold for nonmotor learning? With respect to the "specific role" of the cerebellum in modulating nonmotor behaviors, THACH proposes learning to be the underlying mechanism. This raises a problem that is related to the afferents to the red nucleus. A central feature of the Marr (1969) – Albus (1971) theory of motor learning is the interaction between the climbing fiber and mossy fiber systems derived from the olive and pons respectively, as discussed by both THACH and **HOUK et al.** The cerebral afferents of these two systems are markedly different. In the non-human primate, the inferior olive receives most of its descending input from the parvicellular red nucleus, the afferents of which in turn are derived most heavily from motor, premotor and supplementary motor cortices, and to some extent from the postcentral gyrus and area 5 in the superior parietal lobule, but not to any convincing degree (at least in studies to date) from the associative or paralimbic cortices (Humphrey et al. 1984; Kennedy et al. 1986; Kuypers & Lawrence 1967; Saint-Cyr & Courville 1980). Archambault (1914) reported rubral connections with the infratemporal cortices in humans, but this improbable pathway has not been confirmed, and cannot reliably be used in the current discussion. The zona incerta, which receives some projections from prefrontal cortices (Kuypers & Lawrence 1967; Shammah-Lagnado et al. 1985), projects to the inferior olive (Cintas et al. 1980; Saint-Cyr & Courville 1980), so there may be some indirect prefrontal input to the olivary system. Nevertheless, it appears that these two systems are quite different; the red nucleus – inferior olive system seems to be predominantly motor, whereas the pons (as discussed above) is relevant for motor and sensory as well as for associative and paralimbic information. Given the need for climbing fiber – mossy fiber interaction as the basis for context dependent learning, this raises the question of whether the red nucleus/olive can be involved in cognitive processing, and whether it is reasonable to consider learning to be a basis for the cerebellar contribution to cognition? It is conceivable, perhaps, that cerebellar learning may invoke the rubro-olivary system when there is a motor efferent, but not when the learning is purely cognitive. The issue remains open, however, in that there is still the possibility of a demonstration of associative projections to red nucleus in humans, and there is the loophole of the zona incerta.

In sum. There is a wealth of information still unavailable regarding the specific role of the cerebellum in cognition. Nevertheless the field of cognitive neuroscience as applied to the cerebellum is no longer merely emerging, but appears to have come of age.

How to link the specificity of cerebellar anatomy to motor learning?

Fahad Sultan,[a] Detlef Heck,[b] and Harold Bekkering[c]
[a]*Division of Biology, California Institute of Technology, Pasadena, CA 91101;* [b]*Max-Planck-Institut fuer biologische Kybernetik, 72076 Tuebingen, Germany;* [c]*Max-Planck-Institut fuer psychologische Forschung, 80802 Munich, Germany.* **kbekker@mpipf-muenchen.mpg.de**

Abstract: The target article by Thach outlines the Brindley-Marr-Albus-Gilbert (BMAG) learning theory and extends its validity to cognitive processes. We provide here an alternative explanation for the positron emission tomography (PET) data cited by Thach in support of the BMAG model. We also comment on the anatomical and physiological basis of the BMAG model and Houk's adjustable pattern generator (APG), both models being based on similar assumptions. [HOUK et al.; THACH]

In contrast to many other brain structures the cerebellar cortex is highly anisotropic (Braitenberg & Atwood 1958). Two functional interpretations of this strikingly regular structure came out in the fifties and sixties. In the first one the cerebellar cortex was interpreted as a structure universally transforming time into space and vice versa, with the parallel fibers being interpreted as delay lines (Braitenberg 1961; Braitenberg & Atwood 1958). According to the second interpretation, the anisotropic structure emerges as an epiphenomenon of the fact that a single Purkinje cell has 150,000 synapses (rat Purkinje cell, Napper & Harvey 1988) and that so many synapses require a special way of packaging (Albus 1971; Marr 1969). According to the latter model, and also to the BMAG/APG models, the operation performed by the Purkinje cell is to recognize certain activity patterns in its parallel fiber input. The cell would respond to only a small fraction of all possible patterns, whereby the selection of effective patterns would be a result of learning mediated by climbing fiber input.

One crucial point about the Brindley-Marr-Albus-Gilbert model (BMAG) and **HOUK et al.**'s adjustable pattern generator (APG) is the necessity of sparse coding and the assumption that a suitable mechanism for data reduction at the input would be provided by the Golgi cells. Several observations lead us to be cautious about the assumption of sparse coding: (1) A given rosette (the mossy fibers presynaptic specialization) may establish synapses with up to 50 granule cell dendrites in the rat (Jakab & Hàmori 1988), that is, there is a clear anatomical possibility of activating a large number of granule cells simultaneously. Also, a given granule cell's dendrite makes multiple synapses with the same mossy fiber rosette (Jakab & Hàmori 1988), which could ensure a reliable transmission of mossy fiber activity: (2) There are very few GABA-ergic neurons in the pontine nuclei (rat), that is, the massive cortico-pontine input to the cerebellum is barely inhibited locally (Brodal et al. 1988): (3) Electrophysiological recordings reveal pronounced activity in the granular layer after somatosensory stimulation (Bower & Woolston 1983; Shambes et al. 1978), which was most likely of postsynaptic origin, that is, originating from the granule cell dendrites. Thus, delays in a negative feedback system and summation of activities in the molecular layer could bypass the Golgi cells inhibition and lead to the simultaneous activation of a large number of parallel fiber/Purkinje cell synapses and thus hamper the prerequisites for sparse coding.

A major criticism of the first or the "delay line" hypothesis is that parallel fibers are much shorter than assumed at the time (Braitenberg & Atwood 1958) and as a consequence the delays which could be produced would be an order of magnitude too short to be of any use in motor control. In spite of this problem, the notion of the cerebellum being a timing device is still an issue of current debate and various experimental efforts (e.g., Hore et al. 1991; Keele & Ivry 1991).

The so-called "tidal wave" theory of the cerebellar cortex (Braitenberg 1983) is a logical extension of the "delay line" hypothesis, although tidal wave theory works with the actual parallel fiber length. In essence it says that sequential activation of granule cells with the right order in time and space will result in a tidal wave of spikes traveling in the parallel fiber system. In this scenario the parallel fibers are still interpreted as delay lines and the Purkinje cells as sensitive coincidence detectors. The parallel fibers transform sequential mossy fiber input into synchronous input to Purkinje cells. The charm of this idea is that it gives a convincing functional explanation for the peculiar anatomical arrangement of parallel fibers and Purkinje cells: the parallel fibers run in parallel in order to preserve the relative timing of action potentials they conduct. The flat and fan-shaped dendritic tree of Purkinje cells enables these cells to detect "tidal waves" of spikes traveling in the parallel fiber system as synchronous and thus most effective inputs. In a set of experiments the responses of the cerebellar cortex *in vitro* to sequential input have been measured and all theoretical predictions tested found clear experimental confirmation (Heck 1993; 1995).

The assumption that the cerebellar cortex is a sequence detector, reacting mainly to dynamic rather than static mossy fiber inputs, offers a new perspective which lets us think differently about the recent PET studies on the cerebellum.

THACH argues that the BMAG model could explain the PET data, showing increased activity in the prefrontal and cerebellar cortex during the learning phase and decreasing again as the task becomes more automated. We would like to suggest an explanation different from **THACH**'s: an increase in the activity of the fronto-ponto-cerebellar mossy fiber system during the learning phase leads to the PET findings and reflects the search for the optimal sequence of muscle activation (i.e., motor program) to accomplish the task. We assume that neo-cortical cell assemblies represent single elements of a motor program and that a motor program thus consists of a number of consecutively activated cell assemblies (i.e., a "synfire chain" as proposed by Abeles 1991). During the learning phase many different assembly-sequences would be produced in a trial and error fashion, leading to increased neo-cortical activity. When, after several trials, the "best" sequence is found, neo-cortical and cerebellar activity is much less because of the much smaller number of neurons and mossy fibers involved in the generation of the movement. The "best" sequence would produce one or several locally restricted "tidal waves" and thus activate only a few Purkinje cells, with little net effect on the overall cerebellar nucleo-olivary inhibition.

In sum, although the BMAG/APG models seems to provide a sufficient explanation for some of the peculiarities of the cerebellum, the computation suggested could just as well be realized in a neo-cortex-like structure. Several anatomical characteristics of the cerebellar cortex we consider as keys to cerebellar function go unexplained, such as that the parallel fibers run in parallel or that the dendrites of Purkinje cells are flat and fan-shaped. The idea, however, that the cerebellum analyses spatio-temporal activity patterns in the mossy fiber input, that is, the tidal wave theory, offers an explanation for the peculiar cerebellar anatomy. Because of this, the investigation of dynamic processes is certainly a most fruitful approach to the understanding of the cerebellum.

We know a lot about the cerebellum, but do we know what motor learning is?

Stephan P. Swinnen,[a] Charles B. Walter,[b] and Natalia Dounskaia[a,c]
[a]*Motor Control Lab, Department of Kinesiology, Catholic University of Leuven, Tervuurse Vest 101, 3001 Heverlee, Belgium;* [b]*Motor Control Lab, School of Kinesiology, University of Illinois at Chicago, 901 W. Roosevelt Road, Chicago, IL 60608;* [c]*Institute of Control Sciences, Russian Academy of Science, Moscow.* **stephan.swinnen@flok.kuleuven.ac.be**

Abstract: In the behavioral literature on human movement, a distinction is made between the learning of parameters and the learning of new movement forms or topologies. Whereas the target articles by Thach, Smith, and Houk et al. provide evidence for cerebellar involvement in parametrization learning and adaptation, the evidence in favor of its involvement in the generation of new movement patterns is less straightforward. A case is made for focusing more attention on the latter issue in the future. This would directly help to bridge the gap between current neurophysiological approaches to the role of the cerebellum and the behavioral expressions of human motor learning. [**HOUK et al.; SMITH; THACH**]

THACH, SMITH and **HOUK et al.** are to be commended for their efforts to come to grips with the question of what motor learning is before advancing evidence in favor of the idea that the cerebellum is a critical substrate for motor learning. Even though there may be implicit agreement among these authors with respect to the most general characterization of motor learning, defined as a set processes associated with practice that results in relatively permanent changes in behavior, there appears to be a general lack of agreement concerning the concrete operationalization of this concept within the context of diverse experimental paradigms. Some physiologists and neurobiologists have studied motor learning in the framework of the gain or adaptation of reflex patterns within the larger context of classical conditioning (for examples of eyeblink and other discrete responses, see **THACH, SMITH,** and **HOUK** et al.). Others have pointed to the acquisition of novel acts, requiring the building of new synergies (**THACH**, sect. 6.7, **SMITH**, sect. 1.1) and this is at the heart of the present commentary.

Within the behavioral literature, a distinction has emerged in the past decade between metrical and structural features of movement. This has important implications for views of motor learning (Newell 1985; Schmidt 1988). Metrical aspects refer to the scaling of movement without modifying its structural features, that is, the heart of the movement pattern. For example, one can reach for a nearby cup or for a pencil located at the edge of the table using the same movement pattern but with different parameter specifications. Scaling is thus required to comply with (relatively minor) changes in the task context. Across these particular variations of movement from trial to trial, the structural aspects are argued to remain invariant (within boundaries).

For various reasons, past behavioral studies have predominantly focused on changes in scaling with practice, as in the production of a movement with a particular movement time or amplitude. This only requires the learner to acquire the appropriate application of a scaling factor to an already existing movement form or topology. Evidence advanced by **THACH, SMITH,** and **HOUK et al.** suggests a cerebellar involvement in parameter learning. However, learning is far richer and more elaborate when a new movement form or topology has to be acquired in addition to movement scaling, and this creation or building of new movement patterns has become the focus of more recent behavioral studies. In this respect, it has been argued that learning does not occur *de novo* but should be considered against the background of preferred coordination modes or synergies that are part of the preexisting motor repertoire of the performer (Swinnen et al. 1993; Zanone & Kelso 1992). Such preexisting synergies may just as well refer to reflex patterns (e.g., the tonic neck reflex) as to the production of voluntary coordination patterns, such as the "natural" tendency to produce mirror-image movements in the upper limbs (Swinnen et al. 1991) or to generate isodirectional movements in the upper and lower limbs in the sagittal plane (Swinnen et al. 1995). Accordingly, learning not only requires the building of a new coordination pattern but may also involve the suppression of preexisting coordination modes in case these do not converge with the new modes. The aforementioned behavioral biases often give rise to persisting errors during skill acquisition (Walter & Swinnen 1994). It is conceivable that inhibitory processes play a dominant role in this respect. For example, when trying to produce bimanual movements with different spatiotemporal features, a general tendency emerges to synchronize both limb motions initially, that is, the mirror-image symmetry constraint is very powerful and needs to be overcome to build new movement synergies (Swinnen et al. 1991). The study of the building of novel coordination patterns is currently a dominant theme in behavioral approaches to motor learning.

The latter form of learning is most reminiscent of the changes that occur when acquiring new skills in dance, gymnastics, athletics, and so on, which require the building of new action patterns that give rise to simultaneous and sequential movements of limbs and body segments. It is also the type of learning for which a cerebellar involvement has not yet been established convincingly. Nevertheless, recent work by Welsh et al. (1995) suggests that the cerebellum may also be critically involved in this more elaborate form of motor learning. Their studies with animals indicate that motor coordination requires the integrity of the inferior olive which is a major cerebellar afferent. This neural substrate is suggested as a principal locus for the generation of new movement synergies. As Welsh et al. (1995, p. 456) state: "Dynamic rearrangement of electronic coupling within the olivary nucleus, controlled by the deep cerebellar nuclei, may permit neuronal clusters representing different muscles to be selectively coupled and to fire synchronously when those muscles must be simultaneously engaged during movement." The deterioration of interlimb coordination, as observed in Lurcher mice which suffer from cerebellar cortical atrophy (see Smith, sect. 2.2), indirectly supports this contention. Additional work along these lines may eventually provide convincing evidence that an intact cerebellum is indispensable for building new movement synergies against the backdrop of preexisting coordination modes and to help consolidate these synergies across longer time periods.

Even though the aforementioned evidence in favor of a cerebellar involvement in sculpturing new synergies appears promising, a process as complex and multilateral as motor learning undoubtedly requires the intact and concerted operation of many brain structures and this has been stated explicitly by **THACH, SMITH,** and **HOUK et al.** As a consequence, reductionistic approaches bear limitations. It is reasonable to assume that motor learning and the memories it lays down are not both consolidated within one structure. Rather, these functions are widely distributed in the central nervous system. As a result, damage to subcortical as well as cortical structures will result in a deterioration of these functions. For example, Parkinsonian patients show considerable difficulties in forming new motor synergies, even though augmented visual information feedback aids performance (Verschueren et al. 1995). On the other hand, Asanuma (1989) has argued for a considerable involvement of higher cortical structures in shaping new movement patterns. Accordingly, attempts to model the convergent operations of the subcortical and cortical brain structures during the sculpturing of new motor synergies within the context of skill acquisition should receive more attention in the future.

Motor learning and synaptic plasticity in the cerebellum

Richard F. Thompson

Neuroscience Program, University of Southern California, Los Angeles, CA 90089-2520. **thompson@neuro.usc.edu**

Abstract: For reasons I have never understood, some students of the cerebellum have been unwilling to accept the now overwhelming evidence that the cerebellum exhibits lasting synaptic plasticity and plays an essential role in some forms of learning and memory. With a few exceptions (e.g., target article by **SIMPSON et al.**) this is no longer the case, as is clear in the excellent target articles on cerebellar LTD and the excellent target review by **HOUK et al.** [**CRÉPEL et al.; HOUK et al.; KANO; LINDEN; SIMPSON et al.; SMITH; VINCENT**]

The target articles I am reviewing fall into two groups, one concerned with processes of synaptic plasticity and the other, at the systems level, concerned with cerebellar functions and learning. I begin with the papers by **LINDEN, CRÉPEL et al., VINCENT,** and **KANO** on mechanisms of synaptic plasticity. The discovery of long-term depression (LTD) in cerebellar cortex by Ito and associates is one of the major findings in modern neuroscience. Initially, there was some controversy, but thanks to the extensive and elegant studies by Ito and associates and work from many other laboratories, the phenomenon is now firmly established. The original observations were that contiguous stimulation of mossy/parallel fibers and climbing fibers resulted in a pronounced and prolonged depression of the activated parallel fiber synapses on Purkinje neuron dendrites.

In a series of already classic studies, Linden and Conner made extraordinarily effective use of embryonic tissue culture of mouse cerebellar cortex to study mechanisms of LTD. Functionally, this culture system resembles *in vitro* (slice) and *in vivo* cerebellar cortex in several ways. The results of these studies and many others using the *in vitro* slice are in close agreement in identifying the conditions necessary to induce LTD in Purkinje neurons. In brief, three processes appear necessary for normal induction of cerebellar LTD: Ca influx through voltage gated channels, Na influx through AMPA receptor-associated channels or voltage gated Na channels, and (less certain) activation of the metabotropic glutamate receptor mGluR1, this last leading to activation of protein kinase C (PKC). In the cerebellar slice or *in vivo*, Ca influx is produced by climbing fiber activation (or depolarization) and activation of AMPA and metabotropic receptors by glutamate released at parallel fiber synapses (or applied) (see also **CRÉPEL et al.** and **VINCENT**).

Specific contributions by these papers include the following. **LINDEN:** LTD induction is input specific, appears due in part to spatially constrained activation of PKC in Purkinje neuron dendrites and is therefore post-synaptic. **CRÉPEL et al.:** using the cerebellar slice preparation and inducing Ca influx by direct depolarization of Purkinje neurons, two different mechanisms for LTD induction appeared to be the case, one dependent primarily on AMPA receptors and the other on strong activation of the metabotropic receptors, involving the PKC activation that leads ultimately to changes in the functional characteristics of AMPA receptors at the parallel fiber-Purkinje cell synapses. **VINCENT:** examines the hypothesis that NO might play a key role in induction of LTD. He concludes that Purkinje neurons do not generate NO whereas granule and basket cells do (in response to NMDA activation). He finds that this increased NO release is associated with a large increase in extracellular cyclic GMP, which also occurs in response to activation of AMPA and metabotropic glutamate receptors (probably involving Purkinje neurons), suggesting a role for cGMP as an intercellular messenger in cerebellar cortex. He concludes that NO cannot play an essential role in the induction of LTD in cerebellar cortex. **KANO:** describes a new form of plasticity in cerebellar cortex. In brief, repeated activation of climbing fibers induces a "rebound potentiation" of inhibitory potentials in Purkinje neurons (whole-cell patch-clamp recording). This would result in prolonged depression of Purkinje neurons due to increased inhibition (IPSCs measured). He presents evidence suggesting that the mechanism involves up-regulation of Purkinje neuron $CABA_A$ receptors by PKA via phosphorylation of (perhaps) the receptor protein. As he notes, potentiation of inhibition has been reported in other systems as well, for example, on Mauthner cells in goldfish (Korn et al. 1992) and IPSCs of deep cerebellar nuclei neurons (Morishita & Sastry 1993). Note that Kano's rebound activation of inhibition would be synergistic with LTD in Purkinje neurons.

These four papers are all most impressive cellular analyses of mechanisms of experience (i.e., activation) induced changes in processes of cellular plasticity in cerebellar cortex. There is general agreement by all these authors on basic common mechanisms and a refreshing lack of controversy. One has the impression of an important and exciting field making rapid progress. My only complaint about these papers is the authors' unwillingness to speculate about the possible functional/behavioral implications of their findings (see Ito 1989; Linden 1994; Thompson & Krupa 1994 for such discussions).

The target article by **SMITH** is interesting and somewhat puzzling. He focuses on the role of the cerebellum in motor learning but defines motor learning in a rather idiosyncratic manner, for example, "it adjusts joint and limb mechanics" a definition that would seem very narrow in that it excludes a number of learned behaviors, such as, head turn, eyelid closure, even limb flexion, that are clearly cerebellar dependent. It is puzzling that he discusses the possible role of the cerebellum in conditioned reflexes (sect. 2.5) and ignores virtually all the relevant literature in this field.

SMITH's focus in the article is on possible cerebellar involvement in learning optimal time – varying control of joint stiffness. His review of this literature is comprehensive and he builds a reasonable case, with appropriate cautions. The Purkinje neuron and nuclear neuron recordings he shows are for already-trained animals (except for Fig. 6). It is not clear in his presentation how typical these unit responses are. Without some knowledge of the proportion of neurons he studied in each paradigm that responded as in the examples, and the proportion of neurons that showed differing responses, and the regions from which they are recorded, it is not possible to draw any meaningful conclusions. Perhaps he can provide this information in his response.

The title of **SIMPSON et al.**'s target article ought to be "The function of climbing fibers in the context of the visual climbing fiber input to the rabbit's vestibulocerebellum." By this extreme narrowing of the topic they are able to exclude much of the relevant evidence "on climbing fibers and their consequences." To their credit they do cite the important studies by Houk and associates (Gellman et al. 1983; 1985) showing that in the awake cat complex spike responses to a passively applied stimulus generally failed to occur when a similar stimulus was produced by a voluntary movement.

Part of the problem appears to be that **SIMPSON et al.** are quite unfamiliar with the broad field of learning and memory and the possible importance of the climbing fiber system in this context. In the case of classical conditioning of discrete responses (eyeblink, limb flexion) appropriate lesions of the inferior olive made before training completely prevent learning and if made after training result in extinction and abolition of the learned response in the face of continued training (McCormick et al. 1985; Voneida et al. 1990; Yeo et al. 1986). Further, electrical neurostimulation of the DAO can serve as a very effective US; the exact movement so elicited can be trained to any neutral stimulus (Mauk et al. 1986; Steinmetz et al. 1989). To our knowledge this is the only system in the brain, other than reflex afferents, that can serve as an effective US for the learning of discrete responses. Thus, these findings assume considerable importance in the broad context of brain substrates of learning and memory. Finally, in Purkinje neuron recordings in such learning paradigms, the aversive US consistently evokes complex spikes at the beginning of training but not in trained animal (Foy & Thompson 1986). In an important study, Sears and Steinmetz (1991) recorded unit activity from neurons in the effective region of the DAO (as determined above by lesion and simulation). The CS (tone) did not influence unit activity at all. US onset evoked a burst of spikes at the beginning of training but this decreased to no response in trained animals (but the same evoked response on US alone trials). All these lines of evidence argue very strongly that in classical conditioning of discrete responses the DAO-climbing fibers system serves as the US reinforcing or teaching input for learning. We submit that in the context of learning, this is the only instance where the reinforcing system has been fully identified and as such is important.

Part of the problem seems to be that some students of the cerebellum (e.g., **SIMPSON**) feel that a structure like the cerebellum or a system like the climbing fibers can have only one function. It would be very surprising if this were the case. Evolution has a way of making economical use of what is available to solve new problems.

Finally, the target article by **HOUK et al.** is superb. It provides an extensive, detailed and balanced review of the broad field concerned with models of the cerebellum and motor learning. This paper is a most refreshing example of an objective and unbiased review of the field (and they do not always agree with me).

Limitations of PET and lesion studies in defining the role of the human cerebellum in motor learning

D. Timmann and H. C. Diener
Department of Neurology, University of Essen, 45122 Essen, Germany.
tnk0b0@sp2.power.uni-essen.de

Abstract: PET studies using classical conditioning paradigms are reported. It is emphasized that PET studies show "cerebellar involvement" and not "specific function" in learning paradigms. The importance of dissociating motor performance and learning deficits in human lesions studies is demonstrated in two exemplary studies. The different role of the cerebellum in adaptation of postural reflexes and learning of complex voluntary arm movements is discussed. [THACH]

1. Human PET studies. A number of PET studies, which have been reviewed in great detail in THACH's target article, provide evidence that the cerebellum is involved in motor learning. More recent PET studies demonstrated involvement of the human cerebellum in classical conditioning of the eyeblink reflex using bolus injection of $H_2{}^{15}O$ (Molchan et al. 1994) or 2-[^{18}F] flourodeoxyglucose (FDG) (Logan & Grafton 1995). In these studies cerebellar activity prior conditioning was compared with cerebellar activity after acquisition of eyeblink conditioning using subtraction analysis. A recent study from our laboratory analyzed changes in rCBF during the process of flexion reflex conditioning using correlation analysis (Timmann et al. 1995a). Regional cerebral blood flow of each scan was correlated with the incidence of conditioned responses (CR) of the corresponding scan. The correlation analysis showed that rCBF in the ipsilateral cerebellar cortex correlated with the process of classical conditioning (= CR incidence). The changes in the cerebellar cortical activity might well reflect increased climbing fiber activity during task acquisition (Gilbert & Thach 1977). These results agree with previous PET studies investigating learning sequences of single finger movements which demonstrated most extensive activation of the cerebellum during task acquisition (Friston et al. 1992; Jenkins et al. 1994). However, it has to be emphasized that PET studies themselves do not solve the question if the specific role of the cerebellum in classical conditioning is the storage of engrams of the learned behavior ("learning hypothesis," Thompson & Krupa 1994) or the capacity to execute conditioned and unconditioned responses properly ("performance hypothesis," Bloedel & Bracha 1995).

Furthermore – as pointed out in THACH's target article – the question of the functional role of the cerebellum during acquisition of the learned behavior is not resolved by human PET studies themselves. In fact, different theories of cerebellar function in motor performance and learning based on animal data and the unique cerebellar anatomical structure can be used for interpretation of similar PET data. For example, the changes in rCBF activity may reflect the monitoring of peripheral events (Kolb et al. 1987). Thach suggested that "cerebellar pathways are used to build through trial and error learning behavioral context-response linkages, and to build up appropriate responses from simpler constitutive elements." Bower (1995) proposed that the main function of the cerebellum is to control the acquisition of sensory information so that the highest quality sensory data is available for computations performed by the rest of the nervous system.

In conclusion, PET studies provide increasing evidence that the cerebellum is involved in motor learning among other neural structures. However, the question of the specific role of the cerebellum in motor learning cannot be answered by PET studies themselves.

2. Human lesion studies. Motor learning paradigms involve motor performance. Therefore, differences in performance cannot be taken as a demonstration of a learning deficit. For example, in the mirror drawing study of Sanes et al. (1990) the findings show that cerebellar patients can dramatically improve their performance. Nevertheless, the authors concluded that the patient group had a substantial learning impairment based on intergroup comparisons with normal subjects. In contrast, Timmann et al. (1994) demonstrated that cerebellar patients could improve their performance in a series of two-dimensional tracing tasks. Cerebellar subjects were capable of improving substantially their performance of a complex motor task involving the recall of memorized shapes and the visuomotor control of a tracing movement. However, these observations did not indicate that the cerebellum was uninvolved in task acquisition. The study demonstrated clear differences between the patients and control subjects in their percent improvement.

Horak and Diener (1994) have previously reported that patients with cerebellar deficits were unable to scale the magnitude of their early automatic postural responses to the predicted amplitudes of surface translations based on central set from prior experience. However, findings of a more recent study showed that both control and cerebellar subjects overresponded when they expected, on the basis of prior experience, a larger displacement than they actually received and underresponded when they expected a smaller displacement than they received (Timmann & Horak 1995b). The most hypermetric patients showed the least amount of amplitude scaling but were not affected in their ability to predict displacement amplitudes based on prior experience. The large trial-by-trial response magnitude variability in cerebellar patients did not allow for significant scaling when only 5–7 sequential trials were provided. These results clearly demonstrate that the underlying cause of predictive scaling deficits in cerebellar patients is due to difficulty in precisely modifying response gain based on set and not disability to develop predictions based on prior experience.

In fact, the cerebellum's main role in automatic postural responses may be gain control. A recent study demonstrated that latencies and relative spatial-temporal pattern of postural responses to surface displacement are preserved in patients with cerebellar deficits. The ability to use on-line velocity feedback to scale the magnitude of early automatic postural responses was preserved in patients with cerebellar dysfunction (Horak & Diener 1994). Thus, the spatio-temporal organization of automatic postural synergies may be more hardwired by brainstem or spinal mechanisms.

In conclusion, the majority of human lesion studies provide no clear evidence that the role of the cerebellum in motor learning goes beyond motor performance. In particular, the cerebellar role in adaptation of automatic postural responses might well be restricted to gain control and hence much more limited compared to voluntary arm movements.

ACKNOWLEDGMENTS

Parts of this work were supported by grants from the Deutsche Forschungsgemeinschaft (Ti 239/1-1, Ti 239/2-1) to Timmann and (Di 327/6-1) to Diener.

Sensorimotor learning in structures "upstream" from the cerebellum

Paul van Donkelaar
University Laboratory of Physiology, Oxford OX1 3PT, United Kingdom.
paul.vandonkelaar@physiol.ox.ac.uk

Abstract: Thach, following the Marr-Albus theory, suggests that the cerebellum contributes to sensorimotor learning through a process of context-response linkage. However, perceptual learning studies demonstrate that the context itself is subject to adaptive modifications and recent evidence has shown that motor responses derived from such perceptions display appropriate changes which may be quite independent of cerebellar processing. [HOUK et al.; SMITH; THACH]

THACH and several of the other authors in this issue (**HOUK et al.**; SMITH) rightly suggest that the cerebellum plays a role in the learning of new motor skills and adaptive modifications observed in previously-learned responses. Indeed, as reviewed by THACH there is extensive evidence to implicate the cerebellum in this process. However, there is also a growing body of literature which demonstrates changes associated with sensorimotor learning and adaptation in other areas of the CNS, some of which can be classified as "upstream" from and potentially independent of any modulatory influence from the cerebellum. For example, neurons in the extrastriate cortical areas concerned with visual motion processing have recently been shown to display a remarkable degree of plasticity (Zohary et al. 1994). In particular, when repeatedly exposed to their preferred direction of visual motion in a random dot display, such cells become more sensitive to this direction, firing at higher frequencies than normal. This change in neuronal sensitivity is accompanied by analogous improvements in perceptual sensitivity: the monkey can make correct psychophysical judgments about this direction of movement at lower thresholds of motion coherence. A similar improvement in perceptual sensitivity is observed in human subjects exposed to the same conditions (Ball & Sekular 1987).

One could hypothesize that any type of motor output that uses information derived from the processing occurring in these or similar cells would also be affected by such changes. We recently tested this notion with respect to target speed by adaptively modifying the output of the ocular smooth pursuit system (van Donkelaar et al. 1994). This is accomplished by electronically adding a percentage of the eye motion signal to the movement of the target as the subject attempts pursuit (Carl & Gellman 1986). After repeated exposure to such target motion, the subject increases the output of the smooth pursuit system for a given amount of retinal image motion. In other words, the same amount of movement of the target image on the retina elicits a higher velocity smooth pursuit response than before the adaptation.

Now, the mechanism(s) underlying this change could reside at various sensory and/or motor levels in the smooth pursuit system, including within the cerebellum. However, evidence that at least some of the change occurs at a sensory level comes from the fact that the gain of open-loop manual tracking movements performed while the subject visually fixates is also increased following the smooth pursuit adaptation (van Donkelaar et al. 1994). Because the eyes are fixating one must conclude that the changes in the manual responses are due to a misperception of target motion induced by the adaptation procedure: the target appears to move faster than it actually does after the smooth pursuit system has been adapted. As noted above, changes in the perception of motion are thought to be the result of modulations in neuronal activity within the cortical motion processing areas (Tootell et al. 1995). Therefore, it is likely that similar sensory changes underlie the modifications we observed following smooth pursuit adaptation.

Thus, unlike the cerebellar-dependent motor adaptation that THACH and colleagues have studied, our task is an example of sensorimotor learning that is at least partially based on modifications in areas of the CNS that are "upstream" from the cerebellum. In theory, the learning that we have observed could bypass this structure altogether. The extrastriate motion processing areas are part of the dorsal visual processing stream which sends corticocortical projections to frontal motor areas of the cortex (Cavada & Goldman-Rakic 1993). At some of these latter sites there are cells which appear to combine limb and eye movement information (Boussaoud 1995; Gottlieb et al. 1994), and, thus, would be ideally suited to influence the motor output of each of these systems following adaptation. One prediction arising from this hypothesis is that subjects with cerebellar dysfunction should also display increases in smooth pursuit and manual tracking gain after exposure to the adaptive procedure (although these may be partially hidden by performance deficits). We hope to test this prediction in the near future.

Not all sensorimotor learning may be dependent on the cerebellum and the structures it subsequently innervates. When the changes in motor output are partly based on sensory modifications, as appears to occur during smooth pursuit adaptation, the cerebellum may not necessarily be directly involved. Furthermore, this implies that not all sensorimotor learning is the result of new linkages between context and response, as suggested by **THACH.** Rather, the context itself (in our case, the perception of target motion) may be altered during the learning process. The reason that the cerebellum has enjoyed a central role in the controversy surrounding sensorimotor learning is because the tasks that have been used have emphasized the motor side of the process. However, sensory processes obviously make significant contributions to our ability to make accurate movements in everyday life. Thus, in addition to the cerebellum, primary, supplementary, and premotor cortex, basal ganglia, red nucleus, and prefrontal cortex I would add to the list of CNS structures involved in sensorimotor learning those areas that initially process the sensory information on which subsequent motor output is based.

What behavioral benefit does stiffness control have? An elaboration of Smith's proposal

Gerard P. Van Galen, Angelique W. Hendriks, and Willem P. DeJong
Nijmegen Institute for Cognition and Information (NICI), P.O. Box 9104, 6500 HE Nijmegen, The Netherlands. **vangalen@nici.kun.nl; hendriks @nici.kun.nl; dejong@nici.kun.nl**

Abstract: It is argued that stiffness is, as Smith assumes, a likely movement-control parameter for the brain, as is supported by the results of several behavioral experiments. However, without further elaboration of what specific type and implementation of stiffness is meant it is perhaps premature to speculate on its possible locus of control. [SMITH]

In terms of classical mechanics, stiffness is an inherent parameter of any (bio)mechanical system and any constellation of forces is equivalent to a unique stiffness value. Some might argue, therefore, that asking for the role of stiffness control or the locus of its control center(s) is an unnecessary question because stiffness is an inherent implication of the recruited force level. **SMITH,** like us, apparently argues differently, although his definition of stiffness as a behavioral parameter is rather underspecified. Stiffness is indeed a crucial parameter of behavior and it is, together with other biomechanical parameters like dampening, viscosity, and limb inertia are too often neglected in behavioral neuroscience. In our opinion, the concept of stiffness should be better defined and linked up with more elaborate models of movement control before the full role of the cerebellum in this can be assessed.

SMITH does not make any explicit statement about what he means by limb stiffness but from the article at least two variants – and two possible agents for the role of the cerebellum – are to be recognized. One is *phasic* stiffness, which is defined by the sum of the (ant-)agonistic forces acting over time, and the other is *static* stiffness, which is defined by the background cocontraction of agonists and antagonists. An implicit assumption is that the possible role of the Purkinje cells of the cerebellum is to inhibit either the static cocontraction or the phasic contraction of the antagonist during agonist activity. Although we do not deny or have evidence *against* this idea, we would like to propose an additional and – in part – alternative role for stiffness, which may possibly help to clarify the role of the cerebellum. The hypothesis we propose is that stiffness control is an effective means of *accuracy control.* Especially in biophysical and neurological literature, movement control has usually been considered as a displacement task only. In most human movement, however, accuracy control or even static stability is of much greater importance. SMITH, who apparently works in this tradition, touches upon stability control and its possible relation to stiffness, but he never explicitly mentions accuracy control as a possible objective for the cerebellum.

In contrast, the behavioral sciences have reserved a much greater role for accuracy control (Fitts 1954), but they have nearly completely overlooked the constraints of the physical set-up of the human moving limb and its neural implementation. Because we have pursued such a role in the near past, we give a summary of this research with the intention to help formulate more specific hypotheses about the cerebellum's role in movement control.

Van Galen and colleagues (Van Galen & Schomaker 1992; Van Galen et al. 1990) recorded human aiming movements to targets of different sizes. Apart from measuring usual kinematics, they also calculated the degree of noisiness of the movement's velocity signal by applying power spectral density analysis to the raw movement signal (which was recorded by a high resolution digitizer tablet). Their findings suggest that more precise movements were much more heavily filtered from the very outset of the movement, which argues against the feedback control hypothesis and which supports the view that higher degrees of static stiffness were applied during the movement. The authors proposed that cocontraction might be the "filtering" agent to control movement accuracy. In a more direct test of the hypothesis that cocontraction is an instrument of the brain's accuracy control. Enright and Hendriks (1994) measured eye-ball retraction in viewing conditions of varying accuracy demands. The experiment revealed that scrutinizing a visual stimulus leads to a significant retraction of the eyeball which cannot be attributed to accommodation, suggesting that the orientation of the eye is stabilized by simultaneous contraction of the antagonistic muscles. Such cocontraction can be expected to alter the responsiveness of the whole system to changes in motorneuron activation.

We modeled the role of stiffness and, more particularly, the role of background cocontraction by implementing a formal model for the recruitment of muscle force into a biomechanical model of the human arm (Van Galen & DeJong 1995). Essential aspects of this model are (1) that muscle force is inherently noisy because of the stochastic nature of the recruitment process, and (2) that limb stiffness, as defined by the background cocontraction regime, has a filtering, error-reducing effect on the endpoint accuracy of movement. In the model, the cocontraction regime is supposed to be independently set by the brain, which might be the specific contribution of the Purkinje cells. The error reducing capacities of cocontraction are demonstrated in Figure 1 for a set of movements of varying lengths and degrees of cocontraction.

Conclusion. Stiffness is a likely control parameter for the brain. However, before attempts are made to definitively associate the locus of control with either the cerebellum or the motor cortex and its sensory projecting areas, we should have a better specified model for the *role* of stiffness in movement control in general. From behavioral experiments it is concluded that cocontraction, if studied as a measure independent from the overall driving force regime, is an effective means for the brain to provide control over movement endpoint accuracy without the necessity of resorting to on-line feedback control, as is traditionally assumed.

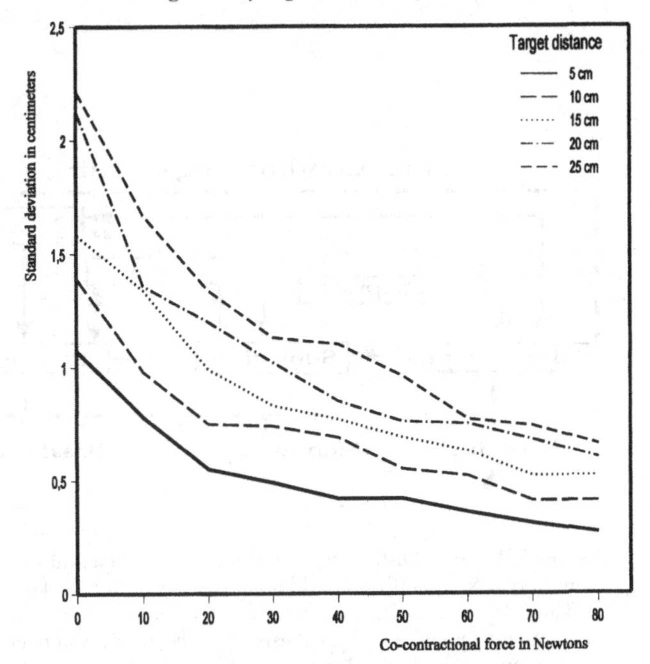

Figure 1 (G. P. Van Galen et al.). The potential ability of cocontraction to improve the accuracy of movement. Force recruitment was simulated by the model of Van Galen & DeJong (1995). The standard deviations of the simulated movement endpoints decrease with level of cocontraction and increase with target distance.

Eyeblink conditioning, motor control, and the analysis of limbic-cerebellar interactions

Craig Weiss and John F. Disterhoft

Department of Cell & Molecular Biology, Northwestern University, Chicago, IL 60611. **cweiss@nwu.edu**

Abstract: Several target articles in this *BBS* special issue address the topic of cerebellar and olivary functions, especially as they pertain to motor learning. Another important topic is the neural interaction between the limbic system and the cerebellum during associative learning. In this commentary we present some of our data on olivo-cerebellar and limbic-cerebellar interactions during eyeblink conditioning. [**HOUK et al.; SIMPSON et al.; THACH**]

Several target articles in this *BBS* special issue address cerebellar and olivary functions in motor learning. The locus of the essential associative site(s) for motor learning is a particularly interesting question to us. Work from several laboratories suggest that this site is within the cerebellum (cortex and/or deep nuclei) for simple conditioning (Thompson 1990, for review), and that the cerebellum and limbic system are required for more demanding conditioning paradigms, e.g., trace conditioning and discrimination reversal (Berger & Orr 1983; Moyer et al. 1990). Other lines of evidence suggest that the cerebellum modulates associative sites within the brainstem [**HOUK et al.**]. Continuing progress on these questions would yield a better understanding of the functional role of the cerebellum and hence of motor control and motor learning in general.

The experimental paradigm that we and others are using to investigate the neurobiology of learning and sensorimotor integration is eyeblink conditioning (e.g., Disterhoft et al. 1977). This paradigm was developed initially as a model behavioral system to investigate systematically the laws of associative learning in the human (Gormezano 1966). As mentioned in several of the target articles, the demonstration that eyeblink conditioning critically engages the cerebellum was a major step forward in a more complete appreciation of the role of the cerebellum in associative motor learning.

Another important issue is the neural interaction between the hippocampus and the cerebellum during associative learning. Our experiments using hippocampally-dependent eyeblink conditioning are extending the analysis of motor learning to include the role of the limbic system (Disterhoft et al. 1995). Multiple-unit hippocampal recordings and lesions of the dentate-interpositus nucleus demonstrate an interdependence between alterations in hippocampal neuronal firing and mediation of the conditioned response via the cerebellar deep nuclei (Clark et al. 1984; Sears & Steinmetz 1990). The neural substrate for this interaction includes the hippocampus, subiculum, retrosplenial cortex, and pontine nuclei (Berger & Bassett 1992). Feedback control from the cerebellum to the hippocampus and other forebrain regions is then mediated by the ventral anterior thalamus. This circuitry is indicated in Figure 1.

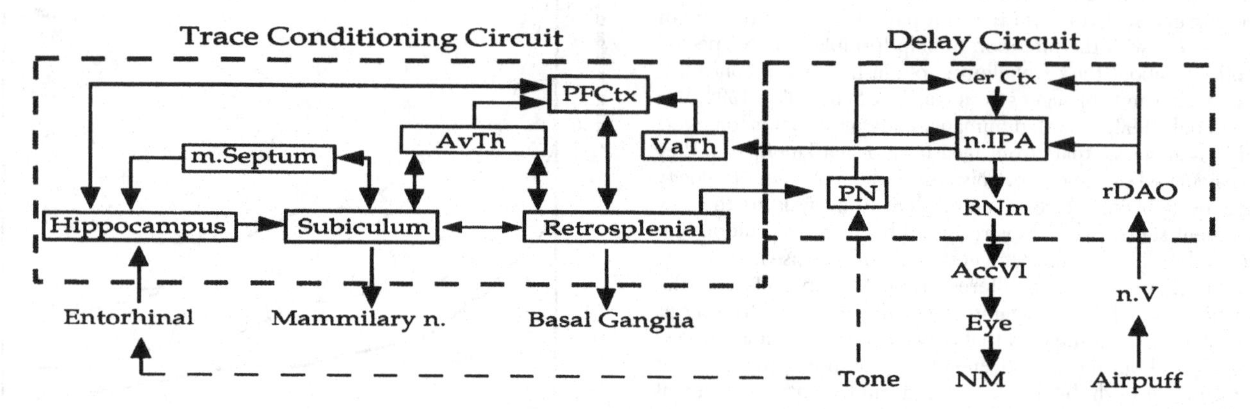

Figure 1 (Weiss & Disterhoft). A block diagram of a limbic-cerebellar circuit that may mediate eyeblink conditioning (adapted from Berger & Bassett 1992 and Thompson 1986). Not all of the connections are indicated, including several discussed by Houk et al. The delay circuit mediates simple forms of conditioning. The trace circuit requires limbic connections which interface with the delay circuit. Our working hypothesis is that the limbic system facilitates the tone input to the cerebellum to permit a sensory-motor association, or conditioned response, during more complex tasks such as trace or discrimination reversal conditioning. AccVI, accessory abducens nucleus; AvTh, ventral part of anterior thalamic nucleus; PN, pontine nuclei; IPA, anterior interpositus nucleus; NM, nictitating membrane; nV, trigeminal nucleus; PFCtx, prefrontal cortex; rDAO, rostral dorsal accessory olive; RNm, magnocellular red nucleus; VaTh, anterior part of ventral thalamic nucleus.

We are carrying out ensemble recordings from simultaneously recorded single hippocampal neurons (Weiss et al. 1996) during and after a hippocampally-dependent version of the trace eyeblink conditioning task (Moyer et al. 1990). The direction and nature of the interdependence between the hippocampus and cerebellum is a particularly interesting area of inquiry. Regardless of the direction of influence between the cerebellum and hippocampus, localized changes in the biophysical properties of hippocampal pyramidal cells are indisputably learning related, even after delay conditioning which does not depend upon the hippocampus for its acquisition (Coulter et al. 1989; Disterhoft et al. 1986).

We have also been using the detailed anatomical and functional information gathered about the eyeblink conditioning circuit in the rabbit to plan and interpret neurobiologically oriented experiments in the human. We demonstrated that amnesic subjects, who have damage to the temporal lobe/hippocampal system, are able to learn the hippocampally-independent delay conditioned eyeblink response at the same rate as control subjects (Gabrieli et al. 1995). The learning was not impaired as was predicted from the rabbit literature (Akase et al. 1989; Berger & Orr 1983; Schmaltz & Theios 1972). In a second study, we have shown that the same delay conditioning task is impaired in amnesic Korsakoff's patients and recovered alcoholics (McGlinchey-Berroth et al. 1995). We assume that this impairment is attributable to the cerebellar deterioration which is a well-documented consequence of alcoholism. A deficit which is correlated with damage to the cerebellum is, of course, what would be anticipated from several of the focus articles [HOUK et al.; SIMPSON et al.; THACH] as well as the rabbit experiments (Thompson 1990, for review). It is also possible that the learning deficit in Korsakoff's patients and alcoholics was due to a malfunction of the limbic system instead of, or in addition to, cerebellar damage (Solomon et al. 1983). This finding is being further examined in ongoing studies.

The most profound and permanent learning deficits have resulted from lesions of the deep cerebellar nuclei. Reports by Thompson and colleagues (1990) and Yeo et al. (1985) have demonstrated permanent abolition of CRs following these lesions. These authors would likely argue that the deficits seen by Welsh et al. (1989) can be explained by the effects of partial lesions. The experiment by Krupa et al. (1993) however, clearly demonstrates that learning did not take place during muscimol inactivation of the deep nuclei (in contrast to THACH, sect. 3.4).

The question of whether or not a cerebellar dependency involves the cerebellar cortex was also discussed by THACH (sect. 3.1). THACH cites Thompson (1990) for cortical ablation studies that have prevented or eliminated CRs. This is surprising since Thompson's group has repeatedly shown functional recovery following cortical lesions. Only the most recent study of Gruart and Yeo (1995) confirmed the necessity of cerebellar cortex for eyeblink conditioning. That study utilized bilateral cerebellar cortex lesions and strengthens the argument for bilateral interactions during learning (e.g., Lavond et al. 1994). These bilateral lesion effects need to be more fully examined before the functional relationship between the cerebellar cortex and deep nuclei is fully understood.

The question of the substrate for learning was also raised by THACH (sect. 3.4), that is, Are new or latent pathways utilized during associative motor learning? The most likely answer is that the substrate is based on latent pathways. A pathway generated *de novo* during learning would be difficult to explain. Conversely, data have been reported which argue for the utilization of existing pathways. Swain et al. (1992) paired a tone CS with electrical stimulation of climbing fibers (CFs) as a US. Those rabbits established CRs that were based upon whatever movement was generated by the climbing fiber stimulation. Similarly, electrical stimulation of the inferior olive (IO) established conditioned responses (CRs) according to whatever movement was evoked by the stimulus when paired with a tone CS (Mauk et al. 1986; Steinmetz et al. 1989). These data suggest that CRs are specified by the somatotopic linkage of CFs through the olivo-cerebellar-rubral system (Gibson et al. 1987).

Functional roles of the IO were discussed by SIMPSON et al. He began with a review of accepted wisdom about the olivocerebellar system. It is interestingly to note that the data upon which current assumptions are based are for the most part approximately 20 years old. In these past 20 years a definitive role for the IO has remained elusive. Progress has been made in regard to the anatomical connections of the IO (e.g., De Zeeuw et al. 1990), but the physiological data are not yet unified by a common functional interpretation.

THACH states that the IO detects and corrects errors in performance (sect. 2.4). **SIMPSON et al.** presents the major arguments for the different proposed functions of the inferior olive in a nonbiased manner that summarizes the data and their limitations. The goal of achieving a consensus among the different theories is admirable. Unfortunately, a consensus for olivary function does not exist, and one was not presented in any of the focus articles.

Progress in the physiological realm has been made by expanding analyses from the single neuron level to the ensemble level (e.g., Llinas & Sasaki 1989). We are also using ensemble analyses to examine the physiology of hippocampal neurons during eyeblink conditioning in the rabbit. This type of analysis should reveal processing that has remained elusive, especially for slowly firing cells such as those within the IO and silent cells within the hippocampus (Deadwyler & Hampson 1995). As indicated by **SIMPSON,** a better understanding of cerebellar function (and of hippocampal function) may depend on an analysis of inhibitory interneurons.

One of the roles proposed for the IO is that of a comparator between intended position and achieved position, or of an error in posture. The exquisite sensitivity of the footpads to external stimuli during rest, but not during movement (Gellman et al. 1985) is partial support for this theory. Receptive fields that are more proximal to the body as well as on the face and head are also exquisitely sensitive in both the awake and anesthetized cat (Gellman et al. 1985; Weiss et al. 1990; 1993). Another potential function of the climbing fiber may be to rapidly change the calcium concentration of Purkinje cell dendrites. The necessity of calcium entry with conjoint stimulation of PFs and CFs during learning (**HOUK et al.,** sect. 4.1, **SIMPSON et al.,** sect. 3.3) is also interesting for the understanding of eyeblink conditioning. Van der Zee et al. (1995) observed bilateral changes in calcium-dependent PKC binding in those regions of cerebellar cortex HVI that have been implicated in EBC.

Another approach to analyze macroscopic changes among regions of the nervous system, as ensemble analyses do within a region, is imaging studies. **THACH** summarizes many of the PET motor learning studies. Eyeblink conditioning has also been observed with PET (Blaxton et al., submitted; Zeffiro et al. 1993). In these studies, young adult humans were trained in the PET scanner during the time periods when the scans were being made. The learning paradigm utilized paired presentations of tones and airpuffs to the eye while the control paradigm presented unpaired tones and puffs. Our data suggest that the neural systems involved in the mediation of eyeblink conditioning in the human are similar to those in the rabbit. Our data are strikingly convergent with what would be anticipated from the PET motor learning studies reviewed by **THACH.**

We observed *increased* activation in structures predicted by the rabbit work, especially in the hippocampus and the cerebellum. In addition, very large and somewhat unexpected changes in the frontal cortex and in the caudate were observed after conditioning. More specifically, and in agreement with the empirical data and theoretical considerations reviewed by **THACH,** we observed that cerebellar cortical changes showed increases which paralleled the behavioral acquisition curve, that is, there were increases in cerebellar blood flow which likely reflected the increase in the ability of humans to perform the eyeblink conditioned response. What was quite striking was the pattern of changes which occurred simultaneously in frontal cortex. The frontal cortex showed a striking enhancement in blood flow during the first block of conditioning which declined as the conditioning proceeded and the eyeblink conditioned response was acquired. The change during the first block of conditioning trials might reflect the registration of the change in contingency from unpaired pseudo-conditioning control trials to the paired conditioning trials. This pattern in frontal cortex was inversely related to the blood flow changes we observed in cerebellum, and of course to the behavioral acquisition, that is, an increase in blood flow which correlated with behavioral learning.

The frontal cortical regions were presumably activated by projection pathways from the cerebellum as demonstrated by Strick (Middleton & Strick 1994; Strick 1994), anticipated by the Leiners (Leiner et al. 1986), and reviewed extensively by **THACH** in his target article. The frontal cortex may then mediate a behavioral "context" that is linked with the cerebellar output. This context would serve to facilitate or rehearse motor responses. We have proposed a similar linkage to account for the hippocampal dependency of trace eyeblink conditioning. In our proposal the hippocampus is required to facilitate the CS input to the cerebellar circuitry to permit a sensory-motor association within Purkinje cells and/or the deep nuclei.

In conclusion, the role of the IO and cerebellum in motor learning and motor control is justifiably still open to debate. The two regions are obviously involved, but the mechanism for this integration is unresolved, as is the mechanism for the interaction between the cerebellar and the limbic systems. The data gathered from several different lines of research have not been fitted together into one unified hypothesis. A fruitful approach to reconciling the data has been to model the system. **HOUK et al.** discuss this issue, and they present a model based on an adjustable pattern generator. Whether or not the model is correct, it provides the valuable service of being testable and modifiable. Models should predict the shape and latency of CRs and the rate at which they develop. A model should also predict the rate and accuracy of VOR adaptation. Modelers, behaviorists, physiologists, and anatomists working in combination should be able to arrive at a unifying theory of cerebellar function which will provide a major step forward in understanding motor control as a whole.

ACKNOWLEDGMENT

This work was supported by NIMH R01MH47340 and NIDA RO1DA07633.

Plasticity of cerebro-cerebellar interactions in patients with cerebellar dysfunction

Karl Wessel

Department of Neurology, Medical University of Luebeck, D-23538 Leubeck, Germany. **wessel@medinf.mu-luebeck.de**

Abstract: Studies comparing movement-related cortical potentials, post-excitatory inhibition after transcranial magnetic brain stimulation, and PET findings in normal controls and patients with cerebellar degeneration demonstrate plasticity of cerebro-cerebellar interactions and hereby support Thach's theory that the cerebellum has the ability to play a role in building behavioral context-response linkages and to build up appropriate responses from simpler constitutive elements. [**THACH**]

The important role of the cerebellum in motor control, motor learning, and cognition is made possible by cerebellar inputs and outputs from side loops of transcortical projections (Allen & Tsukahara 1974; Thach 1970). **THACH** provides evidence from PET and ablation studies that especially in motor learning the cerebellum is but one part of a larger system that includes primary, supplementary, and premotor cortex, basal ganglia, red nuclei, and the prefrontal cortex. Cerebro-cerebellar interactions can also be studied by recording movement-related cortical potentials (MRCPs). Deecke et al. (1976) have proposed that impulse conduction through cortico-cerebellar-cortical loops plays a role, particularly for the parts of the Bereitschaftspotential (BP, readiness potential) immediately preceding the movement onset.

Movement-related cortical potentials (MRCPs) in patients with cerebellar ataxia showed earlier onset and/or depressed amplitude as well as abnormal topographical patterns close to movement onset (Shibasaki et al. 1978; Tarkka et al. 1993). Using sequential and goal-directed motor tasks, which were expected to be more sensitive to cerebellar deficits, and performing a more detailed analysis by differentiating early (NS1; mean amplitude 600–800 msec before movement onset) and late (NS2) components of the potential, we observed even more extensive changes of MRCPs in patients with cerebellar ataxia (Wessel et al. 1994). Peak amplitude of the potential and NS2 amplitude (difference between peak amplitude and amplitude of NS1) were reduced and NS1 amplitude was increased in patients. In addition, the onset of the BP was earlier in the patients than in the controls. The reduced peak amplitude and NS2 could reflect that the primary motor cortex lacks adequate input from the cerebellum in patients.

Using transcranial magnetic brain stimulation we found prolonged postexcitatory inhibition in patients with cerebellar degeneration, which may be a consequence of transient facilitation of cortical inhibitory interneurones, resulting in decreased excitability of primary motor cortex (M1) in patients (Wessel et al. 1996). The increased NS1 and the earlier onset of the BP could be a result of the larger effort through which the patients try to compensate for their motor deficits by extended cortical activation preceding voluntary movements. All these findings are consistent with THACH's idea of context-response linkage, which requires extensive cerebro-cerebellar interaction, as it is demonstrated to occur in the above mentioned studies in patients with cerebellar dysfunction.

Concerning interactions of the cerebellum with prefrontal cortices and the hypothesis of the specific cerebellar contribution to context linkage and shaping of the response through trial and error learning, we can also contribute findings from a PET study in patients with cerebellar degeneration (Wessel et al. 1995). We measured changes in regional cerebral blood flow (rCBF) in relation to a self-paced sequential finger opposition task with the right hand, completing a sequence of movements every 4–6 sec. The reduced activity of cerebellar neurones in the patients produced a complex pattern of rCBF increases and decreases in other brain regions. Our results suggest that for the preparation and execution of sequential finger movements, patients with cerebellar degeneration use a medial premotor system, including the SMA and caudal cingulate motor area (CMA), as well as the M1 and putamen, rather than the ventral premotor area, prefrontal cortex, lobus parietalis inferior, and rostral CMA.

THACH worked out that prefrontal cortex and cerebellum show parallel changes with increased activity during motor and "non-motor" learning and decreased activity during the over-learned or automatic performance. Our study did not specifically address motor learning; regarding adaptation to the movement, control subjects and patients showed no significant difference in rCBF increases during movement between the first and the last (10th) scan. With regard to prefrontal cortices, widespread areas of the frontal lobe contribute to basal ganglia and cerebellar loops (Brooks & Thach 1981).

Since in our study the activation of prefrontal cortex during movement was reduced in patients with cerebellar degeneration, we conclude that activation of prefrontal cortex in relation to arm movements depends partly on cerebellar inputs. Our findings in patients suffering from cerebellar degeneration, using MRCPs, transcranial magnetic stimulation, and PET, demonstrate plasticity of cerebrocerebellar interactions and thereby support THACH's theory that the cerebellum can play a role in building (through trial and error learning) behavioral context-response linkages and can build up appropriate responses from simpler constitutive elements.

Authors' Responses

A cerebellar long-term depression update

David J. Linden

Department of Neuroscience, The Johns Hopkins University School of Medicine, Baltimore, MD 21205. **david.linden@qmail.bs.jhu.edu**

Abstract: Two major themes have emerged in the commentaries elicited by the target articles that concern cerebellar long-term depression (LTD) (CRÉPEL, VINCENT, and LINDEN). First, is a lively debate concerning the potential role of a nitric oxide/cGMP cascade in cerebellar LTD induction. Second is a much broader issue relating to the interchange of information between cerebellar physiologists concerned with mechanisms at a cellular and synaptic level and those working at the level of systems physiology, behavior, or modeling. What contributions can cellular physiologists make to the study of motor learning? Cellular physiologists can provide testable hypotheses to help determine if these synaptic phenomena do underlie particular behaviors (e.g., if cerebellar LTD underlies vestibulo-ocular reflex [VOR] adaptation or eyeblink conditioning, then blockade of cerebellar LTD via, say, mGluR1 inhibition, should interfere with these forms of motor learning). In addition, we can provide descriptive parametric information about basal synaptic function and use-dependent synaptic modifications that can constrain the range of models proposed to underlie a given behavior (e.g., are the timing constraints on LTD induction consistent with VOR adaptation or eyeblink conditioning?).

Writing a target article in 1994 and responding to commentaries in 1996 provides the sensation of engaging in interstellar dialogue with colleagues at the Alpha Centauri Institute of Cerebellar Physiology. My first task is to summarize the major advances that have occurred in the field of cerebellar long-term depression (LTD) since the target article was written, as many of these speak directly to issues raised in the commentaries.

The identification of mGluR1 as the isoform of metabotropic glutamate receptor that contributes to LTD induction, which was originally made using specific inactivating antibodies (Shigemoto et al. 1994), has since been confirmed in two different laboratories using mGluR1 knockout mice (Aiba et al. 1994; Conquet et al. 1994). The impairment of LTD in these mice was originally claimed not to be accompanied by gross alterations in the anatomy of the cerebellum. However, a more recent reexamination by one group has claimed that approximately 25% of Purkinje neurons (PNs) in a mGluR1 mutant mouse are multiply innervated by climbing fibers (CFs) (Chen et al. 1995). Basic properties of evoked parallel fiber (PF) synaptic transmission (sensitivity to AMPA and NMDA receptor antagonists, paired pulse facilitation) were not altered when assessed in brain slice, nor were voltage-gated Ca currents (Aiba et al. 1994). These findings suggest that the impairment of LTD in mGluR1 knockout mice may be attri-

Table 2. Authors' Responses

Distribution of commentators responded to in Authors' Responses

LINDEN: A cerebellar long-term depression update	CRÉPEL: Cellular mechanisms of long-term depression: From consensus to open questions	KANO: A bridge between cerebellar long-term depression and discrete motor-learning: Studies on gene knockout mice
Arbib	**Baudry**	**Arbib**
Bindman		**De Schutter**
De Schutter		**Fiala & Bullock**
Fiala & Bullock		**Houk**
Hartell		**Jaeger**
Hirano		**Thompson**
Houk		
Mori-Okamoto & Okamoto		
Okada		

VINCENT: NO more news from the cerebellum	HOUK & BARTO: More models of the cerebellum	SIMPSON, WYLIE, & DE ZEEUW: More on climbing fiber signals and their consequence(s)
Arbib	**Arbib**	**Arbib**
Baudry	**Bekkering, Heck, & Sultan**	**De Schutter**
Bindman	**Bower**	**Dufossé**
Calabresi, Pisani, & Bernardi	**Dean**	**Feldman & Levin**
Fiala & Bullock	**De Schutter**	**Fiala & Bullock**
Hartell	**Feldman & Levin**	**Hepp**
Hirano	**Fiala & Bullock**	**Jaeger**
Kiedrowski	**Gilbert**	**Kawato**
Mori-Okamoto & Okamoto	**Hepp**	**Miall, Malkmus, & Robertson**
Okada	**Hesslow**	**Roberts, McCollum, & Holly**
	Houk & Alford	**Thompson**
	Jaeger	**Weiss & Disterhoft**
	Kawato	
	Latash & Latash	
	Miall, Malkmus, & Robertson	
	Paulin	
	Schmahmann	
	Sultan, Heck, & Bekkering	
	Swinnen, Walter, & Dounskaia	
	Thompson	
	Weiss & Disterhoft	

SMITH: Resilient cerebellar theory complies with stiff opposition	THACH: Q: Is the cerebellum an adaptive combiner of motor and mental/motor activities? A: Yes, maybe, certainly not, who can say?
Arbib	**Arbib**
Bower	**Bower**
De Schutter	**De Schutter**
Feldman & Levin	**Feldman & Levin**
Gomi	**Fiala & Bullock**
Hore	**Flament & Ebner**
Houk & Alford	**Gielen**
Jaeger	**Gilbert**
Latash & Latash	**Hallett**
Paulin	**Jaeger**
Van Galen, Hendriks, & De Jong	**Latash & Latash**
	Miall, Malkmus, & Robertson
	O'Mara
	Paulin
	Schmahmann
	Sultan, Heck, & Bekkering
	Swinnen, Walter, & Dounskaia
	Timmann & Diener
	van Donkelaar
	Weiss & Disterhoft

buted directly to a deficit in this gene and not to a side effect on some other process known to contribute to LTD induction, such as the operation of voltage-gated Ca channels or AMPA receptors. However, it should be noted that not all of the processes that have been suggested to be necessary for LTD induction have been screened. It would be instructive, for example, to assess properties of the protein kinase C and nitric oxide/cGMP signaling systems in this mutant.

As mGluR1 is coupled via phospholipase C to the production of inositol-1,4,5-trisphosphate (IP3), it has become worthwhile to examine the potential role of this compound in LTD. Unfortunately, specific inhibitors of the IP3 receptor (IP3R) have not yet been developed. Application of heparin, a nonspecific inhibitor of the IP3R, blocked LTD induced by glutamate/depolarization conjunction in cultured PNs (Kasono & Hirano 1995; see also commentary by **Hirano**). Application of thapsigargin, a drug that depletes internal Ca stores through inhibition of the endoplasmic reticulum Ca-ATPase, was similarly effective (Kohda et al. 1995) (although in the slice preparation it was found to block depression induced by bath application of the mGluR agonist ACPD together with depolarization, but not depression induced by PF/depolarization conjunction [Hemart et al. 1995]). Thapsigargin would be expected to deplete Ca stores gated by both the IP3R and the ryanodine receptor, the latter mediating Ca-induced Ca release. As such, it is interesting to note that ryanodine receptor antagonists were also found to block LTD induction in the cell culture preparation (Kohda et al. 1995). A complementary approach to this problem was taken by Kasono and Hirano (1994; 1995), who reported that AMPA pulses will induce LTD when paired with photolytic release of caged Ca loaded into a cultured PN. This analysis was extended to show that photolysis of caged IP3 could induce LTD when paired with both AMPA pulses and depolarization, but that any two of these three signals were not sufficient. These authors suggest a model in which the Ca signals mediated by voltage-gated channels and IP3Rs, possibly boosted by ryanodine receptor mediated Ca-induced Ca release, synergize to contribute to LTD induction. Furthermore, they propose that activation of the other limb of the mGluR1 pathway, production of 1,2-diacylglycerol, is unnecessary for LTD induction. There are several notes of caution that should be sounded about the experiments that support this model. First, heparin has been shown to interfere with G-protein coupling (Khodakhah & Ogden 1993) and thus has the potential to interfere with the linkage between mGluR1 and phospholipase C directly. Second, the authors have shown that photolysis of caged Ca, which occurs throughout the PN, is different than depolarization-induced Ca in terms of its ability to contribute to LTD induction. Analogously, photolysis of caged IP3 cannot be presumed to mimic the spatial pattern of IP3 production driven by synaptic stimulation and might differentially contribute to LTD induction as well. Third, a report using a similar preparation has shown that application of a synthetic diacylglycerol could substitute for mGluR activation when combined with AMPA pulses and depolarization, and that the LTD so induced was specific to the site of diacylglycerol application (Linden 1994). Fourth, while several groups have shown robust Ca responses of PNs to exogenous mGluR agonists, at present, a synaptically driven Ca response mediated by mGluRs has yet to be directly demonstrated (this point is also relevant to the model of **Fiala & Bullock**). The recent development of a knockout mouse that lacks the type I IP3R (Matsumoto et al. 1996), which is the dominant isoform in PNs, may be helpful in addressing these questions.

Until very recently, all models of cerebellar LTD induction contained within them two assumptions: (1) either CF activation or direct PN depolarization sufficient to open voltage-gated Ca channels is necessary for LTD induction, and (2) LTD induction involves, at least in part, an increase in Ca concentration in the dendritic spine head, the compartment underlying the PF synapse. Two recent imaging studies of cerebellar slices have shown that activation of a group of PFs with a stimulating electrode will give rise to transient increases in Ca in dendritic spine heads. One report (Eilers et al. 1995) indicates that Ca increases driven by PF activation are mediated solely by activation of voltage-gated Ca channels as a consequence of AMPA receptor activation. These responses take the form of activated spines together with activated portions of intervening dendritic shaft. Activated spine heads were never seen without activated dendritic shaft, leading these authors to suggest that the Ca response was generated at the level of the dendritic shaft: either the depolarization invaded the shaft resulting in activation of Ca channels in the spine head or Ca ions themselves invaded the spine head. Another report (Denk et al. 1995) suggests that PF-mediated Ca increases can occur from two sources, either activation of Ca-permeable ligand-gated receptors or voltage gated Ca channels. This report, however, finds Ca signals restricted to dendritic spine heads and suggests that Ca flux through these two channels is triggered by ion channels located in the membrane of the spine head itself. In comparing these reports, it should be noted that Denk et al. (1995) used voltage-clamp recording, a manipulation that might be expected to limit the extent of postsynaptic excitation, while Eilers et al. (1995) used current clamp recording in the presence of a $GABA_A$ receptor antagonist, which would tend to promote postsynaptic excitation. In both cases, however, stimulation of PFs in a manner that was subthreshold for activation of dendritic Ca spiking produced an increase in spine head Ca. These observations suggest three main models for the role of CF-mediated Ca signals in LTD induction:

Model 1. Activation of PFs alone results in an increase in Ca in spine heads, but this increase is not sufficient to contribute to LTD induction. Activation of PFs and CFs together results in a larger increase in spine Ca, which is sufficient. This has been the most conventional model, and is suggested by the experiments of Kasono and Hirano (1994; 1995) mentioned above.

Model 2. Activation of CFs and PFs together results in Ca increases in the spine head that are not significantly different in amplitude. The CF contributes a signal to LTD that is dependent upon recruitment of some additional signal other than spine Ca. As CF activation may be replaced by direct PN depolarization, this is unlikely to involve direct release of compounds from the CF terminal or direct activation of interneurons by CFs as suggested by **Okada.** Some alternative details of this model:

(a) There are enzymes that are intimately associated with shaft Ca channels and selectively activated by them. Hence, they are preferentially activated by CF stimulation. Once activated, they result in some signal that is conveyed to the spine head to contribute to LTD induction.

(b) CF-mediated PN depolarization results in Ca influx leading to K efflux and depolarization of PF terminals (see ***Crépel***) or basket cells/Bergman glia (see **Okada**) and the consequent activation of nitric oxide (NO) synthase, leading to an anterograde NO signal that contributes to LTD induction. This model would predict that loading PNs with a Cs-based internal saline should block induction of LTD, a finding that has not been confirmed (Konnerth et al. 1992; Hemart et al. 1995).

(c) PF activation alone *is* sufficient to induce LTD, as recently proposed by **De Schutter** (1995 and this volume) and Llinas (1995). In fact, some recent experiments have suggested that strong stimulation (sufficient to evoke an EPSP > 10 mV recorded at the soma) of PFs at 1 Hz is sufficient to induce LTD, and that this LTD is blocked by postsynaptic Ca chelation and is occluded by LTD induced by conjunctive stimulation (**Hartell,** 1996 and this volume). It is not entirely clear why strong PF EPSP stimulation alone has failed to induce LTD in previous reports.

Another receptor protein has been implicated in LTD induction. The δ2 receptor, which is weakly homologous to NMDA and AMPA type glutamate receptors, is strongly and selectively expressed in PN dendrites. Morphologically, the cerebella of δ2 knockout mice exhibit several alterations including multiple CF innervation of PNs and an approximately two-fold reduction in the number of PF-PN synapses (Kashiwabuchi et al. 1995). Although synaptic transmission appears to be normal at the PF-PN synapse in slices from δ2 knockout mice, cerebellar LTD is completely blocked. This lack of LTD was also observed when glutamate/depolarization conjunctive stimulation was applied to PNs cultured from δ2 deficient mice (Hirano et al. 1995). In culture, the application of δ2 antisense oligonucleotides, which effectively suppressed the expression of the δ2 subunit, also inhibited the induction of LTD (Hirano et al. 1994). In my laboratory, we have recently confirmed that suppression of δ2 expression by antisense treatment produces a complete blockade of LTD induction, and have also extended these observations by measuring a set of processes known to be important for LTD induction including AMPA receptor function, voltage-gated Ca^{2+} influx and phosphoinositide-linked mGluR function, all of which are unaltered by δ2 antisense treatment (Jeromin et al. 1996). The δ2 subunit together with δ1 is believed to form a new family of glutamate receptor subunits, although heterologous expression studies have yet to assign them a functional role. Immunoprecipitation experiments also failed to demonstrate any interaction with other glutamate receptor subunits (Mayal et al. 1995). It is possible that δ2 forms a unique non-NMDA receptor complex that is dominantly expressed at the PF-PN synapses. Screening of cerebellar libraries with probes representing conserved domains of this new family might help to identify new members and thereby help us gain an understanding of how δ2 could potentially contribute to LTD induction.

Certainly the most contentious issue in the field of LTD induction has been the potential role of a NO/cGMP cascade, and recent years have seen a continuation of this debate. In cerebellar cultures made from neuronal nitric oxide synthase (nNOS) knockout mice, LTD induced by glutamate/depolarization conjunctive stimulation was indistinguishable from that in cultures from wild type mice (Linden et al. 1995). In wild type cultures, bath application of cGMP analogs produced a large, transient attenuation of glutamate-gated inward currents. However, application of an activator of soluble guanylate cyclase or an inhibitor of type V cGMP-phosphodiesterase did not mimic the effect of cGMP analogs, and inclusion of cGMP analogs in the patch pipette did not give rise to a slowly developing attenuation, suggesting that these compounds exert their effects at the cell surface. Free Ca was measured in the distal dendritic arbor of single PNs by fura-2 microfluorimetry. Cyclic ADP-ribose (cADPr), which has been suggested to be formed by a cGMP/PKG-dependent process and to be an endogenous ligand of the type 2 ryanodine receptor, increased the Ca response to a 3-second depolarizing step from −80 to +10 mV when included in the patch pipette. This effect was completely antagonized by the coinclusion of a competitive inhibitor ($8\text{-}NH_2$-cADP-r). Induction of LTD was not blocked by inclusion in the patch pipette of three different PKG inhibitors or $8\text{-}NH_2$-cADPr. These results suggest that, while cGMP/cADPr signaling may be important in neuronal Ca regulation, a NO/cGMP cascade is not required for cerebellar LTD induction in culture. However, these results are at odds with those of **Mori-Okamoto & Okamoto,** who found that a long-lasting attenuation of responses to AMPA pulses could be produced by external application of NO donors or internal application of a cGMP/IBMX cocktail. While the cause of this discrepancy is not obvious (it could be a species difference), it, together with the previously mentioned failure of Glaum et al. (1992) to observe effects of NO donors on PF EPSPs in slice, highlights the fact that the differences between labs with regard to NO/cGMP cannot be attributed solely to the use of a culture versus slice preparation.

Using the slice preparation, several laboratories have obtained further evidence consistent with a role of the NO/cGMP pathway in LTD induction. Hartell (1994) reports that injection of 8Br-cGMP into PNs, together with PF activation, is sufficient to induce a depression that occludes LTD produced by PF/CF conjunction, and that inhibitors of either PKC or PKG can block depression induced by either of these manipulations. Another approach has been taken by Lev-Ram et al. (1995), who found that photolysis of caged NO loaded into PNs could substitute for PF activation in LTD induction. When NO photolysis was followed by direct PN depolarization within a 50-msec window, LTD of PF EPSCs was produced. LTD induced in this manner could be blocked by an internal application of a Ca chelator or NO scavenger, but not external application of a NOS inhibitor or an NO scavenger. In contrast, LTD produced by PF/depolarization conjunction could be blocked by either an internally or externally applied NO scavenger, or an externally applied NOS inhibitor, but not an internally applied NOS inhibitor. This pattern of results suggests a model in which activation of PFs causes an anterograde NO signal that acts inside the PN. Furthermore, these findings suggest that the sole function of PF activation in LTD induction is the production of NO, in contrast to a large number of studies indicating a role of PF-mediated activation of AMPA and mGluR1 receptors. Based on the < 50-millisecond lifetime of NO in this preparation, Lev-Ram et al. (1995) have proposed that NO has a space constant for radial decay of < 13 μm, sufficient to preserve at least a rough degree of input specificity. However, it should be cautioned that both this estimate, and the more general set of observations using caged NO, come from experiments in which this compound is uncaged simultaneously throughout the PN, a spatial aspect of the signal that may differ considerably from synaptically released NO.

The contradictory nature of experiments with NO/cGMP reagents has lead several investigators to examine the possibility that multiple induction mechanisms exist for cerebellar LTD, some NO-dependent and some NO-independent (see ***Crépel*, Hartell, Mori-Okamoto & Okamoto,** and **Okada**). Hemart et al. (1995) found that while induction of LTD by PF/depolarization conjunction was blocked by a NOS inhibitor, LTD induced by pairing depolarization with a bath-applied mGluR agonist was not. In contrast, LTD induced by both protocols was blocked by a PKC inhibitor, leading the authors to suggest that while LTD induction in the former case required activation of both NO/cGMP and PKC-mediated pathways, the latter required PKC, but not NO/cGMP – similar to results seen in culture (Linden & Connor 1992; Linden et al. 1995). Another multiple induction mechanism model is presented by **Hartell** (1996 and this volume) in which strong activation of PFs alone results in two forms of LTD, an NO-dependent, postsynaptic Ca-independent form that spreads to sites distant from the locus of PF activation, and a local form which is Ca-dependent and NO-independent. Although Hartell's model is similar to that of Lev-Ram et al. (1995) in that they both hypothesize an anterograde action of NO from PF terminals to PN dendrites, they differ in that Lev-Ram et al. (1995) found no effect of caged NO photolysis in the absence of coincident depolarization/Ca influx.

While considerable attention has been paid to the molecular mechanisms of cerebellar LTD induction, much less effort has been focused upon the mechanisms of LTD expression. A widely accepted notion has been that LTD is expressed, at least in part, as a down-regulation of the postsynaptic sensitivity to AMPA, because LTD may be detected using AMPA or glutamate test pulses in both slice and culture preparations. This notion has received further support from a study showing that the coefficient of variation of PF EPSCs was altered by manipulations known to act presynaptically (such as transient synaptic attenuation produced by addition of adenosine) but was not altered by induction of LTD (Crépel et al. 1995). This finding speaks to the suggestion by **Bindman** that a presynaptic component of cerebellar LTD expression may exist. One form of postsynaptic model for LTD expression is that phosphorylation of AMPA receptors by an enzyme such as PKC (or possibly another kinase downstream from PKC) results in faster desensitization of AMPA currents. Recently, two forms of evidence have emerged that are consistent with this model. First, using an antibody that recognizes the AMPA receptor subunit GluR2 phosphorylated at serine-696 (and possibly some corresponding sites on other subunits), it was shown that bath application of AMPA prior to 8Br-cGMP, a manipulation that is claimed to cause an LTD-like effect in grease gap recordings (Ito & Karachot 1990), produced a persistent (> 30 minutes) increase in immunoreactivity in PN dendrites (Nakazawa et al. 1995). Second, an interaction was found between the induction of LTD and the effects of the drug Aniracetam, which potentiates and prolongs AMPA responses though an inhibitory effect on desensitization. Hemart et al. (1994) report that Aniracetam prevented LTD induction by PF/depolarization in slice and that the effect on Aniracetam was larger on depressed synapses than on naive ones. While the authors contend that these findings constitute evidence that LTD expression involves a desensitization of postsynaptic AMPA receptors, it should be cautioned that since desensitization of AMPA receptors is not proven to be a unitary phenomenon (and in fact, is likely to be a process with several states), this experiment should be viewed as indirect evidence at best. Direct evidence will require actual measurements of desensitization with fast agonist application following LTD in isolated cells or membrane patches.

Behavioral studies in both humans and animals have shown that in a number of learning tasks, information appears to be initially stored as an easily-disrupted short-term form, and then is consolidated into a more stable long-term form. A parallel, and perhaps related dissociation has been demonstrated in persistent changes of synaptic strength: In addition to LTP and LTD, certain patterns of stimulation can give rise to short-term potentiation (STP) and short-term depression (STD), phenomena that typically last from 15 minutes to 2 hours. LTD may be reliably induced in the cultured mouse PN when glutamate pulses and PN depolarization are applied together six times. When the number of these conjunctive stimuli was reduced to two, STD lasting 20 to 40 minutes was induced in 4/12, no change occurred in 3/12, and LTD was induced in 5/12 cells. To understand the molecular events that govern STD versus LTD induction, the enzyme phospholipase A2 (PLA2) was examined (Linden 1995). PLA2 catalyzes cleavage of the 2-acyl group from membrane phospholipids resulting in the liberation of free unsaturated fatty acids such as oleate and arachidonate, which in turn synergistically activate protein kinase C (PKC) when present with diacylglycerol and Ca. Application of oleate with two conjunctive stimuli resulted in an apparent conversion of STD cases to LTD. Application of PLA2 inhibitors during six pairings converted LTD to STD, and this effect could be completely reversed by coapplication of either free arachidonate or oleate. These findings suggest a model in which liberation of unsaturated fatty acids by PLA2 contributes to a synergistic activation of PKC, the full activation of which results in LTD induction and the partial activation of which results in STD induction.

Perhaps the most interesting result in cerebellar synaptic plasticity is the convincing demonstration of a mechanism for LTP at the PF-PN synapse. Although it had been shown in scattered reports that LTP can occur at PF-PN synapses following PF stimulation alone and that this phenomenon is not blocked by postsynaptic Ca chelators, a mechanistic description has only recently emerged. Salin et al. (1996) show that LTP of this synapse may be induced by an 8 Hz × 15 second stimulation, and that induction of this LTP is not blocked by postsynaptic Ca chelation or blockade of ionotropic glutamate receptors (with kynurenate). This LTP, which was associated with a decrease in paired-pulse facilitation, was blocked by removal of external Ca, an inhibitor of PKA, or occlusion via activation of adenylate cyclase. These findings are suggestive of a model identical to that proposed for LTP at the mossy fiber-CA3 synapse in hippocampus (Weisskopf et al. 1994), in which presynaptic Ca entry during stimulation results in stimulation of a Ca-sensitive adenylate cyclase and the consequent activation of PKA.

What aspects of this molecular detail of LTP and LTP induction mechanisms may be of use to systems and computational physiologists? First and foremost, these molecular details provide tools with which to test the hypothesis that cerebellar LTD and LTP as studied in slice or culture are necessary for certain motor learning tasks. Application in the cerebellar molecular layer of drugs that interfere with the function of mGluR1, PKC, PLA2, and perhaps a NO/cGMP system would be instructive in this regard. At present, only a small number of studies have made use of this strategy and these have chosen to focus upon NO/cGMP signaling, the most disputed aspect of the LTD induction process. Nagao and Ito (1991) have shown that VOR adaptation may be blocked by application of hemoglobin, which functions as an extracellular nitric oxide trap, to the subdural space overlying the ipsilateral flocculus in monkey and rabbit. Similarly, Li et al. (1995) found that injection of a nitric oxide synthease (NOS) inhibitor into the vestibulocerebellum of the goldfish also blocked VOR adaptation, and that the effect of this NOS inhibitor could be overcome by simultaneous administration of L-arginine (the substrate of NOS). They found that adaptive gain increases, but not decreases, were blocked by this process and that administration of the NOS inhibitor after VOR adaptation had no effect on retention of this form of motor learning.

A similar approach is provided by the generation of transgenic mice that lack proteins thought to be required for cerebellar LTD. While mice are not the ideal animal model for the analysis of a wide range of motor behaviors, there are nonetheless certain tasks that can be evaluated in them; these include basal tests of motor coordination such as beam walking and tests of conditioning of discrete motor learning such as eyeblink conditioning. Production of a transgenic mouse in which the γ isoform of PKC had been rendered null, did not result in significant alteration of LTD induction by PF/depolarization conjunction (Chen et al. 1995; Kano et al. 1995). While PKCγ is a major isoform in PNs, it is not the sole isoform, so this result is not entirely unexpected. Eyeblink conditioning was not impaired in these mice, but was partially impaired in two knockout mice that lack cerebellar LTD, the mGluR1 knockout (Aiba et al. 1994) and the glial fibrillary acidic protein knockout (Shibuki et al. 1996). There are some obvious problems that complicate the analysis of behavior and physiology in "knockout" mice. First, knockout mice have the gene of interest deleted from the earliest stages of development. As a result, these mice often have a complex developmental phenotype. For example, PKCγ, mGluR1, and $\delta 2$ (but not GFAP) knockout mice all have PNs that fail to undergo the normal developmental conversion from multiple to mono CF innervation. Chen et al. (1995) claim that, in this particular case, this information is useful in providing a double dissociation: mutant mice with multiple CF innervation have impaired motor coordination but may have either normal or impaired eyeblink conditioning. Mutant mice that have impaired cerebellar LTD have impaired eyeblink conditioning, but may have either normal or impaired motor coordination. Another developmental issue that arises is compensation. Knockout of one gene sometimes produces upregulation of the expression of related genes during development. In the PKCγ knockout mouse, a PKC inhibitor is only partially effective in blocking LTD but it is completely effective in the wild type, suggesting a compensatory effect on expression of other kinases (Chen et al. 1995). A second complicating factor is that knockout mice have the gene of interest deleted in every cell of the body, not just the neurons of interest, making it more difficult to ascribe the knockout effects to dysfunction in any one particular structure. For example, eyeblink conditioning seems to be sensitive to both hippocampal and cerebellar lesions depending on the exact details of the task employed (delay vs. trace conditioning), complicating the analysis of behavior in a knockout. Finally, at present, all publications analyzing the physiology of presumed cerebellar motor learning knockout mice have relied solely on in vitro preparations of cerebellar tissue. This form of investigation is limited in that analysis of, say, PF synaptic properties in a slice may be able to tell you about receptor function, short-term facilitation, and so on, but it will not tell you about other potentially important measures, such as the firing rate of PFs.

The second way in which cellular physiologists can contribute to a systems level understanding of the cerebellum is by providing parametric information that constrains the possible patterns of activity that could give rise to long-term synaptic changes. **De Schutter** takes both the systems and cellular communities to task, the former for not considering the known properties of cerebellar synaptic plasticity in building models of cerebellar function, and the latter for not generating the sort of parametric data that would be most useful for systems considerations. While this scolding has a very legitimate point, I believe that there are important corollaries to be added. First, it does no good to generate parametric data if the systems physiologists do not consider it as a whole, but merely pick through it to find bits that may be fit into preexisting models. For example, while the studies cited by De Schutter suggest that CF stimulation must precede PF stimulation to produce LTD (thus making it difficult for the CF to function as an error signal), there are other studies using slightly different stimulation protocols (small trains instead of single pulses in one case) that show that PF before CF pairing may also be effective in inducing LTD (Chen & Thompson 1995; Schreurs et al. 1996). Second, sometimes, even in the absence of extensive parametric data, the knowledge that a particular form of synaptic change *exists* can cause one to completely rethink a given model. For example, the recent demonstration in cell culture that there is a postsynaptic switch between STD and LTD induction that is dependent upon number of conjunctive pairings (Linden 1995) can provide an additional mechanism for modeling information storage in the cerebellum. Likewise, the observation that stimulation of PFs alone at moderate frequency can induce a presynaptically-mediated LTP has important systems-level implications. It complicates models based on the notion that PF stimulation alone will invariably result in LTD (see **Hartell, De Schutter** 1995 and this volume), while providing a resetting mechanism of the kind required for the models of **Arbib** and **Houk.** In addition, the presence of a presynaptically expressed LTP mechanism, but a postsynaptically expressed LTD mechanism, creates the unusual situation in which a synapse could change synaptic weight without being truly "reset": maximal expression of both LTP and LTD could result in a nonplastic synapse at an intermediate synaptic gain.

Cellular mechanisms of long-term depression: From consensus to open questions

F. Crépel
CNRS and University Paris-Sud, Bât. 441, 91405 Orsay Cedex, France.
crepelf@dialup.francenet.fr

Abstract: The target article on cellular mechanisms of long-term depression appears to have been well received by most authors of the relevant commentaries. This may be due to the fact that this review aimed to give a general account of the topic, rather than just describe previous work of the present author. The present response accordingly only raises questions of major interest for future research.

Although the commentaries often emphasize specific points of particular interest to their authors, the overall feeling that emerges from them is an agreement on the scheme proposed in the target article for the induction and expression of long-term depression (LTD). This is noteworthy, since this review was written more than two years ago. In particular, concerning the controversy on the role of nitric oxide (NO) in LTD, it seems reasonable to think – as stressed by many commentators – that cell cultures are not appropriate in this respect, since it is likely that NO is produced outside Purkinje cells by neighboring elements, which are poorly represented in such reduced preparations (see Crépel et al. 1994).

A recent paper by Linden (1995), has introduced a new player in the game, namely, phospholipase A_2 (PLA_2). Indeed, during pairing protocols in cell culture experiments, activation of this enzyme by calcium would allow full activation of protein kinase C (PKC) and LTD induction, whereas only partial activation of PKC would be achieved when PLA_2 was blocked, which would in turn lead to short-term depression (STD) instead of LTD. This is a very important finding even though, as stressed by Linden, the presence of this new cascade must be confirmed in intact tissues before being incorporated in a general scheme of LTD induction. The fact that the transition from STD to LTD occurred in a rather all-or-none manner in these experiments is also puzzling. Indeed, one would have expected that, depending on the number of pairings, the proportion of partly and fully activated molecules of PKC would vary in a continuous rather than all-or-none manner, thus leading to a progressive transition between STD and LTD.

Finally, the commentary by **Baudry** addresses a still completely unsolved question, namely, the fact that in the LTD field, it is believed that phosphorylation of postsynaptic AMPA receptors contributes to the observed long-lasting increase of AMPA receptor-mediated synaptic currents, whereas in the LTD field it is believed that exactly the opposite occurs. Although one might argue that such opposite effects result, for instance, from a different subunit composition of AMPA receptors in hippocampal neurons and in Purkinje cells, or from different phosphorylation sites, there is clearly still something missing from our understanding of LTP and LTD in this respect.

A bridge between cerebellar long-term depression and discrete motor learning: Studies on gene knockout mice

Masanobu Kano
Laboratory for Neuronal Signal Transduction, Frontier Research Program, Riken, Wako-shi, Saitama 351-01, Japan. **mkano@postman.riken.go.jp**

Abstract: Several commentators argue that there remains a gap between cerebellar long-term depression (LTD) studied in vitro and cerebellar functions such as motor learning. Recent developments in a gene-targeting technique for producing mutant mice defective in particular gene products has opened a new era for the study of synaptic plasticity and learning. There are three mutants in which cerebellar LTD are clearly deficient in vitro. Two of these mice show clear impairment in the conditioned eyeblink response, a paradigm known to involve the cerebellum. In contrast, one mutant with apparently normal cerebellar LTD showed no impairment in the conditioned eyeblink response. These new lines of evidence support the view that cerebellar LTD is a cellular basis for discrete motor learning.

De Schutter points out that cellular neuroscientists who study cerebellar long-term depression (LTD) do not address the issue of the functional significance of synaptic plasticity. The same criticism is also raised by **Thompson.** De Schutter claims that the parameters used to induce LTD in vitro are not relevant to behavioral motor learning. He assumes that 100 pairings of parallel fiber and climbing fiber stimulation at 1 to 4 Hz, which is often used to induce LTD in cerebellar slices, are not likely to occur in real life. He claims that lower frequency of pairings (around 0.01 to 0.10 Hz), which nobody has ever tried in cerebellar slices, would be relevant to motor learning.

Jaeger points out that we have little knowledge about how motor control works in the structures with which the cerebellum interacts. The cerebellum receives inputs from various sources, and the outputs project to various structures that we do not understand. He also suggests that the physiology of cerebellar activity patterns is poorly understood. For example, we have little knowledge about the natural pattern of parallel fiber activation and the behavior of intrinsic inhibitory circuits.

The issues raised by these two commentators are not easy to tackle with conventional experimental techniques. For example, stable recordings from cerebellar slices lasting several hours or even days are necessary to reproduce the situations that **De Schutter** claims occur in motor learning in vivo. In practice it is very difficult, however, to achieve such recordings. It is also almost impossible to identify all the structures that receive cerebellar output. The difficulty and technical limitations on answering questions like those raised by the commentators leave a gap between cellular and molecular studies of cerebellar LTD and the behavioral studies of motor control and learning. I suggest that the application of gene-targeting technique to the production of knockout mice defective in particular gene products is one important breakthrough. This new technique appears useful for bridging the gap between cellular/molecular studies and behavioral studies. Here I introduce four examples in which the gene knockout has been used with some success to link cerebellar LTD and motor learning.

R1. mGluR1 deficient mice. Aiba et al. (1994) and Conquet et al. (1994) independently produced mutant mice that lack mGluR1, a subtype of metabotropic glutamate receptors that is abundant in Purkinje cells and has been shown to be required for LTD induction. The mGluR1 mutant mice are viable but clearly ataxic. They show characteristic cerebellar symptoms such as intention tremor, dysmetria, and ataxic gait. The anatomy of the cerebellum is not apparently disturbed. Electrophysiological examinations in our laboratory revealed that voltage-gated Ca^{2+} currents of Purkinje cells are normal and excitatory synaptic transmission from parallel fibers to Purkinje cells and from climbing fibers to Purkinje cells are functional. Both parallel fiber and climbing fiber synapses display normal short-term synaptic plasticity to paired stimuli, suggesting that presynaptic functions of these excitatory synapses are normal. Response to glutamate receptor antagonists is normal in both parallel fiber-mediated and climbing fiber-mediated excitatory postsynaptic currents (EPSCs), suggesting that the postsynaptic receptors are normal. However, about one-third of Purkinje cells in mature mGluR1 mutant mice are innervated by multiple climbing fibers, which is very rare ($< 5\%$) in wild type mice. Moreover, LTD is clearly deficient in the mGluR1 mutant. At the behavioral level, mGluR1 mutant mice display clear impairment in motor coordination, as assessed by the ability of the animals to stay on the roller (rotorod test) and on the rod inclined by 30° to the horizon (inclined rod test). In addition, the mGluR1 mutant mice show partial impairment in the conditioned eyeblink response.

R2. PKCγ deficient mice. The γ isoform of protein kinase C (PKCγ) is the major PKC isoform in cerebellar Purkinje cells. Several lines of evidence have suggested the involvement of PKC in LTD induction. Thus, we examined PKCγ deficient mice that had been produced in Tonegawa's laboratory. The PKCγ mutant mice are viable and display ataxia that is much milder than that of the mGluR1 mutant mice. The anatomy of the cerebellum is not apparently disturbed. Electrophysiologically, voltage-gated Ca^{2+} currents and both parallel fiber- and climbing fiber-mediated EPSCs appear normal. Behavior to paired stimuli and response to glutamate receptor antagonists are normal in both parallel fiber- and climbing fiber-mediated EPSCs. Similar to the mGluR1 mutant mice, about one-third of Purkinje cells in mature PKCγ mutant mice remain multiply innervated by climbing fibers (Kano et al. 1995). LTD can be induced apparently normally, however, in the PKCγ mutant mice. Behaviorally, motor coordination is clearly impaired in both rotorod and thin rod test. In marked contrast, the conditioned eyeblink response is not impaired in PKCγ mutant mice (Chen et al. 1995).

R3. GluRδ2 deficient mice. The ionotropic glutamate receptors consist of six families among which the functions of the δ subfamily are poorly understood. The GluRδ2 is selectively localized in cerebellar Purkinje cells, suggesting its possible involvement in cerebellar LTD. Kashiwabuchi et al. (1995) produced GluRδ2 deficient mice. Morphologically, the number of parallel fiber–Purkinje cell synapses in GluRδ2 mutant mice is reduced to less than one-half of that of the wild type mice. Yet the basic electrophysiology of parallel fiber-Purkinje cell synapses appears normal. Climbing fiber synapses are functional, but again about half of the Purkinje cells in mature GluRδ2 mutant mice remain multiply innervated by climbing fibers. LTD is deficient in GluRδ2 mutant mice in both slice and culture preparations. The GluRδ2 mutant mice are impaired in motor coordination in the rotorod, runway, and rope-climbing tests. Eyeblink conditioning has not been examined. However, GluRδ2 mutant mice show retardation in vestibular compensation (Funabiki et al. 1995).

R4. GFAP deficient mice. Shibuki et al. (1996) analyzed mutant mice deficient in glial fibrillary acidic protein (GFAP) and found unexpected results. The anatomy of the cerebellum is not apparently disturbed. Electrophysiologically, dendritic Ca^{2+} spikes of Purkinje cells and both parallel fiber- and climbing fiber-mediated EPSPs appear normal. Behavior to paired stimuli and response to glutamate receptor antagonists are normal in both parallel fiber- and climbing fiber-mediated EPSPs. Unlike the other three mutant mice mentioned above, Purkinje cells of mature GFAP mutant mice are mono-innervated by climbing fibers. In contrast, LTD is clearly deficient in GFAP mutant mice. Behaviorally, motor coordination is normal in the rotorod, runway, and rope-climbing tests. In marked contrast, the conditioned eyeblink response is clearly impaired in GFAP mutant mice.

The results presented above strongly support the view that cerebellar LTD is a cellular basis for learning about elementary movement such as the conditioned eyeblink response. In mGluR1, GluRδ2, and GFAP mutant mice, cerebellar LTD are clearly deficient in vitro. The mGluR1 and GFAP mutant mice show clear impairment in the conditioned eyeblink response. In contrast, PKCγ mutant mice have apparently normal cerebellar LTD and show no impairment in the conditioned eyeblink response. As pointed out by **De Schutter,** the parameters used to induce LTD in slices may not be relevant to behavioral motor learning. However, the parallelism between LTD in vitro and motor learning in these mutant mice indicates that LTD induced by the conventional parameters is a good performance indicator of motor learning.

The timing of parallel and climbing fiber inputs is an important issue addressed by **Arbib, Fiala & Bullock,** and **Houk.** In slices, LTD is most effectively induced when parallel fibers and climbing fibers are repetitively stimulated at nearly the same time (Karachot et al. 1995). On the other hand, in behavioral learning, the motor error signals from the climbing fiber system must come several hundreds of milliseconds after motor execution signals through the mossy fiber–parallel fiber system. Filia & Bullock and Houk have constructed models by which they try to bridge the cellular and behavioral data. I agree that the mGluR1-mediated response is a good candidate for bridging the parallel fiber and climbing fiber signals. Repetitive activation of parallel fibers induces depolarization of Purkinje cells that lasts several hundreds of milliseconds. This long-lasting depolarization is mediated by mGluR1, since it is resistant to ionotropic glutamate receptor antagonists (CNQX or AP5), but it is blocked by a metabotropic glutamate receptor antagonist (MCPG; Batchelor et al. 1994). In other respects, both models seem to be based largely on assumptions with insufficient experimental support. To judge the validity of these models, we must wait until more experimental data are accumulated.

NO more news from the cerebellum

Steven R. Vincent
Division of Neurological Sciences, Department of Psychiatry, The University of British Columbia, Vancouver, B.C., V6T 1Z3 Canada. **sru@unixg.ubc.ca**

Abstract: Although a number of further studies have now been published exploring possible roles for the NO-cGMP signal transduction system in various models of LTD, there still appears to be confusion regarding the physiological functions of these molecules in the cerebellar cortex. What seems clear is that NO can be produced in granule cells and basket cells, and can stimulate soluble guanylyl cyclase, thereby activating cGMP-dependent protein kinase in Purkinje cells. However, substrates for this kinase are still poorly defined, and its role in the modulation of synaptic transmission remains to be determined. The recent development of novel transgenic models and new pharmacological agents with which to dissect this signal transduction system, gives one confidence that our understanding of the cerebellar NO system will quickly advance. With its unique properties as a signaling molecule, it is clear that NO news has been good news for those studying synaptic plasticity in the cerebellum.

Since the target article was written, a number of interesting studies that further examine the possible role of NO in synaptic plasticity in the cerebellar cortex have been published. In addition, the commentators have raised some relevant issues that need to be addressed. First, with regard to the cellular localization of the NO/cGMP system, the data indicating that NO synthase (NOS) is present in granule and basket cells and not in Purkinje neurons still appear valid. Perhaps this has been most elegantly demonstrated using the single cell PCR approach by Crépel et al. (1994). **Kiedrowski** correctly points out that cells in addition to Purkinje neurons may contain guanylyl cyclase and respond to NO with an increase in cGMP. However, as noted previously, "nervous" mutant mice, which lack mostly Purkinje cells, have decreased guanylyl cyclase activity and cGMP levels (Mao et al. 1975; Schmidt & Nadi 1977). These animals also show a loss of CO-induced cGMP formation, which is presumably based upon activation of soluble guanylyl cyclase (Nathanson et al. 1995). These animals are also reported to show a loss in NOS activity in the cerebellar cortex (Ikeda et al. 1993). However, Wood et al. (1994) did not detect a decrease in cGMP or a loss in response to harmaline in these animals. These results are somewhat confusing, but are consistent with the major role of phosphodiesterase relative to guanylyl cyclase in controlling cerebellar cGMP levels. Furthermore, the observation that harmaline was still effective is consistent with our suggestion that this drug increases cGMP levels in the cerebellar cortex via the activation of olivary projections to the deep cerebellar nuclei and subsequent mossy fiber activation, and not via Purkinje cell activation. As pointed out by **Okada,** this indirect pathway might not account for the activation of NOS seen by Shibuki (1990) following white matter stimulation in cerebellar slices; instead, depolarization of granule and basket nerve terminals by increased extracellular potassium might be responsible.

With regard to granule cells, the data of Biggio et al. (1978) show an almost complete loss of guanylyl cyclase activity and cGMP levels in the cerebellar cortex following kainate lesions, which largely spare granule and glial cells. In addition, Bunn et al. (1986) reported an increase in guanylyl cyclase activity after depletion of granule cells. This would suggest that if granule cells contribute to cGMP increases, their contribution would be small compared to that of Purkinje cells.

Differences in LTD between in vivo studies, acute slice experiments, and Purkinje cell cultures have been noted (**Calabresi et al.; Hartell**). It is known that Purkinje cells can be induced to express NOS under certain conditions (Chen & Aston-Jones 1994; O'Hearn et al. 1995; Saxon & Beitz 1994), and NOS has been reported to be transiently expressed during early postnatal life (Bruning 1993). Thus, the neurochemical and electrophysiological properties of cultured Purkinje cells may well be quite different from the in situ situation. However, Linden et al. (1995) have found that in their culture preparation the Purkinje cells do not express NOS, but do express cGMP-dependent protein kinase, as they do in vivo.

The issue of cell cultures is also raised by **Kiedrowski,** who reported that sodium nitroprusside-induced cGMP increases in cultured cerebellar astrocytes (Kiedrowski et al. 1992). A wide range of studies has also demonstrated that astrocytes express NOS in culture or in response to injury (Murphy et al. 1993). However, it has not been demonstrated that quiescent astrocytes in situ express either NOS or soluble guanylyl cyclase. Indeed, the early immunohistochemical results of Zwiller et al. (1981) must be viewed with caution, since they appear to stain all cells and have not been confirmed by subsequent studies with well-characterized antibodies. Indeed, recent studies have demonstrated increased cGMP immunofluorescence in Purkinje cells in response to various stimuli (Nathanson et al. 1995).

The idea of a transcellular metabolic pathway for citrulline and arginine is of course unproven, but the well-documented glutamate–glutamine pathway between neurons and glia may provide a precedent. Indeed, in the periphery, arginine and citrulline metabolism is a complex process involving distinct portions of the liver, kidney, and intestine. In other brain regions, we have observed examples of neurons that express both NOS and argininosuccinate synthetase (ASS), while in other regions, these enzymes appear in distinct neuronal populations (Arnt-Ramos et al. 1992). In the enteric nervous system, ASS and ASL have been found to colocalize with NOS and appear to recycle citrulline to arginine (Shuttleworth et al. 1995). The levels of arginine and citrulline in the extracellular fluid have been monitored by in vivo microdialysis and appear to vary with NOS activity (Ohta et al. 1994; Sorkin 1993), consistent with results from cerebellar slices (Hansel et al. 1992). **Kiedrowski** correctly questions whether results obtained with antibodies raised to liver urea cycle enzymes may miss other isoforms of these proteins. However, similar observations have been obtained using ASS antibodies developed independently by two groups, and Southern analysis indicates a single ASS gene, mutations of which are associated with a loss of brain ASS activity. Indeed, the development of ASS knockout animals by O'Brien and colleagues, might allow this question to be tested more directly. Although these animals are not viable, perhaps inducible knockouts are on the way.

A number of commentators agree with the postulate advanced by CRÉPEL that there exist two forms of long-term depression (LTD), one dependent upon large calcium increases and the other on nitric oxide (**Hartell; Okada**). However, there still appears to be considerable disagreement on what NO, if anything, might be doing to affect LTD. Indeed, in contrast to the knockout of the metabotropic receptor mGluR1 or the GluRd2 protein, which produces profound changes in motor learning and blocks LTD, knockout of the neuronal NOS gene does not appear to affect motor behavior or LTD induction, at least in isolated Purkinje cells (Huang et al. 1993; Linden et al. 1995).

In acute slices, bath application, but not intra-Purkinje cell application of NOS inhibitors, prevents LTD induced by parallel fiber–climbing fiber costimulation (Daniel et al. 1993; Lev-Ram et al. 1995). It is well known that parallel fiber activation paired with calcium spikes in Purkinje cells will evoke LTD. The biochemical neuroanatomy indicates that NOS is contained in the granule cells, and one would predict that it would be activated by voltage-dependent calcium channel opening following parallel fiber stimulation. This is consistent with our findings with tetrodotoxin (TTX) and w-conotoxin (Luo et al. 1994). Climbing fiber activation would give rise to a large increase in intracellular calcium in the Purkinje cells. Thus one might predict that NO would be able to substitute for parallel fiber stimulation, but not climbing fiber stimulation. However, Shibuki and Okada (1991) found that bath-applied SNP paired with parallel fiber stimulation evoked LTD. Recently, in perhaps the most elegant study examining the role of NO in LTD, the photoactivated release of caged NO inside individual Purkinje cells in slices induced LTD only when paired with depolarization (Lev-Ram et al. 1995). Given these observations, one would predict that NOS inhibitors should have no effect on LTD evoked by combined depolarization and glutamate in cultured neurons, and indeed they do not (Linden & Connor 1992). However, one would also predict that intracellular NO or cGMP increases would induce LTD when combined with depolarization, yet they do not (Linden et al. 1995).

Hartell presents data indicating that strong parallel fiber stimulation alone can induce LTD. This strong stimulation appears to induce a dendritic calcium increase similar to that usually evoked by climbing fiber stimulation or direct depolarization of the Purkinje cells. The relevance of this phenomenon to LTD is not clear, however, since in this preparation a depression of nonstimulated inputs was also induced in a calcium-independent, NOS-dependent manner, implying that NO or cGMP on their own applied into Purkinje cells should depress all parallel fiber inputs. In fact **Mori-Okamoto & Okamoto** have found, in cultured chick Purkinje cells, that intracellular cGMP or NO donors alone will suppress AMPA responses. However others have not observed such an action in slices or cultures (Lev-Ram et al. 1995; Linden et al. 1995).

Fiala & Bullock put forward a scheme in which parallel fiber-dependent, metabotropic receptor-mediated activation of protein kinase C (PKC), and subsequent phosphorylation of the AMPA receptor, is combined with an increased cGMP in Purkinje cells following climbing fiber activation. This results in the PKG-dependent phosphorylation of G-substrate, which then inhibits protein phosphatase-1, preventing dephosphorylation of the AMPA receptor. Calcineurin is suggested to antagonize this pathway by dephosphorylating the G-substrate. However, since climbing fiber activation produces a large increase in calcium in the Purkinje cells, one would predict that calcineurin would be potently activated by this stimulation, particularly given the high affinity of this enzyme for Ca^{2+}/calmodulin. Furthermore, as discussed below, metabotropic receptor activation by parallel fibers does not appear to be associated with increased intracellular calcium in Purkinje cell dendrites.

There is much evidence pointing to a critical role for metabotropic glutamate receptor activation in LTD induction. Indeed, inactivation or knockout of mGluR1 receptors blocks LTD (Aiba et al. 1994; Conquet et al. 1994; Shigemoto et al. 1994). However, it is not clear whether the metabotropic receptors involved are activated by climbing fibers or parallel fibers. In cultured Purkinje cells, depolarization must be combined with both AMPA and metabotropic receptor activation to induce LTD (Linden & Connor 1991). In slices, although activation of mGluR receptors paired with calcium spikes is able to induce LTD (Daniel et al. 1992; Hemart et al. 1995), it is not clear if blockade of these receptors can prevent LTD induction following parallel fiber pairing with calcium spikes. Hemart et al. (1995) found it did not, while Hartell (1994b) found that it did. The main difference in these two studies appears to be the presence of a GABA antagonist in the study by Hemart et al. IPSPs in the preparation of Hartell (1994b) might well inhibit local, voltage-dependent calcium increases in Purkinje dendrites (Callaway et al. 1995), making the preparation dependent upon metabotropic receptor-mediated events.

There have been suggestions that the role of metabotropic receptors in LTD depends upon the release of calcium from intracellular stores following IP3 generation. However, Eilers et al. (1995) have found that the increase in dendritic calcium induced by parallel fiber stimulation does not involve metabotropic receptor activation, and instead depends on the activation of voltage-gated calcium channels following AMPA receptor activation. Likewise, application of t-ACPD alone does not consistently produce detectable change in free-calcium concentration in Purkinje cell dendrites (Hartell 1994b; Llano et al. 1991; Vranesic et al. 1991). However, Kasono and Hirano (1995) found that heparin, which can block IP3 receptors, suppressed LTD, and combined AMPA and depolarization of

Purkinje cells induced LTD only when paired with intracellular IP3 release. Kohda et al. (1995) found that LTD was blocked by thapsigargin or inhibition of the ryandodine receptor, suggesting a role for calcium-dependent calcium release in LTD in cultured Purkinje cells. However, others have not observed such inhibition with thapsigargin (Hemart et al. 1995); Linden et al. (1995) found that inclusion of an antagonist of the suggested ligand for the ryanodine receptor, cyclic ADP-ribose, did not block LTD induction. Finally it should be noted, particularly by those advocating synergistic roles for both NO/cGMP and IP3 inside Purkinje cells to induce LTD, that in the systems in which it has been examined, including cerebellar granule cells, cGMP appears to inhibit IP3 formation (Oliva & Garcia 1995). Thus, it would be premature to say with confidence that t-ACPD-induced suppression of AMPA responses is mediated by elevated intracellular cGMP and activated cGMP-dependent protein kinase (**Mori-Okamoto & Okamoto**).

As pointed out by **Baudry,** phosphorylation of AMPA receptors by PKC, protein kinase A (PKA), or CaMKII is usually associated with increased rather than decreased activity. It may be that phosphorylation of the GluRd2 subunit is important here, since, as noted by **Hirano,** this protein is selectively localized in Purkinje cells and appears from various forms of knockout studies to be essential for LTD (Hirano et al. 1995; Kashiwabuchi et al. 1995). These results must be viewed with some caution, however, since the lack of this protein also affects parallel fiber and climbing fiber synapse formation. Clearly, it will be of great importance to determine the functional significance of this protein.

Bindman suggests that the injection into Purkinje cells of the oxadiazoloquinozaline derivative, ODQ, a selective inhibitor of soluble guanylyl cyclase, may clarify the role of cGMP in LTD. This is clearly a significant advance, given the various actions of the currently used guanylyl cyclase inhibitors LY83583 and methylene blue (Luo et al. 1995). However, the injections into Purkinje cells of various activators of soluble guanylyl cyclase, as well as inhibitors of cGMP-dependent protein kinase, have not resolved this issue.

An exciting recent avenue of research has revealed the presence of a number of previously unrecognized targets for cGMP within neurons. Various ion channels that are directly gated by intracellular cGMP have been identified (El-Husseini et al. 1995a; Leinders-Zufall et al. 1995; Yao et al. 1995). Furthermore, a type II cGMP-dependent protein kinase has been shown to be expressed in the cerebellum (El-Husseini et al. 1995b), and the pharmacology and substrate specificity of this kinase are likely to be distinct from the type I cGMP-dependent protein kinase previously described in Purkinje cells. Other possible targets for cGMP-dependent kinase action have also been recently proposed (Nathanson et al. 1995). Finally, although the idea that extracellular cGMP has a role as a messenger molecule is interesting, recent work by Linden et al. (1995) has shown that only relatively high levels of cGMP have extracellular actions, and it is not clear that such levels could be obtained physiologically. Indeed, **Mori-Okamoto & Okamoto** found no extracellular effects of cGMP on AMPA responses in cultured chick Purkinje cells.

In response to the request of **Arbib,** I would conclude by saying that our prospects for developing precise descriptions of the kinetics of the neurochemical mechanisms underlying LTD are good. However, as to whether the mechanisms underlying physiologically relevant synaptic plasticity in the cerebellum are understood at present, we must still answer NO.

More models of the cerebellum

James C. Houk[a] and Andrew G. Barto[b]
Department of Physiology, Northwestern University Medical School, Chicago, IL 60611[a]; Department of Computer Science, University of Massachusetts, Amherst, MA 01003.[b] **houk@casbah.acns.nwu.edu; barto@cs.umass.edu**

Abstract: Neural models are needed at many levels, ranging from the biophysically based to connectionist abstractions. The adjustable pattern generator (APG) model of the cerebellum discussed in the target article by Houk, Buckingham, and Barto operates at an intermediate level and is constrained by a large body of anatomical and physiological data. The associative memory model originally proposed by Marr (1969) and Albus (1971) on the basis of the microcircuitry of the cerebellar cortex is still relevant today, although it needs to be coupled to the problem of motor pattern generation. In considering the problem of cerebellar learning, one needs some source of training information to adjust network performance so that it performs fruitful interactions with the environment; however, it is not true that this "teacher" needs to be particularly smart. We address these and other important issues raised by the commentators.

In our target article on models of the cerebellum and motor learning, we attempted to provide a reasonably detailed and balanced review of this broad topic, understandably emphasizing the adjustable pattern generator (APG) model that has been the focus of our own research. Whereas some commentators expressed appreciation of this comprehensive coverage (e.g., **Thompson**), others seem anxious to dismiss the entire body of prior work (e.g., **Paulin**). The following reply addresses mainly the specific statements and criticisms as opposed to general attacks. However, we begin by responding to one general issue in neural modeling studies, raised by several of the commentators, but especially sounded by **Jaeger** and **Hepp.**

Jaeger argues that high-level models of the cerebellum are underconstrained and are therefore likely to be of little utility in contributing to an understanding of cerebellar function. What is needed, he states, are "realistic models, which map physiological variables 1:1 with model parameters." **Hepp** likewise argues, though less emphatically, for large-scale cellular models. Although we agree that detailed cellular models are sometimes useful, they are not a panacea. Models are also needed at many other levels. For example, we need models that attempt to bridge between biophysical knowledge, on the one hand, and systems-level concepts, on the other (e.g., Yuen et al. 1995). We also need models comprised of simplified neurons to explore purely connectionist principles (e.g., Rumelhart et al. 1987). Modeling at any of these higher levels is, by its very nature, underconstrained, but *underconstrained* does not mean *unconstrained.* Our target article provides extensive examples of how the APG model is constrained by key anatomical, physiological, and behavioral observations. "Key" is the important word here, since not every detail is included. As was effectively stated by Koch and Segev (1989) in their introductory chapter to an influential book on this general topic, a network model "that incorporates all the known biophysical details is out of the question" and "such a simulation with a vast number of parameters will be as poorly understood as the brain itself."

Sultan et al. raise concerns about the suitability of the cerebellar architecture for pattern recognition. This is a fundamental tenet in the Marr-Albus model, and, as they point out, we also rely on pattern recognition in the APG model. These commentators challenge the usual assumption that granular layer processing achieves a sparse coding of the input to Purkinje cells (PCs); sparse coding is important since it is known to enhance the capacity for pattern recognition. Their anatomical assumptions appear to be at variance with the assumptions presented by Tyrrell and Willshaw (1992) in their definitive analysis of these anatomical design issues. Sultan et al. further suggest that physiological data are not compatible with the sparse coding concept, quoting earlier multiunit observations from the granular layer, which suggested that granule cells are massively activated by the onset of a cutaneous stimulus. Our recent single unit studies in awake monkeys led to the conclusion that most of the electrical activity picked up by an extracellular electrode in the granular layer is attributable to mossy fibers, as opposed to granule cells (Van Kan et al. 1993). If there were a massive activation of mossy fibers (rare in our experience with awake animals), it would nevertheless be compatible with the occurrence of sparse coding at the next stage of processing, in granule cells (Van Kan et al. 1993). Granule cells have very small axons and are extremely difficult to sample with extracellular electrodes, which leaves us relatively uninformed about the nature of coding in parallel fibers – at present, we are forced to rely on theoretical explorations of this issue. Sultan et al. also suggest that pattern recognition theories do not give a role to the fact that parallel fibers are parallel and that the PC dendrite is flat and fan-shaped. They seem unaware of Albus's (1971) point that this architecture provides for maximal convergence onto PCs, which improves the capability for pattern recognition.

Gilbert emphasizes the improved storage capacity that accrues from incorporating longitudinal zones of PCs into computational modules, which is an issue that he addressed in 1974 and that we fully embrace and have incorporated in the APG model. (**Hepp** wonders why this enormous storage capacity is required, pointing out that vestibular-ocular reflex (VOR) control is a relatively simple problem. However, most of the control problems confronted by the cerebellum are likely to be much more complex than simple VOR control.) An important additional feature that we have incorporated in the APG model is the coupling of this associative memory mechanism to the problem of pattern generation. Thus, Gilbert's model follows the Marr-Albus model in treating the cerebellar cortex as an associative memory capable of learning complex transformations, but this theory is restricted to nondynamical, memoryless transformations. Temporal features of motor programs produced in this manner have to be derived from time variations already present in the parallel fiber inputs to the cerebellum, which is inconsistent with single-unit data from the intermediate cerebellum. A comparison of the signals recorded from mossy fibers (Van Kan et al. 1993) and PCs (Harvey et al. 1977) indicates that neural signals undergo a major sensory-to-motor transformation between input and output, a transformation that is clearly dynamical. For example, movement-related mossy fibers are well modulated by either passive or active joint rotation, whereas PCs fire well only during active motion. This is one important reason why we introduced bistable elements to model PC dendritic regions; the flat segments of the bistable function can explain the weakness of sensory responses in PCs (Houk et al. 1990). Further, we emphasize that our hypothesis is not that entire PCs are bistable, but rather that PC dendritic subregions are bistable. This would make entire PCs into more complex multistable systems. A second important justification for bistability of dendritic subregions is the existence of plateau potentials in dendritic recordings from PCs, as discussed in Yuen et al. (1995). (These findings also answer Jaeger's concern for evidence of refractoriness after switching.) Gilbert, along with **Hepp** and several other commentators, suggests that this kind of bistability is not compatible with single-unit recordings, but, as we argue in our target article and in the Yuen paper, multistability with many levels of output clearly is compatible with the recording data. Dendritic bistability, and the resultant PC multistability, would provide a form of short-term memory that in the APG model contributes to the capacity to generate temporal patterns of output that are not time-locked to the input (Houk & Barto 1992). Positive feedback in the recurrent premotor networks that are regulated by PCs also contributes to pattern generation in the APG model. The capacity for temporal pattern generation is an important difference between our model and the Marr-Albus and Gilbert formulations.

Hesslow reviews an experimental observation that challenges our hypothesis that positive feedback in the cortical-rubral-cerebellar recurrent network is an important driving force for generating patterned motor commands. He electrically stimulated the region of the cerebellar cortex that controls one of the eye muscles that contributes to conditioned eyeblink. A brief electrical stimulus produced a brief pause in electromyographic (EMG) activity, rather than terminating the EMG burst. If the brief stimulus succeeded in terminating the burst, this would have provided support for the positive feedback hypothesis. In disagreement with Hesslow, we do not believe that the negative finding is strong evidence against the positive feedback hypothesis. One can think of many mechanisms that might have interfered with a simple outcome. For example, there are multiple loops in this system that control many muscles in a distributed fashion. The PCs stimulated by Hesslow may have inhibited only one of these loops. Furthermore, recent evidence suggests two stages in the output network, as reviewed in our target article. Although Hesslow's stimulus may have curtailed discharge in a subset of loops controlling the monitored EMG, one needs to record from the brainstem in order to determine the extent to which activity persisted in other loops. Hesslow's finding is important and needs to be pursued with studies designed to clarify the mechanism.

Miall et al. suggest that the cerebellum is not a motor pattern generator but instead a sensory mechanism for estimating system state based on motor commands and sensory feedback. We start by responding to their specific criticisms of the APG model and then express our opinion of the sensory theory espoused by them and by others. First, Miall et al. point out that positive feedback can be hard to control. While this is true, we refer them to our work showing how networks of neurons with realistic (nonlinear) relationships between synaptic current and firing rate are relatively easy to control with simulated PC input (Eisenman et al. 1991). While positive feedback in linear systems is hard to control, positive feedback in nonlinear networks is potentially advantageous (Hopfield 1982). Their second comment is a suggestion that transcranial magnetic stimulation might be used to artificially terminate movement commands (related to the **Hesslow** commentary discussed earlier). Werhahn et al. (1993) in fact demonstrated that this procedure reduces motor cortical excitability for 5 to 8 msec. Third, they point out that many PCs increase their discharge during movement, whereas pauses in discharge are required in the APG model to permit the buildup of positive feedback. We have discussed this in several articles; we postulated that both pauses and increased discharge are needed to shape positive feedback (e.g., see sect. 3.3 and Fig. 5). Pauses would permit the buildup of positive feedback in modules that innervate agonist muscles, whereas increased discharge would inhibit positive feedback from spreading to modules that innervate antagonist muscles.

In commenting on Thach's target article, **Miall et al.** suggest that the APG model is not capable of coordinating synergies. This is mirrored by **Kawato**'s suggestion that the APG model does not provide a mechanism for learning invariant features of multijoint arm movements on the basis of only target-error information. While we agree that we have not yet demonstrated how the APG model can coordinate complex synergies for 3-D multijoint arm movements, we have published work suggesting how this might be accomplished for a 2-D movement by an array of APG modules (Berthier et al. 1993). This issue was also discussed in a recent *Trends in Neuroscience* article (Houk et al. 1993). According to our view, individual APG modules are responsible for controlling single synergies, so that a set of synergies would be coordinated as a function of behavioral context by selecting the appropriate set of APGs and suitably weighting their influence on the overall motor command. This mechanism could very easily result in invariant features of arm movement. Although our modular pattern-generator approach is unusual from an engineering perspective, we do not believe that orthodox engineering practices exhaust the possibilities for achieving efficient control of complex systems. A major goal of our ongoing work is to demonstrate the control competence of the APG model.

Miall et al. share with **Bower** and with **Paulin** the hypothesis that the cerebellum functions as a sensory predictor or state estimator. This claim is difficult to reconcile with the salient motor deficits produced by lesions of the intermediate and medial cerebellum, but offers a potential explanation for the lack of clear motor deficits with lesions of the lateral cerebellum. A sensory designation may actually be more appropriate for the cerebellar-like structures that are phylogenetic elaborations of the dorsal cochlear nucleus (Mugnaini & Maler 1993). Apparently, the same neuronal architecture that works well for controlling movements can be used to advantage in controlling nonmotor functions. In the dorsal cochlear nucleus, these nonmotor functions probably fit a sensory designation, whereas in the lateral cerebellum the designation "cognitive function" may be more appropriate than "sensory," as discussed later. These commentators, instead, propose that all parts of the cerebellum perform sensory functions. Miall et al. point out that those parts of the inferior olive that innervate motor regions of the cerebellum have clear sensory properties. On this basis they suggest that the cerebellum has to learn a sensory function, because it is being taught by sensory signals. This reasoning is quite erroneous: Sensory signals are widely used to train control systems, and it has been known for some time that sensory feedback is important in shaping motor behavior.

Feldman & Levin review their concept of sensorimotor integration in limb movement control and suggest that cerebellar models should conform to it. We urge these authors to study the APG model, since it could provide an anatomical and physiological bridge between their abstract "frame of reference hypothesis" and the actual biology of the system. Their hypothesis relates in specific and relatively straightforward ways to our suggestion that PCs learn to recognize patterns of sensory and efference copy input that indicate when the endpoint of a movement is about to be achieved, thus terminating motor commands. While **Latash & Latash** suggest that PC synapses are not economical for storing such information, they provide no rationale for their unorthodox opinion. The concept of endpoint recognition in the APG model also bears obvious relations to the theory of equilibrium point control in force fields (Bizzi et al. 1995), an issue that was brought up by **Jaeger.** We are currently exploring how PCs might use delayed sensory feedback to predictively control endpoint positioning of a nonlinear spring-mass system (Barto et al. 1995).

Dean suggests that advances in understanding the role of the cerebellum in low-level control of saccadic eye movements could be used to refine and extend the APG model. We wholeheartedly support this suggestion in an article on this subject (Houk et al. 1992).

Bekkering et al. list three reasons why the APG model is incompatible with saccadic control, but it is difficult to understand their logic. First, they point out that saccadic commands must be recalibrated frequently throughout life. However, this presents no challenge to the APG model, since its adaptive mechanism has a continuously operating, low-magnitude potentiation process that counterbalances the depression triggered by climbing fiber activity. We presume that this adaptive mechanism operates throughout life. Second, they point out that saccadic eye movements often undershoot or overshoot and are then followed by corrective movements. Instead of being a problem for the APG model, precisely this source of variation is normally present in our implementation of the limb control model (Houk et al. 1990; Fig. 13.8B). The same mechanisms were suggested as important operational features in our conceptual extension of the APG model to saccadic gaze control (Houk et al. 1992). Third, these commentators suggest that the APG model would not adapt to displaced targets, whereas we made extensive use of the displaced-target data in developing our saccade model (Houk et al. 1992). We also note that the APG model does not operate open-loop as they claim; instead, it operates in a quasi-feedforward mode (Houk et al. 1990).

Kawato comments on several of our interpretations of his work. In section 1, he raises no objection to our conclusion that inferior olivary activity is basically a sensory signal, whereas his theory requires a motor representation. He simply states his faith that somehow there must be a sensory-to-motor transformation in the climbing fiber pathway. This is a hard requirement for his theory and difficult to rationalize physiologically. Although a precise transformation to motor coordinates would also be convenient for the APG theory, as discussed by Berthier et al. (1993), it is not a requirement. This is because PCs in the APG model learn simply when to switch their activity so as to terminate the motor command, whereas Kawato's theory requires PCs to respond in a continuous manner to inputs that designate a complete desired trajectory. In section 2 he responds to our hypothesis that desired trajectory signals are not generally present in the brain and are available only in special cases (e.g., from the semicircular canals for use controlling the VOR and conceivably from the parietal cortex for the control of pursuit eye movements). His comment that only a small portion of parietal neurons have firing patterns suitable for the construction of pursuit commands is very interesting. We agree that it indicates that a major sensory-to-motor transformation occurs in the cerebellar cortex. Kawato points out that Kalaska described signals in area 5 that encode the velocity profile, but these may originate as back-projections from the motor cortex, which is known to contain signals that can code movement velocity. His theory further requires that the desired trajectory include a specified acceleration.

The APG model is rejected by **Arbib** because he interprets it as suggesting that motor control is being mediated exclusively by the cerebellum. Rather, the cerebellar microzone is only one component of the APG and is responsible for sculpting the activity patterns that are set up within the recurrent premotor networks mentioned earlier. We have further proposed that these premotor networks are capable of mediating automatic actions in the absence of any PC regulation (Houk & Barto 1992). The output of an APG can thus be either an elemental command (downward activation of a muscle synergy), or an activation of higher cortical regions as discussed later. In either case, the role of the cerebellum posited by the APG model is consistent with Arbib's notion of schema modulation. In fact, one could view it as providing a biological mechanism whereby schema modulation might be implemented.

Arbib also wonders "Who teaches the teacher?" suggesting that "it is implausible that evolution should provide innate olivary circuitry to signal the position of a dart relative to a bullseye." Of course we agree that this is a critical question (Houk & Barto 1992), but we do not regard it as being as troublesome as Arbib suggests. We know from computational studies, for example, that learning is possible despite great variations in the quantity and quality of training information (Barto 1994). It is not true that a "teacher" needs to be particularly smart to provide information that a learning system can use to improve its behavior: The learning system needs to be correspondingly more sophisticated in its ability to take advantage of low-quality training information. In our target article we outline our current working hypothesis as to how simple sensory responses at the level of the inferior olive could be readily shaped into effective climbing fiber training signals, without any need of invoking an intelligent teacher.

De Schutter suggests that long-term depression (LTD) can only be induced when a climbing fiber signal precedes activity in a parallel fiber; if this were true, it would interfere with learning since the signals transmitted by climbing fibers are delayed indications of error. Apparently, he is not aware of the recent article by Chen and Thompson (1995) reporting the opposite timing. Commentaries by **Houk & Alford,** by **Arbib,** and by **Fiala & Bullock** discuss the intracellular signal transduction mechanisms that may be available to mediate this "trace" mechanism. Furthermore, the cellular mechanisms underlying the cerebellar learning rule have been discussed extensively in our target article and also in several prior publications (Houk et al. 1990; Houk & Barto 1992). Rather than repeating this material here, we refer De Schutter to these writings.

Swinnen et al. point out that the generation of new movement patterns is more complex than the adaptive adjustment of movement parameters addressed by most cerebellar models. Pattern generation has several aspects in addition to the generation of the bursts that control individual movement segments. We have proposed that spatial pattern generation, for controlling movement synergies, can be accommodated by an array of APG modules as discussed earlier in this reply and elsewhere (Berthier et al. 1993). In contrast, the pattern generation required for controlling movement sequences may be upstream from the cerebellar circuitry that controls limb movements. While sequence generation certainly involves circuits through the supplementary motor cortex and basal ganglia, it may include control loops through the lateral cerebellum as well.

Weiss & Disterhoft point out that a conditioned reflex in response to a conditioned stimulus (CS) that is physically presented requires an intact hippocampus in addition to an intact cerebellum. They review a variety of data indicating that a number of higher structures are required to support this form of learning, called trace conditioning. Their recent PET studies in human subjects indicate increased activity in the hippocampus, prefrontal cortex, and caudate nucleus, in addition to the cerebellum, even though the trace paradigm was not used. Might this activity nevertheless be related to the potential for forming traces?

Schmahmann comments on the potential role of the cerebellum in cognitive function. He reviews his and other data indicating that the cerebellum not only receives from but also controls output to cognitive processing regions such as area 46 of the prefrontal cortex. Fostering a cognitive function for the lateral cerebellum are behavioral deficits that could be designated "dysmetria of thought," in analogy with the classic motor deficits designated "dysmetria" (of movement) associated with deficits of intermediate and medial cerebellum. However, Schmahmann raises a problem that arises when one compares the climbing and mossy fiber inputs to the lateral regions of the cerebellum. Mossy fiber inputs originate from a considerably broader group of higher cortical areas that do the climbing fiber inputs. He believes that there is more correspondence between the sources of climbing and mossy fiber input to the motor areas of the cerebellum. This may not be the case. As reviewed in section 4.3 of our target article, excitatory input to the inferior olive appears to be dominated by somatosensory signals, with efference copy signals providing predominantly an inhibitory gating of the sensory input. Mossy fiber inputs instead show a much richer mixture of sensory, efference copy, and high-level features in their signaling. One interpretation of this is that the cerebellum uses training signals that are quite basic in comparison to the operations that it then coordinates, by using a wider diversity of mossy fiber input. This principle could apply as well to cognitive regions in species that have an expansion of the lateral cerebellum. This might allow training information derived from the more primary regions of the cerebral cortex to be used to train the cerebellum to use more complex signals deriving from associative regions of the cerebral cortex.

More on climbing fiber signals and their consequence(s)

J. I. Simpson,[a] D. R. Wylie,[b] and C. I. De Zeeuw[c]
[a]*Department of Physiology and Neuroscience, New York University Medical Center, New York, NY 10016.* **simpsjø1@popmail.med.nyu.edu;**
[b]*Department of Psychology, University of Alberta, Edmonton, Alberta, Canada, T6G 2E1;* [c]*Department of Anatomy, Erasmus University Rotterdam, 3000 DR Rotterdam, Postbus 1738, The Netherlands.*

Abstract: Several themes can be identified in the commentaries. The first is that the climbing fibers may have more than one function; the second is that the climbing fibers provide sensory rather than motor signals. We accept the possibility that climbing fibers may have more than one function – hence "consequence(s)" in the title. Until we know more about the function of the inhibitory input to the inferior olive from the cerebellar nuclei, which are motor structures, we have to keep open the possibility that the climbing fiber signals can be a combination of sensory and motor signals.

R1.1. More than one role for the climbing fibers. In asserting that we feel that climbing fibers can have only one function, **Thompson** seems to have failed to see the "s" in parentheses in our title, indicating that there could be several consequences to climbing fiber signals. We undertook our experiments in the spirit of trying to bring about some consensus by seeing what support could be found for several hypotheses in one particular system. Indeed, in trying to reach a consensus, which does not mean that only one function will prevail but rather that we simply reach agreement, we acknowledge the possibility that there could be more than one role for the climbing fibers in cerebellar function. In keeping with that view, we illustrate a Purkinje cell whose activity shows rhythm, post-complex spike facilitation of the simple spikes, and reciprocal modulation of complex and simple spikes in response to retinal image movement (Fig. 1). **Weiss & Disterhoft** suggest, as do we and other commentators, that the climbing fibers are not necessarily performing only one job. Thompson's comment that our title does not reflect our context overlooks the basic tenet that the cerebellum functions similarly in all its parts. We subscribed to that view in choosing our title and believe that investigations of the vestibulocerebellum in relation to compensatory eye movements in the rabbit will be generalizable to other parts of the cerebellum and to other species.

Contrary to **Thompson**'s claim that we are unfamiliar with the possible importance of the climbing fiber system in the context of learning and memory, we are familiar enough with the cerebellar learning and memory literature to know that it is rife with contradiction and replete with claims and counterclaims. The putative role of climbing fibers in motor learning was not given as much attention as other hypothesized roles because we have no recent experimental data of our own on this subject. It is surprising that Thompson did not cite the work of Ito's laboratory (Karachot et al. 1994) on the effect of the order of presentation of the complex and simple spikes on long-term depression. Ito and his colleagues reported that repeated application (every 20 sec for 10 min) of a parallel fiber stimulus train immediately followed by a climbing fiber stimulus train, as in classical conditioning, did not induce significant long-term depression. From other experiments done with pairing stimuli at frequencies of 0.25 to 4 Hz, Ito and colleagues concluded that long-term depression was inhibited when parallel fiber stimulation preceded climbing fiber stimulation by 10 to 100 milliseconds. Elsewhere, Thompson (Chen & Thompson 1995) points out that when pairing is done at higher frequencies, it is ambiguous which stimulus precedes and which follows. There seems, however, to be some possibility of sorting out this chicken-and-egg problem by using low frequencies and a greater range of parallel fiber–climbing fiber stimulus intervals.

Arbib acknowledges that he has one foot in the learning camp and the other in the real-time role camp. Like us, he could be seen either as a fence sitter or as open-minded. Arbib's question of who teaches the teacher is an important one. In the case of the visual climbing fibers to the flocculus, the experiments by Soodak et al. (1988) indicate that vision itself is not the teacher, because rabbits deprived of vision for several months from birth, upon eye opening, immediately show normal climbing fiber responses to retinal slip as well as reciprocity with the simple spikes. Either the reciprocal relation between simple and complex spikes is wired in, or there are "instruction" signals from the vestibular system and/or eye muscle proprioceptors that are conveyed, most likely through the prepositus hypoglossi, the ventral dentate, and the *y* group, and that produce climbing fiber signals that possibly lead to a reciprocal relation with the simple spikes when retinal slip is present.

Kawato points out that, during the ocular following response, the modulation of the visual complex spike activity in the monkey's ventral paraflocculus is in the wrong direction to produce an eye movement that would reduce retinal slip. This point was made in 1976 by Gonshor and Melville Jones on the basis of data then available from the rabbit. We hasten to point out, however, that because of the postinhibitory rebound burst (Llinás & Mühlethaler 1988), the net

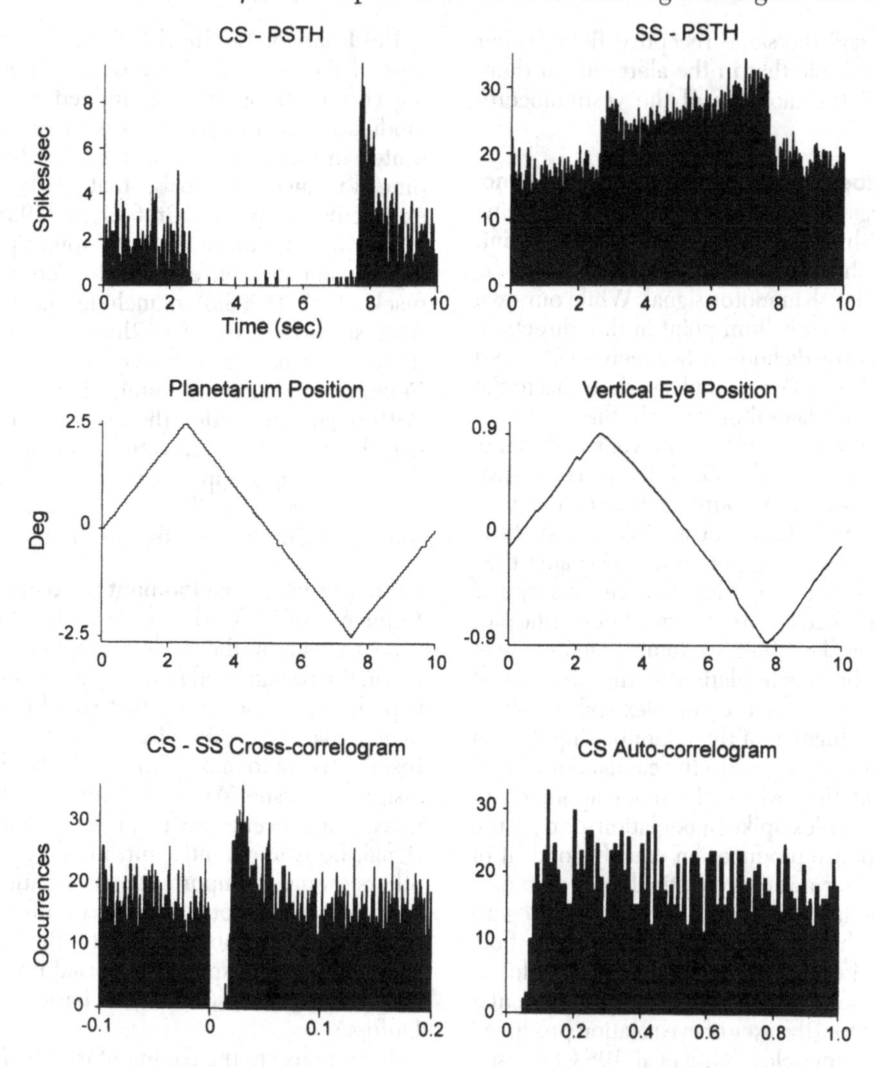

Figure R1. A floccular Purkinje cell, recorded in the awake rabbit, that has it all – rhythm, post-complex spike facilitation of the simple spikes, and reciprocal modulation of complex and simple spikes in response to retinal image movement. The complex and simple spike PSTHs (top row) were compiled from 60 stimulus cycles (binwidth 50 msec). A whole-field optokinetic stimulus was presented monocularly to the eye contralateral to the recorded flocculus. The stimulus consisted of constant speed rotation at 1°/second with reversal of the direction of rotation every 5 seconds (middle row, left). The axis of the planetarium projector was horizontal and at 45° contralateral azimuth. The vertical component of the eye movement is shown in the middle row, right. The complex spike-simple spike cross-correlogram (bottom row, left; binwidth 1 msec) shows that this cell was of the pause-facilitation type. Because of the near absence of the complex spike activity during half of the stimulus cycle, the increase in simple spikes after the pause occurred during that half of the stimulus cycle in which the simple spike activity was less than the spontaneous level, which was about 27 spikes/second. The three clear cycles of oscillation in the complex spike autocorrelogram (bottom row, right; binwidth 10 msec) show the presence of rhythm at about 9 Hz in the complex spike activity. The correlograms were obtained from the same recordings used for the PSTHs. (The data for this figure were obtained in collaboration with Sebastiaan Koekkoek).

effect of the complex spike on the activity of the recipient vestibular nuclei neurons could be an increase, and not a decrease, in firing. We also note that during vestibular stimulation with the rabbit in darkness, a minority of Purkinje cells show complex spike and simple spike modulations that increase and decrease together (De Zeeuw et al. 1995b). Furthermore, recordings made in other areas of the cerebellum (e.g., Kolb & Rubia 1980; Ojakangas & Ebner 1994) have revealed that the modulation of complex and simple spike activity is not necessarily reciprocal.

The commentary by **Roberts et al.** is thoughtful and provides an attractive way of looking at the seemingly disparate characterizations of climbing fiber function by drawing a parallel to the wave-particle duality of light. With regard to coupling modules we can only point out that, as stated in our article, synchrony was restricted to complex spikes that had the same receptive field proper ties. However, it is possible that in the alert animal there could be coupling of the modules of the vestibulocerebellum.

R1.2. Are climbing fiber signals sensory rather than motor? Several investigators of the cerebellum, including some represented in this issue by their commentary (**Miall et al.** and **Kawato**), have suggested that climbing fibers provide a sensory rather than motor signal. While our own studies in the vestibulocerebellum point in that direction, we do not believe that the dichotomy between sensory and motor is firmly established. For example, with regard to the dorsal cap and the ventrolateral outgrowth, the inhibitory input from the prepositus hypoglossi, the ventral dentate, and the *y* group (De Zeeuw et al. 1993; 1994) can influence both the excitability and the coupling between olivary neurons (e.g., Lang 1995; Llinás et al. 1974), and these nuclei are quite likely conveying motor signals because they have numerous neurons that are related to eye velocity and eye position. As mentioned above, in a small percentage of Purkinje cells in the awake rabbit, we found (De Zeeuw et al. 1995b) with vestibular stimulation in the absence of vision a modest modulation of the complex spikes, which likely is due to the modulation of the inhibitory input from the nucleus prepositus hypoglossi to the caudal dorsal cap.

Kawato notes that the inverse dynamics approach to characterizing the complex spike modulation during the ocular following response produces an equally good fit of the modulation whether a sensory (retinal slip) or motor (eye movement kinematics) reference is used. With regard to reference frames, the visual climbing fibers of the flocculus are organized in a reference frame that could be interpreted as being sensory (the axes of the semicircular canals) or as being motor (the axes of eye rotation produced by particular sets of eye muscles) (Graf et al. 1988; Simpson et al. 1981; Wylie & Frost 1993; in press). There may be some difficulty in using the inverse dynamics approach for some parts of the cerebellum outside the ventral paraflocculus and flocculus. For example, in the case of the vestibular climbing fiber signals carried to the uvula and nodulus (Barmack et al. 1993a), the spatial arrangement between the sensory axes (the axes of the semicircular canals) and the axes of the motor-command coordinates may be much more complicated than for the flocculus, because the uvula and nodulus control not only the eyes, but also most likely neck muscles. In line with this concern is the finding in the rabbit that the good spatial alignment of the best axes for complex and simple spike modulation found in the flocculus with optokinetic stimulation is not nearly so prevalent in the nodulus (Kano et al. 1991).

Feldman & Levin also draw attention to the usefulness of the concept of reference frames for understanding cerebellar operations. Indeed, the finding of visual modulation of complex spikes in the flocculus was presented in the context of a reference frame, which at that time was suggested to be that of the canals and/or the extraocular muscles (Graf et al. 1988; Simpson et al. 1981). The notion of complex spikes providing signals in a particular reference frame has been extended by Barmack et al. (1993a) to include the vestibular climbing fiber signals conveyed to the nodulus and uvula. While these reference frames may not be the ones to which Feldman and Levine refer, their notion is intriguing. With regard to shifting the origins of reference frames, it may be that the prepositus hypoglossi and the ventral dentate and *y* group act, if not to shift the origin of the reference frame as the eye moves in the head, then to shift its orientation relative to the head.

R1.3. The credit assignment problem and other issues. **Fiala & Bullock** address what has become a common concern, namely, the credit assignment problem. This issue was in the background for many years, so it is important that hypotheses are advanced that could lead to telling experiments. We have difficulty understanding the point **Dufossé** is trying to make. Perhaps he is addressing the credit assignment issue. We are not sure. While he agrees that an unexpected event can result in an olive-mediated error signal, he is apparently untroubled by what to us and to others seems an important consideration, namely, that the complex spike error signals occur after those simple spike signals that presumably led to the error. **Miall et al.** distinguished unexpected external events from motor errors, although we see them as basically the same, as does Dufossé.

In response to the commentaries by **Hepp, Jaeger,** and **De Schutter,** we can but agree that there is the need for more cooperation between systems neurophysiologists, biophysicists, and modelers in order to arrive at some agreement as to how the cerebellum operates. We also agree that we need to know the natural patterns of parallel fiber activity as well as the activity patterns of the intrinsic inhibitory cortical neurons. Like Hepp, we have often wondered why, indeed, there are so many Purkinje cells, especially in the rabbit, where there seems to be only one general class of Purkinje cell in each zone of the flocculus. All we can say at this time is that there is a range of eye position and eye velocity signals on these cells, but the significance of the range for controlling eye movements is unknown.

Resilient cerebellar theory complies with stiff opposition

Allan M. Smith
Centre de Recherche en Sciences Neurologiques, Département de Physiologie, Université de Montréal, Montréal, Québec, Canada H3C 3J7.

Abstract: In response to several requests from commentators, an unambiguous definition of time-varying joint stiffness is provided. However, since a variety of different operations can be used to measure stiffness, a problem for quantification admittedly still exists. Several commentaries pointed out the advantage of controlling joint stiffness in optimizing the speed–accuracy trade-off known as Fitts's law. The deficit in rapid reciprocal movements and the impact on joint stiffness inhibition caused by cerebellar lesions is clarified here, as the target article was apparently misinterpreted by some readers. In response to the challenge that there is little consensus among cerebellar physiologists, several areas of tacit agreement with other theories of cerebellar function are enumerated. An alternative interpretation of studies showing a transient activation of the cerebellum in motor learning is suggested. Finally, the relationship between the command signals generated by supraspinal centers such as the cerebellum and spinal interneuron networks controlling muscle synergies is discussed.

As intended, the suggestion that the cerebellum plays an important role in controlling time-varying joint stiffness seems to have evoked a variety of reactions ranging from some mitigated agreement to diametric opposition. However, regardless of the merits of this hypothesis (or the lack thereof), I will have achieved my ultimate satisfaction if the target article successfully provokes even a single fresh experiment yielding some new perspectives on the cerebellum.

Perhaps the most fundamental criticism of the target article concerns whether the cerebellum plays *any* role in controlling joint stiffness or whether this parameter simply emerges as a dependent function of individual muscle activation patterns. The target article freely admitted this lacuna (reiterated by **Paulin**) and was intended to issue the challenge of finding evidence for or against this hypothesis. In addition, several commentators (**Feldman & Levin, Latash & Latash, Van Galen et al.**) pointed out a general need for specifying an unambiguous operational definition of joint stiffness. Other commentators (**De Schutter, Jaeger, Paulin, Van Galen et al.**) felt that the diversity of opinions expressed by HOUK, THACH, and myself reflected the unfettered imaginations of the authors unconstrained by a sufficiently confining number of facts. Since I firmly believe this is not the case, I tried to identify a few aspects of cerebellar function about which there is a general consensus, if not full agreement. In addition, some comments were directed to specific issues such as the cerebellar control of spinal interneurons and the significance of studies implying a transient and cognitive function for the cerebellum in motor learning. I hope the following remarks will sharpen the focus on substantive and potentially resolvable issues rather than merely perpetuating a polemic on matters of opinion.

R1. The definition of time varying joint stiffness. Offering a general definition of joint stiffness as the relationship between change in torque and angular position (described in the target article as the product of a displacement vector and a stiffness matrix) is relatively straightforward, although arriving at an operational measurement of joint stiffness is considerably more complicated. A practical obstacle to studying joint or limb stiffness is the need to apply repeated force-pulse perturbations, which by themselves provoke modifications in subsequent voluntary movements. These perturbations may be either discrete force pulses of fixed amplitude, rate, and plane of application, or they may be continuous and randomly varying. Even though both methods may be thought to measure joint stiffness, the actual results may differ depending on whether or not feedback is included. It should also be pointed out that the term *joint* does not imply any special anatomical or physiological significance (except as a place where muscles can influence the mechanical impedance of the limb) and the fact that cocontracting biarticular muscles can increase the stiffness of several limb segments is an inherent feature of the physical plant.

Despite the suggestion by **Latash & Latash** that such a definition of joint stiffness presents an insurmountable conceptual difficulty (because joints rotate and therefore cannot store and release potential energy), there appears to be ample precedent in mechanics for describing torsion or rotational stiffness. Moreover, if individual muscles have stiffness as a result of their elastic properties, then it follows that muscles stretching across a joint must impart some of that stiffness to the joint. Stiffness, therefore, appears to be a useful term to describe the complex mechanics of multisegmented limbs. Nevertheless, I do agree with **Feldman & Levin** that joint stiffness is an emergent property comprised of many mechanical components. This emergent property is what Bennett et al. (1992) have called "a lumped-parameter model" of muscle, tendon, and proprioceptive feedback dynamics. In this model, the short-range muscle stiffness, the length-tension properties, the force-velocity relationship, and reflex feedback are amalgamated into a single convolved parameter to describe joint or limb stiffness.

R2. The advantages of controlling joint stiffness. All joints have some associated stiffness at rest and by virtue of the torques developed by the muscles that stretch across them, and their mechanical stiffness varies throughout each movement. The essential and fundamental question raised by the target article is whether the nervous system, and in particular the cerebellum, actually regulates limb or joint stiffness or whether this parameter is simply an incidental byproduct of the overlapping activity of agonist and antagonist muscles. In addition, several authors (Hasan 1986; Hogan 1990; Kawato & Gomi 1992; Lacquaniti et al. 1992; Milner & Cloutier 1993) have described numerous mechanical advantages to controlling joint stiffness, and I am particularly grateful for the comments by **Gomi** and **Van Galen et al.** for adding the important parameter of movement accuracy to this growing list. Van Galen and Schomaker (1992) have indicated how modulating joint stiffness is a more likely means to modulate Fitts's speed–accuracy trade-off law than peripheral feedback modulations. However, the difficulty, as **Paulin** correctly pointed out, concerns how stiffness control differs from simply controlling the onset and offset of muscle activity.

R3. The cerebellar syndrome and its relation to joint stiffness. An independent control of stiffness might be demonstrated in several ways, and some of these were mentioned in the target article. In my experience, the activity of single cerebellar cells is usually related to single muscles or muscles with similar mechanical actions rather than to muscles with mechanically antagonistic or opposing actions. As stated in the target article, the activity of Purkinje cells appears to be inversely related to whether the muscles are performing in a reciprocal or cocontracting mode. It follows, therefore, that single-cells cannot control joint stiffness, which can only be achieved by the simultaneous activity of populations of neurons related to pairs of antagonist muscles. Contrary to the suggestion by **Hore,** there is little if any disagreement between the proposal made in the target article and his suggestion that the onset of agonist and antagonist EMG is delayed after cerebellar lesions. The target article never suggested that cocontraction is a major feature of cerebellar *ataxia.* Instead, it stated that poorly modulated cocontraction is a fundamental characteristic of the difficulty in performing rapid successions of alternating movements and that this deficit is exhibited by many cerebellar patients. Joint stiffness is the essential issue, however, since delays in muscle contraction and relaxation times can create distortions between the timing of EMG patterns and the actual joint stiffness changes. The primary deficit following cerebellar lesions appears to be in the timing and scaling of rapid voluntary muscle activation, which is ataxia.

Ideally, the time-varying modulation of joint or limb stiffness could be investigated in unilateral cerebellar-damaged patients and compared to the stiffness profiles in the unaffected limb, which is matched for its passive mechanical properties. A recent study by Bastian and Thach (1995) indicates that cerebellar-damaged patients compensate poorly for the intersegmental torques generated during multijoint movements, but whether this implies a direct cerebellar control over joint stiffness remains to be proven.

R4. Areas of tacit agreement about cerebellar function. In my opinion, there is more agreement than disagreement about cerebellar function than might be supposed from reading the accounts by HOUK, THACH, and myself. Cerebellar physiology is now a vastly wider domain than it was even a few decades ago, and areas of tacit agreement are often inadvertently not made explicit. On an abstract level, I believe there is now a general consensus that the cerebellum operates as a feedforward controller as opposed to relying exclusively on sensory feedback and reflex modulation. The ultimate superiority of one anticipatory computational model over another will be determined by the model's predictive merits, regardless of whether this control model takes the form of an adaptable pattern generator (HOUK et al.), a Smith (no-relation) predictor (Miall et al. 1993), a Kalman filter (Paulin 1989) or an inverse dynamics calculator (Kawato & Gomi 1993).

A second area of tacit agreement among the majority (but certainly not all) of cerebellar physiologists is that some sort of synaptic plasticity takes place within the cerebellar cortex. Although the site and mechanisms are currently hotly debated, there appears to be substantial support for the notion that associative learning induces plastic changes somewhere within the cerebellar cortex.

A third aspect, about which I entirely agree with Thach (1992), is that the cerebellum is intimately involved in the formation and coordination of muscles synergies. Moreover, as Thach suggested, the long parallel fiber beams are well suited to simultaneously coordinate synergies in the diverse muscle groups and to unite the somatotopically fractured afferents within the granular cell layer (Bower & Woolston 1983; Shambes et al. 1978). In the same way, the cerebellum may associate teleceptive, particularly visual, information as was suggested in an earlier target article by Stein and Glickstein (1992).

R5. The transient activation of the cerebellum in motor learning. Bower pointed out that recent PET studies of the metabolic activity of the cerebellum in human subjects show transient activations in a surprising range of behaviors, particularly in relation to task difficulty. Similarly, multiple floating wire recordings of single cerebellar nuclear cells in cats learning complex forelimb movements also show that the magnitude of the task-related modulation reaches a peak at the time when the task is first performed reasonably well, but then progressively decreases as the task becomes well practiced (Milak et al. 1995). While these observations are undisputed, they are open to various interpretations. Typically, subjects attempting to perform a difficult visuomotor tracking task are initially hampered by excessive cocontraction, especially if position control is exacting. As performance improves with practice, this stiffness caused by antagonist cocontraction usually subsides. In other cases, where subjects have been given complex puzzles that require planning moves several steps in advance, the transient cerebellar activity may have more to do with the long-range planning of movements. That is, the transient changes in cerebellar activity during the acquisition of motor skills may be explained by the mental rehearsal of potential movements and the progressive attrition of nonessential muscle activation accompanying skill acquisition.

R6. Commanding, integrating, and coordinating muscle synergies through spinal interneurons. Several commentators objected that the role of the cerebellum in sensorimotor integration had been neglected in the target article. **Jaeger,** for example, suggested that given our general ignorance about the spinal circuits controlling muscle synergies and our only rudimentary knowledge of cerebellar activity patterns, it is not surprising that the cerebellar contribution to the control of movement is confusing. Nonetheless, I cannot fully agree with either of these assertions. Neurophysiologists since Sherrington have devoted considerable effort to elaborating the spinal circuits controlling muscles synergies. In addition to the study by Bizzi et al. (1995) cited by **Jaeger,** Swedish scientists led for several decades by Prof. Lundberg have devoted considerable effort to elaborating the spinal interneuron networks controlling muscles synergies; current investigations into the spinal rhythm generator for locomotion and the upper cervical spinal network for reaching movements are prime examples (see recent review by Hultborn & Illert, 1991).

How supraspinal commands select and switch between spinal networks associated with different synergies was the theme of the commentary by **Arbib.** His suggestion that locomotion deficits associated with cerebellar lesions reflect an inability to integrate postural and locomotor synergies seems both reasonable and worthy of further investigation. Specifically, how segmental pathways integrate the commands to cocontract or reciprocally activate antagonist muscles is a very important issue in motor control. However, as **Feldman & Levin** pointed out, in my formulation the control signals produce reciprocal or coactive EMGs that determine the joint or limb stiffness, whereas according to their positional frame of reference theory, the R and C commands are hypothetical constructs that are independent of the EMG but which critically depend on proprioceptive feedback to define the spatial frame of reference. The identification of the Ia inhibitory interneuron (Jankowska & Roberts 1972) and its inhibition by Renshaw cells (Hultborn et al. 1979) were important contributions to our understanding of how peripheral afferents control reciprocal inhibition and antagonist cocontraction. My personal bias is that analogous reciprocal or coactive command circuits may also be found within the motor cortex, basal ganglia, and cerebellum, and to some degree all three structures may contribute (but in different ways) to joint or limb stiffness but without the necessity of proprioceptive feedback. This assertion is supported by evidence from different sources. For example, the inhibition of a tonically active antagonist EMG occurs even before EMG onset in the agonist muscle (Hufschmidt & Hufschmidt 1954), and is clearly seen in patients with peripheral large-fiber neuropathy (Forget & Lamarre 1987). Also, most forms of spasticity (pathological cocontractions) are not caused by hyperreflexia and persist after rhizotomy. Instead, they appear to result from some disturbance of the open-loop reciprocal command signal produced by lesions of the central nervous system.

Single-cell recordings in awake animals have registered the discharge patterns of cerebellar cells in a wide variety of different movements such as locomotion, reaching, grasping, and oculomotor saccades. Moreover, cerebellar responses to diverse perturbations have been studied in all these movements. If our knowledge of cerebellar discharge patterns is still as rudimentary as **Jaeger** suggests, it is certainly not for lack of trying. Neurologizing on the basis of incomplete knowledge is by no means unique to the cerebellum and historically it has proven to be both a necessary and useful endeavor. Finally, although **Paulin** feels that the current views of cerebellar function are both too diverse and less than compelling, I am convinced that the very diversity of opinion will ultimately be the source of better and more convincing theories in the future.

Q: Is the cerebellum an adaptive combiner of motor and mental/motor activities? A: Yes, maybe, certainly not, who can say?

W. Thomas Thach

Department of Anatomy and Neurobiology, Washington University School of Medicine, St. Louis, MO 63110. **thachw@thalamus.wustl.edu**

Abstract: The target article reviews studies of human cerebellar functional imaging and ablation in motor learning. This material is integrated with anatomical and physiological work in laboratory animals. A synthesis is presented as a working hypothesis on how the cerebellum might adapt, learn, and coordinate movement. An attempt is made to extend the motor role to certain mental operations, particularly those associated with mental movement. These notions met with varying enthusiasm. A few commentators seemed pleased with the overall result and offered supportive ideas. Some were critical of the proposed contributions to cognition and motor learning. A surprising number insisted that the cerebellum was more concerned with sensory processing than with movement management. Some thought we know too little to attempt such a synthesis.

R1. Introduction. I am humbled that so many should take such pains to reply. The commentary addressed three main points: (1) the extension of a basic cerebellar function to cognitive operations and the contributions of human studies, (2) the idea that what the cerebellum might contribute is context–response linkage and combination of responding elements through learning, and (3) the question of whether we yet know enough to think about cerebellar control mechanisms generally. Commentary ranged from positive to negative.

R2. A basic cerebellar function in cognitive operations? **Wessell** offers new evidence from studies of human performance of sequences of movements that supports the idea of cerebellar contributions to cognition in a motor domain. Specifically, he argues that the cerebellum participates in the storage of sequences. **Schmahmann,** like the Leiners, believes that the cerebellum should be able to influence many other cerebral operations, including attention, perception, and emotions. His argument is that the cerebellum receives from cerebral areas associated with these activities. My position is that until these cerebral areas are shown to receive from the cerebellum or these activities are shown to be impaired by cerebellar lesion, such arguments are speculative.

Gielen would have the cerebellum engaged in more complex as well as corrective movements. But what is the limit? Do we forget how to tie our shoelaces (or necktie or any other knot) after cerebellar lesions? Are there alternative ways of knowing a knot, and do we have to relearn it after cerebellar lesions? Musical pieces? Driving a car? Is the sequence to any extent stored in the cerebellum? Opinions differ; more work is needed.

Gilbert agrees with the target article on motor learning, but not on "conscious thinking." It may be that our differences are semantic; what I suggested was a cerebellar contribution to frontal cerebral cortical operations to render them automatic. It would be of interest to study mental motor imagery in patients with cerebellar lesions. Ivry et al. (1988), Fiez et al. (1992), and Van Mier et al. (1995) have shown right and left cerebellar hemispheric activity that appears to generalize to abstract functions independent of moving body parts (cf. Thach 1996).

O'Mara compares explicit (involving hippocampus) and

implicit (involving cerebellum?) learning. Opinion is divided over whether performance errors confound assessment of motor learning in cases of cerebellar lesions. We have evidence that performance and adaptation are dissociable (Martin et al. 1996a; 1996b), and that the cerebellum is involved in both. Human cerebellar subjects who have difficulty with adaptation are aware of the difficulty, as O'Mara predicts. Normal subjects who can adapt do so in a way into which they have little or no insight. Once they have adapted, performance is automatic, unconscious. The question is whether the cerebellar output connections to the frontal lobe association cortex cause those areas to be active unconsciously. We have suggested that they might.

R3. A basic role in motor learning? Kawato reports on correlation studies of complex and simple spikes related to the inverse dynamics of moving the eye. The inverse correlation of the two spike forms triggered in the same spatial dimensions are viewed as compatible with the feedback error learning model. He views the proposed cerebellar contributions to cognition as possibly related to the development of a forward model of the controlled body part, a strategy often used in computations. "Mental movement" would presumably reflect the operations of such a forward model.

Arbib argues for the more traditional view of brain motor control including the cerebellar control models of Luciani and Holmes. Thus, the cerebellum tunes downstream motor elements, and there is no need for replacement of a feedback system by a feedforward system. Rather, once an adjustable feedback system is well-tuned, it can expect smaller errors and thus proceed effectively at velocities high enough to appear "ballistic" unless perturbations yield an "unexpectedly" large error. By contrast, the basal ganglia and cerebral cortex are given full executive status for combining the pieces that make up a skilled behavior.

My colleagues and I have taken an opposite view, arguing for the executive and combining functions of the cerebellum for automatic movements (Thach et al. 1992a; 1993a; 1993b) and the permissive activity of the basal ganglia in executive functions of the cerebral cortex (Mink & Thach 1991a; 1991b; 1991c). A cerebellectomy is truly disabling. Damage of the output of the basal ganglia may be of little consequence, and it may actually improve movement in Parkinson's disease.

Weiss & Disterhoft add human studies implicating the cerebellum in conditioned eyeblink, but suggest it may be a latent pathway and therefore an adaptation rather than associative conditioning. **Hallett** addresses the motor learning issue, urging that the cerebellum may participate fully in adaptation of motor patterns but only partially in the acquisition of an entirely new skill, which has to be more widely represented in the nervous system. Haggard's study is given a similar interpretation. **Timmann & Diener** agree with the caveats as to interpretation of imaging data and what little they tell us of cellular events, but believe also that there is really no clear evidence of a cerebellar role in motor learning beyond that of gain control. Finally, **Swinnen et al.,** like Hallett, point out that we lack a definition of what motor learning really is.

To these I would answer that the response to novel perturbation is learned and predictive (Horak 1990; Horak & Diener 1993), that compensation for interactive torques is predictive and probably learned (Bastian et al. 1996), and that gaze-throw calibration in prism-adapted throwing is certainly learned and satisfies definitions of acquired skill (Martin et al. 1996a; 1996b). All are dependent on the cerebellum.

R4. Out with the old: But is the new really new? Flament & Ebner recall the idea of a comparator as the model for cerebellar function. This idea is certainly indebted to the ideas of Weiner (1948) on the comparison of a command and a feedback and on the comparison of two signals generally causing a state-change at the output. What is the evidence for it in the cerebellum? The existence of overlapping motor command and feedback inputs is not unique to the cerebellum. However, the addition of a vision-gated head velocity signal by the inhibitory Purkinje cell onto the medial vestibular nuclear cell to cancel the VOR may well be (Lisberger & Fuchs 1974). The subtraction of the effects of a self-generated electric discharge on the sensory systems of the electric fish may also be unique to the cerebellum (Bell et al. 1995). However, the Purkinje cell seems wired by its many parallel fibers to bring to it many different kinds of information, other than simple command-response signals. Certainly, the firing of the Purkinje cell is not a bistable state, but rather a high-frequency continuum of graded levels of activity. These features favor in my view the idea of a "context recognizer," which could be viewed as a generalized comparator.

As to what the learning-related changes are in the PET scan, I use Ojakangas & Ebner's (1992) evidence as well as our own (Gilbert & Thach 1977) to suggest that they are caused by climbing fiber activity. These may well be comparators, as Oscarsson (1980) first proposed, as well as error detectors.

Sultan et al. propose that the Purkinje cell reads coincidences in parallel fiber activity and is thus able to detect sequences of those events that might cause line-up ("tidal wave") in parallel fiber spike activity. Such a mechanism would seem best able to read "peaks" in parallel fiber activity, whereas Purkinje cell discharge in awake animals varies continuously over a wide range of frequencies.

Miall et al. favor a cerebellar role as a predictor, but not as a motor pattern generator. **Feldman & Levin** say we may have difficulty knowing more about the cerebellum until we know more about the rest of the system and the various possible frames of reference. **Latash & Latash** don't like the narrowness of explaining cerebellar function in terms of stiffness control. They also argue that memory storage must not be at the level of synapses but rather of molecules.

De Schutter has a number of reservations about LTD implementing behavioral plasticity, and points up the need for communication and work between molecular, cellular, and systems scientists (the point of this symposium!). His work with Bower shows amplification of the effects of parallel fiber contacts on the distal dendrites of the Purkinje cell, "so that even inputs contacting the most distal parts of the dendrite have the same access to the soma as more proximally located inputs" (De Schutter & Bower 1994b). This could counterbalance the larger effect of the vertical granule cell axon on the more proximal dendrites.

Fiala & Bullock discuss how metabotropic mechanisms might provide the delays needed to time the movement initiator function with the feedback error modifiability function that would be required for the two to interact.

Jaeger would prefer to have all the information in hand from molecules on up before attempting any kind of synthesis. His comment raises the question of how well one can really collect useful information unless one has some idea of what one is looking for. His view of "wait and see" appears to me too passive and pessimistic; the collection of data and the imagination of what it might mean usually go hand-in-hand.

Bower suggests that the emphasis on motor control has been a big mistake. The cerebellum is viewed as a sensory analyzer and not as a movement generator, much less a motor learner. **Van Donkelaar** suggests that the cerebellum may play a direct role in visual perception. Bower points out that cerebellar cells can be activated by passive somatosensory stimulation more than active movement in both animal and man, and that the sensory analysis is likely to affect sensory perception as well as motor control. He has suggested that the goal of movement may be the acquisition of sensory information! **Paulin** provides a delightful display of metaphor, rhetoric, sarcasm, and punning to the same end.

To these I answer: The cerebellum does indeed receive a lot of sensory information. This is why Sherrington called it the "head ganglion of the proprioceptive system." Snider (Snider & Stowell 1994; Snider & Eldred, 1952) showed that touch, sound, and visual information go there too. Lateral line systems go there, and so do peripheral "c" fibers and, of course, second-order vestibular nerve afferents. In some animals, these inputs comprise most of the input to the cerebellum. But the sensory inputs are not the only inputs, which, especially in man, are dwarfed by the cerebral association area input. There is an emphasis on the motor and premotor cortex at the level of the deep nuclei, but frontal parietal and occipital cortex join these in sending to cerebellar cortex. This was appreciated by Adrian (1943) and extended by Allen and Tsukahara (1974) and their colleagues, and Glickstein et al., 1994. However, there is a lot of sensory input, enough to activate single mossy fiber afferents in unit-recording studies in laboratory animals and to "light up" cerebellar cortex input stages in human PET studies when sensory receptors are stimulated.

Does this mean that the cerebellum is a sensory analyzer? I believe that the reasoning behind this idea hinges on two errors of logic.

(1) *Error of logic in PET interpretation.* First, PET activity is correlated with blood flow, which in turn correlates with activity that is mostly in nerve presynaptic terminals. There are lots of these in the cerebellum. They are easily activated in decerebrate and/or anesthetized animals who neither feel nor move. According to the circuitry, activity in these afferent terminals could lead to excitation, inhibition, or no changes in the activity of the output nuclei. A cerebellar cortical PET "lightup" could mean that the output from the nuclei is turned off rather than on, or that there is no change at the output. The point is that an increase in blood flow in this tissue does not necessarily mean that the cerebellar output is active. [See also multiple book review of Posner & Raichle's *Images of Mind, BBS* 18, 1995.]

The results are of interest, and a caveat to PET and fMR scanners. They suggest to me that one can activate inputs to a major structure and not necessarily lead to its output. It suggests that the structure may make a decision about what to do about the input activity. That's comforting: it gives the structure something to do beyond merely acting as a relay. The next generation of PET interpretation will look at synaptic activity on cerebellar output targets to see if a structure is putting out or not.

(2) *Error of logic in naming neurologic structures.* Despite all the sensory (and other) input to the cerebellum, what is impaired in humans when the cerebellum is damaged is active movement. For a long time, it was thought that this was all that was impaired. The one exception has been weight discrimination, but one has to move an object to know the weight. Recently, Grill et al. (1994) have shown that there are slight but detectable impairments in judgments of acceleration, velocity, and position of passively moved limbs. These in turn are thought to be mediated at least in part by muscle spindle afferents, and it has long been known that the cerebellum controls muscle spindles, as well as other muscles, via gamma motor neurons. No one has ever claimed that cerebellar lesions in humans cause serious or even significant sensory deficits. Humans never complain of major sensory abnormalities. Other parts of the nervous system are quite able to process sensation without the cerebellum. This has been very carefully tested; it is unlikely to be wrong.

Would any one want to rename the alpha motor neuron? It also has an important sensory input – from the Ia muscle spindle afferent. In a micro-PET scan, these active sensory terminals would "light up" the motor neuron pool. But the motor neuron has many other inputs that can also activate it under other conditions. The output of the motor neuron goes only to muscle; its activity causes only movement. That is why we call it a motor neuron. No one has ever proposed that it be called a sensory neuron.

Would anyone except **Paulin** ever propose that a windscreen might be confused with the motor functions of an automobile? If his experimental subjects were human, they could tell him that it is there to keep the weather, fauna, flora, and other noxae out of one's face. Is the cerebellum like a windscreen? Only if one is moving backward.

References

Letters "a" and "r" appearing before authors' initials refer to target article and response, respectively.

Aas, J.-E. & Brodal, P. (1988) Demonstration of topographically organized projections from the hypothalamus to the pontine nuclei: An experimental study in the cat. *Journal of Comparative Neurology* 268:313–38. [JDS]

Abbie, A. A. (1934) The projection of the forebrain on the pons and cerebellum. *Proceedings of the Royal Society of London, Series B* 115:504–22. [JDS]

Abeles, M. (1991) *Corticonics: Neural circuits of the cerebral cortex.* Cambridge University Press. [FS]

Abeliovich, A., Chen, C., Goda, Y., Stevens, C. & Tonegawa, S. (1993a) Modified hippocampal long-term potentiation in PKCγ mutant mice. *Cell* 75:1253–62. [aMKan]

Abeliovich, A., Paylor, R., Chen, C., Kim, J. J., Whener, J. M. & Tonegawa, S. (1993) PKCγ mutant mice exhibit mild deficits in spatial and contextual learning. *Cell* 75:1263–71. [aMKan]

Abrams, R. A., Dobkin, R. & Helfrich, M. (1992) Adaptive modification of saccadic eye movements. *Journal of Experimental Psychology, Human Perception and Performance* 18:922–33. [HB]

Adams, J. A. (1971) A closed-loop theory of motor learning. *Journal of Motor Behavior* 3:111–49. [aJCH, HB]

(1977) Feedback theory of how joint receptors regulate the timing and positioning of a limb. *Psychological Review* 84:504–23. [aJCH]

Adrian, E. D. (1943) Afferent areas in the cerebellum connected with the limbs. *Brain* 66:289–315.

Aggelopoulos, N. C., Duke, C. & Edgley, S. A. (1994) Non-uniform conduction time in the olivocerebellar pathway in the anaesthetized cat. *Journal of Physiology (London)* 476.P:26–27. [aJIS]

Aiba, A., Kano, M., Chen, C., Stanton, M. E., Fox, G. D., Herrup, K., Zwingman, T. A. & Tonegawa, S. (1994) Deficient cerebellar long-term depression and impaired motor learning in mGLuR1 mutant mice. *Cell* 79:377–88. [aJCH, aMKan, MKan, TH, rMKan, rDJL, rSRV]

Akase, E., Alkon, D. L. & Disterhoft, J. F. (1989) Hippocampal lesions impair memory of short-delay conditioned eyeblink in rabbits. *Behavioral Neuroscience* 103:935–43. [CW]

Akazawa, K., Milner, T. E., & Stein, R. B. (1983) Modulation of reflex electromyogram and stiffness in response to stretch of human finger muscle. *Journal of Neurophysiology* 49:16–27. [aAMS]

Akshoomoff, N. A. & Courchesne, E. (1992) A new role for the cerebellum in cognitive operations. *Behavioral Neuroscience* 106:731–38. [aWTT]

Akshoomoff, N. A., Courchesne, E., Press, G. A. & Irague, V. (1992) Contribution of the cerebellum to neuropsychological functioning: Evidence from a case of cerebellar degenerative disorder. *Neuropsychologia* 30:315–28. [aWTT]

Albus J. S. (1971) A theory of cerebellar function. *Mathematical Bioscience* 10:25–61. [aFC, arJCH, aJIS, CG, PFCG, EDS, FS, JDS, aWTT]

(1975) A new approach to manipulator control: The cerebellar model articulation controller (CMAC). *Transactions of the ASME: Journal of Dynamic Systems, Measurement, and Control* 97:220–27. [aJCH]

(1981) *Brains, behavior and robotics*. Byte Books. [aJCH]

Alexander R. M. (1989) Optimization and gaits in the locomotion of vertebrates *Physiological Reviews* 69:1199–1227. [aAMS]

Allen, G. I. & Tsukahara, N. (1974) Cerebrocerebellar communication systems. *Physiological Reviews* 54:957–1006. [KW]

Alley, K. A., Baker, R. & Simpson, J. I. (1975) Afferents to the vestibulo-cerebellum and the origin of the visual climbing fibers in the rabbit. *Brain Research* 98:582–89. [aJIS]

Andersson, G. & Armstrong, D. M. (1985) Climbing fibre input to *b* zone Purkinje cells during locomotor perturbation in the cat. *Neuroscience Letters Supplement* 22:S27. [aJIS]

(1987) Complex spikes in Purkinje cells in the lateral vermis of the cat cerebellum during locomotion. *Journal of Physiology (London)* 385:107–34. [aJIS]

Andersson, G. & Oscarsson, O. (1978) Climbing fiber microzones in cerebellar vermis and their projection to different groups of cells in the lateral vestibular nucleus. *Experimental Brain Research* 32:565–79. [aAMS, aJIS]

Aniksztejn, L. & Ben-Ari, Y. (1991) Novel form of long-term potentiation produced by a K+ channel blocker in the hippocampus. *Nature* 349:67–69. [LJB]

Antziferova, L. I., Arshavsky, Yu. I., Orlovsky, G. N. & Pavlova, G. A. (1980) Activity of neurons of cerebellar nuclei during fictitious scratch reflex in the cat: 1. Fastigial nucleus. *Brain Research* 200:239–48. [aWTT]

Appollonio, I. M., Grafman, J., Schwartz, M. S., Massaquoi, S. & Hallett, M. (1993) Memory in patients with cerebellar degeneration. *Neurology* 43:1536–44. [aWTT]

Arai, A. & Lynch, G. (1992) Factors regulating the magnitude of long-term potentiation induced by theta pattern stimulation. *Brain Research*. 598:1–2. [MB]

Arbib, M. A., Boylls, C. C. & Dev, P. (1974) Neural models of spatial perception and the control of movement. In: *Kybernetik und bionik/cybernetics*, ed. R. Oldenbourg. [aJCH, MAA]

Arbib, M. A., Bischoff, A., Fagg, A. H. & Crafton , S. T. (1995) Synthetic PET: Analyzing large-scale properties of neural netowrks. *Human Brain Mapping* 2:225–33. [MAA]

Arbib, M. A. & Caplan, D. (1979) Neurolinguistics must be computational. *Behavioral and Brain Sciences* 2:449–83. [MAA]

Arbib, M. A., Schweighofer, N. & Thach, W. T. (1995) Modeling the cerebellum: From adaptation to coordination. In: *Motor control and sensory-motor integration: Issues and directions*, ed. D. J. Glencross & J. P. Piek. Elsevier. [MAA]

Archambault, L. (1914–15) Les connexiones corticales du noyau rouge. *Nouvelle Iconographie de la Salpîtrière* 27:187–225. [JDS]

Ariano, M. A., Lewicki, J. A., Brandwein, H. J. & Murad, F. (1982) Immunohistochemical localization of guanylate cyclase within neurons of rat brain. *Proceedings of the National Academy of Sciences of the USA* 79:1316–20. [aDJL, aSRV]

Armstrong, D. M. (1974) Functional significance of the inferior olive. *Physiological Reviews* 54:358–417. [aJIS]

Armstrong, D. M., Campbell, N. C., Edgley, S. A., Schild, R. F. & Trott, J. R. (1982) Investigations of the olivocerebellar and spino-olivary pathways. In: *Cerebellum: New vistas*, eds. S. L. Palay & V. Chan-Palay. Springer-Verlag. [aJIS]

Armstrong, D. M. & Edgley, S. A. (1984) Discharges of Purkinje cells in the paravermal part of the cerebellar anterior lobe during locomotion in the cat. *Journal of Physiology (London)* 352:403–24. [aAMS, aJIS]

Armstrong, D. M., Edgley, S. A. & Lidierth, M. (1988) Complex spikes in Purkinje cells of the paravermal part of the anterior lobe of the cat cerebellum during locomotion. *Journal of Physiology (London)* 400:405–14. [aJIS]

Armstrong, D. M. & Rawson, J. A. (1979) Activity patterns of cerebellar cortical neurons and climbing fibre afferents in the awake cat. *Journal of Physiology (London)* 289:425–48. [aJIS]

Arnt-Ramos, L R., O'Brien, W. E. & Vincent, S. R. (1992) Immunohistochemical localization of argininosuccinate synthetase in the rat brain in relation to nitric oxide synthase-containing neurons. *Neuroscience* 51:773–89. [aSRV]

Arshavsky, Y. I., Berkinblit, M. B., Fuxson, O. I., Gelfand, I. M. & Orlovsky, G. N. (1972a) Recordings of neurones of the dorsal spinocerebellar tract during evoked locomotion. *Brain Research* 43:272–75. [aWTT]

(1972b) Origin of modulation in neurones of the ventral spinocerebellar tract during locomotion. *Brain Research* 43:276–79. [aWTT]

Arshavsky, Y. I., Gelfand, I. M. & Orlovsky, G. N. (1986) Cerebellum and rhythmical movements. In: *Studies of Brain Function*, vol 13., ed. V. Braitenberg. Springer-Verlag. [aJCH]

Arshavsky, Y. I., Orlovsky, G. N., Pavlova, G. A. & Perret, C. (1980) Activity of neurons of cerebellar nuclei during fictitious scratch reflex in the cat: 2. The interpositus and lateral nuclei. *Brain Research* 200:249–58. [aWTT]

Artola, A. & Singer, W. (1987) Long-term potentiation and NMDA-receptors in rat visual cortex. *Nature* 330:649–52. [aFC]

(1990) The involvement of N-methyl-D-aspartate receptors in induction and maintenance of long-term potentiation in rat visual cortex. *The European Journal of Neuroscience* 2:254–69. [aFC]

(1993) Long-term depression of excitatory synaptic transmission and its relationship to long-term potentiation. *Trends in Neuroscience* 16:480–87. [PC]

Asanuma, H. (1989) *The motor cortex*. Raven. [SPS]

Asanuma, C., Thach, W. T. & Jones, E. G. (1983a) Anatomical evidence for segregated focal groupings of efference cells and their terminal ramifications in the cerebellothalamic pathway of the monkey. *Brain Research Review* 5:267–99. [aWTT]

(1983b) Distribution of cerebellar terminations and their relation to other afferent terminations in the ventral lateral thalamic region of the monkey. *Brain Research Review* 5:237–65. [aWTT]

(1983c) Brainstem and spinal projections of the deep cerebellar nuclei in the monkey, with observations on the brainstem projections of the dorsal column nuclei. *Brain Research Review* 5:299–322. [aWTT]

Ashe, J., Taira, M., Smyrnis, N., Pellizzer, G., Georgakopoulos, T., Lurito, J. T. & Georgopoulos, A. P. (1993) Motor cortical activity preceding a memorized movement trajectory with an orthogonal bend. *Experimental Brain Research* 95:118–30. [CG]

Audinat, E., Gahwiler, B. H. & Knopfel, T. (1992) Excitatory synaptic potentials in neurons of the deep nuclei in olivo-cerebellar slice cultures. *Neuroscience* 49:903–11. [aSRV]

Babinski, J. (1899) De l'asynergie cérébelleuse. *Revue Neurologique* (Paris) 7:806–16. [aAMS, aWTT]

(1902) Sur le rôle du cervelet dans les actes volitionnels nécessitant une succession rapide de mouvements (diadococinésie). *Revue Neurologique* 10:1013–15. [aAMS]

(1906) Asynergie et inertie cerebelleuses. *Revue Neurologique* 14:685–86. [aWTT]

Babinski, J. & Tournay, A. (1913) Symptômes des maladies du cervelet. *Revue Neurologique* (Paris) 18:306–22. [aAMS]

Babour, B. (1993) Synaptic current evoked in Purkinje cells by stimulating individual granule cells. *Neuron* 11:759–69. [KH]

Baizer, J. S. & Glickstein, M. (1974) Role of cerebellum in prism adaptation. *Journal of Physiology* 23:34–35. [aWTT]

Baker, P. F. & DiPolo, R. (1984) Axonal calcium and magnesium homeostasis. *Current Topics in Membrane Transport* 22:195–248. [aDLJ]

Balaban, C. D., Billingsley, M. L. & Kincaid, R. L. (1989) Evidence for transsynaptic regulation of calmodulin-dependent cyclic nucleotide phosphodiesterase in cerebellar Purkinje cells. *Journal of Neuroscience* 9:2374–81. [aSRV]

Balaban, C. D. & Henry, R. T. (1988) Zonal organization of olivo-nodulus projections in albino rabbits. *Neuroscience Research* 5:409–23. [aJIS]

Ball, K. & Sekular, R. (1987) Direction-specific improvement in motion discrimination. *Vision Research* 27:953–65. [PVD]

Bandle, E. & Guidotti, A. (1978) Studies on the cell location of cyclic 3',5'-guanosine monophosphate-dependent protein kinase in cerebellum. *Brain Research* 156:412–16. [aSRV]

Bansinath, M., Arbabha, B., Turndorf, H. & Garg, U. C. (1993) Chronic administration of a nitric oxide synthase inhibitor, Nw-nitro-L-arginine, and drug-induced increase in cerebellar cyclic GMP in vivo. *Neurochemical Research* 18:1063–66. [aSRV]

Barmack, N. H., Fagerson, M. & Errico, P. (1993b) Cholinergic projection to the dorsal cap of the inferior olive of the rat, rabbit, and monkey. *Journal of Comparative Neurology* 328:263–81. [aJIS]

Barmack, N. H., Fagerson M., Fredette, B. J., Mugnaini, E. & Shojaku, H. (1993a) Activity of neurons in the beta nucleus of the inferior olive of the rabbit evoked by natural vestibular stimulation. *Experimental Brain Research* 94:203–15. [arJIS]

Barmack, N. H. & Hess, D. T. (1980) Multiple-unit activity evoked in the dorsal cap of inferior olive of the rabbit by visual stimulation. *Journal of Neurophysiology* 43:151–64. [aJIS]

Barmack, N. H., Mugnaini, E. & Nelson, B. J. (1989) Vestibularly-evoked activity of single neurons in the beta nucleus of the inferior olive. In: *The olivocerebellar system in motor control: Experimental brain research series 17,* ed. P. Strata. Springer-Verlag. [aJIS]

Barmack, N. H. & Shojaku, H. (1992) Representation of a postural coordinate system in the nodulus of the rabbit cerebellum by vestibular climbing fiber signals. In:*Vestibular and brain stem control of eye, head and body movements,* eds. H. Shimazu & Y. Shinoda. Japan Scientific Societies Press. [aJIS]

Barnes, C. A., McNaughton, B. L., Bredt, D. S., Ferris, C. D. & Snyder, S. H. (1994). Nitric oxide synthase inhibition in vivo: Lack of effect on hippocampal synaptic enhancement or spatial memory. In: *Long-term potentiation,* vol. 2, ed. M. Baudry & J. L. Davis. MIT Press. [MB]

Barto, A. G. (1994) Reinforcement learning control. *Current Opinion in Neurobiology* 4:888–93. [rJCH]

(1995) Adaptive critics and the basal ganglia. In: *Models of Information Processing in the Basal Ganglia,* ed. J. C. Houk, J. L. Davis & D. G. Beiser. MIT Press. [arJCH]

Barto, A. G., Buckingham, J. T., & Houk, J. C. (1996) A predictive switching model of cerebellar movement control. In: *Advances in Neural Information Processing Systems 8,* eds. D. S. Touretzky, M. C. Mozer & M. E. Hasselmo. MIT Press. [aJCH]

Bastian, A. J., Martin, T. A., Keating, J. G. & Thach, W. T. (in press) Cerebellar ataxia: Abnormal control of interaction torques across multiple joints. *Journal of Neurophysiology.* [rWTT]

Bastian, A. J., Mueller, M. J., Martin, T. A., Keating, J. G. & Thach, W. T. (1994) Control of interaction torques during reaching in normal and cerebellar patients. *Society for Neuroscience Abstracts* 20:933. [aWTT]

Bastian, A. J. & Thach, W. T. (1995) Cerebellar patients made initial directional errors consistent with impaired control of limb dynamics. *Society of Neuroscience Abstracts* 21:1921–1995. [rAMS]

Batchelor, A. M., Madge, D. J. & Garthwaite, J. (1994) Synaptic avtivation of metabotropic glutamate receptors in the parallel fibre-Purkinje cell pathway in rat cerebellar slices. *Neuroscience* 63:911–15. [rMKan]

Baude A., Nusser, Z., Roberts, J. D. B., Mulvihill, E., McIlhinney, R. A. J. & Somogyi, P. (1993) The metabotropic glutamate receptor (mGlurla) is concentrated at perisynaptic membrane of neuronal subpopulations as detected by immunogold reaction. *Neuron* 11:771–87. [aDJL]

Baudry, M. & Davis, J. L., eds. (1994) *Long-term potentiation,* vol. 2. MIT Press. [MB]

Baudry, M. & Lynch, G. (1993) Long-term potentiation: Biochemical mechanisms. In: *Synaptic plasticity: Molecular and functional aspects,* ed. M. Baudry, R. F. Thompson & J. L. Davis. MIT Press. [MB]

Bear, M. F. & Malenka, R. C. (1994) Synaptic plasticity: LTP and LTD. *Current Opinion Neurobiology* 4:389–99. [MB]

Becker, W. (1972) The control of eye movements in the saccadic system. *Bibliography Opthalmology* 82:233–43. [HB]

Becker W. J., Kunesch, E. & Freund, H. J. (1990) Coordination of a multi-joint movement in normal humans and in patients with cerebellar dysfunction. *Canadian Journal of Neurological Sciences* 17:264–74. [aAMS, CG]

Becker, W. J., Morrice, B. L., Clark, A. W. & Lee, R. G. (1991) Multi-joint reaching movements and eye-hand tracking in cerebellar incoordination: Investigation of a patient with complete loss of Purkinje cells. *Canadian Journal of Neurological Sciences* 18:476–87. [aAMS]

Bekkering, H., Abrams, R. A. & Pratt, J. (1995) Transfer of saccadic adaptation to the manual motor system. *Human Movement Science* 14:155–64. [HB]

Bekkers, J. M. & Stevens, C. F. (1989) NMDA and non-NMDA receptors are co-localized at individual excitatory synapses in cultured rat hippocampus. *Nature* 341:230–33. [aFC]

Bell, C. C. (1994) The generation of expectations in cerebellum-like structures. In: *The neurobiology of computation: Proceedings of the annual computational neuroscience meeting.* [aJCH]

Bell, C. C. & Grimm, R. J. (1969) Discharge properties of Purkinje cells recorded on single and double microelectrodes. *Journal of Neurophysiology* 32:1044–55. [aJCH, aJIS]

Bell, C. C. & Kawasaki, T. (1972) Relations among climbing fiber responses of nearby Purkinje cells. *Journal of Neurophysiology* 35:155–69. [aJIS]

Bellugi, U., Bihrle, A., Jernigan, T., Trauner, D. & Doherty, S. (1990) Neuyropsychological, neurological, and neuroanatomical profile of Williams syndrome. *American Journal of Medical Genetics* 6(suppl):115–25. [aWTT]

Ben-Ari, Y. & Aniksztejn, L. (1995) Role of glutamate metabotropic receptors in long-term potentiation in the hippocampus. *Seminar in Neuroscience* 7:127–35. [MB]

Benedetti, F., Montarolo, P. G. & Rabacchi, S. (1984) Inferior olive lesion induces long-lasting functional modifications in the Purkinje cells. *Experimental Brain Research* 55:368–71. [aJIS]

Bennett, D. J. (1993a) Electromyographic responses to constant position errors imposed during voluntary elbow joint movement in human. *Experimental Brain Research* 95:499–508. [aAMS]

(1993b) Torques generated at the human elbow joint in response to constant position errors imposed during voluntary movements. *Experimental Brain Research* 95:488–98. [aAMS, HG]

Bennett, D. J., Hollerbach, J. M., Xu, Y. & Hunter, I. W. (1992) Time-varying stiffness of human elbow joint during cyclic voluntary movement. *Experimental Brain Research* 88:433–42. [arAMS, HG]

Benuck, M., Reith, M. E. A. & Lajtha, A. (1989) Phosphoinositide hydrolysis induced by depolarization and sodium channel activation in mouse cerebrocortical slices. *Neuropharmacology* 28:847–54. [aDJL]

Beppu, H., Nagaoka, M. & Tanaka, R. (1987) Analysis of cerebellar motor disorders by visually guided elbow tracking movement: 2. Contribution of the visual cues on slow ramp pursuit. *Brain* 110:1–18. [PH]

Berger, T. W. & Bassett, J. L. (1992) System properties of the hippocampus. In: *Learning and memory: The behavioral and biological substrates,* ed. I. Gormezano & E. A. Wasserman. Erlbaum. [CW]

Berger, T. W. & Orr, W. B. (1983) Hippocampectomy selectively disrupts discrimination reversal conditioning of the rabbit nictitating membrane response. *Behavioral Brain Research* 8:49–68. [CW]

Berman, A. J., Berman, D. & Prescott, J. W. (1978) The effect of cerebellar lesions on emotional behavior in the rhesus monkey. In: *The Cerebellum, epilepsy and behavior,* ed. I. S. Cooper, M. Riklan & R. S. Snider. Plenum. [aWTT, JDS]

Bernstein, N. A. (1967) *The coordination and regulation of movements.* Pergamon. [aAMS, AGF, LPL]

Bernston, G. G. & Torello, M. W. (1982) The paliocerebellum and the integration of behavioral function. *Physiological Psychology* 10:2–12. [aWTT]

Berthier, N. E. & Moore, J. W. (1986) Cerebellar Purkinje cell activity related to the classically conditioned nictitating membrane response. *Experimental Brain Research* 63:341–50. [aJCH]

Berthier, N. E., Singh, S. P., Barto, A. G. & Houk, J. C. (1993) Distributed representation of limb motor programs in arrays of adjustable pattern generators. *Journal of Cognitive Neuroscience* 5:56–78. [arJCH, PD, JCH]

Berthoz, A. & Pozzo, T. (1988) Intermittent head stabilisation during postural and locomotory tasks in humans. In: *Posture and gait: Development, adaptation and modulation,* ed. B. Amblard, A. Berthoz & F. Clarac. Elsevier. [SMO]

Biel, M., Altenhofen, W., Hullin, R., Ludwig, J., Freichel, M., Flockerzi, V., Dascal, N., Kaupp, U. B. & Hofmann, F. (1993) Primary structure and functional expression of a cyclic nucleotide-gated channel from rabbit aorta. *Federation of European Biological Societies Letters* 329:134–38. [aSRV]

Biggio, G., Brodie, B. B., Costa, E. & Guidotti, A. (1977a) Mechanisms by which diazepam, muscimol and other drugs change the content of cGMP in cerebellar cortex. *Proceedings of the National Academy of Sciences of the USA* 74:3592–96. [aSRV]

Biggio, G., Corda, M. G., Casu, M., Salis, M. & Gessa, G. L (1978) Disappearance of cerebellar cyclic GMP induced by kainic acid. *Brain Research* 154:203–8. [arSRV]

Biggio, G., Costa, E. & Guidotti, A. (1977b) Pharmacologically induced changes in the 3',5'-cyclic guanosine monophosphate content of rat cerebellar cortex: Differences between apomorphine, haloperidol and harmaline. *Journal of Pharmacology and Experimental Therapeutics* 200:207–15. [aSRV]

Biggio, G. & Guidotti, A. (1976) Climbing fiber activation and 3',5'-cyclic guanosine monophosphate (cGMP) content in cortex and deep nuclei of the cerebellum. *Brain Research* 107:365–73. [aSRV, DO]

Bindman, L. J., Murphy, K. P. S. J. & Pockett, S. (1988) Postsynaptic control of the induction of long-term changes in efficacy of transmission at neocortical synapses in slices of rat brain. *Journal of Neurophysiology* 60:1053–65. [aFC]

Bizzi, E., Accomero, N., Chapple, W. & Hogan, N. (1982) Arm trajectory formation. *Experimental Brain Research* 46:139–43. [aAMS]

Bizzi, E., Giszter, S. F., Loeb, E., Mussa-Ivaldi, F. A. & Saltiel, P. (1995)

Modular organization of motor behavior in the frog's spinal cord. *Trends Neuroscience* 10:442–46. [DJ, rJCH, rAMS]

Black, J. E., Isaacs, K. R., Anderson, B. J., Alcantara, A. A. & Greenough, W. T. (1990) Learning causes synaptogenesis, whereas motor activity causes angiogenesis, in cerebellar cortex of adult rats. *Proceedings of the National Academy of Sciences of the USA* 87:5568–72. [aJCH]

Blaxton, T. A., Zeffiro, T. A., Gabrieli, J. D. E., Bookheimer, S. Y., Carrillo, M. C., Theodore, W. H. & Disterhoft, J. F. (submitted) Functional mapping of human learning: A PET activation study of eyeblink conditioning. [CW]

Bles, W., Vianney de Jong, J. M. B. & de Wit, G. (1984) Somatosensory compensation for loss of labyrinthine function. *Acta Otolaryngologica* 97:312–21. [aWTT]

Bliss, T. V. P. & Collingridge, G. L. (1993) A synaptic model of memory: Long-term potentiation in the hippocampus. *Nature* 361:31–39. [aJCH]

Bliss, T. V. P. & Lomo, T. (1973) Long-lasting potentiation of synaptic transmission in the dentate area of the anaesthetized rabbit following stimulation of the perforant path. *Journal of Physiology (London)* 232:331–56. [aMKan]

Bliss, T. V. P. & Lynch, M. A. (1988) Long-term potentiation of synaptic transmission in the hippocampus: Properties and mechanisms. In: *Long-term potentiation, from biophysics to behavior*, ed. P. W. Landfield & S. A. Deadwyler. Alan R. Liss. [aFC]

Bloedel, J. R. (1992) Functional heterogeneity with structural homogeneity: How does the cerebellum operate? *Behavioral and Brain Sciences* 15:666–78. [aJCH, aAMS, aJIS, aWTT, EDS]

Bloedel, R. F. & Bracha, V. (1995) On the cerebellum, cutaneomuscular reflexes, movement control and the elusive engrams of memory. *Behavioral Brain Research* 68:1–44. [DT]

Bloedel, J. R. & Courville, J. (1981) Cerebellar afferent systems. In: *Handbook of Phsyiology: sect. 1. The nervous system: vol. 2. Motor control*, ed. J. Brookhart, V. Mountcastle, V. Brooks, & S. Geiger. American Physiological Society. [aJCH]

Bloedel, J. R. & Ebner, T. J. (1984) Rhythmic discharge of climbing fibre afferents in response to natural peripheral stimuli in the cat. *Journal of Physiology (London)* 352:129–46. [aJIS]

Bloedel, J. R. & Kelly, T. M. (1992) The dynamic selection hypothesis: A proposed function for the cerebellar sagittal zones. In: *The cerebellum revisited*, eds. R. Llinás & C. Sotelo. Springer-Verlag. [aJIS]

Bloedel, J. R. & Roberts, W. J. (1970) Action of climbing fiber in cerebellar cortex of the cat. *Journal of Neurophysiology* 34:17–31. [aJIS, DO]

Blomfield, S. & Marr, D. (1970) How the cerebellum may be used. *Nature* 227:1224–28. [aJCH]

Bloom, F. E., Hoffer, B. J. & Siggins, G. R. (1971) Studies on norepinephrine-containing afferents to Purkinje cells of rat cerebellum: 1. Localization of the fibers and their synapses. *Brain Research* 25:501–21. [aDJL]

Bortolotto, Z. A., Bashir, Z. I., Davies, C. H. & Collingridge, G. L. (1994) A molecular switch activated by metabotropic glutamate receptors regulates induction of long-term potentiation. *Nature* 368:740–43. [MB]

Bossom, J. (1965) The effect of brain lesions on prism adaptation in monkeys. *Psychonomic Science* 45–46. [aWTT]

Bossom, J. & Hamilton, C. R. (1963) Interocular transfer of prism-altered coordinations in split-brain monkeys. *Journal of Comparative Psysiology and Psychology* 56:769–74. [aWTT]

Botez, M. I., Botez, T., Elie, R. & Attig, E. (1989) Role of the cerebellum in complex human behavior. *Italian Journal of Neurological Science* 10:291–300. [aWTT, JDS]

Botez, M. I., Gravel, J., Attig, E. & Vezina, J.-L. (1985) Reversible chronic cerebellar ataxia after phenytoin intoxication: Possible role of cerebellum in cognitive thought. *Neurology* 35:1152–57. [aWTT]

Botez, M. I., Leveille, J. & Botez, T. (1988) Role of the cerebellum in cognitive thought: SPECT and neurological findings. In: *The Australian Society for the Study of Brain Impairment*, ed. M. Matheson & H. Newman. [aWTT]

Boucher, J. L., Genet, A., Vadon, S., Delaforge, M. & Mansuy, D. (1992) Formation of nitrogen oxides and citrulline upon oxidation of N$^{\Omega}$-hydroxy-L-arginine by hemeproteins. *Biochemical and Biophysical Research Communications* 184:1158–64. [DO]

Boulter, J., Hollman, M., O'Shea-Greenfild, A., Hartley, M., Deneris, E., Maron, C. & Heinemann, S. (1990) Molecular cloning and functional expression of glutamate receptor subunit genes. *Science* 249:1033–37. [aFC, aSRV]

Boulton, C. L., Southam, E. & Garthwaite, J. (1995) Nitric oxide-dependent long-term potentiation is blocked by a specific inhibitor of soluble guanyl cyclase. *Neuroscience* 69:699–703. [LJB]

Boussaoud, D. (1995) Primate premotor cortex: Modulation of preparatory neuronal activity by gaze angle. *Journal of Neurophysiology* 73:886–90. [PVD]

Bower, J. M. (1992) Is the cerebellum a motor control device? Commentary on "Function heterogeneity with structural homogeneity: How does the cerebellum operate?" by J. R. Bloedel. *Behavioral and Brain Sciences* 15:714–15. [JMB]

(1995a) The cerebellum as a sensory acquisition controller. *Human Brain Mapping* 2:255–56. [JMB]

(1995b) Is the cerebellum sensory for motor's sake, or motor for sensory's sake? The cerebellum: From structure to control. *Satellite Symposium of the Meeting of European Neuroscience* [Abstracts: p 32], Rotterdam, August 31–September 3. [DT]

(in press) Is the cerebellum sensory for motor's sake, or motor for sensory's sake? *Progress in Brain Research.* [JMB]

Bower, J. M. & Kassel, J. (1990) Variability in tactile projection patterns to cerebellar folia Crus IIA in the Norway rat. *Journal of Comparative Neurology* 302:768–78. [JMB, RCM]

Bower, J. M. & Woolston, D. C. (1983) Congruence of spatial organization of tactile projections to granule cell and Purkinje cell layers of cerebellar hemispheres of the albino rat: Vertical organization of cerebellar cortex. *Journal of Neurophysiology* 49:745–66. [JMB, DJ, FS, rAMS]

Boylls, C. C. (1975a) *A theory of cerebellar function with applications to locomotion*, COINS Technical Report 76–1, Amherst, MA. [aJCH, MGP]

(1975b) Synergies and cerebellar function. In: *Conceptual models of neural organization.* MIT Press. [MAA]

(1980) Contributions to locomotor coordination of an olivo-cerebellar projection to vermis in the cat: Experimental results and theoretical proposals. In:*The inferior olivary nucleus: Anatomy and physiology*, eds. J. Courville, C. de Montigny & Y. Lamarre. Raven. [aJIS]

Braitenberg, A. & Atwood, R. P. (1958) Morphological observations on the cerebellar cortex. *Journal of Corparative Neurology* 109:1–27. [aJCH, aWTT, FS]

Braitenberg, V. (1961) Functional interpretation of cerebellar histology. *Nature* 190:539–40. [FS]

(1983) The cerebellum revisited. *Journal of Theoretical Neurobiology* 2:237–41.

Brand, S., Dahl, A.-L. & Mugnaini, E. (1976) The length of parallel fibers in the cat cerebellar cortex. An experimental light and electron microscope study. *Experimental Brain Research* 26:39–58. [aWTT]

Bracke-Tolkmitt, R., Linden, A., Canavan, G. M., Rockstroh, B., Scholz, E., Wessel, K. & Diener, H. C. (1989) The cerebellum contributes to mental skills. *Behavioral Neuroscience* 103:442–46. [aWTT]

Bredt, D. S., Glatt, C. E., Hwang, P. M., Fotuhi, M., Dawson, T. M. & Snyder, S. H. (1991) Nitric oxide synthase protein and mRNA are discretely localized in neuronal populations of the mammalian CNS together with NADPH diaphorase. *Neuron* 7: 615–24. [aSRV]

Bredt, D. S., Hwang, P. M. & Snyder, S. H. (1990) Localization of nitric oxide synthase indicating a neural role for nitric oxide. *Nature* 347:768–70. [aFC, aDJL, aSRV]

Bredt, D. S. & Snyder, S. H. (1989) Nitric oxide mediates glutamate-linked enhancement of cGMP levels in the cerebellum. *Proceedings of the National Academy of Sciences of the USA* 86:9030–33. [aSRV]

(1990) Isolation of nitric oxide synthetase, a calmodulin-requiring enzyme. *Proceedings of the National Academy of Sciences of the U.S.A.* 87:682–85. [aSRV]

Breese, G. R., Mailman, R. B., Ondrusek, M. G., Harden, T. K. & Mueller, R. A. (1978) Effects of dopaminergic agonists and antagonists on cerebellar guanosine-3',5'-monophosphate (cGMP). *Life Sciences* 23:533–36. [aSRV]

Bridgeman, B., Hendry, D. & Stark, L. (1975) Failure to detect displacement of the visual world during saccadic eye movements. *Vision Research* 15:719–22. [HB]

Briley, P. A., Kouyoumdjian, J. C., Haidamous, M. & Gonnard, P. (1979) Effect of L-glutamate and kainate on rat cerebellar cGMP levels in vivo. *European Journal of Pharmacology* 54:181–84. [aSRV]

Brindley, G. S. (1964) The use made by the cerebellum of the information that it receives from the sense organs. *International Brain Research Organization Bulletin* 3:80. [aWTT, PFCG]

Brodal, P. et al. (1988) GABA-containing neurons in the pontine nuclei of rat, cat and monkey: An immunocytochemical study. *Neuroscience* 25:27–45. [FS]

Brooks, V. B. & Thach, W. T. (1981) Cerebellar control of posture and movement. In: *Handbook of Physiology*, sect. 1, vol. 2, pt. 2, ed. J. M. Brookhart, V. B. Mountcastle & V. B. Brooks. American Physiological Society. [KW]

Brooks, V. B., Kozlovskaya, I. B., Atkin, A., Horvath, F. E. & Uno, M. (1973) Effects of cooling dentate nucleus on tracking-task performance in monkeys. *Journal of Neurophysiology* 36:974–95. [aAMS]

Brüne, B. & Lapetina, E. G. (1989) Activation of a cytosolic ADP-ribosyltransferase by nitric oxide-generating agents. *Journal of Biological Chemistry* 264:8455–58. [aSRV]

Brüning, G. (1993a) Localization of NADPH-diaphorase in the brain of the chicken. *Journal of Comparative Neurology* 334:192–208. [aSRV]

(1993b) NADPH-diaphorase histochemistry in the postnatal mouse cerebellum suggests specific developmental functions for nitric oxide. *Journal of Neuroscience Research* 36:580–87. [arSRV]

Buckingham, J. T., Houk, J. C. & Barto, J. G. (1994) Controlling a nonlinear spring-mass system with a cerebellar model. In: *Proceedings of the eighth Yale workshop on adaptive and learning systems.* [aJCH]

(1995) Adaptive predictive control with a cerebellar model. *Proceedings of the 1995 World Congress on Neural Networks,* Erlbaum. [aJCH]

Bunn, S. J., Garthwaite, J. & Wilkin, G. P. (1986) Guanylate cyclase activites in enriched preparations of neurones, astroglia and a synaptic complex isolated from rat cerebellum. *Neurochemistry International* 8:179–85. [arSRV]

Buonomano, D. V. & Mauk, M. D. (1994) Neural network model of the cerebellum: temporal discrimination and the timing of motor responses. *Neural Computation* 6:38–55. [aJCH]

Burkard, W. P., Pieri, L. & Haefely, W. (1976) In vivo changes of guanosine 3',5'-cyclic phosphate in rat cerebellum by dopaminergic mechanisms. *Journal of Neurochemistry* 27:297–98. [aSRV]

Burnod, Y. & Dufossé, M. (1991) A model for the cooperation between cerebral cortex and cerebellar cortex in movement learning. In: *Brain and space,* ed. J. Paillard. Oxford University Press. [MD]

Caddy, K. W. T. & Biscoe, T. J. (1979) Structural and quantitative studies on the normal C3H and Lurcher mutant mouse. *Philosophical Transactions of the Royal Society of London: Biology* 287:167–200. [aAMS]

Calabresi, P., Pisani, A., Mercuri, N. B. & Bernardi, G. (1994) Post-receptor mechanisms underlying striatal long-term depression. *Journal of Neuroscience* 14:4871–81. [PC]

Callaway, J. C., Lasser-Ross, N. & Ross, W. N. (1995) IPSPs strongly inhibit climbing fiber-activated $[Ca^{2+}]_i$ increases in the dendrites of cerebellar Purkinje neurons. *Journal of Neuroscience* 15:2777–87. [aJIS, rSRV, JCF]

Campbell, N. C., Ekerot, C.-F. & Hesslow, G. (1983) Interaction between responses in Purkinje cells evoked by climbing fibre impulses and parallel fibre volleys in the cat. *Journal of Physiology (London)* 340:225–38. [aJIS]

Cannon, S. C. & Robinson, D.A. (1987) Loss of the neural integrator of the oculomotor system from brainstem lesions in monkey. *Journal of Neurophysiology* 57:1383–1409. [aJCH]

Carl, J. R. & Gellman, R. S. (1986) Adaptive responses in human smooth pursuit. In: *Adaptive processes in the visual and oculomotor systems,* ed. E. L. Keller & D. S. Zee. Pergamon. [PVD]

Carpenter, R. H. S. (1988) *Movements of the eyes.* Pion. [HB]

Carter, C. J., Noel, F. & Scatton, B. (1988) Ionic mechanisms implicated in the stimulation of cerebellar cyclic GMP levels by N-methyl-D-aspartate. *Journal of Neurochemistry* 49:195–200. [aSRV]

Carter, T. L. & McElligott, J. G. (1994) Metabotropic glutamate receptor antagonist (L-AP3) inhibits vestibulo-ocular reflex adaptation when administered into goldfish vestibulo-cerebellum. *Society for Neuroscience Abstracts* 20:17.10. [aJCH]

Cavada, C. & Goldman-Rakic, P. S. (1993) Multiple visual areas in the posterior parietal cortex of primates. *Progress in Brain Research* 95:123–37. [PVD]

Chamberlain, T. J., Halick, P. & Gerrard, R. W. (1963) Fixation of experience in the rat spinal cord. *Journal of Neurophysiology* 26:662–73. [LPL]

Chan-Palay, V. & Palay, S. L. (1979) Immunocytochemical localization of cyclic GMP: Light and electron microscopic evidence for involvement of neuroglia. *Proceedings of the National Academy of Sciences of the USA* 76:1485–88. [aSRV]

Chapeau-Blondeau, F. & Chauvet, G. (1991) A neural network model of the cerebellar cortex performing dynamic associations. *Biological Cybernetics* 65:267–79. [aJCH]

Chen C. & Thompson, R. F. (1992) Associative long-term depression revealed by field potential recording in rat cerebellar slice. *Society for Neuroscience Abstracts* 18:1215. [aJCH, aDJL]

(1995) Temporal specificity of long-term depression in parallel fiber – Purkinje synapses in rat cerebellar slice. *Learning & Memory* 2:185–98. [JCH, rJCH, rDJL, rJIS]

Chen, L. & Huang L.-Y. M. (1992) Protein kinase C reduces Mg^{2+} block of NMDA-receptor channels as a mechanism of modulation. *Nature* 356:521–23. [aDJL]

Chen, C., Kano, M., Chen, L., Bao, S., Kim J. J., Hashimoto, K., Thompson, R. F. & Tonegawa, S. (1995) Impaired motor coordination correlates with persistent multiple climbing fiber innervation in PKCg mutant mice. *Cell* 83:1233–42. [rMKan, rDJK]

Chen, S. & Aston-Jones, G. (1994) Cerebellar injury induces NADPH diaphorase in Purkinje and inferior olivary neurons in the rat. *Experimental Neurobiology* 126:270–76. [rSRV]

Chen, Q. X., Stelzer, A., Kay, A. R. & Wong, R. S. K. (1990) GABAA receptor function is regulated by phosphorylation in acutely dissociated guinea-pig hippocampal neurons. *Journal of Physiology (London)* 420:207–21. [aMKan]

Cheng, H. C., Kemp, B. E., Pearson, R. B., Smith, A. J., Misconi, L., Van Patten, S. M. & Walsh, D. A. (1986) A potent synthetic peptide inhibitor of the cAMP-dependent protein kinase. *Journal of Biological Chemistry* 261:989–92. [aMKan]

Chubb, M. C., Fuchs, A. F. & Scudder, C. A. (1984) Neuron activity in monkey vestibular nuclei during vertical stimulation and eye movements. *Journal of Neurophysiology* 52:724–42. [aJIS]

Cintas, H. M., Rutherford, J. G. & Gwyn, D. G. (1980) Some midbrain and diencephalic projections to the inferior olive in the rat. In: *The inferior olivary nucleus: Anatomy and physiology,* ed. J. Courville, C. de Montigny & Y. Lamarre. Raven. [JDS]

Clark, G. A., McCormick, D. A., Lavond, D. G. & Thompson, R. F. (1984) Effects of lesions of cerebellar nuclei on conditioned behavioral and hippocampal neuronal responses. *Brain Research* 291:125–36. [CW]

Clément, G. & Rézette, D. (1985) Motor behavior underlying the control of an upside-down vertical posture. *Experimental Brain Research* 59:478–84. [aAMS]

Cohen, H., Cohen, B., Raphan, T. & Waespe, W. (1992) Habituation and adaptation of the vestibuloocular reflex: A model of differential control by the vestibulocerebellum. *Experimental Brain Research* 90:526–38. [aJIS]

Colin, F., Manil, J. & Desclin, J. C. (1980) The olivocerebellar system. Delayed and slow inhibitory effects: An overlooked salient feature of the cerebellar climbing fibers. *Brain Research* 187:3–27. [aJIS]

Collingridge, G. L., Kelh, S. J. & McLennan, H. (1983) Excitatory amino acids in synaptic transmission in the Schaffer collateral-commissural pathway of the rat hippocampus. *Journal of Physiology* 334:33–46. [aFC]

Conquet, F., Bashir, Z. I., Davies, C. H., Daniel, H., Ferraguti, F., Bordi, F., Franz-Bacon, K., Reggian, A., Matarerse, V. , Conde, F., Collingridge, G. L. & Crépel, F. (1994) Motor deficit and impairment of synaptic plasticity in mice lacking mGluR1. *Nature* 372:237–42. [aJCH, arMKan, TH, MKan, rDJK, rSRV]

Conrad, B. & Brooks, V. B. (1974) Effects of dentate cooling on rapid alternating arm movements. *Journal of Neurophysiology* 37:792–804. [aAMS]

Cordo, P. J. & Nashner, L. M. (1982) Properties of postural adjustments associated with rapid arm movements. *Journal of Neurophysiology* 47:287–302. [aAMS]

Coulter, D. A., LoTurco, J. J., Kubota, M., Disterhoft, J. F., Moore, J. W. & Alkon, D. L. (1989) Classical conditioning reduces the amplitude and duration of the calcium-dependent afterhyperpolarization in rabbit hippocampal pyramidal cells. *Journal of Neurophysiology* 61:971–81. [CW]

Crépel, F. & Audinat, E. (1991) Excitatory amino acid receptors of cerebellar Purkinje cells: Development and plasticity. *Progress in Biophysic and Molecular Biology* 55:31–46. [aFC]

Crépel, F., Audinat, E., Daniel, H., Hemart, N., Jaillard, D., Rossier, J. & Lambolez, B. (1994) Cellular locus of the nitric oxide-synthase involved in cerebellar long-term depression induced by high external potassium concentration. *Neuropharmacology* 33:1399–1405. [DO, rFC, rSRV]

Crépel, F., Daniel, H., Conde, F., Ferraguit, F. & Conquet, F. (1995) Pre- and postsynaptic mechanisms of cerebellar LTD. *Fourth IBRO World Congress of Neuroscience* [abstract] 13.2. [rDJL]

Crépel, F., Daniel, H., Hemart, N. & Jaillard, D. (1991) Effects of ACPD and AP3 on parallel fibre-mediated EPSPs of Purkinje cells in cerebellar slices *in vitro. Experimental Brain Research* 86:402–6. [aFC, aDJL]

(1993) Mechanisms of synaptic plasticity in the cerebellum. In: *Long-term potentiation: A debate of current issues, vol. 2.*, ed. M. Baudry & J. Davis, MIT Press. [aDJL]

Crépel, F., Dhanjal, S. S. & Sears, T. A. (1982) Effect of glutamate, aspartate and related derivatives on cerebellar Purkinje cell dendrites in the rat: An in vitro study. *Journal of Physiology* 329:297–317. [aFC]

Crépel, F. & Jaillard, D. (1990) Protein kinases, nitric oxide and long-term depression of synapses in the cerebellum. *NeuroReport* 1:133–36. [aFC, arMKan, aDJL, aSRV, NAH, DO, PC]

(1991) Pairing of pre- and postsynaptic activities in cerebellar Purkinje cells induces long-term changes in synaptic efficacy *in vitro. Journal of Physiology (London)* 432:123–41. [aFC, aDJL, PC]

Crépel, F. & Krupa, M. (1988) Activation of protein kinase C induces a long-term depression of glutamate sensitivity of cerebellar Purkinje cells. An *in vitro* study. *Brain Research* 458:397–401. [aFC, aMKan, aDJL]

(1990) Modulation of the responsiveness of cerebellar Purkinje cells to excitatory amino acids. In: *Excitatory amino acids and neuronal plasticity,* ed. Y. Ben-Ari. Plenum. [aFC]

Crick, F. H. C. & Koch, C. (1995) Are we aware of visual activity in the primary visual cortex? *Nature* 375:121–23. [SMO]

Crill, W. E. (1970) Unitary multiple-spiked responses in the cat inferior olive nucleus. *Journal of Neurophysiology* 33:199–209. [aJIS]

Cross, A. J., Misra, A., Sandilands, A., Taylor, M. J. & Green, A. R. (1993) Effect of chlormethiazole, dizocilpine and pentobarbital on harmaline-induced

increase of cerebellar cyclic GMP and tremor. *Psychopharmacology* 111:96–98. [aSRV]
Cuenod, M., Do, K. Q., Vollenweider, F., Zollinger, M., Klein, A. & Streit, P. (1989) The puzzle of the transmitters in the climbing fibers. In: *The olivocerebellar system in motor control [Brain Research Series 17]*, ed. P. Strata. Springer-Verlag. [aFC, aDJL]
Daniel, H., Hemart, N., Jaillard, D. & Crépel, F. (1992) Coactivation of metabotropic glutamate receptors and of voltage-gated calcium channels induces long-term depression in cerebellar Purkinje cells *in vitro*. *Experimental Brain Research* 90:327–31. [aFC, aMKan, aDJL, DO, rSRV]
(1993) Long-term depression requires nitric oxide and guanosine 3'-5' cyclic monophosphate production in cerebellar Purkinje cells. *European Journal of Neuroscience* 5:1079–82. [aFC, aDJL, aSRV, PC, LJB, NAH]
Danysz, W., Wroblewski, J. T., Brooker, G. & Costa, E. (1989) Modulation of glutamate receptors by phencyclidine and glycine in the rat cerebellum: cGMP increase in vivo. *Brain Research* 479:270–76. [aSRV]
Dawson, V. L., Dawson, T. M., London, E. D., Bredt, D. S. & Snyder, S. H. (1991) Nitric oxide mediates glutamate neurotoxicity in primary cortical cultures. *Proceedings of the National Academy of Sciences of the USA* 88:6368–71. [aSRV]
De Camilli, P., MIller, P. E., Levitt, P., Walter, U. & Greengard, P. (1984) Anatomy of cerebellar Purkinje cells in the rat determined by a specific immunohistochemical marker. *Neuroscience* 11:761–817. [aDJL, aSRV]
De Schutter, E. (1995) Cerebellar long-term depression might normalize excitation of Purkinje cells: A hypothesis. *Trends in Neurosciences* 18:291–95. [EDS]
De Schutter, E. & Bower, J. M. (1994a) An active membrane model of the cerebellar Purkinje cell: 1. Simulation of current clamps in slice. *Journal of Neurophysiology* 71:375–400. [JMB, KH, EDS]
(1994b) An active membrane model of the cerebellar Purkinje cell: 2. Simulation of synaptic response. *Journal of Neurophysiology* 71:401–19. [JMB, KH]
(1994c) Simulated responses of cerebellar Purkinje cell are independent of the dendritic location of granule cell synaptic inputs. *Proceedings of the National Academy of Sciences of the USA* 91:4736–40. [JMB, KH, EDS]
de Graaf, J. B., Pelisson, D., Prablanc, C. & Goffart, L. (1955) Modifications in end positions of arm movements following short term saccadic adaptation. *NeuroReport* 6:1733–36. [HB]
de Montigny, C. & Lamarre, Y. (1973) Rhythmic activity induced by harmaline in the olivo-cerebello-bulbar system of the cat. *Brain Research* 53:81–95. [aJIS]
de Vente, J., Bol, J. G. J. M., Berkelmans, H. S., Schipper, J. & Steinbusch H. M. W. (1990) Immunocytochemistry of cGMP in the cerebellum of the immature, adult, and aged rat: The involvement of nitric oxide. A micropharmacological study. *European Journal of Neuroscience* 2:845–62. [aDJL, aSRV]
de Vente, J., Bol, J. G. J. M. & Steinbusch, H. W. M. (1989a) cGMP-producing, atrial natriuretic factor-responding cells in the rat brain. *European Journal of Neuroscience* 1:436–60. [DO]
(1989b) Localization of cGMP in the cerebellum of the adult rat: An immunohistochemical study. *Brain Research* 504: 332–37. [aSRV]
de Vente, J. & Steinbusch, H. W. M. (1992) On the stimulation of soluble and particulate guanylate cyclase in the rat brain and the involvement of nitric oxide as studied by cGMP immunocytochemistry. *Acta Histochemica* 92:13–38. [aDJL, aSRV]
De Zeeuw, C. I. (1990) Ultrastructure of the cat inferior olive. Ph.D. thesis, Erasmus University, Rotterdam. [aJIS]
De Zeeuw, C. I., Gerrits, N. M., Voogd, J., Leonard, C. S. & Simpson, J. I. (1994) The rostral dorsal cap and ventrolateral outgrowth of the rabbit inferior olive receive a GABAergic input from dorsal group y and the ventral dentate nucleus. *Journal of Comparative Neurology* 341:420–32. [arJIS]
De Zeeuw, C. I., Hertzberg, E. & Mugnaini, E. (1995a) The dendritic lamellar body: A new neuronal organelle putatively associated with dendrodendritic gap junctions. *Journal of Neuroscience* 15(2):1587–1604. [aJIS]
De Zeeuw, C. I., Holstege, J. C., Ruigrok, T. J. H. & Voogd, J. (1989) Ultrastructural study of the GABAergic, the cerebellar, and the mesodiencephalic innervation of the cat medial accessory olive: Anterograde tracing combined with immunocytochemistry. *Journal of Comparative Neurology* 284:12–35. [aJIS]
(1990) Mesodiencephalic and cerebellar terminals end up on the same dendritic spines within the glomeruli of the cat and rat inferior olive: An ultrastructural study using a combination of (3H)leucine and WGA-HRP anterograde tracing. *Neuroscience* 34:645–55. [aJIS, CW]
De Zeeuw, C. I. & Ruigrok, T. J. H. (1994) Olivary neurons in the nucleus of Darkschewitsch in the cat receive excitatory monosynaptic input from the cerebellar nuclei. *Brain Research* 653:345–50. [aJIS]
De Zeeuw, C. I., Ruigrok, T. J. H., Holstege, J. C., Jansen, H. J. & Voogd, J. (1990) Intracellular labeling of neurons in the medial accessory olive of the cat: 2. Ultrastructure of dendritic spines and their GABAergic innervation. *Journal of Comparative Neurology* 300:478–94.
De Zeeuw, C. I., Ruigrok, T. J. H., Holstege, J. C., Schalekamp, M. P. A. & Voogd, J. (1990) Intracellular labeling of neurons in the medial accessory olive of the cat: 3. Ultrastructure of the axon hillock and initial segment and their GABAergic innervation. *Journal of Comparative Neurology* 300:495–510. [aJIS]
De Zeeuw, C. I., Wentzel, P. & Mugnaini, E. (1993) Fine structure of the dorsal cap of the inferior olive and its GABAergic and non-GABAergic input from the nucleus prepositus hypoglossi in rat and rabbit. *Journal of Comparative Neurology* 327:63–82. [arJIS]
De Zeeuw, C. I., Wylie, D. R., DiGiorgi, P. L. & Simpson, J. I. (1994b) Projections of individual Purkinje cells of identified zones in the flocculus to the vestiublar and cerebellar nuclei in the rabbit. *Journal of Comparative Neurology* 349:428–47. [aJIS]
De Zeeuw, C. I., Wylie, D. R., Stahl, J. S. & Simpson, J. I. (1995b) Phase relations of Purkinje cells in the rabbit flocculus during compensatory eye movements. *Journal of Neurophysiology* 74:2051–64. [aJIS]
Deadwyler, S. A. & Hampson, R. E. (1995) Ensemble activity and behavior: What's the code? *Science* 270:1316–18. [CW]
Dean, P. (1995) Modelling the role of the cerebellar fastigial nuclei in producing accurate saccades: The importance of burst timing. *Neuroscience* 68:1059–77. [PD]
Dean, P., Mayhew, J. E. W. & Langdon, P. (1994) Learning and maintaining saccadic accuracy: A model of brainstem-cerebellar interactions. *Journal of Cognitive Neuroscience* 6:117–38. [aJCH, MAA, HB, PD]
Decety, J. & Michel, F. (1989) Comparative analysis of actual and mental movement times in two graphic tasks. *Brain and Cognition* 11:87–97. [aWTT]
Decety, J., Sjoholm, H., Ryding, E., Stenberg, G. & Ingvar, D. H. (1990) The cerebellum participates in mental activity: Tomographic measurements of regional cerebral blood flow. *Brain Research* 535:313–17. [aWTT]
Deecke, L., Grözinger, B. & Korhuber, H. H. (1976) Voluntary finger movement in man: Cerebral potentials and theory. *Biological Cybernetics* 23:99–119. [KW]
Demer, J. L., Echelman, D. A. & Robinson, D. A. (1985) Effects of electrical stimulation and reversible lesions of the olivocerebellar pathway on Purkinje cell activity in the flocculus of the cat. *Brain Research* 346:22–31. [aJIS]
Denk, W., Sugimori, M. & Llinas, R. (1995) Two types of calcium response limited to single spines in cerebellar Purkinje cells. *Proceedings of the National Academy of Sciences of the USA* 92:8279–82. [rDJL]
Desclin, J. C. (1974) Histological evidence supporting the inferior olive as the major source of cerebellar climbing fibers in the rat. *Brain Research* 77:365–84. [aJIS]
DeSerres, S. J. & Milner, T. E. (1991) Wrist muscle activation patterns and stiffness associated with stable and unstable mechanical loads. *Experimental Brain Research* 86:451–58. [aAMS]
Desmond, J. E. & Moore, J. W. (1991) Single-unit activity in red nucleus during the classically conditioned rabbit nictitating membrane response. *Neuroscience Research* 10:260–79. [aJCH]
Detre, J. A., Mairn, A. C., Aswad, D. W. & Greengard, P. (1984) Localization in mammalian brain of G-substrate, a specific substrate for guanosine 3',5'-cyclic monophosphate-dependent protein kinase. *Journal of Neuroscience* 4:2843–49. [aSRV]
Deubel, H., Wolf, W. & Hauske, G. (1986) Adaptive gain control of saccadic eye movements. *Human Neurobiology* 5:245–53. [HB]
Deuschl, G., Toro, C., Zeffiro, T., Massaquoi, S. & Hallett, M. (in press) Adaptation motor learning of arm movements in patients with cerebellar disease. *Journal of Neurology, Neurosurgery, and Psychiatry*. [MH]
Di Pellegrino, G., Fadiag, L., Fogassi, L., Gallese, V. & Rizzolatti, G. (1992) Understanding motor events: A neurophysiological study. *Experimental Brain Research* 91:176–80. [aWTT]
Dickie, B. G. M., Lewis, M. J. & Davies, J. A. (1990) Potassium-stimulated release of nitric oxide from cerebellar slices. *British Journal of Pharmacology* 101:8–9. [aSRV]
(1992) NMDA-induced release of nitric oxide potentiates aspartate overflow from cerebellar slices. *Neuroscience Letters* 138:145–48. [aSRV]
Diener, H.-C. & Dichgans, J. (1992) Pathophysiology of cerebellar ataxia. *Movement Disorders* 7:95–109. [JH]
Diener, H. C., Dichgans, J., Guschlbauer, B., Bacher, M., Rapp, H. & Langenbach, P. (1990) Associated postural adjustments with body movements in normal subjects and patients with Parkinsonism and cerebellar disease. *Revue Neurologique (Paris)* 146:555–63. [aAMS]
Diener, H. C., Hore, J., Ivry R. B. & Dichgans, J. (1993) Cerebellar dysfunction of movement and perception. *Canadian Journal of Neurological Sciences* 20(suppl. 3):S62-S69. [aAMS, CG]

Dietrichs, E. (1984) Cerebellar autonomic function: Direct hypothalamo-cerebellar pathway. *Science* 223:591–93. [JDS]
Dinnendahl, V. & Stock, K. (1975) Effects of arecoline and cholinesterase-inhibitors on cyclic guanosine 3',5'-monophosphate in mouse brain. *Naunyn-Schmiedebergs Archives of Pharmacology* 290:297–306. [aSRV]
Disterhoft, J. F., Coulter, D. A. & Alkon, D. L. (1986) Conditioning-specific membrane changes of rabbit hippocampal neurons measured in vitro. *Proceedings of the National Academy of Sciences of the USA* 83: 2733–37. [CW]
Disterhoft, J. F., Kronforst, M. A., Moyer, J. R., Jr., Thompson, L. T., Van der Zee, E. & Weiss, C. (1995) Hippocampal neuron changes during trace eyeblink conditioning in the rabbit. In: *Acquisition of motor behavior in vertebrates,* ed. J. R. Bloedel, T. J. Ebner & S. P. Wise. MIT Press. [CW]
Disterhoft, J. F., Kwan, H. H. & Lo, W. D. (1977) Nictitating membrane conditioning to tone in the immobilized albino rabbit. *Brain Research* 137:127–43. [CW]
Dodson, R. A. & Johnson, W. E. (1979) Effects of ethanol, arecoline, atropine and nicotine, alone and in various combinations, on rat cerebellar cyclic guanosine 3',5'-monophosphate. *Neuropharmacology* 18:871–76. [aSRV]
(1980) Effects of general central nervous system depressants with and without calcium ionophore A23187 on rat cerebellar cyclic guanosine 3',5'-monophosphate. *Research Communications in Chemical Pathology and Pharmacology* 29:265–80. [aSRV]
Dolphin, A. C., Detre, J. A., Schlichter, D. J., Nairn, A. C., Yeh, H. H., Woodward, D. J. & Greengard, P. (1983) Cyclic nucleotide-dependent protein kinases and some major substrates in the rat cerebellum after neonatal X-irradiation. *Journal of Neurochemistry* 40:577–81. [aSRV]
Dom, R., King, J. S. & Martin, G. F. (1973) Evidence for two direct cerebello-olivary connections. *Brain Research* 57:498–501. [aJIS]
Dornay, M., Uno, Y., Kawato, M. & Suzuki, R. (in press) Minimum muscle tension change trajectories. *Journal of Motor Behavior.* [MD]
Dow, R. S. (1942) The evolution and anatomy of the cerebellum. *Biological Reviews* 17:179–220. [aJIS]
Dow, R. S. & Moruzzi, G. (1958) *The physiology and pathology of the cerebellum.* University of Minnesota Press. [JDS]
Drevets, W. C., Videen, T. O., MacLeod, A.-M. K., Haller, J. W. & Raichle, M. E. (1992) PET imagees of blood flow changes during anxiety: Correction [letter]. *Science* 256:1696. [aWTT]
Dufossé, M., Ito, M., Jastreboff, P. & Miyashita, Y. (1978) A neuronal correlate in rabbit's cerebellum to adaptive modification of the vestibulo-ocular reflex. *Brain Research* 150:611–16. [MD]
Dugas, C. & Smith, A. M. (1992) Responses of cerebellar Purkinje cells to slip of a hand-held object. *Journal of Neurophysiology* 67:483–95. [aAMS]
Dunn, M. E. & Mugnaini, E. (1993) Influence of granule cells on the survival and differentiation of Purkinje cells in dissociated cerebellar cultures. *Society for Neuroscience Abstracts* 19:1723. [aDJL]
Dunwiddie, T. V. (1990) *Adenosine and adenosine receptors,* ed. M. Williams. Humana Press. [aDJL]
East, S. J. & Garthwaite, J. (1990) Nanomolar N^G-nitroarginine inhibits NMDA-induced cyclic GMP formation in rat cerebellum. *European Journal of Pharmacology* 184:311–13. [aSRV]
(1992) Actions of a metabotropic glutamate receptor agonist in immature and adult rat cerebellum. *European Journal of Pharmacology* 219:395–400. [aSRV]
Ebner, T. J. & Bloedel, J. R. (1981a) Temporal patterning in simple spike discharge of Purkinje cells and its relationship to climbing fiber activity. *Journal of Neurophysiology* 45:933–47. [aJIS]
(1981b) Role of climbing fiber afferent input in determining responsiveness of Purkinje cells to mossy fiber inputs. *Journal of Neurophysiology* 45:962–71. [aJIS]
(1984) Climbing fiber action on the responsiveness of Purkinje cells to parallel fiber inputs. *Brain Research* 309:1822–186. [aJIS]
Ebner, T. J., Flament, D. & Shanbhag, S. J. (1996) The cerebellum's role in voluntary motor learning: Clinical, electrophysiological, and imaging studies. In: *Acquisition of motor behavior in vertebrates,* ed. J. R. Bloedel, T. J. Ebner & S. P. Wise. MIT Press. [DF]
Ebner, T. J., Yu, Q. & Bloedel, J. R. (1983) Increase in Purkinje cell gain associated with naturally activated climbing fiber input. *Journal of Neurophysiology* 50:205–19. [aJIS]
Eccles, J. C., Ito, M. & Szentagothai, J. (1967) *The cerebellum as a neuronal machine.* Springer-Verlag Berlin. [aFC, aAMS, aJIS, aWTT, MGP]
Eccles, J. C., Llinás, R. & Sasaki, K. (1966) The excitatory synaptic action of climbing fibres on the Purkinje cells of the cerebellum. *Journal of Physiology (London)* 182:268–96. [aJIS]
Eccles, J. C., Sabah, N. H., Schmidt, R. F. & Táboríková, H. (1972) Cutaneous mechanoreceptors influencing impulse discharges in cerebellar cortex: 3. In Purkinje cells by climbing fiber input. *Experimental Brain Research* 15:484–97. [aJIS]
Edwards, F. A., Konnerth, A., Sakmann, B. & Takahashi, T. (1989) A thin slice preparation for patch-clamp recordings from neurones of the mammalian central nervous system. *Pflügers Archiv* 414:600–12. [aMKan]
Eilers, J., Augustine, G. J. & Konnerth, A. (1995) Subthreshold synaptic Ca^{2+} signalling in fine dendrites and spines of cerebellar Purkinje neurons. *Nature* 373:155–58. [NAH, DO, rSRV]
Eisenman, L. N., Keifer, J. & Houk, J. C. (1991) Positive feedback in the cerebro-cerebellar recurrent network may explain rotation of population vectors. In: *Analysis and modeling of neural systems.* ed. F. Eeckman. Kluwer. [arJCH]
Ekerot, C.-R., Garwicz, M. & Schouenborg, J. (1991) Topography and nociceptive receptive fields of climbing fibres projecting to the cerebellar anterior lobe in the cat. *Journal of Physiology (London)* 441:257–74. [aJCH]
Ekerot C. F. & Kano, M. (1985) Long-term depression of parallel fibre synapses following stimulation of climbing fibres. *Brain Research* 342:357–60. [aFC, aMKan, aDJL, aJIS, aWTT]
(1989) Stimulation parameters influencing climbing fibre induced long-term depression of parallel fibre synapses. *Neuroscience Research* 6:264–68. [aJCH, aDJL, CFE, JCF, EDS]
Ekerot, C. F. & Oscarsson, O. (1981) Prolonged depolarization elicited in Purkinje cell dendrites by climbing fibre impulses in the cat. *Journal of Physiology (London)* 318:207–21. [aFC, aJIS]
El-Husseini, A. E.-D., Bladen, C. & Vincent, S. R. (1995a) Expression of the olfactory cyclic nuceotide gated channel (CNG1) in the rat brain. *Neuroreport* 6:1331–35. [rSRV]
(1995b) Molecular characterization of a type II cyclic GMP-dependent protein kinase expressed in the rat brain. *Journal of Neurochemistry* 64:2814–17. [rSRV]
Enright, J. T. & Hendriks, A. W. (1994) To stare or to scrutinize: "Grasping" the eye for better vision. *Vision Research* 34:2039–42. [GPVG]
Escudero, M., de la Cruz, R. R. & Delgado-Garcia, J. M. (1992) A physiological study of vestibular and prepositus hypoglossi neurons projecting to the abducens nucleus in the alert cat. *Journal of Physiology (London)* 458:539–60. [aJIS]
Fagni, L., Bossu, J. L. & Bockaert, J. (1991) Activation of a large-conductance Ca^{2+}-dependent K^+ channel by stimulation of glutamate phosphoinositide-coupled receptors in cultured cerebellar granule cells. *European Journal of Neuroscience* 3:788–96. [aSRV]
Farrant, M. & Cull-Candy, S. G. (1991) Excitatory amino acid receptor-channels in Purkinje cells in thin cerebellar slices. *Proceedings of the Royal Society of London, Series B* 244:179–84. [aSRV]
Feldman, A. G. (1980a) Superposition of motor programs: 2. Rapid forearm flexion in man. *Neuroscience* 5:91–95. [aAMS]
(1980b) Superposition of motor programs: 1. Rhythmic forearm movements in man. *Neuroscience* 5:81–90. [aAMS]
Feldman, A. G., Adamovich, S. V. & Levin, M. D. (1995) The relationship between control, kinematic and electromyographic variables in fast single-joint movements in humans. *Experimental Brain Research* 103:440–50. [AGF]
Feldman, A. G. & Levin, M. F. (1993) Control variables and related concepts in motor control. *Concepts in Neuroscience* 4:25–51. [AGF]
(1995) Positional frames of reference in motor control: Their origin and use. *Behavioral and Brain Sciences* 78:723–806. [AGF]
Ferrendelli, J. A., Chang, M. M. & Kinscherf, D. A. (1974) Elevation of cyclic GMP levels in central nervous system by excitatory and inhibitory amino acids. *Journal of Neurochemistry* 22:535–40. [aSRV]
Ferrendelli, J. A., Kinscherf, D. A. & Kipnis, D. M. (1972) Effects of amphetamine, chlorpromazine and reserpine on cyclic GMP and cyclic AMP levels in mouse cerebellum. *Biochemical and Biophysical Research Communications* 46:2114–20. [aSRV]
Ferster, D. & Spruston, N. (1995) Cracking the neural code. *Science* 270:756–57. [EDS]
Fiala, J. C., Grossberg, S. & Bullock, D. (1995) *Metabotropic glutamate receptor activation in cerebellar Purkinje cells as substrate for adaptive timing of the classically conditioned eye blink response.* Technical Report CSD/CNS-TR-95-029, Department of Cognitive and Neural Systems, Boston University. [JCF]
Fiez, J. A., Petersen, S. E., Cheney, M. K. & Raichle, M. E. (1992) Impaired nonmotor learning and error detection associated with cerebellar damage. *Brain* 115:155–78. [aWTT]
Fitts, P. M. (1954) The information capacity of the human motor system in controlling the amplitude of movement. *Journal of Experimental Psychology* 47:381–91. [aAMS, GPVG]
Flament, D., Ellermann, J., Ugurbil, K. & Ebner, T. J. (1994) Functional magnetic resonance imaging (fMRI) of cerebellar activation while learning

to correct for visuomotor errors. *Society for Neuroscience Abstracts* 20:20. [DF]
Flament, D. & Hore, J. (1986) Movement and electromyographic disorders associated with cerebellar dysmetria. *Journal of Neurophysiology* 55:1221–33. [aAMS]
Flament, D., Lee, J.-H., Ugurbil, K. & Ebner, T. J. (1995) Changes in motor cortical and subcortical activity, during the acquisition of motor skill, investigated usuing functional MRI (4T, echo planar imaging). *Society for Neuroscience Abstracts* 21:1422. [DF]
Flash, T. (1987) The control of hand equilibrium trajectories in multi-joint arm movements. *Biological Cybernetics* 57:257–74. [HG]
Flash, T. & Mussa-Ivaldi, F. A. (1990) Human arm stiffness characteristics during the maintenance of posture. *Experimental Brain Research* 82:315–26. [aAMS]
Floeter, M. K. & Greenough, W. T. 91979) Cerebellar plasticity: Modification of Purkinje cell structure by differential rearing in monkeys. *Science* 206:227–29. [aWTT]
Flourens, P. (1824/1968) *Recherches expérimentales sur les propriétés et les fonctions du systèm nerveux dan les animaux vertébres*. Paris: Cervot. Translated, 1968, in: *The human brain and spinal cord,* ed. E. Clarke & C. D. O'Malley. University of California Press, Berkeley. [aJCH, aWTT, RCM]
Forget, R. & Lamarre, Y. (1987) Rapid elbow flexion in the absence of proprioceptive and cutaneous feedback. *Human Neurobiology* 6:27–37. [rAMS]
Förstermann, U., Gorsky, L E., Pollock, J. S., Schmidt, H. H. H. W., Heller, M. & Murad, F. (1990) Regional distribution of EDRF/NO-synthesizing enzymes(s) in rat brain. *Biochemical and Biophysical Research Communications* 168:727–32. [aSRV]
Fortier, P. A., Kalaska, J. F. & Smith, A. M. (1989) Cerebellar neuronal activity related to whole-arm reaching movements in the monkey. *Journal of Neurophysiology* 62:198–211. [aAMS]
Fortier, P. A., Smith, A. M. & Kalaska, J. F. (1993) Comparison of cerebellar and motor cortex activity during reaching: Directional tuning and response variability. *Journal of Neurophysiology* 69:1136–49. [aAMS, HG]
Fortier, P. A., Smith, A. M. & Rossignol, S. (1987) Locomotor deficits in the mutant mouse Lurcher. *Experimental Brain Research* 66:271–86. [aAMS]
Foy, M. R. & Thompson, R. F. (1986) Single unit analysis of Purkinje cell discharge in classically conditioned and untrained rabbits. *Neuroscience Abstracts* 12:753. [RFT]
Frens M. A., Van Opstal, A. J. (1994) Auditory-evoked saccades in two dimensions: Dynamical characteristics, influence of eye position and sound source spectrum. In: *Information processing underlying gaze control,* ed. J. Delgado-Garc!a, P. Vidal & E. Godaux. Oxford University Press. [CG]
Frick, R. B. (1982) The ego and the vestibulocerebellar system. *Psychoanalytic Quarterly* 51:93–122. [JDS]
Fries, W. (1990) Pontine projection from striate and prestriate visual cortex in the macaque monkey: An anterograde study. *Visual Neuroscience* 4:205–16. [JDS]
Friston, K. J., Frith, C. D., Passingham, R. E., Liddle, P. F. & Frackowiak, R. S. J. (1992) Motor practice and neurophysiological adaptation in the cerebellum: A positron tomography study. *Proceedings of the Royal Society of London* 248:223–28. [aWTT, DT]
Frolov, A. A., Roschin, V. Y. & Biryukova, E. V. (1993) Adaptive neural model of multijoint movement control by working point analysis. *Neural Network World* 4:141–56. [MD]
Frysinger, R. C., Bourbonnais, D., Kalaska, J. F. & Smith, A. M. (1984) Cerebellar cortical activity during antagonist cocontraction and reciprocal inhibition of forearm muscles. *Journal of Neurophysiology* 51(1):32–49. [aAMS, aWTT, HG]
Fuchs, A. F., Robinson, F. R. & Straube, A. (1993) Role of the caudal fastigial nucleus in saccade generation: 1. Neuronal discharge pattern. *Journal of Neurophysiology* 70:1723–40. [aJCH, PD]
Fujita, M. (1982) Adaptive filter model of the cerebellum. *Biological Cybernetics* 45:195–206. [aJCH]
Fukuda, M., Yamamoto, T. & Llinás, R. (1987) Simultaneous recordings from Purkinje cells of different folia in the rat cerebellum and their relation to movement. *Society for Neuroscience Abstracts* 13:603. [aJIS]
Funabiki, K., Mishina, M. & Hirano, T. (1995) Retarded vestibular compensation in mutant mice deficient d2 glutamate receptor subunit. *NeuroReport* 7:189–92. [TH, rMKan]
Furuyama, T., Inagaki, S. & Takagi, H. (1993) Localizations of a1 and b1 subunits of soluble guanylate cyclase in the rat brain. *Molecular Brain Research* 20:335–44. [aSRV]
Fushiki, H., Sato, Y., Miura, A. & Kawasaki, T. (1994) Climbing fiber responses of Purkinje cells to retinal image movement in cat cerebellar flocculus. *Journal of Neurophysiology* 71:1336–50. [aJIS]
Gabbiani, F., Midtgaard, J. & Knöpfel, T. (1994) Synaptic integration in a model of cerebellar granule cell. *Journal of Neurophysiology* 72:999–1009. [KH]
Gabrieli, J. D. E., McGlinchey-Berroth, R., Carrillo, M. C., Gluck, M. A., Cermak, L. S. & Disterhoft, J. F. (1995) Intact delay-eyeblink classical conditioning in amnesics. *Behavioral Neuroscience* 109:819–27. [CW]
Gaffan, D. (1992) The role of the hippocampus-fornix-mammillary system in episodic memory. In: *Neuropsychology of memory,* ed. L. R. Squire & N. Butters. Guilford. [SMO]
Gaffan, D. & Harrison, S. (1989) A comparison of the effects of fornix transection and sulcus principalis ablation upon spatial learning by monkeys. *Behavioral Brain Research* 31:207–20. [SMO]
Galiana, H. L. (1985) Comissural vestibular nuclear coupling: A powerful putative site for producing adaptive change. In: *Adaptive mechanisms in gaze control: Facts and theories*, ed. A. Berthoz & G. Melvill Jones. Elsevier. [aJCH]
(1986) A new approach to understanding adaptive visual-vestibular interactions in the central nervous system. *Journal of Neurophysiology* 55:349–74. [aJCH]
Galiana, H. L. & Guitton, D. (1992) Central organization and modelling of eye-head coordination during orienting gaze shifts. In: *Sensing and controlling motion: Vestibular and sensorimotor function*, vol. 656, ed. B. Cohen, D. L. Tomka & F. Guedry. Annals of the New York Academy of Science. [aJCH]
Galiana, H. L. & Outerbridge, J. S. (1984) A bilateral model for central neural pathways in the vestibuloocular reflex. *Journal of Neurophysiology* 51:210–41. [aJCH]
Galione, A., White, A., Willmott, N., Turner, M., Potter, B. V. L. & Watson, S. P (1993) cGMP mobilizes intracellular Ca^{2+} in sea urchin eggs by stimulating cyclic ADP-ribose synthesis. *Nature* 365:456–59. [aSRV]
Gao, J.-H., Parsons, L. M., Bower, J. M. Xiong, J., Li, J. & Fox, P. T. (in press) Cerebellum implicated in sensory acquisition and discrimination rather than motor control. *Science.* [JMB]
Garthwaite, J. (1991) Glutamate, nitric oxide and cell-cell signalling in the nervous system. *Trends in Neuroscience* 14:60–67. [aSRV]
Garthwaite, J., Charles, S. L. & Chess-Williams, R. (1988) Endothelium-derived relaxing factor release on activation of NMDA receptors suggests role as intercellular messenger in the brain. *Nature* 336:385–88. [aFC, aSRV]
Garthwaite, J. & Brodbelt, A. R. (1989) Glutamate as the principal mossy fibre transmitter in rat cerebellum: Pharmacological evidence. *European Journal of Neuroscience* 2:177–80. [aSRV]
Garthwaite, J. & Garthwaite, G. (1987) Cellular origins of cyclic GMP responses to excitatory amino acid receptor agonists in rat cerebellum *in vitro*. *Journal of Neurochemistry* 48:29–39. [aDJL, LK]
Garthwaite, J., Garthwaite, G., Palmer, R. M. J. & Moncada, S. (1989a) NMDA receptor activation induces nitric oxide synthesis from arginine in rat brain slices. *European Journal of Pharmacology* 172:413–16. [aSRV]
Garthwaite, J., Southam, E. & Anderson, M. (1989b) A kainate receptor linked to nitric oxide synthesis from arginine. *Journal of Neurochemistry* 53:1952–54. [aFC, aSRV]
Gasic, G. P. & Hollman M. (1992) Molecular neurobiology of glutamate receptors. *Annual Review of Physiology* 54:507–36. [aFC]
Gauthier, G. M., Hofferer, J.-M., Hoyt, W. F. & Stark, L. (1979) Visual-motor adaptation: Quantative demonstration in patients with posterior fossa involvement. *Archives of Neurology* 36:155–60. [aWTT]
Gellman, R. S., Gibson, A. R. & Houk, J. C. (1985) Inferior olivary neurons in the awake cat: Detection of contact and passive body displacement. *Journal of Neurophysiology* 54:40–60. [aJCH, aJIS, RCM, RFT, CW]
Gellman, R. S., Houk, J. C. & Gibson, A. R. (1983) Somatosensory properties of the inferior olive in the cat. *Journal of Comparative Neurology* 215:228–43. [aJIS, RFT]
Georgopoulos A. P., Kalaska, J. F., Crutcher M. D., Caminiti R. & Massey J. T. (1984) The representation of movement direction in the motor cortex: Single cell and population studies. In: *Dynamic aspects of neocortical function,* ed. G. M. Edelman, W. M. Cowan & W. E. Gall. Wiley. [CG]
Georgopoulos, A. P. & Massey, J. T. (1987) Cognitive spatial-motor processes: 1. The making of movements at various angles from a stimulus direction. *Experimental Brain Research* 65:361–70. [CG]
Gerrits, N. M., Voogd, J. & Magras, I. N. (1985) Vestibular afferents of the inferior olive and the vestibulo-olivo-cerebellar climbing fiber pathway to the flocculus in the cat. *Brain Research* 332:325–36. [aJIS]
Ghelarducci, B., Ito, M. & Yagi, N. (1975) Impulse discharges from flocculus Purkinje cells of alert rabbit during visual stimulation combined with horizontal head rotation. *Brain Research* 87:66–72. [aJIS]
Ghez, C., Hening, W. & Favilla, M. (1990) Parallel interacting channels in the initiation and specification of motor response features. In: *Attention and performance: 8. Motor representation and control,* ed. M. Jeannerod. Erlbaum. [aJCH]
Gibson, A. R., Horn, K. M. & Van Kan, P. L. E. (1990) Interpositus discharge during reaching. *Society for Neuroscience Abstracts* 16:637. [aAMS]

Gibson, A. R., Robinson, F. R., Alam, J. & Houk, J. C. (1987) Somatotopic alignment between climbing fiber input and nuclear output of the intermediate cerebellum. *Journal of Comparative Neurology* 260:362–77. [aJCH, CW]

Gielen, C. C. A. M. & van Gisbergen, J. A. M. (1990) The visual guidance of saccades and fast aiming movements. *News in Physiological Science* 5:58–63. [aJCH]

Gilbert, P. F. C. (1974) A theory of memory that explains the function and structure of the cerebellum. *Brain Research* 70:1–18. [aJCH, aWTT, PFCG]

(1975) How the cerebellum could memorize movements. *Nature (London)* 254:688–89. [aJCH, PFCG]

Gilbert, P. F. C. & Thach W. T. (1977) Purkinje cell activity during motor learning. *Brain Research* 128:309–28. [aJCH, aJIS, aWTT, MD, PFCG, JCH, DT]

Gilman, S. (1969a) Fusimotor fiber responses in the decerebellate cat. *Brain Research* 14:218–21. [aAMS]

(1969b) The mechanism of cerebellar hypotonia: An experimental study in the monkey. *Brain* 92:621–38. [aAMS]

Glasauer, S., Amoriaum, M. A., Vitte, E. & Berthoz, A. (1994) Goal-directed linear locomotion in normal and labyrinthine-defective subjects. *Experimental Brain Research* 98:323–35. [SMO]

Glaum, S. R., Slater, N. T., Rossi, D. J. & Miller, J. R. (1992) The role of metabotropic glutamate receptors at the parallel-fiber-Purkinje cell synapse. *Journal of Neurophysiology* 68:1453–62. [aFC, aSRV, aDJL, LJB, NAH, DO]

Glickstein, M. (1993) Motor skills but not cognitive tasks. *Trends in Neuroscience* 16:450–51. [PFCG]

(1994) Cerebellar agenesis. *Brain* 117:1209–12. [SMO]

Glickstein, M., Gerrits, N., Kralj-Hans, J., Mercier, B., Stein, J., Voogd, J. (1994) Visual pontocerebellar projections in the macaque. *Journal of Comparative Neurology* 349:51–72.

Glickstein, M., May, J. G. & Mercier, B. E. (1985) Corticopontine projection in the macaque: The distribution of labeled cortical cells after large injections of horseradish peroxidase in the pontine nuclei. *Journal of Comparative Neurology* 235:343–59. [JDS]

Gluck, M. A. & Thompson, R. F. (1990) Adaptive signal processing and the cerebellum: Models of classical conditioning and VOR adaptation. In: *Neuroscience and connectionist theory*, ed. M. A. Gluck & D. E. Rumelhart. Erlbaum. [aJCH]

Goldberg, M. E., Musil S. Y., Fitzgibbon E. J., Smith M. & Olson C. R. (1993) The role of the cerebellum in the control of saccadic eye movements. In: *Role of the basal ganglia and cerebellum in voluntary movements*, ed. N. Mano, I. Hamada & M. R. DeLong. Elsevier. [PD, CG]

Gomi, H. & Kawato, M. (1992) Adaptive feedback control models of the vestibulocerebellum and spinocerebellum. *Biological Cybernetics* 68:105–14. [aJCH, aAMS]

(1995) The change of human arm mechanical impedance during movements under different environmental conditions. In: *Society for Neuroscience 25th Annual Meeting*. San Diego, CA: Society for Neuroscience. [HG, MKaw]

(1996) Mechanical impedance of human arm during multi-joint movemnt in the horizontal plane. *Jounal of the Society of Instrument and Control Engineers* 32(3) [in Japanese]. [HG]

Gonshor, A. & Melvill-Jones, G. (1976) Extreme vestibulor-ocular adaptation induced by prolonged optical reversal of vision. *Journal of Physiology (London)* 256:381–414. [aWTT, rJIS]

Goodkin, H. P., Keating, J. G., Martin, T.A. & Thach, W. T. (1993) Preserved simple and impaired compound movement after infarction in the territory of the superior cerebellar artery. *Canadiam Journal of Neurology Science* 20(suppl.3):S93–104. [aWTT]

Goodman, D. & Kelso, J. (1983) Exploring the functional significance of physiological tremor: A biospectroscopic approach. *Experimental Brain Research* 49:419–31. [aJIS]

Goodman, R. R., Kuhar, M. J., Hester, L. & Snyder, S. H. (1983) Adenosine receptors: Autoradiographic evidence for their location on axon terminals of excitatory neurons. *Science* 220:967–69. [aDJL]

Gorassini, M., Prochazka, A. & Taylor, J. L. (1993) Cerebellar ataxia and muscle spindle sensitivity. *Journal of Neurophysiology* 70:1853–62. [aAMS]

Görcs, T. J., Penke, B., Bóti, Z., Katarova, Z. & Hámori, J. (1993) Immunohistochemical visualization of a metabotropic glutamate receptor. *NeuroReport* 4:283–86. [aSRV]

Gordon, A. M., Huxley, A. F. & Julien, F. J. (1966) The variation in isometric tension with sarcomere length in vertebrate muscle fibres. *Journal of Physiology* 184:170–92. [aAMS]

Gormezano, I. (1966) Classical conditioning. In: *Experimental methods and instrumentation in psychology*, ed. J. B. Sidowski. McGraw-Hill. [CW]

Goslow, G. E. Jr., Reinking, R. M. & Stuart, D. G. (1973) The cat step cycle: Hind limb joint angles and mucles lengths during unrestrained locomotion. *Journal of Morphology* 141:1–42. [aAMS]

Gottlieb, J. P., MacAvoy, M. G. & Bruce, C. J. (1994) Neural responses related to smooth-pursuit eye movements and their correspondence with electrically elicited smooth eye movements in the primate frontal eye field. *Journal of Neurophysiology* 72:1634–53. [PVD]

Graf, W., Simpson, J. I. & Leonard, C. S. (1988) Spatial organization of visual messages of the rabbit's cerebellar flocculus. II. Complex and simple spike responses of Purkinje cells. *Journal of Neurophysiology* 60:2091–2121. [arJIS]

Grafman, J., Litvan, I., Massaquoi, S., Stewart, M., Sirigu, A., & Hallett, M. (1992) Cognitive planning deficity in patients with cerebellar atrophy. *Neurology* 42:1493–1496.

Grafton, S. T., Hazeltine, E. & Ivry, R. (1995) Functional mapping of sequence learning in normal humans. *Journal of Cognitive Neuroscience* 7:497–510. [MH]

Grafton, S. T., Mazziotta, J. C., Presty, S., Friston, K. J., Frackowiak, R. S. J. & Phelps, M. E. (1992) Functional anatomy of human procedural learning determined with regional cerebral blood flow and PET. *Journal of Neuroscience* 12:2542–48. [aWTT]

Granit, R. & Phillips, C. G. (1956) Excitatory and inhibitory processes acting upon individual Purkinje cells of the cerebellum in cats. *Journal of Physiology (London)* 133:520–47. [aJIS]

Grant, S. G., O'Dell, T. J., Karl, K. A., Stein, P. L., Soriano, P. & Kandel, E. R. (1992) Impaired long-term potentiation, spatial learning, and hippocampal development in *fyn* mutant mice. *Science* 258:1903–10. [aMKan]

Graybiel, A. M., Nauta, H. J. W., Lasek, R. J. & Nauta, W. J. H. (1973) A cerebello-olivary pathway in the cat: An experimental study using autoradiographic tracing techniques. *Brain Research* 58:205–11. [aJIS]

Greenberg, L. H., Troyer, E., Ferrendelli, J. A. & Weiss, B. (1978) Enzymatic regulation of the concentration of cyclic GMP in mouse brain. *Neuropharmacology* 17:737–45. [aSRV]

Greengard, P., Jen, J., Nairn, A. C. & Stevens, C. F. (1991) Enhancement of the glutamate receptor response by cAMP dependent protein kinase in hippocampal neurons. *Science* 253:1135–38. [MB]

Grill, S. E., Hallett, M., Marcus, C., McShane, L. (1994) Disturbance of kinaesthesia in patients with cerebellar disorders. *Brain* 117:1433-47.

Groenewegen, H. J. & Voogd, J. (1977) The parasagittal zonation within the olivocerebellar projection: 1. Climbing fiber distribution in the vermis of the cat cerebellum. *Journal of Comparative Neurology* 174:417–88. [aJIS]

Groenewegen, H. J., Voogd, J. & Freedman, S. L. (1979) The parasagittal zonation within the olivocerebellar projection: 2. Climbing fiber distribution in the intermediate and hemispheric parts of cat cerebellum. *Journal of Comparative Neurology* 183:551–602. [aJIS]

Grossberg, S. & Kuperstein, M. (1989) *Neural dynamics of adaptive sensory-motor control*. Pergamon. [aJCH]

Grover, L. M. & Teyler, T. J. (1992) N-methyl-D-aspartate receptor-independent long-term potentiation in area CA1 of rat hippocampus: Input-specific induction and preclusion in a non-tetanized pathway. *Neuroscience* 49:7–11. [MB]

Gruart, A. & Yeo, C. H. (1995) Cerebellar cortex and eyeblink conditioning: Bilateral regulation of conditioned responses. *Experimental Brain Research* 104:431–48. [CW]

Grusser, O. J., Pause, M. & Schreiter, U. (1990) Vestibular neurones in the parietoinsular cortex of monkeys (*macaca fascicularis*): Visual and neck receptor responses. *Journal of Physiology* 430:559–83. [SMO]

Guidotti, A., Biggio, G. & Costa, E. (1975) 3-Acetylpyridine: A tool to inhibit the tremor and increase of cGMP content in cerebellar cortex elicited by harmaline. *Brain Research* 96:201–5. [aSRV]

Guiramand, J., Vignes, M., Mayat, E., Lebrun, F., Sassetti, I. & Recasens, M. (1991) A specific transduction mechanism for the glutamate action on phosphoinositide metabolism via the quisqualate metabotropic receptor in rat brainsynaptoneurosomes: 1. External Na^+ requirement. *Journal of Neurochemistry* 57:1488–1500. [aDJL]

Guitton, D., Munoz, D. P. & Galiana, H. L. (1990) Gaze control in the cat: Studies and modeling of the coupling between orienting eye and head movements in different behavioral tasks. *Journal of Neurophysiology* 64:509–31. [aJCH]

Gusovsky, F., Hollingsworth, E. B. & Daly, J. W. (1986) Regulation of phosphatidylinositol turnover in brain synaptoneurosomes: Stimulatory effects of agents that enhance influx of sodium ions. *Proceedings of the National Academy of Science of the USA* 83:3003–7. [aDJL]

Guzmán-Lara, S. (1993) *Adjusting connections using reflexes as guidance*. NPB Technical Report 8, Northwestern University Institute of Neuroscience. [aJCH]

Haby, C., Lisovoski, F., Aunis, D. & Zwiller, J. (1994) Stimulation of the cyclic GMP pathway by NO induces expression of the immediate early genes *c-fos* and *junB* in PC12 cells. *Journal of Neurochemistry* 62:496–501. [aSRV]

Haggard, P. N., Jenner, J. R. & Wing, A. M. (1994) Kinematic patterns in a case of unilateral cerebellar damange. *Neuropsychologia* 32:827–46. [PH]

Haggard, P. & Wing, A. M. (1995) Coordinated responses following mechanical perturbation of the arm during prehension. *Experimental Brain Research* 102:483–94. [PH]

Haidamous, M., Kouyoumdjuan, J. C., Briley, P. A. & Gonnard, P. (1980) In vivo effects of noradrenaline and noradrenergic receptor agonists and antagonists on rat cerebellar cyclic GMP levels. *European Journal of Pharmacology* 63:287–94. [aSRV]

Haier, R. J., Siegel, B. W. Jr., MacLachlan, A., Soderling, E., Lottenberg, S. & Buchbaum, M. (1992) Regional glucose metabolic changes after learning a complex visuospatial motor task: A positron emission tomography study. *Brain Research* 570:134–43. [aWTT]

Hallett, M., Berardelli, A., Matheson, J., Rothwell, J. & Marsden, C. D. (1991) Physiological analysis of simple rapid movements in patients with cerebellar deficits. *Journal of Neurology, Neurosurgery, and Psychiatry* 53:124–33. [JH]

Hallett, M., Pascual-Leone, A. & Topka, H. (in press) Adaptation and skill learning. Evidence for different neural substrates. In: *Acquisition of motor behavior in vertebrates,* ed. J. R. Bloedel, T. J. Ebner & S. P. Wise. [MH]

Hallett, M., Shahani, B. T. & Young R. R. (1975) EMG analysis of patients with cerebellar deficits. *Journal of Neurology, Neurosurgery and Psychiatry* 38:1163–69. [aAMS, aWTT, CG, JH]

Hansel, C., Batchelor, A., Cuénod, M., Garthwaite, J., Knöpfel, T. & Do, K. Q. (1992) Delayed increase of extracellular arginine, the nitric oxide precursor, following electrical white matter stimulation in rat cerebellar slices. *Neuroscience Letters* 142:211–14. [arSRV]

Harris, C. M. (1995) Does saccadic undershoot minimize saccadic flight-time? A Monte Carlo study. *Vision Research* 35:691–701. [PD]

Harrison, N. L. & Lambert, N. A. (1989) Modification of GABAA receptor function by an analog of cyclic AMP. *Neuroscience Letters* 105:137–42. [aMKan]

Hartell, N. A. (1994a) Induction of cerebellar long-term depression requires activation of glutamate metabotropic receptors. *NeuroReport* 5:913–16. [NAH, MKan, rSRV]

(1994b) cGMP acts within cerebellar Purkinje cells to produce long-term depression via mechanisms involving PKC and PKG. *NeuroReport* 5:833–36. [NAH, rDJL, rSRV]

(in press) Strong activation of paralell fibers produces localized calcium transients and a form of LTD which spreads to distant synapses. *Neuron.* [NAH, rDJL]

Harting, J. K. (1977) Descending pathways from the superior colliculus: An autoradiographic analysis in the rhesus monkey (*Macaca mulatta*). *Journal of Comparative Neurology* 173:583–612. [JDS]

Harvey, R. J., Porter, R. & Rawson, J. A. (1977) The natural discharges of Purkinje cells in paravermal regions of lobules V and VI of the monkey's cerebellum. *Journal of Physiology* 271:515–36. [arJCH]

(1979) Discharges of intracerebellar nuclear cells in monkeys. *Journal of Physiology* 297:559–80. [aJCH]

Hasan, Z. (1986) Optimized movement trajectories and joint stiffness in unperturbed, inertially loaded movements. *Biological Cybernetics* 53:373–82. [arAMS]

Hassler, R. (1950) Uber kleinhirnprojektionen zum mittlehirn und thalamus beim menschen. *Deutsche Zeitscrift fur Nervenheikunde* 163:629–71. [aWTT]

Hawkes, R., Blyth, S., Chockkan, V., Tano, D., Ji, Z. & Mascher, C. (1993) Structural and molecular compartmentation in the cerebellum. *Canadian Journal of Neurological Science* 20:S29-S35. [aJCH]

Heath, R. G. (1977) Modulation of emotion with a brain pacemaker. *Journal of Nervous and Mental Disease* 165:300–17. [JDS]

Hebb, D. O. (1949) *The organization of behavior.* Wiley. [aFC, aAMS]

Heck, D. (1993) Rat cerebellar cortex in vitro responds specifically to moving stimuli. *Neuroscience Letters* 157:95–98. [FS]

(1995) Sequential input to guinea pig cerebellar cortex in vitro strongly affects Purkinje cells via parallel fibers. *Naturwissenschaften* 82:201–3. [FS]

Hecker, M., Sessa, W. C., Harris, H. J., Anggard, E. E. & Vane, J. R. (1990) The metabolism of L-arginine and its significance for the biosynthesis of endothelium-derived relaxing factor: Cultured endothelial cells recycle L-citrulline to L-arginine. *Proceedings of the National Academy of Sciences of the USA* 87:8612–16. [LK]

Hemart, N., Daniel, H., Jaillard, D. & Crépel, F. (1994) Properties of glutamate receptors are modified during long-term depression in cerebellar Purkinje cells. *Neuroscience Research* 19:213–21. [aFC, rDJL]

(1995) Receptors and second messengers involved in long-term depression in rat cerebellar slices in vitro: A reappraisal. *European Journal of Neuroscience* 7:45–53. [aFC, DO, rDJL, rSRV]

Herdnon, R. M. & Coyle, J. T. (1978). Glutaminergic innervation, kainic acid and selective vulnerability in the cerebellum. In: *Kainic acid as a tool in neurobiology,* ed. G. McGeer, J. W. Olney & P. L. McGeer. Raven. [aFC]

Herrmann-Frank, A. & Varsanyi, M. (1993) Enhancement of Ca^{2+} release channel activity by phosphorylation of the skeletal muscle ryanodine receptor. *Federation of European Biochemical Societies Letters* 332:237–42. [aSRV]

Hertz, J., Krogh, A. & Palmer, R. G. (1991) *Introduction to the theory of neural computation.* Addison-Wesley. [EDS]

Hesslow, G. (1994a) Correspondence between climbing fibre input and motor output in eyeblink related areas in cat cerebellar cortex. *Journal of Physiology* 476:229–44. [GH]

(1994b) Inhibition of classically conditioned eyeblink responses by stimulation of the cerebellar cortex in the cat. *Journal of Physiology* 476:245–25. [GH]

Hidaka, H., Tanaka, T., Onoda, K., Hagiwara, M., Watanabe, M., Ohta, H., Ito, Y., Tsurudome, M. & Yoshida, T. (1988) Cell-specific expression of protein kinase C isozymes in the rabbit cerebellum. *Journal of Biological Chemistry* 263:4523–26. [aFC, aDJL]

Hikosaka, O., Matsumara, M., Kojima, J. & Gardiner, T. W. (1993) Role of basal ganglia in initiation and suppression of saccadic eye movements. In: *Role of the cerebellum and basal ganglia in voluntary movement,* ed. N. Mano, I. Hamada & M. R. DeLong. Excerpta Medica. [MAA]

Hirano, T. (1990a) Depression and potentiation of the synaptic transmission between a granule cell and a Purkinje cell in rat cerebellar culture. *Neuroscience Letters* 119:141–44. [aDJL]

(1990b) Effects of postsynaptic depolarization in the induction of synaptic depression between a granule cell and a Purkinje cell in rat cerebellar culture. *Neuroscience Letters* 119:145–47. [aDJL, aMKan]

(1991) Differential pre- and postsynaptic mechanisms for synaptic potentiation and depression between a granule cell and a Purkinje cell in rat cerebellar culture. *Synapse* 7:321–23. [aDJL]

Hirano, T. & Kasono, K. (1993) Spatial distribution of excitatory and inhibitory synapses on a Purkinje cell in rat cerebellar culture. *Journal of Neurophysiology* 70:1316–25. [aDJL]

Hirano, T. Kasono, K., Araki, K. & Mishina, M. (1995) Suppression of LTD in cultured Purkinje cells deficient in the glutamate receptor d2 subunit. *NeuroReport* 6:524–26. [TH, MKan, rDJL, rSRV]

Hirano, T., Kasono, K., Araki, K., Shinozuka, K. & Mishina, M. (1994) Involvement of the glutamate receptor d2 subunit in the long-term depression of glutamate responsiveness in cultured rat Purkinje cells. *Neuroscience Letters* 182:172–76. [TH, MKan, rDJL]

Hirsch, J. C. & Crépel, F. (1990) Use-dependent changes in synaptic efficacy in rat prefrontal neurons in vitro. *Journal of Physiology* 427:31–49. [aFC]

Hoff, B. & Arbib, M. A. (1992) A model of the effects of speed, accuracy, and perturbation on visually guided reaching. In: *Control of arm movement in space: Neurophysiological and computational approaches [Experimental Brain Research Series 22],* ed. R. Caminiti, P. B. Johnson & Y. Burnod. [MAA]

Hoffer, B. J., Siggins, G. R., Oliver, A. P. & Bloom, F. E. (1971) Cyclic AMP mediation of norepinephrine inhibition in rat cerebellar cortex: A unique class of synaptic responses. *Annals of the New York Academy of Sciences* 185:531–49. [aSRV]

Hoffer, J. A. & Andreassen, S. (1981) Limitations in the servo-regulation of soleus muscle stiffness in premammillary cats. *Muscle Receptors & Movement* 308:311–24. [aAMS]

Hofmann, M., Spano, P. F., Trabucchi, M. & Kumakura, K. (1977) Guanylate cyclase activity in various rat brain areas. *Journal of Neurochemistry* 29:395–96. [aSRV]

Hogan, N. (1990) Mechanical impedance of single- and multi-articular systems. In: *Multiple muscle systems: Biomechanics and movement organization,* ed. J. M. Winters & S. L. Woo. Springer-Verlag. [arAMS]

Hogan, N. & Flash, T. (1987) Moving gracefully: Quantitative theories of motor coordination. *Trends in Neuroscience* 10:170–74. [aAMS]

Hollman, M., O'Shea-Greenfield, A., Rogers, S. W. & Heinemann, S. (1989) Cloning by functional expression of a member of the glutamate receptor family. *Nature* 342:643–48. [aFC]

Holmes, G. (1917) The symptoms of acute cerebellar injuries due to gunshot injuries. *Brain* 40:461–535. [aWTT]

(1922a) Clinical symptoms of cerebellar disease and their interpretation. The Croonian lectures 1. *Lancet* 1:1117–82. [aWTT]

(1922b) Clinical symptoms of cerebellar disease and their interpretation. The Croonian lectures 2. *Lancet* 1:1237. [aWTT]

(1922c) Clinical symptoms of cerebellar disease and their interpretation. The Croonian lectures 3. *Lancet* 2:59–65. [aWTT]

(1922d) Clinical symptoms of cerebellar disease and their interpretation. The Croonian lectures 4. *Lancet* 2:111–15. [aWTT]

(1939) The cerebellum of man. *Brain* 62:1–30. [aAMS, aWTT, MAA]

Hope, B. T., Michael, G. J., Knigge, K. M. & Vincent, S. R. (1991) Neuronal

NADPH-diaphorase is a nitric oxide synthase. *Proceedings of the National Academy of Sciences of the USA* 88:2811–14. [aSRV]
Hopfield, J. J. (1982) Neural networks and physical systems with emergent collective computational abilities. *Proceedings of the National Academy of Sciences of the USA* 2554–58. [rJCH]
Horak, F. B. (1990) Comparison of cerebellar and vestibular loss on scaling of postural responses. In: *Disorders of posture and gait*, ed. T. Brandt, W. Paulus, W. Bles, M. Dietrerich, S. Drafczyk & A. Straube. Stuttgart: Georg Thieme Verlag. [aWTT]
Horak, F. B. & Diener, H. C. (1993) Cerebellar control of postural scaling and central set in stance. *Journal of Neurophysiology* 72:479–93. [aAMS, aWTT, JMB, DT]
Horak, F. B., Esselman, P. E., Anderson, M. E. & Lynch, M. K. (1984) The effects of movement velocity, mass displaced and task certainty on associated postural adjustments made by normal and hemiplegic individuals. *Journal of Neurology, Neurosurgery, and Psychiatry* 47:1020–28. [aAMS]
Hore, J. (1993) Arm ataxia: Disorders in cerebellar-cortical function. *Biomedical Research* 14(suppl. l):23–26. [JH]
Hore, J. & Flament, D. (1988) Changes in motor cortex neural discharge associated with the development of cerebellar limb ataxia. *Journal of Neurophysiology* 60:1285–1302. [JH]
Hore, J., Wild, B. & Diener, H.-C. (1991) Cerebellar dysmetria at the elbow, wrist and fingers. *Journal of Neurophysiology* 65:563–71. [JH, FS]
Horn, A. K. E. & Hoffmann, K. P. (1987) Combined GABA immunocytochemistry TMB/HRP histochemistry of pretecto nuclei projecting to the inferior olive in rats, cats, and monkeys. *Brain Research* 409:135–38. [aJIS]
Horn, R. & Marty, A. (1988) Muscarinic activation of ionic currents measured by a new whole-cell recording method. *Journal of General Physiology* 92:145–159. [aDJL]
Houk, J. C. (1989) Cooperative control of limb movements by the motor cortex, brainstem and cerebellum. In: *Models of brain function*, ed. R. M. J. Cotterill. Cambridge University Press. [aJCH]
(1990) Role of cerebellum in classical conditioning. *Society for Neuroscience Abstracts* 16:474. [aJCH]
(1992) Learning in modular networks. In: *Proceedings of the 7th Yale workshop on adaptive and learning systems*, ed. K. S. Narendra. Center for Systems Science. [aJCH]
Houk, J. C., Adams, J. L. & Barto, A. G. (1995) A model of how the basal ganglia generates and uses neural signals that predict reinforcement. In: *Models of Information processing in the basal ganglia*, ed. J. C. Houk, J. L. Davis & D. G. Beiser. MIT Press. [aJCH, JCH]
Houk, J. C. & Barto, A. G. (1992) Distributed sensorimotor learning. In: *Tutorials in motor behavior 2*, ed. G. E. Stelmach & J. Requin. Elsevier. [arJCH, aMS, PD, JCH]
Houk, J. C., Galiana, H. L. & Guitton, D. (1992) Cooperative control of gaze by the superior colliculus, brainstem and cerebellum. In: *Tutorials in motor behavior 2*, ed. G. E. Stelmach & J. Requin. Elsevier. [arJCH, PD]
Houk, J. C. & Gibson, A. R. (1987) Sensorimotor processing through the cerebellum. In: *New concepts in cerebellar neurobiology*, ed. J. S. King. Alan R. Liss. [aJCH]
Houk, J. C., Keifer, J. & Barto, A. G. (1993) Distributed motor commands in the limb premotor network. *Trends in Neuroscience* 16:27–33. [arJCH, PD, JCH]
Houk, J. C., Singh, S. P., Fisher, C. & Barto, A. G. (1990) An adaptive sensorimotor network inspired by the anatomy and physiology of the cerebellum. In: *Neural networks for control*, ed. W. T. Miller, R. S. Sutton & P. J. Werbos. MIT Press. [arJCH, PD]
Houk, J. C. & Rymer, W. Z. (1981) Neural control of length and tension. In: *Handbook of physiology: vol. 2. The nervous system: pt. 1. Motor control*, ed. J. M. Brookhardt & V. B. Mountcastle. American Physiological Society. [aAMS, MD, CG]
Houk, J. C. & Wise, S. P. (1995) Distributed modular architectures linking basal ganglia, cerebellum and cerebral cortex: Their role in planning and controlling action. *Cerebral Cortex* 5:95–110. [aJCH]
Huang, P. L., Dawson, T. M., Bredt, D. S., Snyder, S. H. & Fishman, M. C. (1993) Targeted disruption of the neuronal nitric oxide synthase gene. *Cell* 75:1273–86. [TH, rSRV]
Hudson, B. D., Valcana, T., Bean, G. & Timiras, P. S. (1976) Glutamic acid: A strong candidate as the neurotransmitter of cerebellar granule cell. *Neurochemical Research* 1:73–82. [aFC]
Hufschmidt, H. J. & Hufschmidt, T. (1954) Antagonist inhibition as the earliest sign of a sensory-motor reaction. *Nature* 174:607. [rAMS]
Hultborn, H. & Illert, M. (1991) How is motor behavior reflected in the organization of spinal systems? In: *Motor control: Concepts and issues*, ed. D. R. Humphrey & H.-J. Freund. Wiley. [rAMS]
Hultborn, H., Lindstrom, S. & Wigstrom, H. (1979) On the function of recurrent inhibition in the spinal cord. *Experimental Brain Research* 37:399–403. [rAMS]
Humphrey, D. R., Gold, R. & Reed, D. J. (1984) Sizes, laminar and topographic origins of cortical projections to the major divisions of the red nucleus in the monkey. *Journal of Comparative Neurology* 225:75–94. [JDS]
Humphrey, D. R. & Reed, D. J. (1983) Separate cortical systems for control of joint movement and joint stiffness: Reciprocal activation and co-activation of antagonist muscles. In: *Motor control in health and disease*, ed. J. E. Desmedt. Raven. [aAMS]
Hunter, I. W. & Kearney, R. E. (1982) Dynamics of human ankle stiffness: Variation with mean ankle torque. *Journal of Biomechanics* 15:747–52. [aAMS]
Ignarro, L. J., Ballot, B. & Wood, K. S. (1984) Regulation of soluble guanylate cyclase activity by porphyrins and metalloporphyrins. *Journal of Biological Chemistry* 259:6201–7. [DO]
Ikeda, M., Morita, I., Murota, S.-I., Sekiguchi, F., Yuasa, T. & Miyatake, T. (1993) Cerebellar nitric oxide synthase activity is reduced in nervous and Purkinje cell degeneration mutants but not in climbing fiber-lesioned mice. *Neuroscience Letters* 155:148–50. [arSRV]
Inhoff, A. W., Diener, H. C., Rafal, R. D. & Ivry, R. B. (1989) The role of cerebellar structures in the execution of serial movements. *Brain* 112:565–81. [aAMS, aWTT]
Inoue, M., Oomura, Y., Yakushiji, T. & Akaike, N. (1986) Intracellular calcium ions decrease the affinity of the GABA receptor. *Nature* 324:156–58. [aMKan]
Iriki, A., Pavlides, C., Keller, A. & Asanuma, H. (1989) Long-term potentiation in the motor cortex. *Science* 245:1385–87. [aMKan]
Isaac, J. T. R., Nicoll, R. A. & Malenka, R. C. (1995) Evidence for silent synapses: Implications for the expression of LTP. *Neuron* 15:427–34. [MB]
Isaacson, J. S. & Nicoll, R. A. (1991) Aniracetam reduces glutamate receptor desensitization and slows the decay of fast excitatory synaptic currents in the hippocampus. *Proceedings of the National Academy of Sciences of the USA* 88:10936–40. [aFC]
Ishikawa, J., Kawaguchi, S. & Rowe, M. J. (1972) Actions of afferent impulses from muscle receptors on cerebellar Purkinje cells: 2. Responses to muscle contraction. Effects mediated via the climbing fiber pathway. *Experimental Brain Research* 16:104–14. [aJIS]
Ito, M. (1969) Neurons of cerebellar nuclei. In: *The interneuron*, ed. M.A.B. Brazier. UCLA Forum. [aJCH]
(1970) Neurophysiological aspects of the cerebellar motor control system. *International Journal of Neurology* 7:162–76. [aJCH], aJIS
(1972) Neural design of the cerebellar motor control system. *Brain Research* 40:81–84. [aJIS, aWTT]
(1982) Cerebellar control of the vestibulo-ocular reflex-around the flocculus hypothesis. *Annual Review of Neuroscience* 5:275–96. [aJIS]
(1984) *The cerebellum and neural control*. Raven. [aFC, aJCH, aMKan, aAMS, aJIS, aWTT, PD, EDS]
(1987) Characterization of synaptic plasticity in cerebellar and cerebral neocortex. In: *The neural and molecular bases of learning*, ed. J. P. Changeux & M. Nonishi. Wiley. [aFC]
(1989) Long-term depression. *Annual Review in Neuroscience* 12:85–102. [aFC, aJCH, aMKan, aJIS, aWTT, EDS, RFT]
(1990) A new physiological concept on the cerebellum. *Revue Neurologique* 10:564–69. [aWTT]
(1991) The cellular basis of cerebellar plasticity. *Current Opinion in Neurobiology* 1:616–20. [aDJL]
(1993) Movement and thought: Identical control mechanism by the cerebellum. *Trends in Neurosciences* 16:448–50. [aJCH, MKaw]
Ito, M., Jastreboff, P. J. & Miyashita, Y. (1982b) Specific effects of unilateral lesions in the flocculus upon eye movements in albino rabbits. *Experimental Brain Research* 45:233–42. [aJIS]
Ito, M. & Karachot, L. (1990) Messengers mediating long-term desensitization in cerebellar Purkinje cells. *NeuroReport* 1:129–32. [aFC, aMKan, arDJL, aSRV, DO]
(1989) Long-term desensitization of quisqualate-specific glutamate receptors in Purkinje cells investigated with wedge recording from rat cerebellar slices. *Neuroscience Research* 7:168–71. [aWTT]
(1992) Protein kinases and phosphatase inhibitors mediating long-term desensitization of glutamate receptors in cerebellar Purkinje cells. *Neuroscience Research* 14:27–38. [aFC, aDJL, NAH]
Ito, M., Kawai, N. & Udo, M. (1968) The origin of cerebellar-induced inhibition of Deiters neurones: 3. Localization of the inhibitory zone. *Experimental Brain Research* 4:310–20. [aAMS]
Ito, M., Nisimaru, N. & Yamamoto, M. (1977) Specific patterns of neuronal connections involved in the control of the rabbit's vestibulo-ocular reflexes by the cerebellar flocculus. *Journal of Physiology (London)* 265:833–54. [aJIS]

Ito, M., Obata, K. & Ochi, R. (1966) The origin of cerebellar-induced inhibition of Deiters' neurons: 2. Temporal correlation between the trans-synaptic activation of Purkinje cells and the inhibition of Deiters neurons. *Experimental Brain Research* 2:350–64. [aAMS]

Ito, M., Sakurai, M. & Tongroach, P. (1982) Climbing fibre induced depression of both mossy fiber responsiveness and glutamate sensitivity of cerebellar Purkinje cells. *Journal of Physiology (London)* 324:113–34. [aFC, aDJL, aMKan, aJIS, aWTT, MD, RFT]

Ito, M., Shiida, T., Yagi, N. & Yamamoto, M. (1974) The cerebellar modification of rabbit's horizontal vestibulo-ocular reflex induced by sustained head rotation combined with visual stimulation. *Proceedings of the Japan Academy* 50:85–89. [aWTT]

Ito, M. & Simpson, J. (1971) Discharges in Purkinje cell axons during climbing fibre activation. *Brain Research* 31:215–19. [aJIS]

Ito, M., Tanabe, S., Kohda, A. & Sugiyama, H. (1990) Allosteric potentiation of quisqualate receptors by a nootropic drug aniracetam. *Journal of Physiology (London)* 424:533–43. [aFC]

Ito, M. & Yoshida, M. (1966) The origin of cerebellar-induced inhibition of Deiters' neurons: 1. Monosynaptic initiation of the inhibitory post synaptic potential. *Experimental Brain Research* 2:330–49. [aAMS]

Ivry, R. B. & Diener, H. C. (1991) Impaired velocity perception in patients with lesions of the cerebellum. *Journal of Cognitive Neuroscience* 3:355–66. [aAMS]

Ivry, R. B. & Keele, S. W. (1989) Timing functions of the cerebellum. *Journal of Cognitive Neuroscience* 1:136–52. [aAMS, aWTT]

Ivry, R. B., Keele, S. W. & Diener, H.C. (1988) Dissociation of the lateral and medial cerebellum in movement timing and movement execution. *Experimental Brain Research* 73:167–80. [aAMS]

Jackson, J. H. (1890) A study of convulsions. In: *Selected writings of John Hughlings Jackson* (1932), ed. J. Taylor. Basic. [aWTT]

Jaeger, D. & Bower, J. M. (1994) Prolonged responses in rat cerebellar Purkinje cells following activation of the granule cell layer: An intracellular in vitro and in vivo investigation. *Experimental Brain Research* 100:200–14. [JMB, DJ]

Jahnsen, H. (1986) Extracellular activation and membrane conductances of neurones in the guinea-pig deep cerebellar nuclei *in vitro*. *Journal of Physiology (London)* 372:149–68. [aJIS]

Jakab, R. L. & Hàmori, J. (1988) Quantitative morphology and synaptology of cerebellar glomeruli in the rat. *Anatomy and Embryology* 179:81–88. [FS]

Jami, L. (1992) Golgi tendon organs in mammalians skeletal muscle: Functional properties and central actions. *Physiological Reviews* 72:623–66. [MD]

Jankowska, E. & Roberts, W. J. (1972) An electrophysiological demonstration of the axonal projections of single spinal interneurones in the cat. *Journal of Physiology (London)* 222:597–622. [rAMS]

Jansen, J. & Brodal, A. (1954) *Aspects of cerebellar anatomy*. Oslo: Johan Grundt Tanum Forlag. [aJIS]

Jeannerod, M. (1981) Intersegmental coordination during reaching at natural visual objects. In: *Attention and performance 9*, ed. J. Long & A. Baddeley. Erlbaum. [PH]

(1994) The representing brain: Neural correlates of motor intention and imagery. *Behavioral and Brain Science* 17:187–201. [aWTT]

Jenkins, I. H., Brooks, D. J., Nixon P. D. & Frackowiak, R. S. J. (1994) Motor sequence learning: A study with positron emission tomography. *Journal of Neuroscience* 14:3775–90. [aWTT, DT, JDS]

Jeromin, A., Huganir, R., & Linden, D. J. (1996) Suppression of the glutamate receptor delta-2 subunit produces a specific impairment in cerebellar long-term depression. *Journal of Neurophsiology*, in press. [rDJL]

Jongen, H. A. H., Denier Van Der Gon, J. J. & Gielen, C. C. A. M. (1989) Inhomogeneous activation of motoneurone pools as revealed by co-contraction of antagonistic human arm muscles. *Experimental Brain Research* 75:555–62. [aAMS]

Kalaska, J. F., Cohen, D. A. D., Prud'homme, M. & Hyde, M. L. (1990) Parietal area 5 neuronal activity encodes movement kinematics, not movement dynamics. *Experimental Brain Research* 80:351–64. [MKaw]

Kandel, E. & Schwartz, J. (1982) Molecular biology of learning: Modulation of transmitter release. *Science* 218:433–43. [aMKan]

Kano, M. (1995) Plasticity of inhibitory synapses in the brain: A possible memory mechanism that has been overlooked. *Neuroscience Research* 21:177–82. [aMKan]

Kano, M. Hashimoto, K., Chen, C., Abeliovich, A., Aiba, A., Kurihara, H., Watanabe, M., Inoue, Y. L. & Tonegawa, S. (1995) Impaired synapse elimination during cerebellar development in PKCγ mutant mice. *Cell* 83:1223–31. [rMKan, rDJL]

Kano, M. & Kato, M. (1987) Quisqualate receptors are specifically involved in cerebellar synaptic plasticity. *Nature* 325:276–79. [aFC, aMKan, aDJL]

(1988) Mode of induction of long-term depression at parallel fibre-Purkinje cell synapses in rabbit cerebellar cortex. *Neuroscience Research* 5:544–56. [aDJL, aJIS]

Kano, M., Kano, M.-S., Kusunoki, M. & Maekawa, K. (1990a) Nature of optokinetic response and zonal organization of climbing fiber afferents in the vestibulocerebellum of the pigmented rabbit: 2. The nodulus. *Experimental Brain Research* 80:238–51. [aJIS]

Kano, M., Kano, M.-S. & Maekawa, K, (1991) Optokinetic response of simple spikes of Purkinje cells in the cerebellar flocculus and nodulus of the pigmented rabbit. *Experimental Brain Research* 87:484–96. [rJIS]

Kano, M. & Konnerth, A. (1992a) Potentiation of GABA-mediated currents by cAMP-dependent protein kinase. *NeuroReport* 3:563–66. [aMKan]

(1992b) Cerebellar slices for patch clamp recording. In: *Practical electrophysiological methods,* ed. H. Kettenmann & R. Grantyn. Wiley-Liss. [aMKan]

Kano, M., Rexhausen, U., Dreessen, J. & Konnerth, A. (1992) Synaptic excitation produces a long-lasting rebound potentiation of inhibitory synaptic signals in cerebellar Purkinje cells. *Nature (London)* 356:601–4. [aMKan, aJIS]

Kano, M.-S., Kano, M. & Maekawa, K. (1990) Receptive field organization of climbing fiber afferents responding to optokinetic stimulation in the cerebellar nodulus and flocculus of the pigmented rabbit. *Experimental Brain Research* 82:499–512. [aJIS]

Kapoula, Z. & Robinson, D. A. (1986) Saccadic undershoot is not inevitable: Saccades can be accurate. *Vision Research* 26:735–43. [HB]

Karachot, L., Kado, R. T. & Ito, M. 91994) Stimulus parameters for induction of long-term depression in *in vitro* rat Purkinje cells. *Neuroscience Research* 21:161–68. [EDS, rJIS]

Kasai, H. & Petersen, O. H. (1994) Spatial dynamics of second messengers: IP3 and cAMP as long-range and associative messengers. *Trends in Neurosciences* 17:95–101.

Kashiwabuchi, N., Ikeda, K., Araki, K., Hirano, T., Shibuki, K., Takayama, C., Inoue, Y., Kutsuwada, T., Yagi, T., Kang, Y., Aizawa, S. & Mishina, M. (1995) Impairment of motor coordination, Purkinje cell synapse formation, and cerebellar long-term depression in GluRd2 mutant mice. *Cell* 81:245–52. [MKan, rMKan, TH, rDJL, rSRV]

Kasono, K. & Hirano, T. (1994) Critical role of postsynaptic calcium in cerebellar long-term depression. *NeuroReport* 6:17–20. [PC, TH, rDJL]

(1995) Involvement of inositol triphosphate in cerebellar long-term depression. NeuroReport 6:569–72. [TH, MKan, rDJL, rSRV]

Katayama, S. & Nisimaru, N. (1988) Parasagittal zonal pattern of olivo-nodular projections in rabbit cerebellum. *Neuroscience Research* 5:424–38. [aJIS]

Kawano, K., Shidara, M., Takemura, A., Inoue, Y., Gomi, H. & Kawato, M. (1994) A linear time-series regression analysis of temporal firing patterns of cerebral, pontine and cerebellar neurons during ocular following. *Japanese Journal of Physiology* 44:S219. [MKaw]

Kawano, K., Shidara, M., Watanabe, Y. & Yamane, S. (1994) Neural activity in cortical area MST of alert monkey during ocular following responses. *Journal of Neurophysiology* 71:2305–24. [MKaw]

Kawano, K., Shidara, M. & Yamane, S. (1992) Neural activity in dorsolateral pontine nucleus of alert monkey during ocular following responses. *Journal of Neurophysiology* 67:680–703. [MKaw]

Kawato, M. (1990a) Computational schemes and neural network models for formation and control of multijoint arm trajectory. In: *Neural networks for control*, ed. T. Miller, R. S. Sutton & P. J. Werbos. MIT Press. [aJCH]

(1990b) Feedback-error learning neural network for supervised motor learning. In. *Advanced neural computers* , ed. R. Eckmiller. Elsevier. [HB]

(1995) Analysis of neural firing frequency by a generalized linear model. *Technical Report of The Institute of Electronics, Information and Communication Engineers* NC95–33:31. [MKaw]

Kawato, M. & Gomi, H. (1992a) A computational model of four regions of the cerebellum based on feedback-error learning. *Biological Cybernetics* 68:95–103. [arAMS, aJCH, CG, RCM]

(1992b) The cerebellum and VOR/OKR learning models. *Trends in Neuroscience* 15:445–53. [aJCH, aJIS, DF]

(1993) Feedback-error-learning model of cerebellar motor control. In: *Role of the cerebellum and basal ganglia in voluntary movement*, ed. N. Mano, I. Hamada & M. R. DeLong. Elsevier. [aJCH, DF, rAMS]

Keating, J. G. & Thach, W. T. (1993) Complex spike activity in the awake behaving monkey: Non-clock-like discharge. *Society of Neuroscience Abstracts* 19:980. [aJIS]

(1995) Nonclock behavior of inferior olive neurons: Interspike interval of Purkinje cell complex spike discharge in the awake behaving monkey is random. *Journal of Neurophysiology* 73:1329–40. [aWTT]

Keele, S. W. (1981) Behavioral analysis of movement. In: *Handbook of Physiology: sect. 1. The nervous system: vol. 2. Motor control, pt. 2.*, ed. J. M. Brookhart, V. B. Mountcastle & V. B. Brooks. American Physiological Society, [aWTT]

Keele, S. W. & Ivry, R. (1990) Does the cerebellum provide a common computation for diverse tasks? A timing hypothesis. *Annals of the New York Acadamy of Science* 608:179–211. [aWTT, FS]

Keifer, J. & Houk, J. C. (1995) In vitro classical conditioning of abducens nerve discharge in turtles. *Journal of Neuroscience.* [aJCH]

Keinänen, K., Wisden, W., Sommer, B., Werner, P., Herb, A., Verdoorn, T. A., Sakmann, B. & Seeburg, P. H. (1990) A family of AMPA-selective glutamate receptors. *Science* 249:556–60. [aFC]

Keller, E. L. (1989) The cerebellum. In: *The neurobiology of saccadic eye movements* , ed. R. H. Wurtz & M. E. Goldberg. Elsevier. [HB, PD]

Kelso, J. A. S. (1995) *Dynamic patterns of the self-organization of the brain and behavior.* MIT Press. [AGF]

Kennedy, P. R. (1990) Corticospinal, rubrospinal, and rubroolivaary projections: A unifying hypothesis. *Trends in Neuroscience* 13:474–79. [aWTT]

Kennedy, P. R., Gibson, A. R. & Houk, J. C. (1986) Functional and anatomic differentiation between parvicellular and magnocellular regions of red nucleus in the monkey. *Brain Research* 364:124–36. [JDS]

Kennelly, P. J. & Krebs, E. G. (1991) Consensus sequences as substrate specificity determinants for protein kinases and protein phosphatases. *Journal of Biological Chemistry* 266:15555–58. [aSRV]

Khater, T. T., Quinn, K. J., Pena, J., Baker, J. F. & Peterson, B.W. (1993) The latency of the cat vestibulo-ocular reflex before and after short- and long-term adaptation. *Experimental Brain Research* 94:16–32. [aJCH]

Khodakhah, K. & Ogden, D. (1993) Functional heterogeneity of calcium release by inositol trisphosphate in single Purkinje neurones, cultured cerebellar astrocytes, and peripheral tissues. *Proceedings of the National Academy of Sciences of the USA* 90:4976–80. [rDJL]

Kiedrowski, L., Costa, E. & Wroblewski, J. T. (1992a) Glutamate receptor agonists stimulate nitric oxide synthase in primary cultures of cerebellar granule cells. *Journal of Neurochemistry* 58:335–41. [arSRV, LK]

(1992b) *In vitro* interaction between cerebellar astrocytes and granule cells: A putative role for nitric oxide. *Neuroscience Letters* 135:59–61. [LK]

Kim, J. H., Wang, J.-J. & Ebner, T. J. (1987) Climbing fiber afferent modulation during treadmill locomotion in the cat. *Journal of Neurophysiology* 57:787–802. [aJIS]

(1988) Alterations in simple spike activity and locomotor behavior associated with climbing fiber input to Purkinje cells in a decerebrate walking cat. *Neuroscience* 25:475–89. [aJIS]

Klann, E., Chen, S. J. & Sweatt, J. D. (1993) Mechanism of protein kinase C activation during the induction and maintenance of long-term potentiation probed using a selective peptide substrate. *Proceedings of the National Academy of Sciences of the USA* 90:8337–41. [aFC]

Klein, P. S., Sun, T. J., Saxe, C. L. III, Kimmel, A. R., Johnson, R. L. & Devreotes, P. N. (1988) A chemoattractant receptor controls development in *Dictyostelium discoideum. Science* 241:1467–72. [aSRV]

Klopf, A. H. (1982) *The hedonistic neuron: A theory of memory, learning and intelligence.* Harper and Row Hemispheres. [aJCH]

Knöpfel, T., Vranesic, I., Staub, C. & Gahwiler, B. H. (1991) Climbing fibre responses in olivo-cerebellar slice cultures: 2. Dynamics of cytosolic calcium in Purkinje cells. *European Journal of Neuroscience* 3:343–48. [aMKan]

Knowles, R. G., Palacios, M., Palmer, R. M. & Moncada, S. (1989) Formation of nitric oxide from L-arginine in the central nervous system: A transduction mechanism for stimulation of the soluble guanylate cyclase. *Proceedings of the National Academy of Sciences of the USA* 86:5159–62. [aFC, aSRV]

Kobayashi, Y., Kawano, K., Takemura, A., Inoue, Y., Kitama, T., Gomi, H. & Kawato, M. (1995) Inverse-dynamics representation of complex spike discharges of Purkinje cells in monkey cerebellar ventral parafloculus during ocular following responses. *Society for Neuroscience Abstracts* 21:140. [MKaw]

(submitted) Climbing fiber discharges can convey information sufficient for motor learning with their ultra-low firing rates. [MKaw]

Koch, C. & Segev, I. (1989) *Methods in neuronal modeling.* MIT Press. [rJCH]

Kocsis, J. D., Eng, D. L. & Bhisitkul, R. B. (1984). Adenosine selectively blocks parallel fiber-mediated synaptic potentials in rat cerebellar coretx. *Proceedings of the National Academy of Sciences of the USA* 81:6531–34. [aDJL]

Kohda, K., Inoue, T. & Mikoshiba, K. (1995). Ca^{2+} release from Ca^{2+} stores, particularly from ryanodine-sensitive Ca^{2+} stores, is required for the induction of LTD in cultured cerebellar Purkinje cells. *Journal of Neurophysiology* 74:2184–88. [MKan, rDJL, rSRV]

Kolb, F. P. & Rubia, F. J. (1980) Information about peripheral events conveyed to the cerebellum via the climbing fiber system in the decerebrate cat. *Experimental Brain Research* 38:363–73. [arJIS]

Kolb, F. P., Rubia, F. J. & Bauswein, E. (1987) Cerebellar unit responses of the mossy fibre system to passive movements in the cerebrate cat: 1. Responses to static parameters. *Experimental Brain Research* 68:234–43. [DT]

Komatsu, Y., Toyama, K., Maeda, J. & Sakaguchi, H. (1981) Long-term potentiation investigated in a slice preparation of striate cortex of young kittens. *Neuroscience Letters* 26:269–74. [aMKan]

Konnerth, A. (1990) Patch-clamping in slices of mammalian CNS. *Trends in Neuroscience* 13:321–23. [aMKan]

Konnerth, A., Dreessen, J. & Augustine, G. J. (1992) Brief dendritic calcium signals initiate long-lasting synaptic depression in cerebellar Purkinje cells. *Proceedings of the National Academy of Sciences of the USA* 89:7051–55. [aFC, aMKan, arDJL, aSRV]

Konnerth, A., Llano, I. & Armstrong, C. M. (1990) Synaptic currents in cerebellar Purkinje cells. *Proceedings of the National Academy of Sciences of the USA* 87:2662–65. [aMKan, aDJL]

Korn, H., Oda, Y. & Faber, D. S. (1992) Long-term potentiation of inhibitory circuits and synapses in the central nervous system. *Proceedings of the National Academy of Sciences of the USA* 89:440–43. [aMKan, RFT]

Kowall, N. W. & Mueller, M. O. (1988) Morphology and distribution of nicotinamide adenine dinucleotide phosphate (reduced form) diaphorase reactive neurons in human brainstem. *Neuroscience* 26:645–54. [aSRV]

Krommenhoek, K. P., Van Opstal, A. J., Gielen, C. C. A. M. & Van Gisbergen, J. A. M. (1993) Remapping of neural activity in the motor colliculus: A neural network study. *Vision Research* 33:1287–98. [aJCH]

Krupa, M. & Crépel, F. (1990) Transient sensitivity of rat cerebellar Purkinje cells to N-methyl-D-aspartate during development. A voltage-clamp study in *in vitro* slices. *European Journal of Neuroscience* 2:312–16. [aDJL]

Krupa, D. J., Thompson, J. K. & Thompson, R. F. (1993) Localization of a memory trace in mammalian brain. *Science* 260:989–91. [aWTT, CW]

Kuhnt, U. & Voronin, L. L. (1994) Interaction between paired-pulse facilitation and long-term potentiation in area CA1 of guinea-pig hippocampal slices: Application of quantal analysis. *Neuroscience* 62:392–97. [LJB]

Kullman, D. M. (1994) Amplitude fluctuations of dual-component EPSCs in hippocampal pyramidal cells: Implications for long-term potentiation. *Neuron* 12:1111–20. [LBJ]

Kusunoki, M., Kano, M., Kano, M.-S. & Maekawa, K. (1990) Nature of optokinetic response and zonal organization of climbing fiber afferents in the vestibulo-cerebellum of the pigmented rabbit: 1. The flocculus. *Experimental Brain Research* 80:225–37. [aJIS]

Kuypers, H. G. J. M. & Lawrence, D. G. (1967) Cortical projections to the red nucleus and the brainstem in the rhesus monkey. *Brain Research* 4:151–88. [JDS]

Künzle, H. & Akert, K. (1977) Efferent connections of cortical area 8 (frontal eye field) in *Macaca fascicularis.* A reinvestigation using the autoradiographic technique. *Journal of Comparative Neurology* 173:147–64. [JDS]

Lackner, J. R. & DiZio, P. (1994) Rapid adaptation to coriolis force perturbations of arm trajectory. *Journal of Neurophysiology* 72:299–313. [aAMS]

Lacquaniti, F., Carrozzo, M. & Borghese, N. A. (1993) Time-varying mechanical behavior of multijointed arm in man. *Journal of Neurophysiology* 69:1443–64. [arAMS]

Lacquaniti, F. & Maioli, C. (1989) The role of preparation in tuning anticipatory and reflex responses during catching. *Journal of Neuroscience* 9:134–48. [aAMS]

Lainé, J. & Axelrad, H. (1994) The candelabrum cell: A new interneuron in the cerebellar cortex. *Journal of Comparative Neurology* 339:159–173. [aSRV]

Lamarre, Y., Montigny, C., Dumont, M. & Weiss, M. (1971) Harmaline-induced rhythm activity of cerebellar and lower brain stem neurons. *Brain Research* 32:246–50. [aJIS]

Lambolez, B., Audinat, E., Bochet, P., Crépel, F. & Rossier, J. (1992) AMPA receptor subunits expressed by single Purkinje cells. *Neuron* 9:247–58. [aFC]

Lang, E. J. (1995) Synchronicity, rhythmicity, and movement: The role of the olivocerebellar system in motor coordination. Ph.D. thesis, New York University. [arJIS]

Lang, E. J., Sugihara, I. & Llinás, R. (1989) Intraolivary injection of picrotoxin causes reorganization of complex spike activity. *Society of Neuroscience Abstracts* 15:77.5. [aJIS]

(1990) Lesions of the cerebellar nuclei, but not of the mesencephalic structures, alters the spatial pattern of complex spike synchronicity as demonstrated by multiple electrode recordings. *Society Neuroscience Abstracts* 16:370.3. [aJIS]

(1992) The ability of motor cortex stimulation to evoke vibrissal movements is modulated by a 10 Hz signal arising in the inferior olive. *Society of Neuroscience Abstracts* 18:178.6. [aJIS]

Larkman, A., Hannay, T., Stratford, K. & Jack, J. (1992) Presynaptic release probability influences the locus of long-term potentiation. *Nature* 360:70–73. [LBJ]

Lashley, K. S. (1951) The problem of serial order in behavior. *Cerebral mechanisms in behavior: The Hixon Symposium* 36:506–28. [aAMS]

Latash, L. P. (1979) Trace changes in the spinal cord and some basic problems of the neurophysiology of memory. In: *Seventh Gagra talks: The neurophysiological basis of memory,* ed. T. N. Oniani. Tbilisi: Metsniereba. [LPL]

Latash, M. L. & Gottlieb, G. L. (1991) Reconstruction of shifting elbow joint compliant characteristics during fast and slow movements. *Neuroscience* 43:697–712. [aAMS]

Latash, M. L. & Zatsiorsky, V. M. (1993) Joint stiffness: Myth or reality? *Human Movement Sciences* 12:653–92. [LPL]

Lavond, D. G., Kanzawa, S. A., Ivkovich, D. & Clark, R. E. (1994) Transfer of learning but not memory after unilateral cerebellar lesion in rabbits. *Behavioral Neuroscience* 108:284–93. [CW]

Lechtenberg, R. & Gilman, S. (1978) Speech disorders in cerebellar disease. *Annals of Neurology* 3:285–90. [aWTT]

Lee, W. A. (1984) Neuromotor synergies as a basis for coordinated intentional action. *Journal of Motor Behavior* 16:135–70. [aAMS]

Lee, W. A., Buchanan, T. S. & Rogers, M. W. (1987) Effects of arm acceleration and behavioral condition on the organization of postural adjustments during arm flexion. *Experimental Brain Research* 66:257–70. [aAMS]

Leicht, R., Rowe, M. J. & Schmidt, R. F. (1977) Mossy and climbing fiber inputs from cutaneous mechanoreceptors to cerebellar Purkyne cells in unanesthetized cats. *Experimental Brain Research* 27:459–77. [aJIS]

Leinders-Zufall, T., Rosenboom, H., Barnstable, C. J., Shepherd, G. M. & Zuff, F. (1995) A calcium-permeable cGMP-activated cation conductance in hippocampal neurons. *NeuroReport* 1761–65. [rSRV]

Leiner, H. C. & Leiner, A. L. (1989) Reappraising the cerebellum: What does the hindbrain contribute to the forebrain? *Behavioral Neuroscience* 103:998–1008. [aWTT]

Leiner, H. C., Leiner, A. L. & Dow, R. S. (1986) Does the cerebellum contribute to mental skills? *Behavioral Neuroscience* 100:443–53. [aWTT, JDS, CW]

(1987)Cerebro-cerebellar learning loops in apes and humans. *Italian Journal of Neurological Science* 425–436.

(1989) Reappraising the cerebellum: What does the hindbrain contribute to the forebrain? *Behavioral Neuroscience* 103:998–1008. [aJCH]

(1991) The human cerebrocerebellar system: Its computing, cognitive, and language skills. *Behavioral and Brain Research* 44:113–28. [aWTT]

(1993) Cognitive and language functions of the human cerebellum. *Trends in Neuroscience* 16:444–54. [PFCG, JDS]

Lemij, H. G. (1990) Asymetrical adaptation of human saccades to anisometropic spectacles. Thesis, Erasmus University, Rotterdam. [HB]

Leonard, C. S. & Simpson, J. I. (1986) Simple spike modulation of floccular Purkinje cells during the reversible blockade of their climbing fiber afferents. In: *Adaptive processes in visual and oculomotor systems*, ed. E. Keller & D. Zee. Pergamon. [aJIS]

Leonard, C. S., Simpson, J. I. & Graf, W. (1988) Spatial organization of visual messages of the rabbit's cerebellar flocculus: 1. Typology of inferior olive neurons of the dorsal cap of Kooy. *Journal of Neurophysiology* 60:2073–90. [aJIS]

Leranth, C. & Hamori, J. (1981) Quantitative electron microscope study of synaptic terminals to basket neurons in cerebellar cortex of rat. *Journal of Mikroskopik-Anatomy Forsch.* 95:1–14. [aSRV]

Lev-Ram, V., Makings, L. R., Keitz, P. F., Kao, J. P. Y. & Tsien, R. Y. (1995) Long-term depression in cerebellar Purkije neurons results from coincidence of nitric oxide and depolarization-induced Ca^{2+} transients. *Neuron* 15:407–15. [JCH, NAH, TH, LBJ, MKan, DO, rDJL, rSRV]

Lev-Ram, V., Miyakawa, H., Lasser-Ross, N. & Ross, W. N. (1992) Calcium transients in cerebellar Purkinje neurons evokes by intracellular stimulation. *Journal of Neurophysiology* 68:1167–77. [aMKan]

Levi, G., Gordon, R. D., Gallo, V., Wilkin, G. P. & Balazs, R. (1982) Putative amino acid transmitters in the cerebellum: 1. Depolarization induced release. *Brain Research* 239:425–45. [aDJL]

Levin, M. F., Feldman, A. G., Milner, T. E. & Lamarre, Y. (1992) Reciprocal and coactivation commands for fast wrist movements. *Experimental Brain Research* 89:669–77. [aAMS, AGF]

Li, J., Smith, S. S. & McElligott, J. G. (1995) Cerebellar nitric oxide is necessary for vestibulo-ocular reflex adaptation, a sensorimotor model of learning. *Journal of Neurophysiology* 74:489–94. [rDJL]

Liao, D. Z., Hessler, N. A. & Malinow, R. (1995) Activation of postsynaptically silent synapses during pairing-induced LTP in cal region of hippocampal slice. *Nature* 375:400–4. [MB, LBJ]

Lieberman, P. (1969) Primate vocalizations and human linguistic ability. *Journal of the Acoustical Society of America* 44:1574–84. [aWTT]

Lincoln, T. M. & Cornwell, T. L. (1993) Intracellular cyclic GMP receptor proteins. *Federation of American Societies of Experimental Biology Journal* 7:328–38. [aSRV]

Linden, D. J. (1994a) Input-specific induction of cerebellar long-term depression does not require presynaptic alteration. *Learning and Memory* 1:121–28. [aDJL, NAH, LBJ]

(1994b) Long-term synaptic depression in the mammalian brain. *Neuron* 12:457–72. [aJCH, aMKan, aDJL, RFT]

(1995) Phospholipase A_2 controls the induction of short-term versus long-term depression in the cerebellar Purkinje neuron in culture. *Neuron* 15:1393–1401. [rDJL]

Linden, D. J. & Connor, J. A. (1991) Participation of postsynaptic PKC in cerebellar long-term depression in culture. *Science* 254:1656–59. [aFC, aMKan, aDJL, rSRV]

(1992) Long-term depression of glutamate currents in cultured cerebellar Purkinje neurons does not require nitric oxide signalling. *European Journal of Neuroscience* 4:10–15. [aFC, aDJL, aSRV, NAH, DO]

(1993) Cellular mechanisms of long-term depression in the cerebellum. *Current Opinion in Neurobiology* 3:401–6. [aDJL]

(1995) Long-term synaptic depression. *Annual Review of Neuroscience* 18:319–57. [aDJL, RFT]

Linden, D. J., Dawson, T. M. & Dawson, V. L. (1995) An evaluation of the nitric oxide/cGMP-dependent protein kinase cascade in the induction of cerebellar long-term depression in culture. *Journal of Neuroscience* 15(7):5098–5105. [NAH, rDJL, rSRV]

Linden, D. J., Dickinson, M. H., Smeyne, M. & Connor J. A. (1991) A long-term depression of AMPA currents in cultured cerebellar Purkinje neurons. *Neuron* 7:81–89. [aFC, aJCH, aMKan, aDJL]

Linden, D. J. & Routtenberg, A. (1989) The role of protein kinase C in long-term potentiation: A testable model. *Brain Research Reviews* 14:279–96. [MB]

Linden, D. J., Smeyne, M. & Connor, J. A. (1993) Induction of cerebellar long-term depression in culture requires postsynaptic action of sodium ions. *Neuron* 10:1093–1100. [aDJL, PC]

(1994) *Trans*-ACPD, a metabotropic receptor agonist, produces calcium mobilization and an inward current in cultured cerebellar Purkinje neurons. *Journal of Neurophysiology* 71:1992–98. [aDJL]

Linden, D. J., Smeyne, M., Sun, S. C. & Connor, J. A. (1992) An electrophysiological correlate of protein kinase C isozyme distribution in cultured cerebellar neurons. *Journal of Neuroscience* 12:3601–8. [aDJL]

Lisberger, S. G. (1988) The neuronal basis for learning of simple motor skills. *Science* 242:728–35. [aJIS]

(1994) Neural basis for motor learning in the vestibuloocular reflex of primates: 3. Computational and behavioral analysis of the sites of learning. *Journal of Neurophysiology* 72:974–98. [aJCH]

Lisberger, S. G. & Fuchs, A. F. (1978) Role of primate flocculus during rapid behavioral modication of vestibuloocular reflex: 1. Purkinje cell activity during visually guided horizontal smooth pursuit eye movements and passive head rotation. *Journal of Neurophysiology* 41:733–63. [KH]

Lisberger, S. G. & Pavelko, T. A. (1988) Brain stem neurons in modified pathways for motor learning in the primate vestibulo-ocular reflex. *Science* 242:77173. [aJIS]

Lisberger, S. G. & Sejnowski, T. J. (1992) Motor learning in a recurrent network model based on the vestibulo-ocular reflex. *Nature (London)* 360:159–61. [aJIS]

Llano, I., DiPolo, R. & Marty, A. (1994) Calcium-induced calcium release in cerebellar Purkinje cells. *Neuron* 12:663–73. [PC, rSRV]

Llano, I., Dreessen, J., Kano, M. & Konnerth, A. (1991) Intradendritic release of calcium induced by glutamate in cerebellar Purkinje cells. *Neuron* 7:577–83. [aFC, aMKan, aDJL]

Llano, I., Leresche, N. & Marty, A. (1991) Calcium entry increases the sensitivity of cerebellar Purkinje cells to applied GABA and decreases inhibitory postsynaptic currents. *Neuron* 6:565–74. [aMKan]

Llano, I., Marty, A., Armstrong, C. & Konnerth, A. (1991) Synaptic- and agonist-induced excitatory currents of Purkinje cells in rat cerebellar slices. *Journal of Physiology (London)* 424:183–213. [aMKan]

Llinás, R. (1964) Mechanisms of supraspinal actions upon spinal cord activities differences between reticular and cerebellar inhibitory actions upon alpha extensor motorneurons. *Journal of Neurophysiology* 27:1117–26. [aAMS]

(1970) Neuronal operations in cerebellar transactions. In: *The neurosciences: Second study program*, ed. F. O. Schmitt. Rockefeller University Press. [aJIS]

(1974) Eighteenth Bowditch lecture: Motor aspects of cerebellar control. *Physiologist* 17:19–46. [aJIS]

(1981) Electrophysiology of cerebellar networks. In: *Handbook of physiology*, sect. 1, vol 2, part 2, ed. V. B. Brooks. American Physiological Society. [aWTT]

(1985) Functional significance of the basic cerebellar circuit in motor coordination. In: *Cerebellar functions*, ed. J. R. Bloedel, J. Dichgans & W. Precht. Springer-Verlag. [aJIS]

(1991) The noncontinuous nature of movement execution. In: *Motor control: Concepts and issues*, ed. D. R. Humphrey & H.-J. Freund. Wiley. [aJIS]

(1995) Thorny issues in neurons [News and Views}. *Nature* 373:107–8. [rDJL]

Llinás, R., Baker, R. & Sotelo, C. (1974) Electrotonic coupling between neurons in cat inferior olive. *Journal of Neurophysiology* 37:560–71. [arJIS]

Llinás, R. & Mühlethaler, M. (1988a) An electrophysiological study of the *in vitro*, perfused brain stem-cerebellum of adult guinea-pig. *Journal of Physiology (London)* 404:215–40. [aJIS]

(1988b) Electrophysiology of guinea-pig cerebellar nuclear cells in the *in vitro* brain stem-cerebellar preparation. *Journal of Physiology (London)* 404:241–58. [arJIS, aSRV]

Llinás, R. & Sasaki, K. (1989) The functional organization of the olivo-cerebellar system as examined by multiple Purkinje cell recordings. *European Journal of Neuroscience* 1:587–602. [aJIS, CW]

Llinás, R. & Sugimori, M. (1980) Electrophysiological properties of *in vitro* Purkinje cell dendrites in mammalian cerebellar slices. *Journal of Physiology (London)* 305:197–213. [aJIS]

Llinás, R. & Volkind, R. A. (1973) The olivo-cerebellar system: Functional properties as revealed by harmaline-induced tremor. *Experimental Brain Research* 18:69–87. [aJIS]

Llinás, R., Walton, K., Hillman, D. E. & Sotelo, C. (1975) Inferior olive: Its role in motor learning. *Science* 190:230–31. [aWTT]

Llinás, R. & Welsh, J. P. (1993) On the cerebellum and motor learning. *Current Opinion in Neurobiology* 3:958–65. [aJCH, aJIS]

Llinás, R. & Yarom, Y. (1981a) Electrophysiology of mammalian inferior olivary neurones *in vitro*. Different types of voltage-dependent ionic conductances. *Journal of Physiology (London)* 315:549–67. [aJIS]

(1981b) Properties and distribution of ionic conductances generating electroresponsiveness of mammalian inferior olivary neurones *in vitro* . *Journal of Physiology (London)* 315:569–84. [aJIS]

(1986) Oscillatory properties of guinea-pig inferior olivary neurones and their pharmacological modulation: An *in vitro* study. *Journal of Physiology (London)* 376:163–82. [aJIS]

Logan, C. G. & Grafton, S. T. (1995) Functional anatomy of human eyeblink conditioning determined with regional cerebral glucose metabolism and positron-emission tomography. *Proceedings of National Academy of Sciences of the USA* 92:7500–4. [DT]

Lohmann, S. M., Walter, U., Miller, P. E., Greengard, P. & Camilli, P. D. (1981) Immunohistochemical localization of cyclic GMP-dependent protein kinase in mammalian brain. *Proceedings of the National Academy of Sciences of the USA* 78:653–57. [aFC, aDJL, aSRV]

Lopez-Barneo, J., Darlot, C., Berthoz, A. & Baker, R. (1982) Neuronal activity in prepositus nucleus correlated with eye movement in the alert cat. *Journal of Neurophysiology* 47:329–52. [aJIS]

Lou, J.-S. & Bloedel, J. R. (1986) The responses of simultaneously recorded Purkinje cells to the perturbations of the step cycle in the walking ferret: A study using a new analytical method – the real time post synaptic response (RTPR). *Brain Research* 365:340–44. [aJIS]

(1992a) Responses of sagittally aligned Purkinje cells during perturbed locomotion: synchronous activation of climbing fiber inputs. *Journal of Neurophysiology* 68:570–80. [aJIS]

(1992b) Responses of sagittally aligned Purkinje cells during perturbed locomotion: relation of climbing fiber activation to simple spike modulation. *Journal of Neurophysiology* 68:1820–33. [aJIS]

Luciano, L. (1891) *Il cervelletto: Nuovi studi di fisiologia normale e patolgica.* Firenze: Le Monnier. [aWTT]

(1915) The hindbrain. In: *Human physiology*, trans. F. A. Welby. Macmillan. [aWTT]

Luebke, A. E. & Robinson, D. A. (1992) Climbing fiber intervention blocks plasticity of the vestibuloocular reflex. *Annals of the New York Academy of Science* 656:428–30. [aJCH]

Lum-Ragan, J. T. & Gribkoff, V. K. (1993) The sensitivity of hippocampal long-term potentiation to nitric oxide synthse inhibitors is dependent upon the pattern of conditioning stimulation. *Neuroscience* 57:973–83. [aDJL]

Luo, D., Knezevich, S. & Vincent, S. R. (1993) N-methyl-D-aspartate-induced nitric oxide release: An *in vivo* microdialysis study. *Neuroscience* 57:897–900. [aSRV]

Luo, D., Leung, E. & Vincent, S. R. (1994) Nitric oxide-dependent efflux of cGMP in rat cerebellar cortex: An *in vivo* microdialysis study. *Journal of Neuroscience* 14:263–71. [arSRV, LK]

Luo, D. & Vincent, S. R. (1994) Metalloporphyrins inhibit nitric oxide-dependent cGMP formation in vivo. *European Journal of Pharmacology* 267:263–67. [DO]

Luthi, A., Laurent, J. P., Figurov, A., Muller, D. & Schachner, M. (1994) Hippocampal long-term potentiation and pleural cell-adhesion molecules L1 and NCAM. *Nature* 372:777–79. [MB]

Lynch, J. C., Hoover, J. E. & Strick, P. L. (1992) The primate frontal eye field is the target of neural signals from the substantia nigra, superior colliculus, and dentate necleus. *Society for Neuroscience Abstracts* 18:855. [aWTT]

MacKay, W. A. (1988) Unit acitivity in the cerebellar nuclei related to arm reaching movements. *Brain Research* 442:240. [aAMS]

MacKay, W. A. & Murphy, J. T. (1979) Cerebellar modulation of reflex gain. *Progress in Neurobiology* 13:1410–23. [aAMS, aWTT]

Macklis, R. M. & Macklis, J. D. (1992) Historical and phrenologic reflections on the nonmotor functions of the cerebellum: Love under the tent? *Neurology* 42:928–32. [aWTT]

Macpherson, J. M. (1988a) Strategies that simplify the control of quadrupedal stance: 1. Forces at the ground. *Journal of Neurophysiology* 60:204–17. [aAMS]

(1988b) Strategies that simplify the control of quadrupedal stance: 2. Electromyographic activity. *Journal of Neurophysiology* 60:218–31. [aAMS]

(1991) How flexible are muscle synergies? In: *Motor control: Concepts and issues*, ed. D. R. Humphrey & H.-J. Freund. Wiley. [aAMS]

Maekawa, K. & Simpson, J. I. (1973) Climbing fiber responses evoked in vestibulo-cerebellum of rabbit from visual system. *Journal of Neurophysiology* 36:649–66. [aJIS]

Mahamud, S., Barto, A. G., Kettner, R. E. & Houk, J. C. (1995) A model of prediction in smooth eye movements. *Fourth Annual Computation and Neural Systems Conference.* [aJCH]

Mai, N., Bolsinger, P., Avarello, M., Diener, H. C. & J. Dichgans (1988) Control of isometric finger force in patients with cerebellar disease. *Brain* 111:973–98. [aAMS]

Mailman, R. B., Mueller, R. A. & Breese, G. R. (1978) The effect of drugs which alter GABA-ergic function on cerebellar guanosine-3',5'-monophosphate content. *Life Sciences* 23:623–28. [aSRV]

Malenka, R. C., Kauer, J. A., Perkel, D. J., Mauk, M. D., Kelly, P. T., Nicoll, R. A. & Waxham, M. N. (1989) An essential role for postsynaptic calmodulin and protein kinase activity in long-term potentiation. *Nature* 340:554–57. [MB]

Malkmus, M., Miall, R. C. & Stein, J. F. (in preparation) A model of the cerebellar cortex: Learning sensory predictions. [RCM]

Mallorga, P., Tallman, J. F., Henneberry, R. C., Hirata, F., Strittmatter, W. T. & Axelrod, J. (1980). Mepacrine blocks beta adrenergic agonist-induced desensitization in astrocytoma cells. *Proceedings of the National Academy of Sciences of the USA* 77:1341–45. [aDJL]

Mano, N., Kanazawa I. & Yamamoto, K. (1986) Complex-spike activity of cerebellar Purkinje cells related to wrist tracking movement in monkey. *Journal of Neurophysiology* 56:137–58. [aJCH, aJIS, MD]

(1989) Voluntary movements and complex-spike discharges of cerebellar Purkinje cells. In: *The olivocerebellar system in motor control: Experimental brain research series 17*, ed. P. Strata. Springer-Verlag. [aJIS]

Mano, N.-I. & Yamamoto, K. I. (1980) Simple-spike activity of cerebellar Purkinje cells related to visually guided wrist tracking movement in the monkey. *Journal of Neurophysiology* 43:713–28. [aAMS]

Manzoni, O. J., Weisskopf, M. G. & Nicoll, R. A. (1994) MCPG antagonizes metabotropic glutamate receptors but not long-term potentiation in the hippocampus. *European Journal of Neuroscience* 6:1050–54. [MB]

Mao, C. C., Guidotti, A. & Costa, E. (1974a) The regulation of cyclic guanosine monophosphate in rat cerebellum: Possible involvement of putative amino acid neurotransmitters. *Brain Research* 79:510–14. [aSRV]

(1974b) Interactions between g-aminobutyric acid and cyclic guanosine 3',5'-monophosphate in rat cerebellum. *Molecular Pharmacology* 10:736–45. [aSRV]

Mao, C. C., Guidotti, A. & Landis, S. (1975) Cyclic GMP: Reduction of cerebellar concentrations in 'nervous' mutant mice. *Brain Research* 90:335–39. [arSRV, LK]

Marcinkiewicz, M., Morcos, R. & Chretien, M. (1989) CNS connections with the median raphe nucleus: Retrograde tracing with WGA-apoHRP-gold complex in the rat. *Journal of Comparative Neurology* 289:11–35. [JDS]

Maren, S. & Baudry, M. (1995) Properties and mechanisms of long-term synaptic plasticity in the mammalian brain: Relationships to learning and memory. *Neurobiology of Learning & Memory* 63:1–18. [MB]

Marquis, M. & Green, E. J. (1994) Cortical representation of motion during unrestrained spatial navigation in the rat. *Cerebral Cortex* 7:27–39. [SMO]

Marr, D. (1969) A theory of cerebellar cortex. *Journal of Physiology (London)* 202:437–70. [aFC, arJCH, aJIS, aWTT, MD, CG, PFCG, DJ, EDS, FS, JDS]

Martin, L. J., Blackstone, C. D., Huganir, R. L. & Price, D. L. (1992) Cellular localization of a metabotropic glutamate receptor in rat brain. *Neuron* 9:259–70. [aFC, aDJL]

Martin, T. A., Keating, J. G., Goodkin, H. P., Bastian, A. J. & Thach, W. T. (1993) Storage of multiple gaze-hand calibrations. *Society for Neuroscience Abstracts* 19:980. [aWTT]

(1995) Localization of specific regions of the cerebellar system involved in prism adaptation. *Society for Neuroscience Abstracts.* [aWTT]

(in press a) Throwing while looking through prisms: 1. Focal olivocerebellar lesions impair adaptation. *Brain.* [rWTT]

(in press b) Throwing while looking through prisms: 2. Specificity and storage of multiple gaze-throw calibrations. *Brain.* [rWTT]

Marwaha, J., Palmer, M. R., Woodward, D. J., Hoffer, B. J. & Freedman, R. (1980) Electrophysiological evidence for presynaptic actions for

phencyclidine on noradrenergic transmission in rat cerebellum. *Journal of Pharmacology and Experimental Therapeutics* 215:606–13. [aSRV]
Massicotte, G. & Baudry, M. (1991) Triggers and substrates of hippocampal synaptic plasticity. *Neurobiology and Biobehavioral Review* 15:415–23. [MB]
Massion, J. (1973) Intervention des voies cérébello-corticales et cortico-cérébelleuses dans l'organisation et la régulation du mouvement. *Journal of Physiology (Paris)* 67:117A-170A. [aAMS]
(1992) Movement, posture and equilibrium: Interaction and coordination. *Progress in Neurobiology* 38:35–56. [aAMS]
Masu, M., Tanabe, Y., Tsuchida, K., Shigemoto, R. & Nakanishi, S. (1991) Sequence and expression of a metabotropic glutamate receptor. *Nature* 349:760–65. [aDJL]
Matsumoto, M., Nakagawa, T., Inoue, T., Nagata, E., Tanaka, K., Takano, H., Minowa, O., Kuno, J., Sakakibara, S., Yamada, M., Yoneshima, H., Miyawaki, A., Fukuuchi, Y., Furuichi, T., Okano. H., Miloshita, K. & Noda, T. (1996) Ataxia and epileptic seizures in mice lacking type 1 inositol-1,4,5-triphosphate receptor. *Nature* 379:168–71. [rDJL]
Matsumoto, T., Nakane, M., Pollock, J. S., Kuk, J. E. & Forstermann, U. (1993) A correlation between soluble brain nitric oxide synthase and NADPH-diaphorase is only seen after exposure of the tissue to fixative. *Neuroscience Letters* 155:61–64. [aSRV]
Matsuoka, I., Giuili, G., Poyard, M., Stengel, D., Parma, J., Guellaen, G. & Hanoune, J. (1992) Localization of adenylyl and guanylyl cyclase in rat brain by in situ hybridization: Comparison with calmodulin mRNA distribution. *Journal of Neuroscience* 12:3350–60. [aSRV]
Mauk, M. D., Steinmetz, J. E. & Thompson, R. F. (1986) Classical conditioning using stimulation of the inferior olive as the unconditioned stimulus. *Proceedings of the National Academy of Sciences of the USA* 83:5349–53. [RFT, CW]
May, J. G. & Anderson, R. A. (1986) Different patterns of corticopontine projections from separate cortical fields within the inferior parietal lobule and dorsal prelunate gyrus of the macaque. *Experimental Brain Research* 63:265–78. [JDS]
May, P. J., Hall, W. C., Porter, J. D. & Sakai, S. T. (1993) The comparative anatomy of nigral and cerebellar control over tectally initiated orienting movements. In: *Role of the cerebellum and basal ganglia in voluntary movement,* ed. N. Mano, I. Hamada & M. R. DeLong. Excerpta Medica. [aWTT]
Mayer, B., John, M. & Bohme, E. (1990) Purification of a Ca^{2+}/calmodulin-dependent nitric oxide synthase from porcine cerebellum. *Federation of European Biochemical Societies Letters* 277:215–19. [aSRV]
Mayer, B., Klatt, P., Bohme, E. & Schmidt, K. (1992) Regulation of neuronal nitric oxide and cyclic GMP formation by Ca^{2+}. *Journal of Neurochemistry* 59:2024–29. [aSRV, LK]
Mayer, M. L. & Westbrook, G. L. (1987) The physiology of excitatory amino acids in the vertebrate central nervous system. *Progress in Neurobiology* 28:197–276. [aFC]
Mays, L. E. & Sparks, D. L. (1980) Saccades are spatially, not retinocentrally, coded. *Science* 208:1163–65. [HB]
Mazziotta, J. C., Grafton, S. T. & Woods, R. C. (1921) The human motor system studied with PET measurements of cerebral blood flow: Topography and motor learning. In: *Brain work and mental activity* (Alfred Benzen Symposium 31), ed. N. A. Lassen, D. H. Ingvar, M. E. Raichle & L. Friberg. Copenhagen: Munksgaard. [aWTT]
McCollum, G. (1992) Rules of combination that generate climbing fiber tactile receptive fields. *Neuroscience* 50(3):707–25. [aJIS]
(submitted) Climbing fiber ensemble activity indicated by receptive fields. [PDR]
McCormick, D. A., Lavond, D. G., Clark, G. A., Kettner, R. E., Rising, C. E. & Thompson, R. F. (1981) The engram found? Role of the cerebellum in classical conditioning of nictitating membrane and eyelid responses. *Bulletin of the Psychonomic Society* 18:105–15. [aWTT]
McCormick, D. A., Steinmetz, J. E. & Thompson, R. F. (1985) Lesions of the inferior olivary complex cause extinction of the classically conditioned eyeblink response. *Brain Research* 359:120–30. [RFT]
McCormick, D. A. & Thompson, R. F. (1984) Cerebellum: Essential involvement in the classically conditioned eyelid response. *Science* 223:296–99. [aWTT]
McCrea, D. A. (1992) Can sense be made of spinal interneurons? *Behavioral and Brain Sciences* 15:633–43. [aAMS]
McCrea, R. A. & Baker, R. (1985). Anatomical connections of the nucleus prepositus of the cat. *Journal of Comparative Neurology* 237:377–407. [aJIS]
McDevitt, C. J., Ebner, T. J. & Bloedel, J. R. (1982) The changes in Purkinje cell simple spike activity following spontaneous climbing fiber inputs. *Brain Research* 237:484–91. [aJIS]
McDonald L. J. & Moss, J. (1993) Stimulation by nitric oxide of an NAD linkage to glyceraldehyde-3-phosphate dehydrogenase. *Proceedings of the National Academy of Sciences of the USA* 90:6238–41. [aSRV]
McFarland, J. L. & Fuchs, A. F. (1992) Discharge patterns in nucleus prepositus hyposglossi and adjacent medial vestibular nucleus during horizontal eye movement in behaving macaques. *Journal of Neurophysiology* 68:319–32. [aJIS]
McGlade-McCulloh, E., Yamamoto, H., Tan, S. E., Bricjey, D. A. & Soderling, T. R. (1993) Phosphorylation and regulation of glutamate receptors by calcium/calmodulin-dependent protein kinase II. *Nature* 362:640–42. [MB]
McGlinchey-Berroth, R., Cermak, L. S., Carrillo, M. C., Armfield, S., Gabrieli, J. D. E. & Disterhoft, J. F. (1995) Impaired delay eyeblink conditioning in amnesic Korsakoff's patients and recovered alcoholics. *Alcoholism: Clinical Experimental Research* 19:1127–32. [CW]
McIlwain, J. (1986) Effects of eye position on saccades evoked electrically from superior colliculus of alert cats. *Journal of Neurophysiology* 55:97–112. [aJCH]
McLaughlin, S. (1967) Parametric adjustment in saccadic eye movements. *Perception & Psychophysics* 2:359–62. [HB]
McNaughton, B. L. & Nadel, L. (1990) *Hebb-Marr networks and the neurobiological representation of action of space.* MIT Press. [SMO]
Meffert, M. K., Haley, J. E., Schuman, E., Schulman, H. & Madison, D. V. (1994) Inhibition of hippocampal heme oxygenase, nitric oxide synthase, and long-term potentiation by metalloporphyrins. *Neuron* 13:1225–33. [DO]
Melis, B. J. M. & van Gisbergen, J. A. M. (1995) Short-term adaptation of electrically-induced saccades in monkey superior colliculus. Submitted manuscript. [CG]
Melvill-Jones, G. & Watt, D. G. D. (1971) Observations on the control of stepping and hopping movements in man. *Journal of Physiology* 40:1038–50. [aWTT]
Meyer-Lohman, J., Hore, J. & Brooks, V. B. (1977) Cerebellar participation in generation of prompt arm movements. *Journal of Neurophysiology* 40:1038–50. [aWTT]
Miall, R. C., Weir, D. J. & Stein, J. F. (1987) Visuo-motor tracking during reversible inactivation of the cerebellum. *Experimental Brain Research* 65:455–64. [aAMS]
Miall, R. C., Weir, D. J., Wolpert, D. M. & Stein, J. F. (1993) Is the cerebellum a Smith predictor? *Journal of Motor Behavior* 25:203–16. [aJCH, RCM, rAMS]
Miall, R. C. & Wolpert, D. M. (1996) Forward models in physiological motor control. Submitted manuscript. [RCM]
Middleton, F. A. & Strick, P. L. (1994) Anatomical evidence for cerebellar and basal ganglia involvement in higher cognitive function. *Science* 266:458–61. [aJCH, JDS, CW]
Midtgaard, J. (1992) Membrane properties and synaptic responses of Golgi cells and stellate cells in the turtle cerebellum in vitro. *Journal of Physiology (London)* 457:329–54. [KH]
Milak, M. S., Bracha, V. & Bloedel, J. R. (1995) Relationship of simultaneously recorded cerebellar nuclear neuron discharge to the acquisition of a complex, operantly conditioned forelimb movement in cats. *Experimental Brain Research* 105:325–30. [rAMS]
Miles, F. A., Braitman, D. J. & Dow, B. M. (1980) Long-term adaptive changes in primate vestibuloocular reflex: 4. Electrophysiological observations in flocculus of adapted monkeys. *Journal of Neurophysiology* 43:1477–93. [aJIS]
Miles, F. A. & Lisberger, S. G. (1981) Plasticity in the vestibulo-ocular reflex: A new hypothesis. *Annual Review of Neuroscience* 4:273–99. [aJCH, aJIS]
Miller, S. & Oscarsson, O. (1970) Termination and functional organization of spinoolivocerebellar paths. In: *The cerebellum in health and disease,* ed. W. S. Fields & W. D. Willis. Green. [DF]
Miller, W. T. (1987) Sensor-based control of robotic manipulators using a general learning algorithm. *IEEE Journal of Robotics & Automation* RA-3:157–65. [aJCH]
Milner, T. E. (1993) Dependence of elbow viscoelastic behavior on speed and loading in voluntary movements. *Experimental Brain Research* 93:177–80. [aAMS]
Milner, T. E. & Cloutier, C. (1993) Compensation for mechanically unstable loading in voluntary wrist movement. *Experimental Brain Research* 94:522–32. [arAMS]
Mink, J. W. & Thach, W. T. (1991a) Basal ganglia motor control: 1. Nonexclusive relation of pallidal discharge to five movement modes. *Journal of Neurophysiology* 65:273–300. [rWTT]
(1991b) Basal ganglia motor control: 2. Late pallidal timing relative to movement onset and inconsistent pallidal coding of movement parameters. *Journal of Neurophysiology* 65:201–29. [rWTT]
(1991c) Basal ganglia motor control: 3. Pallidal ablation: Normal reaction time, muscle cocontraction, and slow movement. *Journal of Neurophysiology* 65:330–51. [rWTT]

Minsky, M. L. (1963) Steps toward artificial intelligence. In: *Computers and thought*, ed. E. A. Feigenbaum & J. Feldman. McGraw-Hill. [JCH]

Minsky, M. & Papert, S. (1969) *Perceptrons.* MIT Press.

Mitoma, H., Kobayashi, T., Song S.-Y. & Konishi, S. (1994) Enhancement by serotonin of GABA-mediated inhibitory currents in cerebellar Purkinje cells. *Neuroscience Letters* 173:127–30. [aMKan]

Miyakawa, H., Lev-Ram, V., Lasser-Ross, N. & Ross, W. N. (1992) Calcium transients evoked by climbing fiber and parallel fiber synaptic inputs in guinea pig cerebellar Purkinje neurons. *Journal of Neurophysiology* 4:1178–89. [aFC, aMKan]

Miyashita, E. & Tamai, Y. (1989) Subcortical connections to frontal "oculomotor" areas in the cat. *Brain Research* 502:75–87. [aJIS]

Mizukawa, K., McGeer, P. L., Vincent, S. R. & McGeer, E. G. (1989) Distribution of reduced-nicotinamide-adenine-dinucleotide phosphate diaphorase positive cells and fibers in the cat central nervous system. *Journal of Comparative Neurology* 279:281–311. [aSRV]

Molchan, S. E., Sunderland, T., McIntosh, A. R., Herscovitch, P. & Schreurs, B. G. (1994) A functional anatomical study of associative learning in humans. *Proceedings of National Academy of Sciences of the USA* 91:8122–26. [DT]

Montarolo, P. G., Palestini, M. & Strata, P. (1982) The inhibitory effect of the olivocerebellar input on the cerebellar Purkinje cells in the rat. *Journal of Physiology (London)* 332:187–202. [aJIS]

Moore, J. W., Desmond, J. E. & Berthier, N. E. (1989) Adaptively timed conditioned responses and the cerebellum: A neural network approach. *Biological Cybernetics* 62:17–28. [aJCH]

Mori-Okamoto, J., Okamoto, K. & Tatsuno, J. (1993) Intracellular mechanisms underlying the suppression of AMPA responses by *trans*-ACPD in cultured chick Purkinje neurons. *Molecular and Cellular Neurosciences* 4:375–86. [aSRV, JMO]

Morishita, W. & Shastry, B. R. (1993) Long-term depression of IPSPs in rat deep cerebellar nuclei. *NeuroReport* 4:719–22. [aMKan, RFT]

Mortimer, J. A. (1973) Temporal sequence of cerebellar Purkinje and nuclear activity in relation to the acoustic startle response. *Brain Research* 50:457–62.

Moss, S. J., Doherty, C. A. & Huganir, R. L. (1992) Identification of the cAMP-dependent protein kinase and protein kinase C phosphorylation sites within the major intracellular domains of the β1, γ2S, and γ2L subunits of the γ-aminobutyric acid type A receptor. *Journal of Biological Chemistry* 267:14470–76. [aMKan]

Moss, S. J., Smart, T. G., Blackstone, C. D. & Huganir, R. L. (1992) Functional modulation of GABAA receptors by cAMP-dependent protein phosphorylation. *Science* 257:661–65. [aMKan]

Moyer, J. R., Deyo, R. A. & Disterhoft, J. F. (1990) Hippocampectomy disrupts trace eye-blink conditioning in rabbits. *Behavioral Neuroscience* 104:243–52. [CW]

Mugnaini, E. (1983) The length of cerebellar parallel fibers in chicken and rhesus monkey. *Journal of Comparative Neurology* 220:7–15. [aWTT]

Mugnaini, E. & Maler, L. (1993) Comparison between the fish electrosensory lateral line lobe and the mammalian dorsal cochlear nucleus. In: *Contributions of electrosensory systems to neurobiology and neuroethology*, vol. 173, ed. C. C. Bell, C. D. Hopkins & K. Grant. *Journal of Comparative Physiology A.* [arJCH]

Mulkey, R. M., Herron, C. E. & Malenka, R. C. (1993) An essential role for protein phosphatases in hippocampal long-term depression. *Science* 261:1051–55. [MB]

Mulle, C., Choquet, D., Korn, H. & Changeux, J. P. (1992) Calcium influx through nicotinic receptor in rat central neurons: Its relevance to cellular regulation. *Neuron* 8:135–43. [aMKan]

Murphy, J. T. & Sabah, N. H. (1970) The inhibitory effect of climbing fiber activation and cerebellar Purkinje cells. *Brain Research* 19:486–90. [aJIS, MAA]

Murphy, S., Simmons, M. L., Agullo, L., Garcia, A., Feinstein, D. L., Galea, E., Reis, D. J., Minc-Golomb, D. & Schwartz, J. P. (1993) Synthesis of nitric oxide in CNS glial cells. *Trends in Neuroscience* 8:323–28. [DO, rSRV]

Mussa-Ivaldi, F. A., Hogan, N. & Bizzi, E. (1985) Neural, mechanical, and geometric factors subserving arm posture in humans. *Journal of Neuroscience* 5:2732–43. [aAMS]

Nagao, S. (1983) Effects of vestibulo-cerebellar lesions upon dynamic characteristics and adaptation of vestibulo-ocular and optokinetic responses ain pigmented rabbits. *Experimental Brain Research* 53:36–46. [aJIS]

(1988) Behavior of floccular Purkinje cells correlated with adaptation of horizontal optokinetic eye movement response in pigmented rabbits. *Experimental Brain Research* 73:489–97. [aJIS]

Nagao, S. & Ito, M. (1991) Subdural application of hemoglobin to the cerebellum blocks vestibuloocular reflex adaptation. *NeuroReport* 2:193–96. [aFC, aMKan, aSRV, aDJL]

Nairn, A. C. & Greengard, P. (1983) Cyclic GMP-dependent protein phosphorylation in mammalian brain. *Federation of American Societies of Experimental Biology, Proceedings* 42:3107–13. [aSRV]

Nakamura, H., Saheki, T., Ichiki, H., Nakata, K. & Nakagawa, S. (1991) Immunocytochemical localization of argininosuccinate synthetase in the rat brain. *Journal of Comparative Neurology* 312:652–79. [aSRV, LK]

Nakamura, H., Saheki, T. & Nakagawa, S. (1990) Differential cellular localization of enzymes of L-arginine metabolism in the rat brain. *Brain Research* 530:108–12. [aSRV, LH]

Nakane, M., Ichikawa, M. & Deguchi, T. (1983) Light and electron microscopic demonstration of guanylate cyclase in rat brain. *Brain Research* 273:9–15. [aSRV, LK]

Nakanishi, N. (1992) Molecular diversity of glutamate receptors and implications for brain function. *Science* 258:597–603. [aFC]

Nakanishi, N., Sneider, N. A. & Axel, R. (1990) A family of glutamate receptor genes: Evidence for the formation of heteromultimeric receptors with distinct channel properties. *Neuron* 5:569–81. [aFC]

Nakazawa, K., Mikawa, S., Hashikawa, T. & Ito, M. (1995) Transient and persistent phosphorylation of AMPA-type glutamate receptor subunits in cerebellar Purkinje cells. *Neuron* 15:697–709. [rDJL]

Nakazawa, K. & Sano, M. (1974) Studies on guanylate cyclase. A new assay method for guanylate cyclase and properties of the cyclase from rat brain. *Journal of Biological Chemistry* 249:4207–11. [aSRV]

Napper, R. M. A. & Harvey, R. J. (1988) Number of parallel fiber synapses on an individual Purkinje cell in the cerebellum of the rat. *Journal of Comparative Neurology* 274:168–77. [FS]

Nathanson J. A., Scavone, C., Scanlon, C. & McKee, M. (1995) The cellular Na^+ pump as a site of action for carbon monoxide and glutamate: A mechanism for long-term modulation of cellular activity. *Neuron* 14:781–94. [DO, rSRV]

Nelson, M. E. & Bower, J. M. (1990) Brain maps and parallel computers. *Trends in Neuroscience* 13:403–8. [aWTT]

Nelson, B. & Mugnaini, E. (1989a) Origins of the GABAergic inputs to the inferior olive. *Experimental Brain Research* 17:86–107. [aAMS]

(1989b) Origins of GABAergic inputs to the inferior olive. In: *The olivocerebellar system in motor control: Experimental brain research series 17*, ed. P. Strata. Springer-Verlag. [aJIS]

Newell, K. M. (1985) Coordination, control and skill. In: *Differing perspectives in motor learning, memory, and control,* ed. D. Goodman, R. B. Wilberg & I. M. Franks. North-Holland. [SPS]

Newsome, W. T., Wurtz, R. H. & Komatsu, H. (1988) Relation of cortical area MT and MST to pursuit eye movements: 2. Differentiation of retinal from extraretinal inputs. *Journal of Neurophysiology* 60:604–20. [MKaw]

Nichols, T. R. (1994) A biomechanical perspective on spinal mechanisms of coordinated muscular action: An architectural principal. *Acta Anatomica* 151:1–13. [AGF]

Nichols, T. R. & Houk, J. C. (1976) Improvement of linearity and regulation of stiffness that results from action of the stretch reflex. *Journal of Neurophysiology* 39:119–42. [aAMS]

Nicoletti, F., Meek, J. M., Iadora, M. J., Chuang, D. M., Roth, B. L. & Costa, E. (1986) Coupling of inositol phospholipid metabolism with excitatory amino acid recognition sites in rat hippocampus. *Journal of Neurochemistry* 46:40–46. [aFC]

Nicoletti, F., Wroblewski, J. T., Novelli, A., Guidotti, A. & Costa, A. (1986) The activation of inositol phospholipid metabolism as a signal-transducing system for excitatory amino acids in primary cultures of cerebellar granule cells. *Journal of Neurosciences* 6:1905–11. [aSRV, LK]

Nicoll, R. A. & Malenka, R. C. (1995) Contrasting properties of two forms of long-term potentiation in the hippocampus. *Nature* 377:115–18. [MB]

Nishizuka, Y. (1986) Studies and perspectives of protein kinase C. *Science* 233:305–11. [aFC]

(1992). Intracellular signaling by hydrolysis of phospholipids and activation of protein kinase C. *Science* 258:607–14. [aDJL]

Noda, H. (1991) Cerebellar control of saccadic eye movements: Its neural mechanisms and pathways. *Japanese Journal of Physiology* 41:351–68. [PD]

Novelli, A. & Henneberry, R. C. (1987) cGMP synthesis in cultured cerebellar neurons is stimulated by glutamate via a Ca^{2+}-mediated, differentiation-dependent mechanism. *Developmental Brain Research* 34:307–10. [aSRV]

Nuñes-Cardozo, B. & Van der Want, J. J. L. (1990) Ultrastructural organization of the retinal-pretecto-olivary pathway. A combined WGA-HRP retrograde/GABA immunohistochemical study. *Journal of Comparative Neurology* 291:313–27. [aJIS]

O'Dell, T. J., Hawkins, R. D., Kandel, E. R. & Arancio, O. (1991) Tests of the roles of two diffusible substances in long-term potentiation: Evidence for nitric oxide as a possible early retrograde messenger. *Proceedings of the National Academy of Sciences of the USA* 88:11285–89. [MB]

O'Dell, T. J. & Kandel, E. R. (1994) Low-frequency stimulation erases LTP

through an NMDA receptor-mediated activation of protein phosphatases. *Learning & Memory* 1:129–39. [MB]
O'Hearn, E., Zhang, P. & Molliver, M. E. (1995) Excitotoxic insult due to ibogaine leads to delayed induction of neuronal NOS in Purkinje cells. *NeuroReport* 6:1611–16. [rSRV]
O'Mara, S. M. (1995) Spatially selective firing properties of hippocampal formation neurons in rodents and primates. *Progress in Neurobiology* 45:253–74. [SMO]
O'Mara, S. M., Rolls, E. T., Berthoz, A. & Kesner, R. P. (1994) Neurons responding to whole-body motion in the primate hippocampus. *Journal of Neuroscience* 14:6511–23. [SMO]
Ohata, K., Shimazu, K., Komatsumoto, S., Araki, N., Shibata, M. & Fukuuchi, Y. (1994) Modification of striatal arginine and citrulline metabolism by nitric oxide synthase inhibitors. *NeuroReport* 5:766–68. [rSRV]
Ohtsuka, K. & Noda, H. (1995) Discharge properties of Purkinje cells in the oculomotor vermis during visually guided saccades in the macaque monkey. *Journal of Neurophysiology* 74:1828–40. [PD]
Ojakangas, C. L. & Ebner, T. (1992) Purkinje cell complex spike changes during a voluntary arm movement learning task in the monkey. *Journal of Neurophysiology* 68:2222–36. [aJCH, aJIS, aWTT, PD]
(1994) Purkinje cell complex spike activity during voluntary motor learning: relationship to kinematics. *Journal of Neurophysiology* 72:2617–30.[arJIS]
Okada, D. (1992) Two pathways of cyclic GMP production by glutamate receptor-mediated nitric oxide synthesis. *Journal of Neurochemistry* 59:1203–9. [aSRV, DO]
Okada, Y. & Miyamoto, T. (1989) Formation of long-term potentiation in superior colliculus slices from the guinea pig. *Neuroscience Letters* 96:108–13. [aMKan]
Oliva, A. M. & Garcia, A. (1995) Cyclic GMP inhibition of stimulated phosphoinositide hydrolysis in neuronal cultures. *NeuroReport* 6:565–68. [rSRV]
Optican, L. M. (1985) Adaptive properties of the saccadic system. In: *Adaptive mechanisms in gaze control,* ed. A. Berthoz & G. Melvill Jones. Elsevier. [PD]
Optican, J. M. & Robinson, D. A. (1980) Cerebellar adaptive control of primate saccadic system. *Journal of Neurophysiology* 44:1058–76. [aJCH, HB, PD, CG]
Orioli, P. J. & Strick, P. L. (1989) Cerebellar connections with the motor cortex and the arcuate premotor area: An analysis employing retrograde transneuronal transport of WGA-HRP. *Journal of Comparative Neurology* 288:612–26. [aWTT]
Orlovsky, G. N. (1972) The effect of different descending systems on flexor and extensor activity during locomotion. *Brain Research* 40:359–71. [MAA]
Oscarsson, O. (1969) The sagittal organization of the cerebellar anterior lobe as revealed by the projection patterns of the climbing fiber system. In: *Neurobiology of cerebellar evolution and development*, ed. R. Llinás. AMA. [aJIS]
(1973) Functional organization of spinocerebellar paths. In: *Handbook of sensory physiology,* vol. 2, ed. A. Iggo. Springer-Verlag. [aJIS, DF]
(1979) Functional units of the cerebellum-sagittal zones and microzones. *Trends in Neuroscience* 2:143–45. [aJIS]
(1980) Functional organization of olivary projection to the cerebellar anterior lobe. In: *The inferior olivary nucleus: Anatomy and physiology*, ed. J. Courville, C. de Montigney & Y. Lamarre. Raven. [aJCH, aJIS, DF]
Parfitt, K. D., Hoffer, B. J., Bickford-Wimer, P. C. (1990) Potentiation of gamma-aminobutyric acid-mediated inhibition by isoproterenol in the cerebellar cortex: Receptor specificity. *Neuropharmacology* 29:909–16. [aMKan]
Parsons, L. M., Fox, P. T., Downs, J. H., Glass, T., Hirsch, T. B., Martin, C. C., Jerabek, P. A. & Lancaster, J. L. (1995) Use of implicit motor imagery for visual shape discrimination as revealed by PET. *Nature* 375:54. [JDS]
Partsalis, A. M., Zhang, Y. & Highstein, S. M. (1993) The *y* group in vertical visual-vestibular interactions and VOR adaptation in the squirrel monkey. *Society of Neuroscience Abstracts* 19:138. [aJIS]
(1995) Dorsal Y group in the squirrel monkey: 2. Contribution of the cerebellar flocculus to neuronal responses in normal and adapted animals. *Journal of Neurophysiology* 73:632–50. [aJCH]
Pascual-Leone, A., Cammarota, A., Wassermann, E. M., Brasil-Neto, J. P., Cohen, L. G. & Hallett, M. (1993a) Modulation of motor cortical outputs to the reading hand of Braille readers. *Annals of Neurology* 34:33–37. [MH]
Pascual-Leone, A., Cohen, L. G., Dang, N., Brasil-Neto, J. P., Cammarotta, A. & Hallett, M. (1993b) Acquisition of fine motor skills in humans is associated with the modulation of cortical motor outputs [abstract]. *Neurology* 43(suppl. 2):A157. [MH]
Pascual-Leone, A., Grafman, J., Clark, K., Stewart, M., Massaquoi, S., Lou, J.-S. et al. (1993c) Procedural learning in Parkinson's disease and cerebellar degeneration. *Annals of Neurology* 34:594–602. [MH]
Pascual-Leone, A., Grafman, J. & Hallett, M. (1994) Modulation of cortical motor output maps during development of implicit and explicit knowlege. *Science* 263:1287–89. [MH]
Pastor, A. M., de la Cruz, R. R. & Baker, R. (1994) Cerebellar role in adaptation of the goldfish vestibuloocular reflex. *Journal of Neurophysiology* 72:1383–94. [KH]
Paulin, M. G. (1989) A Kalman filter theory of the cerebellum. In: *Dynamic interactions in neural networks: Models and data,* ed. M. A. Arbib & S. Amari. Springer-Verlag. [aJCH, RCM, rAMS]
(1993) The role of the cerebellum in motor control and perception. *Brain, Behavior and Evolution* 41:39–50. [RCM, MGP]
Paulin, M. G., Nelson, M. E. & Bower, J. M. (1989) Dynamics of compensatory eye movement control: An optimal estimation analysis of the vestibulo-occular reflex. *International Journal of Neural Systems* 1:23–29. [JMB]
Perkel, D. J., Hestrin, S., Sah, P. & Nicoll, R. A. (1990) Excitatory synaptic currents in Purkinje cells. *Proceedings of the Royal Society of London* 241:116–21. [aDJL]
Petersen, S. E., Fox, P. T., Posner, M. I., Mintum, M. A. & Raichle, M.E . (1988) Positron emission tomographic studies of the cortical anatomy of single-word processing. *Nature* 331:585–89. [JDS]
(1989) Positron emission tomographic studies of the processing of single words. *Journal of Cognitive Neuroscience* 1:153–70. [aWTT]
Peterson, B. W., Baker, J. F. & Houk, J. C. (1991) A model of adaptive control of vestibuloocular reflex based on properties of cross-axis adaptation. *Annals of the New York Academy of Science* 627:319–37. [aJCH]
Peterson, B. W. & Houk, J. C. (1991) A model of cerebellar-brainstem interaction in the adaptive control of the vestibuloocular reflex. *Acta Otolaryngology (Stockholm)* 481:428–32. [aJCH]
Peterson, D. A., Peterson, D. C., Archer, S. & Weir, E. K. (1992) The nonspecificity of specific nitric oxide synthase inhibitors. *Biochemical and Biophysical Research Communications* 187:797–801. [aSRV]
Petralia, R. S. & Wenthold, R. J. (1992) Light and electron immunocytochemcial localization of AMPA-selective glutamate receptors in the rat brain. *Journal of Comparative Neurology* 318:329–54. [aSRV]
Porter, C. M., Van Kan, P. L. E., Horn, K. M., Bloedel, J. R. & Gibson, A. R. (1993) Functional divisions of cat rMAO. *Society for Neuroscience Abstracts* 19:1216. [aJIS]
Porter, N. M., Twyman, R. E., Uhler, M. D. & MacDonald, L. (1990) Cyclic AMP-dependent protein kinase decreases GABAa receptor current in mouse spinal neurons. *Neuron* 5:789–96. [aMKan]
Prablanc, C., Tzavaras, A. & Jeannerod, M. (1975) Adaptation of the two arms to opposite prism displacements. *Quarterly Journal of Experimental Psychology* 27:667–71. [aWTT]
Prochazka, A. (1989) Sensorimotor gain control: A basic strategy of motor systems? *Progress in Neurobiology* 33:281–307. [aJCH]
Quadroni, R. & Knöpfel, T. (1994) Compartmental model of type A and type B guinea pig medial vestibular neurons. *Journal of Neurophysiology* 72:1911–24. [KH]
Rack, P. M. H. & Westbury, D. R. (1969) The effects of length and stimulus rate on tension in the isometric cat soleus muscle. *Journal of Physiology* 204:443–60. [aAMS]
(1974) The short range stiffness of active mammalian muscle and its effect on mechanical properties. *Journal of Physiology* 240:331–50. [aAMS]
Raichle, M. E., Fiez, J. A., Videen, T. O., MacLeod, A. K., Pardo, J. V., Fox, P. T & Petersen, S. E. (1994) Practice-related changes in human brain functional anatomy during nonmotor learning. *Cereberal Cortex* 4:8–26. [aJCH, aWTT]
Rao, T. S., Contreras, P. C., Cler, J. A., Emmett, M. R., Mick, S. J., Iyengar, S. & Wood, P. L. (1991) Clozapine attenuates N-methyl-D-aspartate receptor complex-mediated responses *in vivo*: Tentative evidence for a functional modulation by a noradrenergic mechanism. *Neuropharmacology* 30:557–65. [aSRV]
Rawson, J. A. & Tilokskulchai (1981) Suppression of simple spike discharges of cerebellar Purkinje cells by impulses in climbing fibre afferents. *Neuroscience Letters* 25:125–30. [aJIS]
(1982) Climbing fibre modification of cerebellar Purkinje cell responses to parallel fibre inputs. *Brain Research* 237:492–97. [aJIS]
Rawson, N. R. (1932) The story of the cerebellum. *Canadian Medical Association Journal* 26:220–25. [aWTT]
Recasens, M., Sassetti, I., Nourigat, A., Sladeczek, F. & Bockaert, J. (1987) Characterization of subtypes of excitatory amino acid receptors involved in the stimulation of inositol phosphate synthesis in rat-brain synaptoneurosomes. *European Journal of Pharmacology* 141:87–93. [aFC]
Reis, D. J., Doba, N. & Nathan, M. A. (1973) Predatory attack, grooming and consummatory behaviors evoked by electrical stimulation of cat cerebellar nuclei. *Science* 182:845–47. [JDS]
Rekate, H. L., Grubb, R. L., Aram, D. M., Hahn, J. F. & Ratcheson, R. A.

(1985) Muteness of cerebellar origin. *Archives of Neurology* 697–98. [aWTT]
Richard, E. A., Sampat, P. & Lisman, J. E. (1995) Distinguishing between roles for calcium in Limulus photoreceptor excitation. *Cell Calcium* 18:330–40. [JCF]
Riley, H. A. (1928) Mammalian cerebellum: Comparative study of abor vitae and folial patterns. *Archives of Neurology and Psychiatry* 20:898:1–34. [RCM]
Rispal-Padel, L., Cicirata, F. & Pons, C. (1982) Cerebellar nuclear topography of simple and synergistic movements in the alert baboon (*Papio papio*). *Experimental Brain Research* 47:365–80. [aAMS]
Roberts, P. (in press) Classification of temporal patterns in dynamic biological networks. *Physics Review E.* [PDR]
Robertson, L. T. (1984) Topographic features of climbing fiber input in the rostral vermal cortex of the cat cerebellum. *Experimental Brain Research* 55:445–54. [aJIS]
Robertson, L. T. & Laxer, H. D. (1981) Localization of cutaneously elicited climbing fibre responses in lobule V of the monkey cerebellum. *Brain Behavior and Evolution* 18:157–68. [aJIS]
Robertson, L. T., Laxer, K. D. & Rushmer, D. S. (1982) Organization of climbing fiber input from mechanoreceptors to lobule V vermal cortex of the cat. *Experimental Brain Research* 46:281–91. [aJIS]
Robertson, L. T. & McCollum, G. (1989) Ensembles of climbing fiber tactile receptive fields encode distinct information for various cerebellar regions. In: *The olivocerebellar system in motor control: Experimental brain research series 17,* ed. P. Strata. Springer-Verlag. [aJIS]
Robinson, D. A. (1963) A method of measuring eye movement using a scleral search coil in a magnetic field. *Transactions on Bio-Medical Electronics* BME-10:137–45. [aJIS]
(1964) The mechanics of human saccadic eye movement. *Journal of Physiology* 174:245–64. [HB]
(1975) Oculomotor control signals. In: *Basic mechanisms of ocular motility and their clinical implications,* ed. G. Lennerstrand & P. Bach-y-Rita. Pergamon. [aJCH, HB]
(1976) Adaptive gain control of the vestibulo-ocular reflex by the cerebellum. *Journal of Neurophysiology* 39:954–69. [aJCH, aWTT]
(1981) Control of eye movments. In: *Handbook of physiology: The nervous system: vol. 2. Motor control, part 2,* ed. V. B. Brooks. American Physiological Society. [aAMS]
(1987) Why visuomotor systems don't like negative feedback and how they avoid it. In: *Vision, brain and cooperative computation,* ed. M. A. Arbib. MIT Press. [aJCH]
(1991) Overview. In: *Eye movements,* ed. R. H. S. Carpenter. Macmillan. [PD]
Robinson, F. R., Fuchs, A. & Straube, A. (1995) Saccadic adaptation deficits after muscimol inactivation of the caudal fastigial nucleus in macaque. *Society for Neuroscience Abstracts* 1271. [PD]
Robinson, D. A. & Optican, L. M. (1981) Adaptive plasticity in the oculomotor system. In: *Lesion induced neuronal plasticity in sensorimotor systems,* ed. H. Flohr & W. Precht. Springer-Verlag. [aJCH]
Roland, L. (1809) *Saggio sopra la vera struttura del cervello dell'uomo e dgeli animali e sopra le funzioni del sistema nervoso.* Sassari: Stampeia da S.S.R.M. Priveligiata. [aWTT]
(1823) Experiences sur les fonctions du systeme nerveux. *Journal de Physiologie Experimentale* 3:95–113. [aWTT]
Roland, P. E. (1987) Metabolic mapping of sensorimotor integration in the human brain. In: *Motor areas of the cerebral cortex* (Ciba Foundation Symposium 132), ed. G. Bock, M. O'Connor & J. Marsh. Wiley. [aWTT]
Roland, P. E., Eriksson, L., Widen, L. & Stone-Elander, S. (1988) Changes in regional cerebral oxidative metabolism induced by tactile learning and recognition in man. *European Journal of Neuroscience* 1:3–17. [aWTT]
Rolls, E. T. & O'Mara, S. M. (1993) Neurophysiological and theoretical analysis of how the primate hippocampus functions in memory. In: *Brain mechanisms of perception: From neuron to behavior,* ed. T. Ono, L. R. Squire, M. Raichle, D. Perrett & M. Fukuda. Oxford University Press. [SMO]
Rondot, P., Bathien, N. & Toma, S. (1979) Physiopathology of cerebellar movement. In: *Cerebro-cerebellar interactions,* ed. J. Massion & K. Sasaki. Elsevier/North Holland. [aAMS, JH]
Rosenblatt, F. (1958) The perceptron: A probabilistic model for information storage and organization in the brain. *Psychological Review* 65:386–408. [MD]
Rosenmund, C., Legendre, P. & Westbrook, G. L. (1992) Expression of NMDA channels on cerebellar Purkinje cells acutely dissociated from newborn rats. *Journal of Neurophysiology* 68:1901–5. [aDJL]
Ross, C. A., Bredt, D. & Snyder, S. H. (1990) Messenger molecules in the cerebellum. *Trends in Neurosciences* 13:216–22.
Ross, W. N. & Werman, R. (1987) Mapping calcium transients in the dentrites of Purkinje cells from the guinea pig cerebellum in vitro. *Journal of Physiology (London)* 389:319–36. [aMKan]
Ross, W. N., Lasser-Ross, N. & Werman, R. (1990) Spatial and temporal analysis of calcium dependent electrical activity in guinea pig Purkinje cells dendrites. *Proceedings of the Royal Society of London, Series B: Biological Science* 240:173–85. [aMKan]
Rossignol, S. & Melvill-Jones, G. (1976) Audio-spinal influence in man studied by the H-reflex and its possible role on rhythmic movements synchronised to sound Electroenceph. *Clinical Neurophysiology* 41:83–92. [aAMS]
Rubia, F. J. & Kolb, F. P. (1978) Responses of cerebellar units to a passive movement in the decerebrate cat. *Experimental Brain Research* 31:387–401. [aJIS]
Rumelhart, D. E., Hinton, G. E., McClelland, J. L. (1987) A general framework for parallel distributed processing. In: *Parallel Processing,* ed. D. E. Rumelhart & J. L. McClelland. MIT Press. [rJCH]
Rushmer, D. S., Roberts, W. J. & Augter, G. K. (1976) Climbing fiber responses of cerebellar Purkinje cells to passive movement of the cat forepaw. *Brain Research* 106:1–20. [aJIS]
Ryding, E., Decety, J., Sjoholm, H., Stengerg, G. & Ingvar, D. H. (1993) Motor imagery activates the cerebellum regionally. A SPECT rCBRF study with 99m Tx-HMPAO. *Cognitive Brain Research* 1:94–99. [aWTT]
Sacktor, T. C., Osten, P., Valsamis, H., Jiang, X., Naik, M. U. & Sublette, E. (1993) Persistent activation of the ∂ isoform of protein kinase C in the maintenance of long-term potentiation. *Proceedings of the National Academy of Sciences of the USA* 90: 8342–46. [aFC]
Sainburg, R. L., Poizner, H. & Ghez, C. (1993) Loss of proprioception produced deficits in interjoint coordination. *Journal of Neurophysiology* 70:2136–47. [aWTT]
Saint-Cyr, J. A. & Courville, J. (1980) Projections from the motor cortex, midbrain, and vestibular nuclei to the inferior olive in the cat: Anatomical organization and functional correlates. In: *The inferior olivary nucleus: Anatomy and physiology,* ed. J. Courville, J. C. de Montigny & Y. Lamarre. Raven. [JDS]
Sakaue, M., Kuno, T. & Tanaka C. (1988) Novel type of monoclonal antibodies against cyclic GMP and application to immunocytochemistry of the rat brain. *Japanese Journal of Pharmacology* 48:47–56. [aDJL, aSRV]
Sakimura, K., Kutsuwada, T., Ito, I., Manabe, T., Takayama, C., Kushiya, E., Yagi, T., Aizawa, S., Inoue, Y., Sugiyama, H. & Mishina, M. (1994) Reduced hippocampal LTP and spatial learning in mice lacking NMDA receptor $\varepsilon 1$ subunit. *Nature* 373:151–55. [aMKan]
Sakurai, M. (1987) Synaptic modification of parallel fibre-Purkinje cell transmission in *in vitro* guinea pig cerebellar slices. *Journal of Physiology (London)* 394:463–80. [aJCH, aMKan, aDJL, aJIS]
(1990) Calcium is an intracellular mediator of the climbing fiber in induction of cerebellar long-term depression. *Proceedings of the National Academy of Sciences of the USA* 87:3383–85. [aFC, aMKan, aDJL, aJIS]
Sandoval, M. E. & Cotman, C. W. (1978) Evaluation of glutamate as a neurotransmitter of cerebellar parallel fibers. *Neuroscience* 3:199–206. [aDJL]
Sanes, J. N., Dimitrov, B. & Hallett, M. (1990) Motor learning in patients with cerebellar dysfunction. *Brain* 113:103–20. [aWTT, MH, DT]
Sanes, J. N., Suner, S., Lando, J. F. & Donoghue, J. P. (1988) Rapid reorganization of adult rat motor cortex somatic representation patterns after motor nerve injury. *Proceedings of the National Academy of Sciences of the USA* 85:2003–7. [aJCH]
Sarrafizadeh, R. (1994) Sensory triggering of limb motor programs: Neural correlates of decisions for action. In: *NPB Technical Report 9.* Northwestern University Institute of Neuroscience. [aJCH]
Sarrafizadeh, R., Keifer, J. & Houk, J. C. (1996). Somatosensory and movement-related properties of red nucleus: A single unit study in the turtle. *Experimental Brain Research* 108:1–17. [aJCH]
Sasaki, K., Bower, J. M. & Llinás, R. (1989) Multiple Purkinje cell recording in rodent cerebellar cortex. *European Journal of Neuroscience* 1:572–86. [aJIS]
Sasaki, K. S., Kawaguchi, S., Oka, H., Saki, M. & Mizuno, N. (1976) Electrophysiological studies on the cerebellocerebral projections in monkeys. *Experimental Brain Research* 24:495–507. [aWTT]
Sato, Y., Miura, A., Fushiki, H. & Kawasaki, T. (1992) Short-term modulation of cerebellar Purkinje cell activity after spontaneous climbing fiber input. *Journal of Neurophysiology* 68(6):2051–62. [aJIS]
(1993) Barbiturate depresses simple spike activity of cerebellar Purkinje cells after climbing fiber input. *Journal of Neurophysiology* 69:1082–90. [aJIS]
Satoh, T., Ross, C. A., Villa, A., Supattapone, S., Pozzan, T., Snyder, S. H. & Meldolesi, J. (1990) The inositol 1,4,5-trisphosphate receptor in cerebellar Purkinje cells: Quantitative immunogold labeling reveals concentration in an ER subcompartment. *Journal of Cell Biology* 111:615–24. [aDJL]

Saxon, D. W. & Beitz, A. J. (1994) Cerebellar injury induced NOS in Purkinje cells and cerebellar afferent neurons. *NeuroReport* 5:809–13. [rSRV]

Schell, G. R. & Strick, P. L. (1983) The origin of thalamic inputs to the arcuate premotor and supplementary motor areas. *Journal of Neuroscience* 4:539–60. [aWTT]

Schieber, M. H. & Thach, W. T. (1985a) Trained slow tracking: 1. Muscular production of wrist movemnt. *Journal of Neurophysiology* 55:1213–27. [aWTT]

(1985b) Trained slow tracking: 2. Bidirectional discharge patterns of cerebellar nuclear, motor cortex, and spindle afferent neurons. *Journal of Neurophysiology* 54:1228–70. [aAMS, aWTT]

Schilling, K., Dickinson, M., Connor, J. A. & Morgan, J. I. (1991) Electrical activity in cerebellar cultures determines Purkinje cell dendritic growth patterns. *Neuron* 7:891–902. [aDJL]

Schilling, K., Schmidt, H. H. H. W. & Badder, S. L. (1994) Nitric oxide synthase expression reveals compartments of cerebellar granule cells and suggests a role for mossy fibers in their development. *Neuroscience* 59:893–903. [aDJL, TH]

Schlichter, D. J., Casnellie, J. E. & Greengard, P. (1978) An endogenous substrate for cGMP-dependent protein kinase in mammalian cerebellum. *Nature* 273:61–62. [aSRV]

Schlichter, D. J., Detre, J. A., Aswad, D. W., Chehrazi, B. & Greengard, P. (1980) Localization of cyclic GMP-dependent protein kinase and substrate in mammalian cerebellum. *Proceedings of the National Academy of Sciences of the USA* 77:5537–41. [aSRV]

Schmahmann, J. D. (1991) An emerging concept: The cerebellar contribution to higher function. *Archives of Neurology* 48:1178–87. [aWTT, JDS]

(1992) An emerging concept: The cerebellar contribution to higher function. *Archives of Neurology* 49:1230. [JDS]

(1994) The cerebellum in autism: Clinical and anatomic perspectives. In: *The neurobiology of autism*, ed. M. L. Bauman & T. L. Kemper. Johns Hopkins University Press. [JDS]

Schmahmann, J. D. & Pandya, D. N. (1987) Posterior parietal projections to the basis pontis in rhesus monkey: Possible anatomical substrate for the cerebellar modulation of complex behavior? *Neurology* (suppl. 37):291. [JDS]

(1989) Anatomical investigation of projections to the basis pontis from posterior parietal association cortices in rhesus monkey. *Journal of Comparative Neurology* 289:53–73. [JDS]

(1991) Projections to the basis pontis from the superior temporal sulcus and superior temporal region in the rhesus monkey. *Journal of Comparative Neurology* 308:224–48. [JDS]

(1993) Prelunate, occipitotemporal, and parahippocampal projections to the basis pontis in rhesus monkey. *Journal of Comparative Neurology* 337:94–112. [JDS]

(1995) Prefrontal cortex projections to the basilar pons: Implications for the cerebellar contribution to higher function. *Neuroscience Letters* 199:175–78. [JDS]

Schmaltz, L. W. & Theios, J. (1972) Acquisition and extinction of a classically conditioned response in hippocampectomized rabbits (*Oryctalagous cuniculus*). *Journal of Comparative Physiology and Psychology* 79:328–33. [CW]

Schmidt, K., Klatt, P. & Mayer, B. (1993) Characterization of endothelial cell amino acid transport systems involved in the actions of nitric oxide synthase inhibitors. *Journal of Pharmacology and Experimental Therapeutics* 44:615–21. [aSRV]

Schmidt, M. J. & Nadi, N. S. (1977) Cyclic nucleotide accumulation in vitro in the cerebellum of 'nervous' neurologically mutant mice. *Journal of Neurochemistry* 29:87–90. [arSRV, LK]

Schmidt, R. A. (1988) *Motor control and learning: A behavioral emphasis.* Human Kinetics. [aJCH, SPS]

(1991) *Motor learning & performance.* Human Kinetics . [HB]

Schreurs, B. G. & Alkon, D. L. (1993) Rabbit cerebellar slice analysis of long-term depression and its role in classical conditioning. *Brain Research* 631:235–40. [JCF, aDJL, aJIS, EDS]

Schreurs, B. G., Oh, M. M. & Alkon, D. L. (1996) Pairing-specific long-term depression of Purkije cell excitatory postsynaptic potentials results from a classical conditioning procedure in the rabbit cerebellar slice. *Journal of Neurophysiology* 75:1051–60. [rDJL]

Schulmann, J. A. & Bloom, F. E. (1981) Golgi cells of the cerebellum are inhibited by inferior olive activity. *Brain Research* 210:350–55. [aJIS]

Schulz, P. E., Cook, E. P. & Johnston, D. (1994) Changes in paired-pulse facilitation suggest presynaptic involvement in long-term potentiation. *Journal of Neuroscience* 14:5325–37. [LBJ]

Schuman, E. & Madison, D. V. (1991) A requirement for the intercellular messenger nitric oxide in long-term potentiation. *Science* 254:1503–6. [MB]

Schwartz, D. W. F. & Tomlinson, R. D. (1977) Neurononal responses to eye muscle stretch in cerebellar lobule VI of the cat. *Experimental Brain Research* 27:101–11. [PD]

Schweighofer, N. (1995) *Computational models of the cerebellum in the adaptive control of movements.* PhD dissertation, University of Southern California. [MKaw]

Schweighofer, N. & Arbib, M. A. (submitted) From behavior to second messengers: A multilevel approach to cerebellar learning. [MAA]

Schweighofer, N. & Arbib, M. A. & Dominey, P. F. (in press) A model of adaptive control of saccades. *Biological Cybernetics.* [MAA]

Schweighofer, N., Spoelstra, J., Arbib, M. A. & Kawato, M. (submitted) Role of the cerebellum in reaching quickly and accurately: 2. A detailed model of the intermediate cerebellum. [MKaw]

Scudder, C. S. (1988) A new local feedback model of the saccadic burst generator. *Journal of Neurophysiology* 59:1455–75. [aJCH]

Sears, L. L. & Steinmetz, J. E. (1991) Dorsal accessory inferior olive activity diminishes during acquisition of the rabit classically conditioned eyelid response. *Brain Research* 545:114–22. [RFT]

Sears, L. L. & Steinmetz, J. E. (1990) Acquisition of classically conditioned-related activity in the hippocampus is affected by lesions of the cerebellar interpositus nucleus. *Behavioral Neuroscience* 104:681–92. [CW]

Segal, L. A. (1995) Grappling with complexity. *Complexity* 1:18–25. [PDR]

Seitz, R. J., Roland, P. E., Bohm, C., Greitz, T. & Stone-Elander, S. (1990) Motor learning in man: A positron emission tomography study. *NeuroReport* 1:57–66. [aWTT]

Selig, D. K., Lee, H. K., Bear, M. F. & Malenka, R. C. (1995) Reexamination of the effects of MCPG on hippocampal LTP, LTD, and depotentiation. *Journal of Neurophysiology* 74:1075–82. [MB]

Sessler, F. M., Mouradian, R. D., Cheng, J. T., Yeh, H. H., Liu, W. & Waterhouse, B. D. (1989) Noradrenergic potentiation of cerebellar Purkinje cell responses to GABA: Evidence for mediation through the b-adrenoceptor-coupled cyclic AMP system. *Brain Research* 499:27–38. [aMKan]

Shadmehr, R., Mussa-Ivaldi, F.A. & Bizzi, E. (1993) Postural force fields of the human arm and their role in generating multijoint movements. *Journal of Neuroscience* 13:45–62. [aAMS]

Shambes, G. M., Gibson, J. M. & Welker, W. I. (1978) Fractured somatotopy in granular cell tactile areas of rat cerebellar hemispheres revealed by micromapping. *Brain Behaviour & Evolution* 15:94–140. [FS, rAMS]

Shammah-Lagnado, S. J., Negrao, N. & Ricardo, J. A. (1985) Afferent connections of the zona incerta: A horseradish peroxidase study in the rat. *Neuroscience* 15:109–34. [JDS]

Sharp, A. H., McPherson, P. S., Dawson, T. M., Aoki, C., Campbell, K. P. & Snyder, S. H. (1993) Differential immunohistochemical localization of inositol 1,4,5-triphosphate- and ryanodine-sensitive Ca^{2+} release channels in rat brain. *Journal of Neuroscience* 13:3051–63. [aSRV]

Sherman, J. & Schmahmann, J. D. (1995) The spectrum of neuropsychological manifestations in patients with cerebellar pathology. *Human Brain Mapping* (suppl. 1):361. [JDS]

Sherrington, C. S. (1909) Reciprocal innervation of antagonist muscles. Fourteenth note – on double reciprocal innervation. *Proceedings of the Royal Society of London, Series B* 81:249–68. [aAMS]

(1947) *The integrative action of the nervous system.* Yale University Press. [aAMS]

Shibasaki, H., Shima, F. & Kuroiwa, Y. (1978) Clinical studies of the movement-related cortical potential (MP) and the relationship between the dentatorubrothalamic pathway and readiness potential (RP). *Journal of Neurology* 219:15–25. [KW]

Shibuki, K. (1990) An electrochemical microprobe for detecting nitric oxide release in brain tissue. *Neuroscience Research* 9:69–76. [arSRV, DO]

(1993) Nitric oxide: A multi-functional messenger substance in cerebellar synaptic plasticity. *Seminars in the Neurosciences* 5:217–23. [aDJL]

Shibuki, K., Gomi, H., Chen, L., Bao, S., Kim, J. J., Wakatsuki, H., Fujisaki, T., Fujimoto, K., Katoh, A., Ikeda, T., Chen, C., Thompson, R. F. & Itohara, S. (1996) Deficient cerebellar long-term depression, impaired eyeblink conditioning, and normal motor coordination in GFAP mutant mice. *Neuron* 16:587–99. [rMKan, rDJL]

Shibuki, K. & Okada, D. (1991) Endogenous nitric oxide release required for long-term synaptic depression in the cerebellum. *Nature* 349:326–28. [aFC, aMKan, aDJL, arSRV, JCH, NAH, DO]

(1992) Cerebellar long-term potentiation under suppressed postsynaptic Ca^{2+} activity. *NeuroReport* 3:231–34. [aSRV, NAH]

Shidara, M., Kawano, K., Gomi, H. & Kawato, M. (1993) Inverse dynamics model eye movement control by Purkinje cells in the cerebellum. *Nature* 365(2):50–52. [aJCH, HG, MKaw]

Shigemoto, R., Abe, T., Nomura, S., Nakanishi, S. & Hirano, T. (1994) Antibodies inactivating mGluR1 metobotropic glutamate receptors block long-term depression in cultured Purkinje cells. *Neuron* 12:1245–55. [TH, MKan, rSRV]

Shigemoto, R., Nakanishi, S. & Hirano, T. (1994) Antibodies inactivating mGluR1 metabotropic glutamate receptor block long-term depression in cultured Purkinje cells. *Neuron* 12:1245–55. [aDJL, PC, rDJL]

Shojaku, H., Barmack, N. H. & Mizukoshi, K. (1991) Influence of vestibular and visual climbing fiber signals on Purkinje cell discharge in the cerebellar nodulus of the rabbit. *Acta Otolaryngologica* 481:242–46. [aJIS]

Shuttleworth, C. W. R., Burns, A. J., Ward, S. M., O'Brien, W. E. & Sanders, K. M. (1995) Recycling of L-citrulline to sustain nitric oxide-dependent enteric neurotransmission. *Neuroscience* 68:1295–1304. [rSRV]

Sigel, E., Baur, R. & Malherbe, P. (1991) Activation of protein kinase C results in down-modulation of different recombinant GABAa-channels. *FEBS Letters* 291:150–52. [aMKan]

Siggins, G. R., Henriksen, S. J. & Landis, S. C. (1976) Electrophysiology of Purkinje neurons in the weaver mouse: Iontophoresis of neurotransmitters and cyclic nucleotides, and stimulation of the nucleus locus coeruleus. *Brain Research* 114:53–69. [aSRV]

Siggins, G. R., Hoffer, B. J., Oliver, A. P. & Bloom, F. E. (1971) Activation of a central noradrenergic pathway to cerebellum. *Nature* 233:481–83. [aDJL]

Silva, A. J., Stevens, C. F., Tonegawa, S. & Wang, Y. (1992) Deficient hippocampal long-term potentiation in α-calcium-calmodulin kinase II mutant mice. *Science* 257:201–6. aMKan

Silva, A. J., Paylor, R., Wehner, J. M. & Tonegawa, S. (1992) Impaired spatial learning in α-calcium-calmodulin kinase II mutant mice. *Science* 257:206–11. aMKan

Simpson, J. I. (1984) The accessory optic system. *Annual Review of Neuroscience* 7:13–41. [aJCH]

(1994) Functional and anatomic organization of three-dimensional sye movements in rabbit cerebellar flocculus. *Journal of Neurophysiology* 72:31–46. [aWTT]

Simpson, J. I. & Alley, K. E. (1974) Visual climbing fiber input to rabbit vestibulocerebellum: A source of direction-specific information. *Brain Research* 82:302–8. [aJIS, aWTT]

Simpson, J. I., Graf, W. & Leonard, C. (1981) The coordinate system of visual climbing fibers to the flocculus. In: *Progress in oculomotor research: Developments in neuroscience*, vol. 12, eds. A. Fuchs & W. Becker. Elsevier/North-Holland. [arJIS]

Simpson, J. I., Leonard, C. S. & Soodak, R. E. (1988) The accessory optic system of rabbit: 2. Spatial organization of direction selectivity. *Journal of Neurophysiology* 60:2055–72. [aJIS]

Sinkjaer, T., Wu, C. H., Barto, A. & Houk, J. C. (1990) Cerebellum control of endpoint position – a simulation model. *IJCNN 90* 2:705–10. [aJCH]

Sladeczek, F., Pin, J. P., Recasens, M., Bockaert, J. & Weiss, S. (1985). Glutamate stimulates inositol phosphate formation in striatal neurones. *Nature* 314:717–19. [aFC]

Smirnova, T., Stinnakre, J. & Mallet, J. (1993) Characterization of a presynaptic glutamate receptor. *Science* 262:430–33. [aSRV]

Smith, A. M. (1981) The coactivation of antagonist muscles. *Canadian Journal of Physiology and Pharmacology* 59:733–47. [aAMS]

Smith, A. M. & Bourbonnais, D. (1981) Neuronal activity in cerebellar cortex related to control of prehensile force. *Journal of Neurophysiology* 45:286–303. [aWTT]

Smith, A. M., Frysinger, R. C. & Bourbonnais, D. (1983) Discharge patterns of cerebellar cortical neurons during the co-activation and reciprocal inhibition of forearm muscles. In: *Neural coding of motor performance, Experimental brain research supplementum 7*, edited by J. Massion, J. Paillard, W. Schultz & M. Wiesendanger. Springer-Verlag. [aAMS]

Snider, R. S. (1950) Recent contributions to the anatomy and physiology of the cerebellum. *Archives of Neurology and Psychiatry* 64:196–219. [JDS]

(1975) A cerebellar-ceruleus pathway. *Brain Research* 88:59–63. [JDS]

Snider, R. S. & Eldred, E. (1952) Cerebro-cerebellar relationships in the monkey. *Journal of Neurophysiology* 15:27–40.

Snider, R. S., Maiti, A. & Snider, S. R. (1976) Cerebellar pathways to ventral midbrain and nigra. *Experimental Neurology* 53:714–28. [JDS]

Snider, R. S. & Stowell, A. (1944) Receiving areas of tactile, auditory, and visual systems in the cerebellum. *Journal of Neurophysiology* 7:331-57. [rWTT]

Sobera, L. A. & Morad, M. (1991) Modulation of cardiac sodium channels by cAMP receptors on the myocyte surface. *Science* 253:1286–89. [aSRV]

Solomon, P. R., Solomon, S. D., Vander Schaaf, E. R. & Perry, H. E. (1983) Altered activity in the hippocampus is more detrimental to conditioning than removal of the structure. *Science* 220:329–31. [CW]

Sommer, B., Keinanen, K., Verdoorn, T., Wisden, W., Burnashev, N., Herb, A., Kohler, M., Takagi, T., Sakmann, B. & Seeburg, P. H. (1990) Flip and flop: A cell-specific functional switch in glutamate-operated channels of the CNS. *Science* 249:1580–85. [aFC, aSRV]

Sommer, B. & Seeburg, P. H. (1992) Glutamate receptor channels: Novel properties and new clones. *Trends in Pharmacological Sciences* 13:291–96. [aFC]

Soodak, R. E., Croner, L. J. & Graf, W. (1988) Development of the optokinetic reference frame of floccular Purkinje cells in rabbit. *Society for Neuroscience Abstracts* 14:758. [arJIS]

Soodak, R. E. & Simpson, J. I. (1988) The accessory system of rabbit: 1. Basic visual response properties. *Journal of Neurophysiology* 60:2037–54. [aJIS]

Sorkin, L. S. (1993) NMDA evokes an L-NAME sensitive spinal release of glutamate and citrulline. *NeuroReport* 4:479–82. [rSRV]

Sotelo, C., Gotow, T. & Wassef, M. (1986) Localization of glutamic-acid-decarboxylase-immunoreactive axon terminals in the inferior olive of the rat, with special emphasis on anatomical relations between GABAergic synapses and dendrodendritic gap junctions. *Journal of Comparative Neurology* 252:32–50. [aJIS]

Sotelo, C., Llinás, R. & Baker, R. (1974) Structural study of the inferior olivary nucleus of the cat: Morphological correlates of electrotonic coupling. *Journal of Neurophysiology* 37:541–59. [aJIS]

Southam, E., East, S. J. & Garthwaite, J. (1991) Excitatory amino acid receptors coupled to the nitric oxide-cyclic GMP pathway in rat cerebellum during development. *Journal of Neurochemistry* 56:2072–81. [aSRV]

Southam, E. & Garthwaite, J. (1991a) Comparative effects of some nitric oxide donors on cyclic GMP levels in rat cerebellar slices. *Neuroscience Letters* 130:107–11. [aSRV]

(1991b) Intercellular action of nitric oxide in adult rat cerebellar slices. *NeuroReport* 2:658–60. [aSRV, LK]

Southam, E. & Garthwaite, J. (1991c) Climbing fibers as a source of nitric oxide in the cerebellum. *European Journal of Neuroscience* 3:379–82. [DO]

(1993) The nitric oxide-cyclic GMP signalling pathway in rat brain. *Neuropharmacology* 32:1267–77. [aSRV]

Southam, E., Morris, R. & Garthwaite, J. (1992) Sources and targets of nitric oxide in rat cerebellum. *Neuroscience Letters* 137:241–44. [aDJL, aSRV]

Sparks, D. L. (1988) Neural cartography: Sensory and motor maps in the superior colliculus. *Brain, Behavior & Evolution* 31:49–56. [aJCH]

Spidalieri, G., Busby, L. & Lamarre, Y. (1983) Fast ballistic arm movements triggered by visual, auditory and somesthetic stimuli in the monkey: 2. Effects of unilateral dentate lesion on discharge of precentral cortical neurons and reaction time. *Journal of Neurophysiology* 50:1359–79. [aAMS]

Squire, L. R. (1992) Memory and the hippocampus: A synthesis from findings with rats, monkeys, and humans. *Psychological Review* 99:195–231. [SMO]

Squire, L. R. & Zola-Morgan, S. (1991) The medial temporal lobe memory system. *Science* 253:1380–86. [SMO]

Stahl, J. S. & Simpson, J. I. (1995) Dynamics of rabbit vestibular nucleus neurons and the influence of the flocculus. *Journal of Neurophysiology* 73:1396–1413. [aJIS]

Stamler, J. S., Singel, D. J. & Loscalzo, J. (1992) Biochemistry of nitric oxide and its redox-activated forms. *Science* 258:1898–1902. [DO]

Staub, C., Vranesic, I. & Knöpfel, T. (1992) Responses to metabotropic glutamate receptor activation in cerebellar Purkinje cells: Induction of an inward current. *European Journal of Neuroscience* 4:832–39. [aFC, aDJL]

Stein, J. F. & Glickstein, M. (1992) Role of the cerebellum in visual guidance of movement. *Physiological Reviews* 72:967–1017. [arAMS]

Stein, R. B. (1991) Reflex modulation during locomotion: Functional significance. In: *Adaptability of human gait*, ed. A. E. Patla. Elsevier. [CG]

Steinmetz, J. E. (1990a) Classical nictitaing membrane conditioning in rabbits with varying interstimulus intervals and direct activation of cerebellar mossy fibers as the CS. *Behavioral Brain Research* 38:97–108. [JCF]

(1990b) Neuronal activity in the rabbit interpositus nucleus during classical NM-conditioning with a pontine-nucleus-stimulation CS. *Psychological Science* 1:378–82. [JCF]

Steinmetz, J. E., Lavond, D. G., Ivkovich, I., Logan, C. G. & Thompson, R. F. (1922) Disruption of classical eyelid conditioning after cerebellar lesions: Damage to a memory trace system or a simple performance deficit? *Journal of Neuroscience* 12:4403–26. [aWTT]

Steinmetz, J. E., Lavond, D. G. & Thompson, R. F. (1989) Classical conditioning rabbits using pontine nucleus stimulation as a conditioned stimulus and inferior olive stimulation as an unconditioned stimulus. *Synapse* 3(3):225–32. [RFT, CW]

Stelzer, A., Slater, N. T. & ten Bruggencate, G. (1987) Activation of NMDA receptors blocks GABAergic inhibition in an in vitro model of epilepsy. *Nature* 326:698–701. [aMKan]

Stone, L. S. & Lisberger, S. G. (1990) Visual responses of Purkinje cells in the cerebellar flocculus during smooth-pursuit eye movements in monkeys: 2. Complex spikes. *Journal of Neurophysiology* 63:1262–75. [aJCH, aJIS]

Strahlendorf, J. C., Strahlendorf, H. K. & Barnes, C. D. (1979) Modulation of cerebellar neuronal activity by raphé stimulation. *Brain Research* 169:565–69. [aDJL]

Straube, A., Deubel, H., Spuler, A. & Büttner, U. (1995) Different effect of a bilateral deep cerebellar nuclei lesion on externally and internally triggered saccades in humans. *Neuro-Ophthalmology* 15:67–74. [HB]

Strata, P. (1985) Inferior olive: Functional aspects. In: *Cerebellar functions*, ed. J. R. Bloedel, J. Dichgans & W. Precht. Springer-Verlag. [aWTT]

Strick, P. L. (1994) Input to the primate frontal eye field from substantia nigra, superior colliculus, and dentate nucleus demonstrated by transneuronal transport. *Experimental Brain Research* 100(1):181–86. [CW]

Sugihara, I., Lang, E. J. & Llinás, R. (1993) Uniform olivocerebellar conduction time underlies Purkinje cell complex spike synchronicity in the rat cerebellum. *Journal of Physiology (London)* 470:243–71. [aJIS]

Sugimori, M. & Llinas, R. R. (1990) Real-time imaging of calcium influx in mammalian cerebellar Purkinje cells in vitro. *Proceedings of the National Academy of Sciences of the USA* 87:5084–88. [aMKan]

Sugiyama, H., Ito, I. & Hirono, C. (1987). A new type of glutamate receptor linked to inositol phospholipid metabolism. *Nature* 325:531–33. [aFC]

Suko, J., Maurer-Fogy, I., Plank, B., Bertel, O., Wyskovsky, W., Hohenegger, M. & Hellmann, G. (1993) Phosphorylation of serine 2843 in ryandodine receptor-calcium release channel of skeletal muscle by cAMP-, cGMP- and CaM-dependent protein kinase. *Biochemica et Biophysica Acta* 1175:193–206. [aSRV]

Sutton, R. S. & Barto, A. G. (1981) Toward a modern theory of adaptive networks: Expectation and prediction. *Psychological Review* 88:135–70. [aJCH, JCH]

Suzuki, R. (1987) A hierarchical neural-network model for control and learning of voluntary movement. *Biological Cybernetics* 57:169–85. [MKan]

Swain, R. A., Shinkman, P. G., Nordholm, A. F. & Thompson, R. F. (1992) Cerebellar stimulation as an unconditioned stimulus in classical conditioning. *Behavioral Neuroscience* 106:739–50. [CW]

Swinnen, S. P., Dounskaia, N., Verschueren, S., Serrien, D. J. & Daelman, A. (1995) Relative phase destabilization during interlimb coordination: The disruptive role of kinesthetic afferences induced by passive movement. *Experimental Brain Research* 105: 439–54. [SPS]

Swinnen, S. P., Walter, C. B., Lee, T. D. & Serrien, D. J. (1993) Acquiring bimanual skills: contrasting forms of information feedback for interlimb decoupling. *Journal of Experimental Psychology: Learning, Memory, & Cognition* 19:1328–44. [SPS]

Swinnen, S. P., Young, D. E., Walter, C. B. & Serrien, D. J. (1991) Control of asymmetrical bimanual movements. *Experimental Brain Research* 85:163–73. [SPS]

Szekely, A. M., Barbaccia, M. L., Alho, H. & Costa, E. (1989) In primary cultures of cerebellar granule cells the activation of N-methyl-D-aspartate-sensitive glutamate receptors induces *c-fos* mRNA expression. *Molecular Pharmacology* 35:401–8. [aSRV]

Szentágothai, J. & Rajkovits, K. (1959) Uber den urspring der kletterfasern des kleinjirns. *Zeitschrift für Anatomie und Entwicklungsgeschichte* 121:130–41. [aJIS]

Tachibana, H., Argane, K. & Sugita, M. (1995) Event-related potentials in patients with cerebellar degeneration: Electrophysiological evidence for cognitive impairment. *Cognitive Brain Research* 2:173–80. [JMB]

Tank, D. W., Sugimori, M., Connor, J. A. & Llinas, R. R. (1988) Spatially resolved calcium dynamics of mammalian Purkinje cells in cerebellar slice. *Science* 242:773–77. [aMKan]

Takagi, H., Takimizu, H., de Barry, J., Kudo, Y. & Yoshioka, T. (1992) The expression of presynaptic t-ACPD receptor in rat cerebellum. *Biochemical and Biophysical Research Communications* 189:1287–95. [aDJL, aSRV]

Takemura, A., Inoue, Y., Kawano, K., Shidara, M., Gomi, H. & Kawato, M. (submitted) Characterization of neuronal firing patterns during short-latency ocular following responses by linear time-series regression analysis. [MKaw]

Tan, J., Gerrits, N. M., Nanhoe, R. S., Simpson, J. I. & Voogd, J. (1995a) Zonal organization of the climbing fiber projection to the flocculus and nodulus of the rabbit. A combined axonal tracing and acetylcholinesterase histochemical study. *Journal of Comparative Neurology* 356:1–22. [aJIS]

Tan, J., Simpson, J. I. & Voogd, J. (1995b) Anatomical compartments in the white matter of the rabbit flocculus. *Journal of Comparative Neurology* 356:23–50. [aJIS]

Tang, C. M., Shi, Q. Y., Katchman A. & Lynch, G. (1991) Modulation of the time course of fast EPSPs and glutamate channel kinetics by Aniracetam. *Science* 254:288–90. [aFC]

Tanji, J. (1985) Comparison of neural activities in the monkey supplementary and precentral motor areas. *Trends in Neuroscience* 18:137. [aWTT]

Tanji, J. & Evarts, E. V. (1976) Anticipatory activity of motor cortex in relation to direction of an intended movement. *Journal of Neurophysiology* 39:1062–68. [aWTT]

Tanji, J. & Shima, K. (1994) Role for supplementary motor area cells in planning several movements ahead. *Nature* 371:413–16. [MH]

Tarkka, I. M., Massaquoi, S. & Hallett, M. (1993) Movement-related cortical potentials in patients with cerebellar degeneration. *Acta Neurologica Scandinavica* 88:129–35. [KW]

Tempia, F., Dieringer, N. & Strata, P. (1991) Adaptation and habituation of the vestibulo-ocular reflex in intact and inferior olive-lesioned rats. *Experimental Brain Research* 86:568–78. [aJIS]

ter Haar Romeny, B. M., Denier Van Der Gon, J. J., & Gielen, C. C. A. M. (1984) Relation between location of a motor unit in the human biceps brachii and its critical firing levels for different tasks. *Experimental Neurology* 85:631–50. [aAMS]

Terzuolo, C. A., Soechting, J. F. & Palminteri, R. (1973) Studies on the control of some simple motor tasks: 3. Comparison of the EMG pattern during ballistically initiated movements in man and squirrel monkey. *Brain Research* 62:242–46. [aAMS]

Terzuolo, C. A., Soechting J. F. & Viviani P. (1973) Studies on the control of some simple motor tasks: 2. On the cerebellar control of movements in relation to the formation of intentional command. *Brain Research* 58:217–22. [CG]

Thach, W. T. (1967) Somatosensory receptive fields of single units in the cat cerebellar cortex. *Journal of Neurophysiology* 30:675–96. [aJIS]

(1968) Discharge of Purkinje and cerebellar nuclear neurons during rapidly alternating arm movement in the monkey. *Journal of Neurophysiology* 31:785–97. [aJIS]

(1970a) Discharge of cerebellar neurones related to two maintained postures and two prompt movements: 1. Nuclear cell output. *Journal of Neurophysiology* 35:527–36. [KW, aJIS]

(1970b) Discharge of cerebellar neurons related to two maintained postures and two prompt movements: 2. Purkinje cell output and input. *Journal of Neurophysiology* 33:537–47. [PFCG]

(1980) Complex spikes, the inferior olive, and natural behavior. In: *The inferior olivary necleus*, ed. J. Courville. Raven. [aWTT]

Thach, W. T., Goodkin, H. P. & Keating, J. G. (1991) Inferior olive disease in man prevents learning novel synergies. *Society for Neuroscience Abstracts* 17:1380. [aWTT]

(1992) The cerebellum and the adaptive coordination of movement. *Annual Review of Neuroscience* 15:403–42. [aJCH, arAMS, aJIS, arWTT]

Thach, W. T., Goodkin, H. P., Keating, J. G. & Martin, T. A. (1992) Prism adaptation in throwing is specific for arm and type of throw. *Society for Neuroscience Abstracts* 18:516. [aWTT]

Thach, W. T., Lane, S. A., Mink, J. W. & Goodkin, H. P. (1992) Cerebellar output: Multiple maps and modes of control in movement coordination. In: *The cerebellum revisited*, ed. R. Llinas & C. Sotela. Springer-Verlag. [aWTT]

Thach, W. T., Martin, T. A., Keating, J. G., Goodkin, H. P. & Bastian, A. J. (1995) Schematic model of short- and long-term adjustments of eye-hand coordination in throwing. *Society for Neuroscience Abstracts.* [aWTT]

Thach, W. T., Mink, J. W., Goodkin, H. P. & Keating, J. G. (1993) Combining versus gating motor programs: Differential roles for cerebellum and basal ganglia? In: *Role of cerebellum and basal ganglia in voluntary movement*, ed. N. Mano, I. Hamada & M. R. DeLong. Elsevier. [rWTT]

Thach, W. T. & Montgomery, E. B. (1990) Motor system. In: *Neurobiology of disease*, ed. A. L. Pearlman & R. C. Collins. Oxford University Press. [aWTT]

Thach, W. T., Perry, J. G., Kane, S. A. & Goodkin, H. P. (1993) Cerebellar nuclei: Rapid alternating movement, motor somatotopy and a mechanism for the control of muscle synergy. *Revue Neurologique* 149:607–28. [aAMS, arWTT]

Thelen, E. & Smith, A. (1994) *A dynamic systems approach to the development of cognition and action.* MIT Press. [AGF]

Thompson, R. F. (1986) The neurobiology of learning and memory. *Science* 223:941–47. [aJCH, aJIS, aWTT, CW]

(1988) The neural analysis of basic associative learning of discrete behavioral responses. *Trends in Neurosciences* 11:152–55. [EDS, JDS]

(1990) Neural mechanisms of classical conditioning in mammals. *Philosophical Transactions of the Royal Society of London* B 161–70. [aWTT, CW]

Thompson, R. F. & Krupa, D. J. (1994) Organization of memory traces in the mammalian brain. *Annual Review of Neuroscience* 108:44–56. [DT, RFT]

Timmann, D. & Horak, F. B. (1995b) Prediction and set-dependent postural gain control in cerebellar patients. *Society of Neuroscience Abstracts* 21:270. [DT]

Timmann, D., Kolb, F. P., Rijntjes, M., Diener, H. C. & Weiller, C. (1995a) Classical conditioning of the human flexion reflex: A PET study. *European Journal of Neuroscience* 8(suppl.):195. [DT]

Timmann, D., Shimansky, Yu., Larson, P. S., Wunderlich, D. A., Stelmach, G. E. & Bloedel, J. R. (1994) Visuomotor learning in cerebellar patients. *Society of Neuroscience Abstracts* 20:21. [DT]

Tjörnhammar, M.-L., Lazaridis, G. & Bartfai, T. (1986) Efflux of cyclic guanosine 3',5'-monophosphate from cerebellar slices stimulated by L-glutamate or high K+ or N-methyl-N'-nitro-N-nitrosoguanidine. *Neuroscience Letters* 68:95–99. [aSRV]

Tootell, R. B. H., Reppas, J. B., Dale, A. M., Look, R. B., Sereno, M. I., Malach, R., Brady, T. J. & Rosen, B. R. (1995) Visual motion aftereffect in human coritical area MT revealed by functional magnetic resonance imaging. *Nature* 375:139–41. [PVD]

Topka, H., Massaquoi, S. G., Zeffiro, T. & Hallett, M. (1991) Learning of arm trajectory formation in patients with cerebellar deficits [abstract]. *Society for Neuroscience Abstracts* 17:1381. [MH]

Topka, H., Valls-Solle, J., Massaquoi, S. G. & Hallett, M. (1993) Deficit in classical conditioning in patients with cerebellar degeneration. *Brain* 116(pt.4):961–69. [aWTT]

Toyama, H., Tsukahara, N., Kosaka, K. & Matsunami, K. (1970) Synaptic excitation of red nucleus nerone by fibres from interpositus nucleus. *Experimental Brain Research* 11(2):187–98. [aWTT]

Tremblay, J., Gerzer, R. & Hamet, P. (1988) Cyclic GMP in cell function. In: *Advances in second messengers and phosphoprotein research*, vol. 22, ed. P. Greengard & G. A. Robison. Raven. [aFC]

Trowbridge, M. H. & Cason, H. (1932) An experimental study of Thorndike's theory of learning. *Journal of General Psychology* 7:245–60. [HB]

Tsukahara, N. (1972) The properties of the cerebello-pontine reverberating circuit. *Brain Research* 40:67. [MAA]

Tsukahara, N., Bando, T., Murakami, F., & Oda, Y. (1993) Properties of cerebello-precerebellar reverberating circuits. *Brain Research* 274:249–259. [aJCH]

Tsukahara, N., Hultborn, H., Murakami, F. & Fujito, Y. (1975) Electrophysiological study of formation of new synapses and collateral sprouting in red nucleus neurons after partial denervation. *Journal of Physiology* 38:1359–72. [aMKan]

Tsukahara, N., Korn, H. & Stone, J. (1968) Pontine relay from cerebral cortex to cerebellar cortex and nucleus interpositus. *Brain Research* 10:448–53. [aJCH]

Tusa, R. J. & Ungerleider, L. G. (1988) Fiber pathways of cortical areas mediating smooth pursuit eye movements in monkeys. *Annals of Neurology* 23:174–83. [JDS]

Tyler, A. E. & Hutton, R. S. (1986) Was Sherrington right about co-contractions? *Brain Research* 370:171–75. [aAMS]

Tyrrell, T. & Willshaw, D. J. (1992) Cerebellar cortex: Its simulation and the relevance of Marr's theory. *Proceedings of the Royal Society of London, Series B* 336:239–57. [arJCH]

Udo, M., Matsukawa, K., Kamei, H., Minoda, K. & Oda, Y. (1981) Simple and complex spike activities of Purkinje cells during locomotion in the cerebellar vermal zones of decerebrate cats. *Experimental Brain Research* 41:292–300. [aAMS]

Vallebuona, F. & Raiteri, M. (1993) Monitoring of cyclic GMP during cerebellar microdialysis in freely-moving rats as an index of nitric oxide synthase activity. *Neuroscience* 57:577–85. [aSRV]

Van der Steen, J., Simpson, J. I. & Tan, J. (1994) Functional and anatomical organization of three-dimensional eye movements in rabbit cerebellar flocculus. *Journal of Neurophysiology* 72:31–46. [aJIS]

Van der Want, J. J. L., Wiklund, L., Guegan, M., Ruigrok, T. & Voogd, J. (1989) Anterograde tracing of the rat olivocerebellar system with phaseolus vulgaris leucoagglutinin (PHA-L). Demonstration of climbing fiber collateral innervation of the cerebellar nuclei. *Journal of Comparative Neurology* 288:1–18. [aJIS]

Van der Zee, E. A., Palm, I. F., Kronforst, M. A., Maizels, E. T., Shanmugam, M., Hunzicker-Dunn, M. & Disterhoft, J. F. (1995) Trace and delay eyeblink conditioning induce alterations in the immunoreactivity for PKCg in the rabbit hippocampus. *Society for Neuroscience Abstracts* 21:1218. [CW]

van Donkelaar, P., Fisher, C. & Lee, R. G. (1994) Adaptive modification of oculomotor pursuit influences manual tracking responses. *NeuroReport* 5:2233–36. [PVD]

Van Galen, G. P. & De Jong, W.P. (1995) Fitts' law as the outcome of a dynamic noise filtering model of motor control. *Human Movement Science* 14:539–71. [GPVG]

Van Galen, G. P. & Schomaker, R. B. (1992) Fitts' law as a low–pass filter effect of muscle stiffness. *Human Movement Science* 11:11–21. [GPVG, rAMS]

Van Galen, G. P., Van Doorn, R. R. A. & Schomaker, R. B. (1990) Effects of motor programming on the power spectrum density function of finger and wrist movements. *Journal of Experimental Psychology: Human Perception and Performance* 16:755–65. [GPVG]

Van Gisbergen, J. A. M., Robinson, D. A. & Gielen, S. (1981) A quantitative analysis of saccadic eye movements by burst neurons. *Journal of Neurophysiology* 45:417–42. [aJCH]

Van Gisbergen, J. A. M., Van Opstal, A. J. & Hoeks, B. (1989) The transformation of the collicular motor map into rapid eye movements: Implications of a nonorthogonal muscle system. In: *Neural networks from models to applications*, ed. L. Personnaz & G. Dreyfus. Paris: I.D.S.E.T. [aJCH]

van Ingen Schenau, C. J., Boots, P. J. M., Snackers, R. J. & van Woensel, W. W. L. M. (1991) The constrained control of force and position by multi-joint movements. *Neuroscience* 46:197–207. [aAMS]

Van Kan, P. L. E., Gibson, A. R. & Houk, J. C. (1993) Movement-related inputs to intermediate cerebellum of the monkey. *Journal of Neurophysiology* 69:74–94. [arJCH]

Van Mier, H., Tempel, L., Perlmutter, J. S., Raichle, M. E. & Petersen, S. E. (1995) Generalization of practice-related effects in motor learning using the dominant and nondominant hand measured by PET. *Society for Neuroscience Abstracts* 21:1441. [rWTT]

van Zuylan, E. J., Gielen, C. C. A. M. & Denier Van Der Gon, J. J. (1988) Coordination and inhomogeneous activation of human arm muscles during isometric torques. *Journal of Neurophysiology* 60:1523–48. [aAMS]

Verma, A., Hirsch, D. J., Glatt, C. E., Ronnett, G. V. & Snyder, S. H. (1993) Carbon monoxide: A putative neural messenger. *Science* 259:381–84. [aDJL, aSRV, JMO]

Verschueren, S., Swinnen, S. P. & Dom, R. (1995) Interlimb coordination in patients with Parkinson's disease: Learning capabilities and the importance of augmented visual information. In: *Studies in perception and action 3*, ed. G. Bardy, R. J. Bootsma & Y. Guiard. Erlbaum. [SPS]

Viallet, F., Massion, J., Bonnefois-Kyriacou, B., Aurenty, R., Obadia, A. & Khalil, R. (1994) Approche quantative de l'asynergie posturale en pathologie cérébelleuse. *Revue Neurologique* 150:55–60. [aAMS]

Viallet, F., Massion, J., Massarino, R. & Khalil, R. (1987) Performance of a bimanual load-lifting task by Parkinsonian patients. *Journal of Neurology, Neurosurgery, and Psychiatry* 50:1274–1283. [aAMS]

(1992) Coordination between posture and movement in a bimanual load lifting task: Putative role of a medial frontal region in the supplementary motor area. *Experimental Brain Research* 88:674–84. [aAMS]

Vilensky, J. A. & Van Hoesen, G. W. (1981) Corticopontine projections from the cingulate cortex in the rhesus monkey. *Brain Research* 205:391–95. [JDS]

Vincent, P., Armstrong, C. M. & Marty, A. (1992) Inhibitory synaptic currents in rat cerebellar Purkinje cells: Modulation by postsynaptic depolarization. *Journal of Physiology (London)* 456:453–71. [aMKan]

Vincent, P. & Marty, A. (1993) Neighboring cerebellar Purkinje cells communicate via retrograde inhibition of common presynaptic interneurons. *Neuron* 11:885–93. [aMKan]

Vincent, S. R. & Hope, B. T. (1992) Neurons that say NO. *Trends in Neuroscience* 15:108–13. [aSRV]

Vincent, S. R. & Kimura, H. (1992) Histochemical mapping of nitric oxide synthase in the rat brain. *Neuroscience* 46:755–84. [aDJL, aSRV]

Voneida, T., Christie, D., Boganski, R. & Chopko, B. (1990) Changes in instrumentally and classically conditioned limb-flexion responses following inferior olivary lesions and olivocerebellar tractotomy in the cat. *Journal of Neuroscience* 10:3583–93. [RFT]

Voogd, J. (1964) The cerebellum of the cat: Structure and fibre connections. Thesis, Van Gorcum, Assen, The Netherlands. [aJIS]

Voogd, J. & Bigaré, F. (1980) Topographical distribution of olivary and cortico-nuclear fibers in the cerebellum: A review. In: *The inferior olivary nucleus: Anatomy and physiology*, eds. J. Courville, C. de Montigny & Y. Lamarre. Raven. [aJIS]

Vranesic, I., Batchelor, A., Gahwiler, B. H., Garthwaite, J., Staub, C. & Knopfel, T. (1991) Trans-ACPD-induced Ca^{2+} signals in cerebellar Purkinje cells. *NeuroReport* 2:759–62. [aFC, aDJL, rSRV]

Vyklicky, L., Patneau, D. K. & Mayer, M.L. (1991) Modulation of excitatory synaptic transmission by drugs that reduce desensitization at AMPA/kaïnate receptors. *Neuron* 7:971–84. [aFC]

Wada, Y. & Kawato, M. (1993) A neural network model for arm trajectory formation using forward and inverse dynamics models. *Neural Networks* 6:919–32. [MKaw]

Waespe, W., Cohen, B. & Raphan, T. (1983) Role of the flocculus and paraflocculus in optokinetic nystagmus and visual-vestibular interactions: Effects of lesions. *Experimental Brain Research*. 50:9–33. [aJIS]

(1985) Dynamic modification of the vestibulo-ocular reflex by the nodulus and the uvula. *Science* 228:199–202. [aJIS]

Walmsley, B., Hodgson J. A. & Burke R. E. (1978) The forces produced by medial gastrocnemius and soleus muscles during locomotion in freely moving cats. *Journal of Neurophysiology* 41:1203–16. [aAMS]

Walter, C. B. & Swinnen, S. P. (1994) The formation and dissolution of 'bad habits' during the acquisition of coordination skills. In: *Interlimb coordination: Neural, dynamical, and cognitive constraints*, ed. S. P. Swinnen, H. Heuer, J. Massion & P. Casaer. Academic Press. [SPS]

Wang, J.-J., Kim, J. H. & Ebner, T. J. (1987) Climbing fiber afferent modulation during a visually guided, multi-joint arm movement in the monkey. *Brain Research* 410:323–29. [aJIS]

Watkins, J. C. (1981) Pharmacology of excitatory amino acid transmitters. In: *Advances in biochemical psychopharmacology, vol. 29: Amino acid neurotransmitters*, ed. F. W. DeFeudis, F. W. & P. Mandel. Raven. [aFC]

Watson, P. J. (1978) Nonmotor functions of the cerebellum. *Psychological Bulletin* 85:944–67. [JDS]

Weeks, D. L., Aubert, M.-P., Feldman, A. G. & Levin, M. F. (in press) One-trial adaptation of movement to changes in load. *Journal of Neurophysiology.* [AGF]

Weinberger, D., Kleinman, J., Luchins, D., Bigelow, L. & Wyatt, R. (1980) Cerebellar pathology in schizophrenia. A controlled post-mortem study. *American Journal of Psychiatry* 137:359–61. [aWTT]

Weiner, M. J., Hallett, M. & Funkenstein, H. H. (1983) Adaptation to lateral displacement of vision in patients with lesions of the central nervous system. *Neurology* 33:766–72. [aWTT]

Weiner, N. (1948) *Cybernetics: Control and Communication in the animal and in the Machine.* John Wiley.

Weiss, C., Houk, J. C. & Gibson, A. R. (1990) Inhibition of sensory responses of cat inferior olive neurons produced by stimulation of red nucleus. *Journal of Neurophysiology* 64:1170–85. [CW]

Weiss, C., Disterhoft, J. F., Gibson, A. R. & Houk, J.C. (1993) Receptive fields of single cells from the face zone of the cat rostral dorsal accessory olive. *Brain Research* 605:207–13. [aJCH, CW]

Weiss, C., Houk, J. C., & Gibson, A. R. (1990) Inhibiton of sensory responses of cat inferior olive neurons produced by stimulation of red nucleus. *Journal of Neurophysiology* 64:1170–85. [aJCH]

Weiss, C., Kronforst-Collins, M. A. & Disterhoft, J. F. (1996) Activity of hippocampal pyramidal neurons during trace eyeblink conditioning. *Hippocampus* 6(2). [CW]

Weisskopf, M. G., Castillo, P. E., Zalutsky, R. A. & Nicoll, R. A. (1994) Mediation of hippocampal mossy fiber long-term potentiation by cyclic-amp. *Science* 265:1878–82. [MB, rDJL]

Welsh, J. P., Lang, E. J., Sugihara, I. & Llinas, R.. (1995) Dynamic organization of motor control within the olivocerebellar system. *Nature* 374:453–57. [RCM, EDS]

Welsh, J. P. & Harvey, J. A. (1989) Cerebellar lesions and the nictitating membrane reflex: Performance deficits of the conditioned and unconditioned response. *Journal of Neuroscience* 9:299–311. [aJIS, aWTT, CW]

Welsh, J. P., Lang, E. J. & Llinás, R. (1993) The microstructure of coherence in the olivocerebellar system during rhythmic movement in normal and deafferented rats. *Society of Neuroscience Abstracts* 19: 529.9. [aJIS]

Welsh, J. P., Lang, E. J., Sugihara, I. & Llinás, R. (1992) Rhythmic olivo-cerebellar control of skilled tongue movement in relation to patterned hypoglossal nerve activity. *Society of Neuroscience Abstracts* 18:178.7. [aJIS]

(1995) Dynamic organization of motor control within the olivocerebellar system. *Nature* 374:453–57. [aJIS, RCM, SPS]

Welsh, J. P. & Harvey, J. A. (1989) Cerebellar lesions and the nictitating membrane reflex: Performance deficits of the conditioned and unconditioned response. *Journal of Neuroscience* 9:299–311.

Wenthold, R. J., Yokotami, N., Doi, K. & Wada, K. (1992) Immunochemical characterization of the non-NMDA receptor using subunit-specific antibodies. *Journal of Biological Chemistry* 267:501–7. [aFC]

Werhahn, K. J., Meyer, B. U., Rothwell, J. C., Thompson, P. D., Day, B. L. & Marsden, C. D. (1993) Reduction of motor cortex excitability by transcranial magnetic stimulation over the human cerebellum. *Journal of Physiology (London)* 459:149. [rJCH]

Wessel, K., Tegenthoff, M., Vorgerd, M., Otto, V., Nitschke, M. & Malin, J. P. (1996) Enhancement of inhibitory mechanisms in the mortor cortex of patients with cerebellar degeneration: A study with transcranial magnetic brain stimulation. *Electroencephalography and Clinical Neurophysiology* 101:273–81. [KW]

Wessel, K., Verleger, R., Nazarenus, D., Vieregge, P. & Kömpf, D. (1994) Movement-related cortical potentials preceding sequential and goal-directed finger and arm movements in patients with cerebellar atrophy. *Electroencephalography and Clinical Neurophysiology* 92:331–41. [KW]

Wessel, K., Zeffiro, T., Lou, J. S., Toro, C. & Hallett, M. (1995) Regional cerebral blood flow during a self-paced sequential finger opposition task in patients with cerebellar degeneration. *Brain* 118:379–93. [KW]

Wetts, R. & Herrup, K. (1982a) Interaction of granule, Purkinje and inferior olivary neurons in Lurcher chimeric mice: 2. Granule cell death. *Brain Research* 250:358–62. [aAMS]

(1982b) Interaction of granule, Purkinje and inferior olivary neurons in Lurcher chimaeric mice: 1. Qualitative studies. *Journal of Embryology and Experimental Morphology* 68:87–98. [aAMS]

Wetts, R., Kalaska, J. F. & Smith, A. M. (1985) Cerebellar nuclear cell activity during antagonist cocontraction and reciprocal inhibition of forearm muscles. *Journal of Neurophysiology* 54:231–44. [aAMS, aWTT]

Whiting, P., McKernan, R. M. & Iverson, L. L. (1990) Another mechanism for creating diversity in γ-aminobutyrate type A receptors: RNA splicing directs expression of two forms of γ2 subunit, one of which contains a protein kinase C phosphorylation site. *Proceedings of the National Academy of Sciences of the USA* 87:9966–70. [aMKan]

Wiener, S. I. & Berthoz, A. (1993) Vestibular contributions during navigation. In: *Multisensory control of movement,* ed. A. Berthoz. Oxford University Press. [SMO]

Williams, J. H., Errington, M. L., Lynch, M. A. & Bliss, T. V. (1989) Arachidonic acid induces a long-term activity-dependent enhancement of synaptic transmission in the hippocampus. *Nature* 341:739–42. [MB]

Williams, J. H., Li, Y.-G., Nayak, A., Errington, M. L., Murphy, K. P. S. J. & Bliss, T.V.P. (1993) The suppression of long-term potentiation in rat hippocampus by inhibitors of nitric oxide synthase is temperature and age dependent. *Neuron* 11:877–84. [aDJL]

Windhorst, U., Burke, R. E., Dieringer, N., Evinger, C., Feldman, A. G., Hasan, Z. et al. (1991) What are the ouput units of motor behavior and how are they controlled?. In: *Motor control: Concepts and issues,* ed. D. R. Humphrey & H.-J. Freund. Wiley. [aAMS]

Wing, A. M. Turton, A. & Fraser, C. (1986) Grasp size and accuracy of approach in reaching. *Journal of Motor Behavior* 18:245–60. [PH]

Wood, P. L. (1991) Pharmacology of the second messenger, cyclic guanosine 3',5'-monophosphate, in the cerebellum. *Pharmacological Reviews* 43:1–25. [aSRV]

Wood, P. L., Emmett, M. R. & Wood, J. A. (1994) Involvement of granule, basket and stellate neurons but not Purkinje or Golgi cells in cerebellar cGMP inceases *in vivo. Life Sciences* 54:615–20. [arSRV, LK]

Wood, P. L., Emmett, M. R., Rao, T. S., Cler, J., Mick, S. & Iyengar, S. (1990) Inhibition of nitric oxide synthase blocks N-methyl-D-aspartate-, quisqualate-, kainate-, harmaline-, and pentylenetetrazole-dependent increases in cerebellar cyclic GMP in vivo. *Journal of Neurochemistry* 55:346–48. [aSRV]

Wood, P. L., Richard, J. W., Pilapil, C. & Nair, N. P. V. (1982) Antagonists of excitatory amino acids and cyclic guanosine monophosphate in cerebellum. *Neuropharmacology* 21:1235–38. [aSRV]

Wood, P. L., Ryan, R. & Li, M. (1992) NMDA-, but not kainate- or quisqualate-dependent increases in cerebellar cGMP are dependent upon monoaminergic innervation. *Life Sciences* 51:267–70. [aSRV]

Wood, J. & Garthwaite, J. (1994) Models of the diffusional spread of nitric oxide: Implications for neural nitric oxide signalling and its pharmacological properties. *Neuropharmacology* 33:1235–44. [DO]

Woodward, D. J., Hoffer, B. J., Siggins, G. R. & Bloom, F. E. (1971) The ontogenic development of synaptic junctions. Synaptic activation and responsiveness to neurotransmitter substances in rat cerebellar Purkinje cells. *Brain Research* 34:73–97. [LK]

Wooten, G. F. & Collins, R. C. (1981) Metabolic effects of unilateral lesions of the substantia nigra. *Journal of Neuroscience* 1:285–91. [aWTT]

Wu, G. Y. & Brosnan, J. T. (1992) Macrophages can convert citrulline into arginine. *Biochemical Journal* 281:45–48. [LK]

Wylie, D. R. & Frost, B. J. (1993) Responses of pigeon vestibulocerebellar neurons to optokinetic stimulation: 2. The 3-dimensional reference frame of rotation neurons in the flocculus. *Journal of Neurophysiology* 70:2647–59. [arJIS]

(in press) The pigeon optokinetic system: Visual input in extraocular muscle coordinates. *Visual Neuroscience.* [rJIS]

Wylie, D. R., De Zeeuw, C. I., DiGiorgi, P. L. & Simpson, J. I. (1994) Projections of individual Purkinje cells of identified zones in the ventral nodulus to the vestibular and cerebellar nuclei in the rabbit. *Journal of Comparative Neurology* 349:448–63. [aJIS]

Wylie, D. R., De Zeeuw, C. I. & Simpson, J. I. (1995) Temporal relations of the complex spike activity of Purkinje cell pairs in the vestibulocerebellum of rabbits. *The Journal of Neuroscience* 15:2875–87. [aJIS]

Xiao, P., Bahr, B. A., Staubli, U., Vanderklish, P. W. & Lynch, G. (1991) Evidence that matrix recognition contributes to stabilization but not induction of LTP. *Neuroreport* 2:461–64. [MB]

Yamamoto, C., Yamashita, H. & Chujo, T. (1978) Inhibitory action of glutamic acid on cerebellar interneurons. *Nature (London)* 262:786–87. [aJIS]

Yamamoto, M. (1979) Vestibulo-ocular reflex pathways of rabbits and their representation in the cerebellar flocculus. *Progress in Brain Research* 50:451–57. [aJIS]

Yamamoto, T., Yoshida, K., Yoshikawa, H., Kishimoto, Y. & Oka, H. (1992) The medial dorsal nucleus is one of the thalamic relays of the cerebellocerebral responses to the frontal association cortex in the monkey: Horseradish peroxidase and fluorescent dye double staining study. *Brain Research* 579:315–20. [aWTT]

Yan, X. X., Jen, L. S. & Garey, L. J. (1993) Parasagittal patches in the granular layer of the developing and adult rat cerebellum as demonstrated by NADPH-diaphorase histochemistry. *NeuroReport* 4:1227–30. [aSRV]

Yanagihara, D., Kondo, I. & Yoshida, T. (1994) Nitric oxide-mediated cerebellar synaptic plasticity plays an important role in adaptive interlimb coordination during perturbed locomotion. *Japanese Journal of Physiology* 55(suppl. 1):S222. [aMKan]

Yao, X., Segal, A. S., Welling, P., Zhang, X., McNicholas, C. M., Engel, D.,

Boulpaep, E. L. & Desir, G. V. (1995) Primary structure and functional expression of a cGMP-gated potassium channel. *Proceedings of the National Academy of Sciences of the USA* 92:11711–15. [rSRV]

Yeo, C. H., Hardiman, M. J. & Glickstein, M. (1984) Discrete lesions of the cerebellar cortex abolish classically conditioned nictitating membrane response of the rabbit. *Behavior Brain Research* 13:261–66. [aWTT]

(1986) Classical conditioning of the nictitating membrane response of the rabbit: 4. Lesions of the inferior olive. *Experimental Brain Research* 63:81–92. [RFT]

Yeo, C. H., Hardiman, M. J. & Glickstein, M. (1985) Classical conditioning of the nictitating membrane response of the rabbit: 1. Lesions of the cerebellar nuclei. *Experimental Brain Research* 60:87–98. [CW]

Yi, S.-J., Snell, L. D. & Johnson, K. M. (1988) Linkage between phencyclidine (PCP) and N-methyl-D-aspartate (NMDA) receptors in the cerebellum. *Brain Research* 445:147–51. [aSRV]

Yoshikami, D. & Okun, L. M. (1984) Staining of living presynaptic nerve terminals with selective fluorescent dyes. *Nature* 310:53–56. [aDJL]

Young, L. R. (1977) Pursuit eye movements: What is being pursued? In: *Control of gaze by brain stem neurons*, ed. R. Baker & A. Berthoz. [aJCH]

Yuen, G. L., Hockberger, P. E. & Houk, J. C. (1995) Bistability in cerebellar Purkinje cell dendrites modelled with high-threshold calcium and delayed-rectifier potassium channels. *Biological Cybernetics* 73:375–88. [arJCH, KH, EDS]

Yuzaki, M. & Mikoshiba, K. (1992) Pharmacological and immunocytochemical characterization of metabotropic glutamate receptors in cultured Purkinje cells. *Journal of Neuroscience* 12:4253–63. [aDJL]

Zanchetti, A. & Zoccolini, A. (1954) Autonomic hypothalamic outbursts elicited by cerebellar stimulation. *Journal of Neurophysiology* 17:473–83. [JDS]

Zanone, P. G. & Kelso, J. A. S. (1992) Evolution of behavioral attractors with learning: nonequilibrium phase transitions. *Journal of Experimental Psychology: Human Perception and Performance* 18:403–21. [SPS]

Zeffiro, T. A., Blaxton, T. A., Gabrieli, J. D. E., Bookheimer, S. Y., Carrillo, M. C., Benion, E., Disterhoft, J. F. & Theodore, W. H. (1993) Regional cerebral blood flow changes during classical eyeblink conditioning in man. *Society for Neuroscience Abstracts* 19:1078. [CW]

Zhang, J. & Snyder S. H. (1992) Nitric oxide stimulates auto-ADP-ribosylation of glyceraldehyde-3-phosphate dehydrogenase. *Proceedings of the National Academy of Sciences of the USA* 89:9382–85. [aSRV]

Zhang, N., Walberg, F., Laake, J. H., Meldrum, B. S. & Ottersen, O. P. (1990) Aspartate-like and glutamate-like immunoreactivities in the inferior olive and climbing fibre system: A light microscopic and semiquantitative electron microscopic study in rat and baboon (*Papio anubis*). *Neuroscience* 38:61–80. [aFC, aDJL]

Zhuang, P., Toro, C., Grafman, J., Manganotti, P., Leocani, L., Deiber, M.-P. et al. (1995) Functional topography during procedural learning studied with event-related desynchronization mapping (preliminary finding) [abstract]. *Society for Neuroscience Abstracts* 21:1927. [MH]

Zhuo, M., Hu, Y., Schultz, C., Kandel, E. R. & Hawkins, R. D. (1994) Role of guanylyl cyclase and cGMP-dependent protein kinase in long-term potentiation. *Nature* 368:635–39. [MB]

Zipser, D. & Andersen, R. E. (1988) A back-propagation programmed network that simulates response properties of a subset of posterior parietal neurons. *Nature* 331:679–84. [aJCH]

Zohary, G., Celebrini, S., Britten, K. H. & Newsome, W. T. (1994) Neuronal plasticity that underlies improvement in perceptual performance. *Science* 263:1289–91. [PVD]

Zomlefer, M. R., Zajac, F. E. & Levine, W. S. (1977) Kinematics and muscular activity of cats during maximum height jumps. *Brain Research* 126:563–66. [aAMS]

Zwiller, J., Ghandour, M. S., Revel, M. O. & Basset, P. (1981) Immunohistochemical localization of guanylate cyclase in rat cerebellum. *Neuroscience Letters* 23:31–36. [aDJL, LK, rSRV]

Index

Printed in the United States
By Bookmasters